**Photovoltaic Solar Energy**

# Photovoltaic Solar Energy

From Fundamentals to Applications, Volume 2

*Edited by*

*Wilfried van Sark*
Utrecht University, Utrecht, The Netherlands

*Bram Hoex*
University of New South Wales
Sydney, Australia

*Angèle Reinders*
Eindhoven University of Technology
Eindhoven, The Netherlands

*Pierre Verlinden*
Chief Scientist of Yangtze Institute for Solar Technology (YIST)
Jiangyin, China

*Nicholas J. Ekins-Daukes*
University of New South Wales
Sydney, Australia

WILEY

*Registered Offices*
John Wiley & Sons, Inc., 111 River Street, Hoboken, NJ 07030, USA
John Wiley & Sons Ltd, The Atrium, Southern Gate, Chichester, West Sussex, PO19 8SQ, UK

For details of our global editorial offices, customer services, and more information about Wiley products visit us at www.wiley.com.

*Library of Congress Cataloging-in-Publication Data:*

Names: Sark, Wilfried van, editor. | Hoex, Bram, editor. | Reinders,
    Angèle, editor. | Verlinden, Pierre, editor. | Ekins-Daukes, Nicholas
    J., editor. | John Wiley & Sons, publisher.
Title: Photovoltaic solar energy : from fundamentals to applications,
    volume 2 / edited by Wilfried van Sark, Bram Hoex, Angèle Reinders,
    Pierre Verlinden, Nicholas J. Ekins-Daukes.
Description: Hoboken, NJ : Wiley, 2024. | "Since the publication of the
    book Photovoltaic Solar Energy, from Fundamentals to Applications
    (Reinders, 2017), we have been eager to compile another volume, in fact,
    to start a series that would appear every 5–10 years to showcase the
    advancement of the state-of-the-art in photovoltaics research,
    development and deployment. We are now proud to present the second
    volume, and we are very grateful to the many authors who committed time
    to write a new or updated chapter"–Introduction. | Includes
    bibliographical references and index.
Identifiers: LCCN 2024014706 (print) | LCCN 2024014707 (ebook) | ISBN
    9781119578819 (hardback) | ISBN 9781119578833 (adobe pdf) | ISBN
    9781119578840 (epub)
Subjects: LCSH: Photovoltaic power generation. | Photovoltaic cells. |
    Solar energy.
Classification: LCC TK1087 .P478 2024 (print) | LCC TK1087 (ebook) | DDC
    621.31/244–dc23/eng/20240501
LC record available at https://lccn.loc.gov/2024014706
LC ebook record available at https://lccn.loc.gov/2024014707

Cover Design: Wiley
Cover Images: © Tom Penpark/Shutterstock, gerenme/Getty Images,
© eunikas/Adobe Stock Photos, © Spectral-Design/Adobe Stock Photos,
© dim86b/Getty Images, © Gyuszko/Getty Images, © An Qi Wang/Adobe Stock Photos

Set in 9.5/12.5pt STIXTwoText by Straive, Chennai, India

*Solar is becoming the new king of the world's electricity markets*

*Fatih Birol,*
*IEA Executive Director, 2020*

*We dedicate this book to all the young students and entrepreneurs who are to take up the task of realizing 50+ terawatt PV by mid-century as a major building block of a sustainable society.*

# Contents

# About the Editors

**Prof. Dr. Wilfried van Sark** is a full professor of "Integration of Photovoltaics" at the Copernicus Institute of Sustainable Development at Utrecht University, Utrecht, the Netherlands. He is an experimental physicist by training (MSc/PhD) and has 40 years of experience in the field of photovoltaics. He worked on various material systems such as crystalline and thin film silicon and III–V solar cells, both experimentally and theoretically. His current activities focus on employing spectrum conversion to increase solar cell conversion efficiency for next-generation photovoltaic energy converters such as luminescent solar concentrators, as well as performance analysis of building-integrated and standard PV systems in the field, including offshore floating PV. This links to the integration of PV systems in smart grids in the built environment, in which electrical vehicles, demand response, and self-consumption and self-sufficiency play a major role. He is the author of some 250 publications, cited more than 12 000 times, and various (text)books on photovoltaics. He is an associate editor of *Solar Energy and Frontiers in Energy Research* and a member of the editorial boards of *Renewable Energy*, *Energies*, and *Materials*. He is a member of the International Solar Energy Society, a senior member of the Institute of Electrical and Electronic Engineers (IEEE), and a member of scientific committees of various EU and IEEE PV conferences. He presently is the national representative for the International Energy Agency (IEA) PVPS Task 13 (PV Performance, Operation and Reliability of Photovoltaic Systems) and Task 16 (Solar Resource for High Penetration and Large Scale Applications).

**Prof. Dr. Bram Hoex** completed an MSc degree and a PhD degree in applied physics from the Eindhoven University of Technology (TU/e) in the Netherlands. His PhD work on functional thin films for high-efficiency solar cells was recognized by both the SolarWorld "Junior Einstein" and Leverhulme "Technology Transfer" awards. After completing his PhD degree in 2008, he joined the Solar Energy Research Institute of Singapore (SERIS) at the National University of Singapore (NUS) as head of the Photovoltaic Characterization group. From 2012 to the end of 2014, he was Director of the Silicon Materials and Solar Cells Cluster and Group Leader Monocrystalline Silicon Wafer Solar Cells. In 2015, he joined the School of Photovoltaic and Renewable Energy Engineering (SPREE) at University of New South Wales, Sydney, Australia where he is currently Deputy Head of School (Research). His research group focuses on developing and applying nanoscale thin films to a wide range of renewable energy devices to improve their performance, solar cell/module reliability, as well as financial and performance modeling of gigascale solar farms. He is best known for his groundbreaking work on aluminum oxide for crystalline silicon surface passivation, which is now the de facto standard for industrial PERC and TOPCon solar cells. He also pioneered the application of atomic layer deposition for silicon wafer solar cell manufacturing. During his career, he has raised over A\$ 80 million in competitive research funding of which A\$ 23 million as a lead investigator. He has published over 250 journal and conference papers, and he has an h-index of 42. In 2016, he was awarded the mid-career "IEEE PVSC Young Professional Award" for his significant contributions and his potential as a future leader in the field of photovoltaics, and in 2018, he was selected as one of the "Solar 40 under 40" by Renewable Energy World.

**Prof. Dr. Angèle Reinders** is a full professor of "Design of Sustainable Energy Systems" at the Department of Mechanical Engineering in Eindhoven University of Technology (TU/e), Eindhoven, the Netherlands. In her research, she focuses on the integration of solar energy technologies in buildings and mobility by means of simulation, prototyping, and testing. A strong interest exists in designs that enhance user acceptance and reduce environmental impact. Angèle Reinders studied experimental physics at Utrecht University, where she also received her doctoral degree in chemistry. Next, she developed herself in industrial design engineering. She was an associate professor at the University of Twente and a full professor in energy efficient design at TU Delft. In the past, she also worked at the Fraunhofer Institute of Solar Energy, World Bank, ENEA in Naples, and Center of Urban Energy in Toronto. She is known for her books: *Designing with Photovoltaics* (2020), *The Power of Design,* and *Photovoltaic Solar Energy from Fundamentals to Applications.* In 2010, she co-founded the *Journal of Photovoltaics.* She has also initiated many projects and has an extensive publication record.

**Dr. Pierre Verlinden** is currently Chief Scientist of Yangtze Institute for Solar Technology, China, and Adjunct Professor at the University of New South Wales, Sydney. He received the PhD degree in electrical engineering from the Catholic University of Louvain (UCL in Belgium) and was a visiting scholar at Stanford University (USA) thanks to a NATO Research Fellowship. Dr. Verlinden has been working in the field of photovoltaics since 1979 and has published over 200 technical papers and contributed to a number of books. Among several key positions in the PV industry, he was director of R&D at SunPower Corporation (USA) from 1991 to 2001 and Chief Scientist, Vice President, and Deputy Chair of the Academic Committee of the State Key Laboratory of PV Science and Technology at Trina Solar, Changzhou (China). He established the technical leadership of Trina Solar, allowing the company to become one of the largest PV manufacturers of cells and modules worldwide. Dr. Verlinden is the recipient of the 2016 William Cherry Award, awarded by the Institute of Electrical and Electronic Engineers (IEEE), and the 2019 Becquerel Prize, awarded by the EU Commission, for his dedication over the past three decades at the forefront of PV technology and commercialization, leading technology advances including the interdigitated back contact cell, silicon and multijunction III-V CPV cells, and his overall leadership of key R&D organizations. Dr. Verlinden is also a recipient of the 2017 Chinese Government Friendship Award, the highest award given by China to foreigners. In 2023, Dr. Verlinden was awarded a Doctor Honoris Causa degree from the University of New South Wales, Australia, and the Francqui Chair at the University of Hasselt, Belgium. He is a fellow member of the IEEE.

**Prof. Dr. Nicholas J. Ekins-Daukes** is a professor in the School of Photovoltaic and Renewable Energy Engineering at the University of New South Wales, Sydney, Australia. He holds a PhD and an MSc from Imperial College and an MSc in physics and electronics from the University of St Andrews. From 2008 to 2017, he worked in the physics department at Imperial College London, holding positions as Reader, Senior Lecturer, and Lecturer, as well as a Royal Society Industry Fellowship. From 2005 to 2017, he worked as a lecturer at the School of Physics at the University of Sydney. From 2003 to 2006, he was a JSPS research fellow at the Toyota Technological Institute, Japan. His research aims to fundamentally increase the efficiency of photovoltaic solar cells toward the ultimate efficiency limit for solar power conversion of 87%. This research begins by establishing conceptual thermodynamic boundaries for the processes that can (and cannot) lead to efficiency improvements in solar cells. This is followed by practical work to demonstrate working, proof-of-principle devices and often involves computer modeling and optical spectroscopy in addition to standard photovoltaic device measurements. Several "Advanced Concept" devices have emerged from this work, notably the quantum well solar cell that enabled a 32.9% III–V double junction solar cells to be made, photo-molecular up-conversion on amorphous silicon solar cells, photon ratchet sequential absorption photovoltaic device, and several demonstrations of a hot carrier solar cell. In 2022, he demonstrated electrical power from the thermal emission of light using a thermoradiative diode.

# List of Contributors

*Esther Alarcón-Lladó*
Center for Nanophotonics
AMOLF
Amsterdam
The Netherlands

*Thomas G. Allen*
KAUST Solar Center (KSC), Physical
Sciences and Engineering Division (PSE)
King Abdullah University of Science and
Technology (KAUST)
Thuwal
Kingdom of Saudi Arabia

*Kevin Anderson*
Sandia National Laboratories
1515 Eubank Blvd SE
Albuquerque
USA

*Erkan Aydin*
KAUST Solar Center (KSC), Physical
Sciences and Engineering Division (PSE)
King Abdullah University of Science and
Technology (KAUST)
Thuwal
Kingdom of Saudi Arabia

*Tino Band*
DENKweit GmbH
Halle (Saale)
Germany

*Michele De Bastiani*
Department of Chemistry
INSTM Università di Pavia
Pavia
Italy

*Matthew Berwind*
Fraunhofer Institute for Solar Energy
Systems
Division Photovoltaics
Freiburg
Germany

*Karsten Bittkau*
IEK-5 Photovoltaik
Forschungszentrum Jülich GmbH
Jülich
Germany

*Mathieu Boccard*
École Polytechnique Fédérale de
Lausanne (EPFL)
Institute of Electrical and Micro
Engineering (IEM)
Photovoltaics and Thin Film Electronics
Laboratory (PV-LAB)
Neuchâtel
Switzerland

**Christian Braun**
Fraunhofer Institute for Solar
Energy Systems
Division Photovoltaics
Freiburg
Germany

**Xue Chen**
R&D Department
State Key Laboratory of Photovoltaic
Science and Technology, Trina Solar
Changzhou
Jiangsu
China

**Alison Ciesla**
School of Photovoltaic and Renewable
Energy Engineering
UNSW Sydney
Sydney, NSW
Australia

**Karoline Dapprich**
Sinton Instruments
Boulder, CO
USA

**Kaining Ding**
IEK-5 Photovoltaik
Forschungszentrum Jülich GmbH
Jülich
Germany

**Priya Dwivedi**
The University of New South Wales
Sydney, NSW
Australia

**Nicholas J. Ekins-Daukes**
School of Photovoltaic and Renewable
Energy Engineering
University of New South Wales
Sydney, NSW
Australia

**Jiarui Fan**
Longi, R&D Department
Wafer Business Unit
Xi'An, Shaanxi
China

**Francesco Frontini**
University of Applied Sciences and Arts of
Southern Switzerland (SUPSI)
Mendrisio
Switzerland

**Vasilis Fthenakis**
Earth and Environmental Engineering
Department
Center of Life Cycle Analysis
Columbia University
New York, NY
USA

**Nannan Fu**
Longi, R&D Department
Wafer Business Unit
Xi'An, Shaanxi
China

**Sara M. Golroodbari**
Copernicus Institute of Sustainable
Institute
Utrecht University
Utrecht
The Netherlands

**Ziv Hameiri**
The University of New South Wales
Sydney, NSW
Australia

**Clifford W. Hansen**
Photovoltaics and Materials Technologies
Sandia National Laboratories
Albuquerque, NM
USA

*Anita Ho-Baillie*
School of Physics
The University of Sydney
Sydney, NSW
Australia

and

Sydney Nano
The University of Sydney
Sydney, NSW
Australia

and

Australian Centre for Advanced
Photovoltaics (ACAP), School of
Photovoltaic and Renewable Energy
Engineering
University of New South Wales
Sydney
Australia

*Bram Hoex*
School of Photovoltaic and Renewable
Energy Engineering
University of New South Wales
Sydney, NSW
Australia

*Md. Anower Hossain*
School of Photovoltaic and Renewable
Energy Engineering
University of New South Wales
Sydney, NSW
Australia

*Dirk C. Jordan*
National Renewable Energy Laboratory
(NREL)
Golden, CO
United States

*Bishal Kafle*
Fraunhofer ISE
Freiburg im Breisgau
Germany

*Olga Kanz*
IEK-5 Photovoltaik
Forschungszentrum Jülich GmbH
Jülich
Germany

*Kai Kaufmann*
DENKweit GmbH
Halle (Saale)
Germany

*Dominik Lausch*
DENKweit GmbH
Halle (Saale)
Germany

*Enrica Leccisi*
Center of Life Cycle Analysis
Columbia University
New York, NY
USA

*Frank Lenzmann*
TNO
Energy Transition Studies
Amsterdam
The Netherlands

*Maria Antonietta Loi*
Photophysics and OptoElectronics Group
Zernike Institute for Advanced Materials
University of Groningen
Groningen
The Netherlands

**Md Arafat Mahmud**
School of Physics
The University of Sydney
Sydney, NSW
Australia

and

Sydney Nano
The University of Sydney
Sydney, NSW
Australia

and

Australian Centre for Advanced
Photovoltaics (ACAP), School of
Photovoltaic and Renewable Energy
Engineering
University of New South Wales
Sydney
Australia

**Sander A. Mann**
Photonics Initiative
CUNY ASRC
New York
USA

**Johanna May**
Institute for Electrical Power Engineering
(IET) and Cologne Institute for Renewable
Energy (CIRE)
Cologne University of Applied Sciences
Cologne
Germany

**David Moser**
Institute for Renewable Energy
Eurac Research
Bolzano
Italy

**Neel Patel**
Forschungszentrum Jülich
Jülich
Germany

and

Eindhoven University of Technology
The Netherlands

**Angèle Reinders**
Energy Technology & Fluid Dynamics
Group, Department of Mechanical
Engineering
Eindhoven University of Technology
(TU/e)
Eindhoven, MB
The Netherlands

and

Department of Design Production and
Management, Faculty of Engineering
Technology
University of Twente
Enschede, AE
The Netherlands

**Armin Richter**
Fraunhofer ISE
Freiburg im Breisgau
Germany

**Wilfried van Sark**
Copernicus Institute of Sustainable
Development
Utrecht University
Utrecht
The Netherlands

**Timothy W. Schmidt**
School of Chemistry and Chief Investigator
of the ARC Centre of Excellence in Exciton
Science
University of New South Wales
Sydney, NSW
Australia

**Josefine Selj**
Department of Renewable Energy Systems
Institute for Energy Technology (IFE)
Kjeller
Norway

**Elham Shirazi**
Faculty of Engineering Technology,
Department of Design, Production and
Management
University of Twente
Enschede, AE
The Netherlands

**Ronald A. Sinton**
Sinton Instruments
Boulder, CO
USA

**Lenneke Slooff-Hoek**
Department of Solar Energy
TNO Energy and Materials Transition
Petten, ZG
The Netherlands

**Joshua S. Stein**
Climate Security Center
Sandia National Laboratories
Albuquerque, NM
USA

**Anand S. Subbiah**
KAUST Solar Center (KSC), Physical
Sciences and Engineering Division (PSE)
King Abdullah University of Science and
Technology (KAUST)
Thuwal
Kingdom of Saudi Arabia

**Murad J. Y. Tayebjee**
School of Photovoltaic and Renewable
Energy Engineering
University of New South Wales
Sydney, NSW
Australia

**Michelle Vaqueiro Contreras**
School of Photovoltaic and Renewable
Energy Engineering
University of New South Wales
Sydney, NSW
Australia

**Pierre J. Verlinden**
Yangtze Institute for Solar Technology
(YIST)
Jiangyin
China

and

AMROCK Pty Ltd
McLaren Vale, SA
Australia

and

School of Photovoltaic and Renewable
Energy Engineering
UNSW
Sydney
Australia

and

Yangtze Institute of Solar Technology
Jiangyin
China

**Bart Vermang**
Hasselt University and Imec
Hasselt
Belgium

***Han Wang***
Photophysics and OptoElectronics Group
Zernike Institute for Advanced Materials
University of Groningen
Groningen
The Netherlands

***Stefaan De Wolf***
KAUST Solar Center (KSC), Physical
Sciences and Engineering Division (PSE)
King Abdullah University of Science and
Technology (KAUST)
Thuwal
Kingdom of Saudi Arabia

***Jianmei Xu***
R&D Department
State Key Laboratory of Photovoltaic
Science and Technology, Trina Solar
Changzhou, Jiangsu
China

***Masafumi Yamaguchi***
Toyota Technological Institute
Nagoya
Japan

***Shu Zhang***
R&D Department
State Key Laboratory of Photovoltaic
Science and Technology, Trina Solar
Changzhou, Jiangsu
China

and

College of Materials Science and
Technology
Nanjing University of Aeronautics and
Astronautics
Nanjing, Jiangsu, 211106
China

***Jun Zhao***
PV-Center
Swiss Center for Electronic and
Microengineering (CSEM)
Neuchâtel
Switzerland

***Jianghui Zheng***
School of Physics
The University of Sydney
Sydney, NSW
Australia

and

Sydney Nano
The University of Sydney
Sydney, NSW
Australia

and

Australian Centre for Advanced
Photovoltaics (ACAP), School of
Photovoltaic and Renewable Energy
Engineering
University of New South Wales
Sydney
Australia

***Yan Zhu***
The University of New South Wales
Sydney, NSW
Australia

# Preface

Education on photovoltaic solar energy involves a broad range of disciplines, ranging from physics, electrical engineering, materials science to design engineering. Moreover, since around 1995, the field of photovoltaics has rapidly been developing due to technological advances; significant cost reductions of solar cells, panels, and inverters; and more beneficial regulatory frameworks, such as feed-in tariffs.

As such, when many of us were teaching courses at the bachelor's and master's levels, finding an appropriate, contemporary, and affordable textbook that covered the "full spectrum" of knowledge about photovoltaic solar energy was difficult. Also, a book covering the full spectrum of photovoltaic solar energy was lacking for training young engineers in the industry.

Therefore, in 2014, editor Angèle Reinders took the initiative to compile and edit with Pierre Verlinden, Wilfried van Sark and Alex Freundlich as co-editors the first volume of the book *Photovoltaic Solar Energy – From Fundamentals to Applications*, to which many photovoltaic specialists, our colleagues, and good friends contributed. The PV specialists who have written individual chapters of this book are currently active in R&D, covering PV material sciences, solar cell research, and application. Our initiative was strongly supported by our publisher, John Wiley & Sons, and the photovoltaic community. Since its publication in 2017, the book has been received very well.

Already before the COVID pandemic, we as editors saw the importance of updating the first volume and adding contemporary topics that had not been covered in the first volume and to update chapters due to fast technological developments in the field of photovoltaics. In the past years, since 2020, we have realized this second volume, comprising 35 chapters. It provides fundamental and updated contemporary knowledge about various photovoltaic technologies in the framework of materials science, device physics of solar cells, engineering of PV modules, design aspects of new photovoltaic applications, and sustainability aspects in the terawatt age of PV systems.

This book aims to inform undergraduate and/or post-graduate students, young or experienced professionals, about the basic knowledge of each aspect of photovoltaic technologies and the applications thereof in the context of the most recent advances in science and engineering and to provide insight into possible future developments in the field of photovoltaics.

If used as a university textbook, this book would be suitable for most universities with technical study programs offering master's level and graduate courses in renewables, solar energy, photovoltaics, and photovoltaic systems.

We do not aim to compete with the typical specialists' books, which go deeper into a particular topic, but instead, this book provides a broad scope of information and engages an audience with a more diverse background in science and engineering than physics only.

We hope that students in different engineering disciplines will find the book easy to read and to use during their studies. We tried to keep the size of the book to an affordable level for students while still maintaining an extensive scope and giving numerous references to recent publications to allow students to go deeper into the selected topics. We aimed to bring relevant information about all existing photovoltaic technologies, including solar irradiance basics, crystalline silicon devices, perovskite solar cells, advanced materials and designs for PV, characterization and measurement methods, PV module designs, and modeling together in one comprehensive book of a reasonable length.

We hope that, with its structured setup, this book will reach out to interested students who have already completed a bachelor's degree in exact or natural sciences or in an engineering study but also to PV specialists who would like to gain a better understanding of the topic other than their field in photovoltaics. As such, we wish to reach a wide audience and engage more people in the incredibly fast-growing field of photovoltaic technologies.

October 2023        *Wilfried van Sark*
                         (on behalf of all editors and authors that contributed to this book)
                         Utrecht University

# Acknowledgments

Over the past four years, we, the editors of *Photovoltaic Solar Energy, From Fundamentals to Applications, Volume 2,* have had the great pleasure of working with many colleagues to bring this book to life.

Our sincere thanks go therefore to all contributing authors and respected photovoltaic specialists (in alphabetical order): Esther Alarcón-Lladó, Thomas Allen, Kevin Anderson, Erkan Aydin, Tino Band, Matthew Berwind, Karsten Bittkau, Mathieu Boccard, Christian Braun, Xue Chen, Alison Ciesla, Karoline Dapprich, Michele De Bastiani, Stefaan De Wolf, Kaining Ding, Priya Dwivedi, Jiarui Fan, Francesco Frontini, Vasilis Fthenakis, Nannan Fu, Sara Golroodbari, Ziv Hameiri, Clifford Hansen, Anita Ho-Baillie, Md. Anower Hossain, Dirk Jordan, Bishal Kafle, Olga Kanz, Kai Kaufmann, Dominik Lausch, Enrica Leccisi, Frank Lenzmann, Maria Antonietta Loi, Md Arafat Mahmud, Sander A. Mann, Johanna May, David Moser, Neel Patel, Armin Richter, Timothy Schmidt, Josefine Selj, Elham Shirazi, Ronald Sinton, Lenneke Slooff-Hoek, Joshua Stein, Anand Subbiah, Murad Tayebjee, Michelle Vaqueiro Contreras, Bart Vermang, Han Wang, Jianmei Xu, Masafumi Yamaguchi, Shu Zhang, Jun Zhao, Jianghui Zheng, and Yan Zhu.

We would also like to thank the Wiley team, Nandhini Karuppiah, Sandra Grayson, Becky Cowan, Mustaq Ahamed Noorullah, Kavipriya Ramachandran, Katherine Wong, Juliet Booker, Steven Fassioms, Peter Mitchell, and Michelle Dunckley for their professional support in this endeavor, which was even more challenging with the COVID pandemic hitting the globe.

Thank you to all the solar deities who, through the centuries, represented the beneficial powers of the Sun and who give us humans free solar power every day. Thank you, Surya, Ra, Sol, Aryaman, Helios, Doumo, Magec, and Tonatiuh.

Our hope continues to be that humankind finally realizes that solar power is powering our precious one and only Earth.

Acknowledgments

# About the Companion Website

This book is accompanied by a companion website:

**www.wiley.com/go/PVsolarenergy**

The website includes:

- Figures
- Tables

**Part One**

**Introduction to the Book**

# 1

# Introduction

*Wilfried van Sark[1], Bram Hoex[2], Angèle Reinders[3], Pierre Verlinden[4], and Nicholas J. Ekins-Daukes[2]*

[1] *Copernicus Institute of Sustainable Development, Utrecht University, Utrecht, The Netherlands*
[2] *School of Photovoltaic and Renewable Energy Engineering, University of New South Wales, Sydney, Australia*
[3] *Department of Mechanical Engineering, Energy Technology Group, Eindhoven University of Technology, Eindhoven, The Netherlands*
[4] *Yangtze Institute for Solar Technology, JiangYin, China*

## 1.1 Introduction to Photovoltaic Solar Energy, Volume 2

Since the publication of the book *Photovoltaic Solar Energy, from Fundamentals to Applications* (Reinders et al. 2017), we have been eager to compile another volume, in fact, to start a series that would appear every 5–10 years to showcase the advancement of the state-of-the-art in photovoltaics (PVs) research, development, and deployment. We are now proud to present the second volume, and we are very grateful to the many authors who committed time to write a new or updated chapter.

One of the main materials developments in the past decade relates to using perovskites for PV solar energy harvesting. This originated from the groundbreaking work in dye-sensitized solar cells as gradually the dyes have been replaced by organic/inorganic structures with a perovskite crystal structure, i.e. methylammonium lead iodide ($CH_3NH_3PbI_3$, $MAPbI_3$) (Liu et al. 2023). Perovskite solar cells now have demonstrated a record 26% efficiency, and combining perovskite and silicon in a tandem structure has reached over 33% efficiency. Figure 1.1 presents the best-known efficiency graph, diligently maintained and updated by the US National Renewable Energy Laboratory (NREL).

Another major development in PV technology is the rapid decrease of cost, leading to the fast deployment of PV worldwide (Jäger-Waldau 2023). This is best illustrated using the so-called learning or experience curve (Junginger and Louwen 2020), which has been updated since 1976. It is striking that a learning rate of 24% is continued over nearly five decades, this means that with every doubling of cumulative shipments, which happens about every three years, the module price is reduced by 24% (ITRPV 2023). Fluctuations are due to various market effects, such as supply chain disruption and demand–supply mismatches. Note that the learning rate approached 40% taking 2006–2022 data. For thin film technologies, learning rates are similar, e.g. around 20%, but with a cumulative shipment

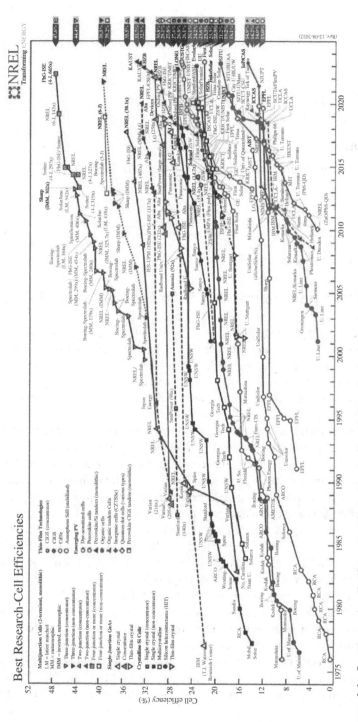

**Figure 1.1** Best research cell efficiencies illustrating that for all PV technologies major efficiency increases have been realized in the past 50 years.
Source: NREL (2023)/Public Domain.

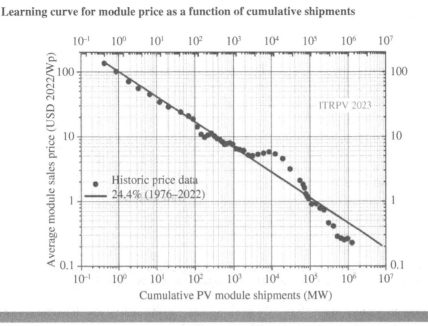

**Learning curve for module price as a function of cumulative shipments**

**Figure 1.2** Learning curve for silicon wafer-based PV technology: module spot market price as a function of cumulative PV module shipments. Source: ITRPV (2023)/International Technology Roadmap for Photovoltaics.

of about 15 times lower than that of silicon wafer-based technology (Philipps and Warmuth 2023) (Figure 1.2).

## 1.2 The First Terawatt

As a consequence of the continued decrease in cost, deployment of PV systems has been booming. Cumulative worldwide shipment has surpassed the 1-terawatt level in early 2022, whereas cumulative installed capacity has reached the 1 TW mark later in the year 2022 (Weaver 2022). Figure 1.3 shows the development of cumulative PV capacity since 2005 at which date only 4 GWp was installed, a level that now is the annual installed capacity of a relatively small country such as the Netherlands (IEA-PVPS 2023). The compound annual growth rate since 2012 has been 28%, and this is expected to continue. Such continued growth, in combination with innovation (Verlinden et al. 2023), is necessary to realize a 100% renewables powered society by mid-century, which would require some 15–75 TWp of PV power (Haegel et al. 2023) in combination with other renewable sources, in particular wind and storage.

PV systems come in a variety of sizes, from the small single-panel solar home system to the large-scale multi GWp PV plant in a large desert area. Using a demarcation of 15 kWp, which is maximum size typically found in residential, grid-connected applications, it is found that about 40% of the PV systems are distributed systems (IEA-PVPS 2022). Of those

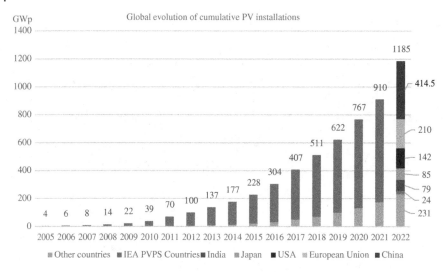

**Figure 1.3** Development of globally installed PV capacity, with a regional breakdown for 2022. Source: IEA-PVPS (2023)/International Energy Agency Photovoltaic Power Systems Programme.

systems, power monitoring is usually not available to grid operators, which may pose problems in ensuring security of supply (Van Sark 2023).

## 1.3 Structure of the Book

After this short introduction, Part 2 presents the basic principles of solar irradiance resources, a very important data for PV modeling. Next, in Part 3, updates on silicon wafer technology compared to Volume 1 are covered in 6 chapters, e.g. on wafering, n-type tunnel oxide passivated contact (TOPCON) cells, heterojunction cells, passivating contacts, degradation, and hydrogenation. Subsequent parts will introduce the reader to other material systems and cell design approaches for PV devices, that is, perovskite solar cells in Part 4, tandem structures and multijunction cells, and nanophotonics approaches in Part 6.

Part 7 covers several characterization techniques that can be used in R&D laboratories and manufacturing lines. Part 8 continues with PV module aspects, in particular, high-efficiency silicon modules in manufacturing context, bifacial modules, smart modules for outdoors partial shading resilience, and colored and building integrated PV modules. This is followed by two chapters on I-V models and thermal aspects of modules.

Part 9 then continues with systems and applications. A performance review is provided, as well as aspects of integration of PV in smart grids with batteries. Recent new applications fields are then covered – building integrated PV, floating PV, PV in agriculture, and PV in and for mobility. This part closes with two chapters on the bankability and reliability of PV systems. Part 10 completes the book and showcases chapters on sustainability aspects on a terawatt scale, life cycle analysis, and recycling.

# References

Haegel, N.M., Verlinden, P., Victoria, M. et al. (2023). Photovoltaics at multi-terawatt scale: waiting is not an option. *Science* 380: 39–42.

IEA-PVPS (2022). Trends in Photovoltaic Applications 2022. Report IEA PVPS T1-43:2022. https://iea-pvps.org/trends_reports/trends-2022/ (accessed 20 February 2024).

IEA-PVPS (2023). Snapshot of Global PV Markets 2023. IEA-PVPS Task 1, report IEA-PVPS-T1-44:2023. https://iea-pvps.org/snapshot-reports/snapshot-2023/ (accessed 20 February 2024).

ITRPV (2023). International Technology Roadmap for Photovoltaic (ITRPV) 2022 Results. https://www.vdma.org/international-technology-roadmap-photovoltaic (accessed 20 February 2024).

Jäger-Waldau, A. (2023). Snapshot of photovoltaics – May 2023. *EPJ Photovoltaics* 14: 23.

Junginger, M. and Louwen, A. (2020). *Technological Learning in the Transition to a Low-Carbon Energy System*. London: Academic Press.

Liu, S., Biju, V.P., Qi, Y. et al. (2023). Recent progress in the development of high-efficiency inverted perovskite solar cells. *NPG Asia Materials* 15: 27.

NREL (2023). Best research-cell efficiency chart. https://www.nrel.gov/pv/cell-efficiency.html (accessed 20 February 2024).

Philipps, S. and Warmuth, W. (2023). Photovoltaics Report (21 February 2023). Fraunhofer Institute for Solar Energy Systems. https://www.ise.fraunhofer.de/en/publications/studies/photovoltaics-report.html (accessed 20 February 2024).

Reinders, A., Verlinden, P., van Sark, W., and Freundlich, A. (ed.) (2017). *Photovoltaic Solar Energy, from Fundamentals to Applications*. Chichester: Wiley.

Van Sark, W. (2023). Photovoltaics performance monitoring is essential in a 100% renewables-based society. *Joule* 7: 1388–1393.

Verlinden, P., Young, D.L., Xiong, G. et al. (2023). Photovoltaic device innovation for a solar future. *Device* 1: 100013.

Weaver, J.F. (2022). World has installed 1TW of solar capacity. *PV Magazine* (15 March 2022). https://www.pv-magazine.com/2022/03/15/humans-have-installed-1-terawatt-of-solar-capacity/ (accessed 20 February 2024).

**Part Two**

**Solar Irradiance Resources**

# 2

# Solar Irradiance Resources

*Joshua S. Stein*

*Climate Security Center, Sandia National Laboratories, Albuquerque, NM, USA*

## 2.1   Introduction

Sunlight is the fuel that is converted into electrical energy by solar photovoltaic systems. The fact that this fuel is free and clean is one of the main advantages of solar energy. To fully understand the availability of sunlight on the surface of the Earth, it is necessary to explore several aspects of the Earth–Sun system and cover some basic orbital dynamics.

## 2.2   Earth–Sun System

The Earth orbits the Sun in an elliptical orbit where the Earth–Sun distance varies by 1.7% during the year (Duffie and Beckman 2006). As shown in Figure 2.1, Earth is closest to the Sun (**perihelion**) around January 3rd and furthest from the Sun (**aphelion**) around July 3rd. These dates vary from year to year due to the mismatch between our calendar and celestial dynamics. The diameter of the Sun is about $10^9$ times larger than that of the Earth, and it is so far away that from the Earth's surface, the Sun's disc appears as a circle approximately 0.51° in diameter. This means that direct light from the sun varies in its direction on the surface of the earth by ±0.255°. By convention, when we define the position of the Sun, we are referring to the position of the center of this circle.

## 2.3   Sun Position Calculations

The position of the Sun relative to an observer on the surface of the Earth is an important quantity for calculating the output of a PV system (Figure 2.2). In this chapter, when we refer to *solar position*, we mean the relative position from the perspective of a horizontal plane on the surface of the Earth. It is relative since the position varies depending on where you are on Earth. For example, when the Sun is rising in California, it is high in the sky in Europe and Africa. We describe solar position with two angles, the **solar elevation angle**

*Photovoltaic Solar Energy: From Fundamentals to Applications, Volume 2*, First Edition.
Edited by Wilfried van Sark, Bram Hoex, Angèle Reinders, Pierre Verlinden, and Nicholas J. Ekins-Daukes.
© 2024 John Wiley & Sons Ltd. Published 2024 by John Wiley & Sons Ltd.
Companion website: www.wiley.com/go/PVsolarenergy

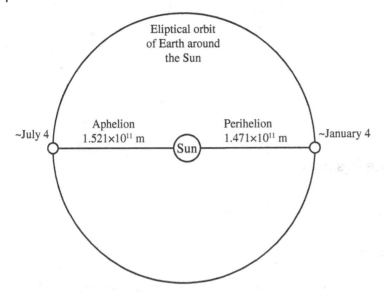

**Figure 2.1** Earth's orbit around the Sun.

**Figure 2.2** Solar position definition.

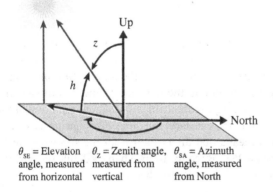

$\theta_{SE}$ = Elevation angle, measured from horizontal

$\theta_{Z}$ = Zenith angle, measured from vertical

$\theta_{SA}$ = Azimuth angle, measured from North

($\theta_{SE}$) and the **solar azimuth angle** ($\theta_{SA}$). The solar elevation angle is the angle between the Sun and the horizon. This same information is also conveyed by the solar zenith angle ($\theta_{Z}$), which is $90° - \theta_{SE}$, and is the angle between the Sun and the zenith, which is the point in the sky that is directly overhead. The solar azimuth angle is the angle of the vector pointing at the Sun projected onto a horizontal surface. In the literature, azimuth angles follow several different conventions. In this chapter, we define the solar azimuth angle with the same convention used for navigation (e.g. North is 0°, East is 90°, South is 180°, and West is 270°).

The solar position is used to determine the angle of incidence of the sunlight on the plane of the PV system (also referred to as plane of array (POA)) and it also affects the magnitude of irradiance hitting standard measurement planes, such as the horizontal plane. Many algorithms, ranging from simple to complex, exist for estimating the Sun's position. Simple

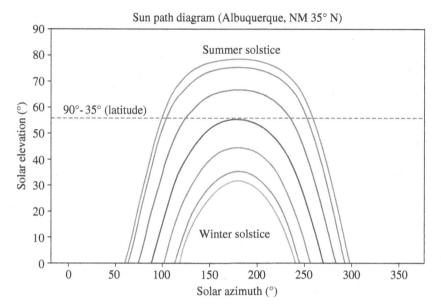

**Figure 2.3** Sun paths for the 21st day of the month for December–June. Sun elevation is equal to 90°-latitude at solar noon on the equinox (March 21st in this figure).

algorithms are appropriate for rough calculations, but there are many reasons to use the most accurate methods available. Perhaps the most obvious is in the case of tracking systems for concentrating photovoltaics (CPV). Simple algorithms, which can have errors as much as 2–3°, are simply not accurate enough for CPV applications. Furthermore, any reduction in the uncertainty of PV output is directly related to the potential economic profits made by the system. For these reasons, it is advisable to use the most accurate algorithms available.

For solar position calculations for solar energy applications, the Solar Position Algorithm (SPA) (Reda and Andreas 2008) is the most accurate available outside the astronomy community. This algorithm calculates the Sun's position to within ±0.0003° and is valid from the year −2000 to 6000. It includes empirical corrections that help match observations over long periods of time, considering the complex orbital and rotation dynamics of the Earth. Thus, for applications where accuracy is important, it is strongly advised to use the most accurate sun position algorithm available. The SPA is included as part of the pvlib-python open-source library (Holmgren et al. 2018).

Figure 2.3 shows a Sun path diagram for seven representative days of the year for Albuquerque, New Mexico, USA, using the SPA.

## 2.4 Extraterrestrial Irradiance

The intensity (in W/m²) of the solar radiation at the top of the Earth's atmosphere, normal to the direction of the Sun's rays, is called extraterrestrial radiation. It varies throughout the year due to the varying Earth–Sun distance from the elliptical orbit. This variation is shown in Figure 2.4.

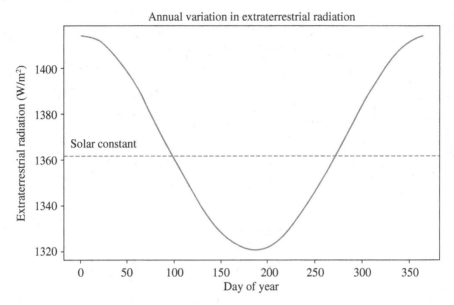

**Figure 2.4** Annual variation in extraterrestrial radiation. *Note*: Calculated using pvlib-python using pvlib.irradiance.get_extra_radiation () function with defaults.

The mean value of extraterrestrial radiation is called the solar constant (approximately $1361\,W/m^2$), which, in fact, varies slightly over time due to changes in the activity of the Sun, but is usually far less than the variation due to the varying Earth–Sun distance. One reason for the fluctuating solar constant is the phenomenon known as sunspots. They are caused by changes in the Sun's magnetic field, resulting in cooler areas on the Sun's surface that emit less light and appear as dark spots. Populations of sunspots tend to appear and disappear in irregular cycles, lasting approximately 11 years. From 1975 to 2005, there were three distinct sunspot cycles that resulted in the solar constant varying by about 0.1% or $1.3\,W/m^2$.

## 2.5 Solar Radiation at the Earth's Surface

While extraterrestrial radiation only varies slightly due to changes in the Earth–Sun distance and the Sun's dynamics, the irradiance measured at the Earth's surface varies significantly due to the Earth's orbit around the Sun, the Earth's rotation and tilted axis, and effects of the atmosphere, including weather patterns and especially clouds.

Extraterrestrial radiation is almost all in the form of direct irradiance, meaning it comes directly from the Sun and is not scattered. As light travels through the atmosphere, it is scattered, and the direct component of the light is reduced as the diffuse component increases. Diffuse light travels in all directions, which allows objects to be visible when they are in shadow. When a cloud obscures the Sun, the direct component is reduced to essentially zero and the remaining light is diffused.

There are three components of irradiance that are typically measured to quantify the solar resource at a site. These include direct normal irradiance (DNI), global horizontal irradiance (GHI), and diffuse horizontal irradiance (DHI). These components are related to each other with the Eq. (2.1). Information about how these quantities are measured is covered in Chapter 3

$$GHI = DNI \times \cos(\theta_Z) + DHI \tag{2.1}$$

## 2.6 Spectral Content of Sunlight

Sunlight at the Earth's surface is composed of a spectrum of wavelengths that reflect the properties of the Sun and the atmosphere. Extraterrestrial radiation has a similar spectrum predicted by a blackbody radiation source, however, the spectrum at the surface of the Earth is significantly affected by absorption in the atmosphere, especially by water, ozone, and carbon dioxide (Figure 2.5).

The effect of the atmosphere on the spectrum changes with the path length of the direct irradiance through the atmosphere. A measure of this path length is referred to as airmass (AM) and it varies with zenith angle and surface elevation as well as air pressure fluctuations. Specific solar reference spectra are defined for testing standards. These include AM0, which refers to the solar spectrum outside of the atmosphere (where

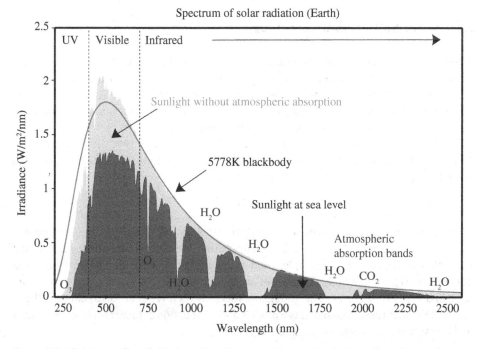

**Figure 2.5** Spectrum of sunlight above the atmosphere and on the Earth surface. Source: By Robert A. Rohde – This file has been extracted from another file, CC BY-SA 3.0, https://commons.wikimedia.org/w/index.php?curid=24648395/Wikimedia Commons/CC BY-SA 3.0.

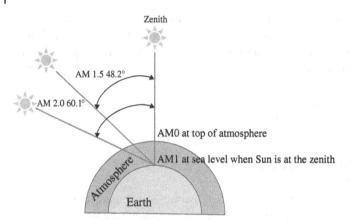

**Figure 2.6** Airmass illustration.

AM = 0); AM1 is the AM at sea level when the solar zenith angle is zero; and AM1.5 is the reference spectrum when the solar zenith angle = 48.2°, and is also the standard spectrum that defines standard test conditions (STC) at which PV modules are to be tested. AM can be approximated by Eq. (2.2). Figure 2.6 illustrates these relationships.

$$AM = 1/\cos(\theta_Z) \tag{2.2}$$

## 2.7 Clear Sky Irradiance Models

Irradiance at a location without the influence of clouds is called "clear sky" irradiance and is useful for comparisons to ground measurements to determine clearness and for simulating the maximum potential performance from a PV system at a given site. There exist several clear sky irradiance models of varying complexity. We will present a few of the simple ones and provide references for other more complex models.

The simplest clear sky model was proposed by Haurwitz (1945, 1948). It estimates clear sky GHI ($GHI_{CS}$) and has two empirical coefficients that were fit to field data in the original paper. $GHI_{CS}$ depends on relative AM or zenith angle ($\theta_Z$):

$$GHI_{CS} = \frac{a}{AM}e^{-b\cdot AM} = 1098\ \cos(\theta_Z)e^{\frac{-0.057}{\cos(\theta_Z)}} \tag{2.3}$$

Despite the simplicity of this model, it performs remarkably well for many applications.

If a more accurate model is needed, and/or if other irradiance components are needed (e.g. DNI & DHI), other more complex models are available. These include Bird (Bird and Hulstrom 1981), Simple Solis (Mueller et al. 2004; Ineichen 2008), Ineichen–Perez (Ineichen and Perez 2002; Perez et al. 2002), and McClear (Lefèvre et al. 2013) among others. The mathematical description of these models and a validation comparison of them against field measurements is available (Mabasa et al. 2021).

## 2.8 Conclusion/Summary

Solar irradiance on the Earth's surface is controlled by the orbital dynamics between the Earth and Sun. These relationships, while complex, can be simplified using approximations and open-source software available for free.

## Author Biography

**Joshua S. Stein** is a Senior Scientist at Sandia National Laboratories where he leads research and development projects in the areas of photovoltaic module and system performance and reliability, specializing in modeling and analysis of new technologies. He is the founder of the PV Performance Modeling Collaborative (PVPMC) and represents the United States in the International Energy Agency PVPS Task 13 on PV Performance and Reliability.

## References

Bird, R.E. and Hulstrom, R.L. (1981). *Simplified Clear Sky Model for Direct and Diffuse Insolation on Horizontal Surfaces*; No. SERI/TR-642-761. Golden, CO, USA: Solar Energy Research Institute.

Duffie, J.A. and Beckman, W.A. (2006). *Solar Engineering of Thermal Processes*, 3e. Wiley.

Haurwitz, B. (1945). Insolation in relation to cloudiness and cloud density. *Journal of Meteorology* 2: 154–166.

Haurwitz, B. (1948). Isolation in relation to cloud type. *Journal of Meteorology* 1948 (5): 110–113.

Holmgren, W.F., Hansen, C.W., and Mikofski, M.A. (2018). pvlib python: a python package for modeling solar energy systems. *Journal of Open Source Software* 3 (29): 884. https://doi.org/10.21105/joss.00884.

Ineichen, P. (2008). A broadband simplified version of the Solis clear sky model. *Solar Energy* 82: 758–762.

Ineichen, P. and Perez, R. (2002). A new airmass independent formulation for the Linke turbidity coefficient. *Solar Energy* 73: 151–157.

Lefèvre, M., Oumbe, A., Blanc, P. et al. (2013). Mcclear: a new model estimating downwelling solar radiation at ground level in clear-sky conditions. *Atmospheric Measurement Techniques* 6: 2403–2418.

Mabasa, B., Lysko, M.D., Tazvinga, H. et al. (2021). The performance assessment of six global horizontal irradiance clear sky models in six climatological regions in South Africa. *Energies* 14: 2583. https://doi.org/10.3390/en14092583.

Mueller, R., Dagestad, K., Ineichen, P. et al. (2004). Rethinking satellite based solar irradiance modelling. The SOLIS clear-sky module. *Remote Sensing of Environment* 91: 160–174.

Perez, R., Ineichen, P., Moore, K. et al. (2002). A new operational model for satellite-derived irradiances: description and validation. *Solar Energy* 73: 307–317.

Reda, I. and Andreas, A. (2008). *Solar Position Algorithm for Solar Radiation Applications*. Golden, Colorado: National Renewable Energy Laboratory.

# 3

# Irradiance and Weather Datasets for PV Modeling

*Joshua S. Stein*

*Climate Security Center, Sandia National Laboratories, Albuquerque, NM, USA*

## 3.1  Introduction

To evaluate the production potential of a solar photovoltaic energy system at a particular site, time series of high-quality solar irradiance, air temperature, wind speed, precipitation, and other weather-related data are required. At a minimum, an entire year of data is necessary to represent all the seasonal and diurnal variations. Developers prefer to have multiple years of data from a site to better predict the effect of interannual variability or the difference in conditions from year to year.

## 3.2  Measurements of Irradiance Data

Irradiance is measured using a radiometer of which there are several types. Direct normal irradiance (DNI) is measured with a pyrheliometer, which only accepts light that is coming directly from the solar disk. This is accomplished by attaching a thermal sensor to a collimator and pointing the instrument at the Sun using a solar tracker. Global measurements include all light hitting the plane of the sensor, regardless of which direction the light is coming from. Global horizontal irradiance (GHI) and diffuse horizontal irradiance (DHI) are both measured with a pyranometer. In the case of a DHI measurement, the direct sunlight is blocked or shaded from hitting the sensor by means of a dark band or ball that is moved with a solar tracker. An alternate method to measure global irradiance is to measure the short circuit current from a calibrated solar cell and convert that to irradiance via a calibration coefficient since current is directly and nearly linearly proportional to irradiance. However, since the solar cell only responds to a portion of the solar spectrum, this method of measuring irradiance should only be used when the effects of the other parts of the spectrum have been deemed unimportant for application. Figure 3.1 shows examples of sensors used to measure these components.

To ensure that the world uses the same reference to calibrate irradiance instruments, the World Radiation Center in Davos, Switzerland, maintains a group of absolute cavity

*Photovoltaic Solar Energy: From Fundamentals to Applications, Volume 2*, First Edition.
Edited by Wilfried van Sark, Bram Hoex, Angèle Reinders, Pierre Verlinden, and Nicholas J. Ekins-Daukes.
© 2024 John Wiley & Sons Ltd. Published 2024 by John Wiley & Sons Ltd.
Companion website: www.wiley.com/go/PVsolarenergy

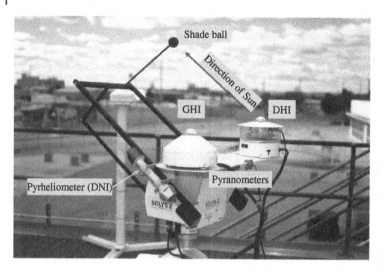

**Figure 3.1** Examples of solar radiometers mounted in the field on a solar tracker.

radiometers (ACRs) that are tracked to point directly at the Sun and measure the DNI. This collection of instruments is the basis of the World Radiometric Reference (WRR). ACRs are the most accurate instruments to measure this radiation and the mean value of the measured DNI from this group of reference instruments is considered the world reference. Periodically, researchers from primary standards laboratories throughout the world come to Davos with their own ACRs and calibrate these instruments to the standard group. These calibrated ACRs are then brought back home and used to calibrate other radiometers, including pyrheliometers and pyranometers during regional calibration intercomparisons. All radiometer calibrations should be tracible back to the WRR.

## 3.3 Satellite Irradiance Datasets

Irradiance can also be estimated from satellite imagery of the Earth. The basic premise of this technique is that the amount of light reflected back to space from a location on Earth is inversely proportional to the amount of irradiance hitting the surface at that location. This is because clouds are more reflective than the ground, thus, if the reflectivity of the ground is characterized at different sun angles, increases in this reflected light are interpreted as increased cloud cover and a corresponding decrease in solar irradiance on the ground. There are a range of models used to convert satellite imagery of the Earth to irradiance at the ground. Physical models (e.g. Emde et al. 2016; Mayer and Kylling 2005) attempt to solve radiative transfer equations but require information about the composition and state of the atmosphere that is rarely available over large areas. These models are being improved but are still in the research stage. Empirical and semiempirical models (e.g. Rigollier et al. 2004; Perez et al. 2013) are actively being used today by industry to generate solar irradiance datasets. These models are based primarily on statistical regressions between measured light intensity from the satellite image, and measured irradiance from ground stations. However, many add additional information such as multi-spectral sensor data that is included on many satellites. Sensors tuned to specific narrow bands of the electromagnetic spectrum provide important information about the local atmospheric and ground cover. For example, snow on the ground can look like clouds and confuse the regression, however, snow is easy to distinguish from clouds using infrared imagery and this results in a more accurate irradiance product. Other spectral bands are sensitive to different kinds of clouds, ozone, and $CO_2$. This information can be used to improve the regression results and deliver more accurate irradiance data. One limitation of satellite irradiance models is the temporal resolution of the imagery. Current satellites used for this purpose typically provide images every 15–30 minutes. Newer satellites are planned for deployment that will provide images every five minutes or less. Data is generally available from 1999, depending on the region. Spatial resolution of the irradiance data is offered as low as 250 m × 250 m, but this varies by source.

There are several organizations and companies that generate and disseminate satellite-derived irradiance data. The community that generates these datasets is competitive and offerings are updated and improved regularly. Data provided by public organizations include:

- **National Solar Radiation Database** (NSRDB) from the US National Renewable Energy Laboratory (NREL). It focuses on the United States but also has data available from Mexico, Central America, and parts of South Asia.
- **PVGIS** from the European Commission includes data from Europe, Africa, and parts of western Asia.
- **NASA** provides lower spatial and temporal resolution for the entire world.

Commercial sources of data include some free data sets as examples, or on a trial basis, but most of the data requires payment and includes licensing agreements. Some of the main commercial sources of global coverage of irradiance data are: Solargis,[1] Clean Power Research,[2] Solcast,[3] and Vaisala.[4] An advantage of using commercial sources is that most of them have merged other weather measurements from ground-based networks to include collocated air temperature, wind speed, and precipitation data. Each of these companies uses different analysis methods and offers a range of data products (including varying spatial and temporal resolutions) and documentation of their data validation methods and results.

## 3.4 Modeled or Processed Irradiance Datasets

While measured data represents the conditions on the ground at the time of the measurement, it does have some limitations when the goal is to estimate the expected energy yield of a PV system before it is built, a necessary step in obtaining financing for the project. For this application, one can choose a single year of data and assume that it is representative of future years or run the yield assessment using all the available years of data and try and

---

1 https://solargis.com
2 https://solaranywhere.com
3 https://solcast.com
4 https://www.vaisala.com/en/digital-and-data-services/renewable-energy

estimate the expected yield from the varying results. Another option that is popular is to process the measured data and assemble what is referred to as a typical meteorological year (TMY).

The methodology for developing TMY datasets was originally developed at Sandia National Laboratories as early as 1978 (Hall 1978; Hall et al. 1978; Menicucci and Femandez 1988). The methods for generating the files were further refined by NREL with the release of TMY2 files (Marion and Urban 1995). TMY2 files were created for 239 sites by analyzing data from 1961–1990, standardizing the output format from solar time to local time, and setting standards for reporting units. Since the release of TMY2, NREL has regularly updated the files to include more site locations as well as adjusted the time period analyzed to create the files. NREL released a major update in 2008 with the release of TMY3 files (Wilcox and Marion 2008), which again changed the formatting standards but included a total of 1020 sites in the United States and its territories, from measurements from 1991–2005. Later, NREL released a gridded TMY product for the United States based on decades of satellite irradiance data (Habte et al. 2014).

TMY datasets contain hourly values of irradiance components, air temperatures, wind speed, precipitation, as well as a number of other details. They include data from every hour of a standard 365-day year (8760 hours). Most of the data in a TMY file is either measured or filled in using well-documented methods designed to minimize bias errors. The basic methodology is to divide up the period of measured data (e.g. 30-year record) into monthly bins and examine each month of the year separately. For instance, to choose data to represent a typical January, all Januarys in the 30-year period are compared and one is selected as being typical.

Over time, the details of the selection method for each type of TMY file have evolved and are explained in the user's guides. In general, the most typical month is chosen by means of assigning weights to each of the data fields and evaluating weighted sums in relation to long-term means and medians. For TMY3 data, 50% of the weighting was split between GHI and DNI, 40% of the weight was split between temperature, and 10% of the weight was assigned to wind speed. The exact choice of the weights has always been a source of discussion and disagreement, but TMY files have historically been used for both PV and concentrated solar power (CSP) purposes, and this has resulted in some issues since CSP plants respond to DNI irradiance while being less sensitive to GHI, especially during cloudy conditions.

## 3.5  Conclusion

Accurate solar irradiance, air temperature, wind speed, precipitation, and other weather-related data are required for the prediction of energy yield for a solar photovoltaic plant. The current best practice is to collect high-quality field measurements of irradiance at the proposed site of the plant for at least a year or more. This data is used to adjust or calibrate historical satellite irradiance data which is then used as input to annual simulations of energy yield. Alternatively, TMY data is used if a single year of predicted performance is desired.

## Author Biography

**Joshua S. Stein** is a Senior Scientist at Sandia National Laboratories where he leads research and development projects in the areas of photovoltaic module and system performance and reliability, specializing in modeling and analysis of new technologies. He is the founder of the PV Performance Modeling Collaborative (PVPMC) and represents the United States in the International Energy Agency PVPS Task 13 on PV Performance and Reliability.

## References

Emde, C., Buras-Schnell, R., Kylling, A. et al. (2016). The libRadtran software package for radiative transfer calculations (version 2.0.1). *Geoscientific Model Development* 9: 1647–1672.

Habte, A., Lopez, A., Sengupta, M., and Wilcox, S. (2014). *Temporal and Spatial Comparison of Gridded TMY, TDY, and TGY Data Sets*. Golden, CO, USA: National Renewable Energy Laboratory TP-5D00-60886.

Hall, I.J. (1978). Generation of a typical meteorological year. *Proceedings of the 1978 Annual Meeting of AS ISES, Denver, CO, USA 2.2*.

Hall, I., Prairie, R., Anderson, H., and Boes, E. (1978). *Generation of Typical Meteorological Years for 26 SOLMET Stations*. SAND78-1601. Albuquerque, NM, USA: Sandia National Laboratories.

Marion, W. and Urban, K. (1995). *User's Manual for TMY2s Typical Meteorological Years*. Golden, CO, USA: National Renewable Energy Laboratory.

Mayer, B. and Kylling, A. (2005). Technical note: the libRadtran software package for radiative transfer calculations – description and examples of use. *Atmospheric Chemistry and Physics* 5: 1855–1877.

Menicucci, D. and Femandez, J. (1988). *A Comparison of Typical Year Solar Radiation Information with the SOLMET Data Base*. SAND87-2379. Albuquerque, NM, USA: Sandia National Laboratories.

Perez, R., Cebecauer, T., and Suri, M. (2013). Semi-empirical satellite models. In: *Solar Energy Forecasting and Resource Assessment* (ed. J. Kleissl), 21–48. Academic Press.

Rigollier, C., Lefevre, M., and Wald, L. (2004). The method Heliosat-2 for deriving shortwave solar radiation from satellite images. *Solar Energy* 77: 159–169.

Wilcox, S. and Marion, W. (2008). *Users Manual for TMY3 Data Sets*. Golden, CO, USA: National Renewable Energy Laboratory TP-581-43156.

# 4

# Irradiance on the Plane of the Array

*Joshua S. Stein[1] and Clifford W. Hansen[2]*

[1]*Climate Security Center, Sandia National Laboratories, Albuquerque, NM, USA*
[2]*Photovoltaics and Materials Technologies, Sandia National Laboratories, Albuquerque, NM, USA*

## 4.1 Introduction

After the solar resource data has been compiled for the selected site (see Chapter 3), the next step in the modeling process is to estimate the plane-of-array (POA) irradiance for the system design. POA irradiance is defined as the irradiance available in a plane parallel to the PV modules. Factors that affect POA irradiance include Sun position, array orientation, tracking, ground reflections, and sky conditions. Models for calculating this quantity are called transposition models; code for many of these models can be found in the pvlib-python open-source library (Holmgren et al. 2018).

## 4.2 Modeling POA Irradiance

POA irradiance ($E_{POA}$) is the sum of three terms:

$$E_{POA} = E_b + E_{sd} + E_g \qquad (4.1)$$

where $E_b$ is the beam irradiance on the array, $E_{sd}$ is the sky diffuse irradiance in the view of the array, and $E_g$ is the ground-reflected irradiance in the view of the array. The array orientation and Sun position are inputs to the calculations of these POA irradiance components.

An important concept for understanding the $E_{sd}$ and $E_g$ components of POA irradiance is the view area of the array. In the case of a standard flat plate PV array, the view area is defined as an infinite hemispherical dome extending above the top of the array. When the array is horizontal, the view area includes all of the sky dome. When the array is tilted, the view area includes a portion of the sky dome and portion of the ground surface. For a tilted

*Photovoltaic Solar Energy: From Fundamentals to Applications, Volume 2*, First Edition.
Edited by Wilfried van Sark, Bram Hoex, Angèle Reinders, Pierre Verlinden, and Nicholas J. Ekins-Daukes.
© 2024 John Wiley & Sons Ltd. Published 2024 by John Wiley & Sons Ltd.
Companion website: www.wiley.com/go/PVsolarenergy

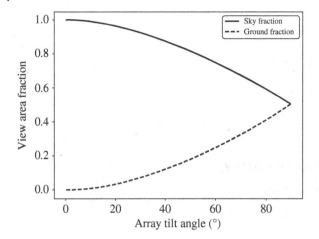

**Figure 4.1** Relative factions of the array view area that are "seeing" the sky and ground, respectively, calculated with Eqs. (4.2) and (4.3).

array, the view area fractions (VAFs) that include the sky and ground for array tilt angle ($\theta_T$) are as follows:

$$\text{VAF}_s = \frac{1 + \cos(\theta_T)}{2} \text{ (sky view area fraction)} \tag{4.2}$$

$$\text{VAF}_g = \frac{1 - \cos(\theta_T)}{2} \text{ (ground view area fraction)} \tag{4.3}$$

Figure 4.1 shows the effect of tilt angle on the VAFs. Notice that for a vertical array at 90° tilt angle, the view area is split evenly between sky and ground. For more typical arrays in the United States and Europe, with tilt angles between 20° and 40°, the fraction of the view area "seeing" the ground is less than 10%. These fractions are important because they provide an indication of how significant the reflective properties of the ground surface (albedo) are to the array performance. An array with a higher tilt angle will be affected by the albedo of the ground to a greater degree. This can be important in situations where the ground albedo can change. For example, ground albedo can increase significantly when there is snow on the ground, and this can lead to significant increases in energy yield for these arrays. In addition, since snow fall and optimal tilt angle usually increase with latitude, these effects are common in high latitudes.

## 4.3 Array Orientation

The orientation of the PV array affects the total irradiance available to the system as well as the shape of the daily irradiance profile. For example, a west-facing PV system will have an irradiance profile skewed toward the afternoon. Figure 4.2 shows examples of clear-sky GHI and POA irradiance profiles for latitude tilts facing South, East, and West in Albuquerque, NM (lat = 35° N).

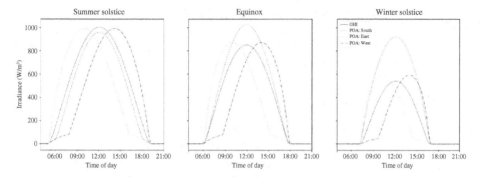

**Figure 4.2** Clear-sky POA irradiance for latitude tilt in Albuquerque, New Mexico, at different times of the year. Calculated using pvlib-python.

### 4.3.1 Fixed Tilt Array Orientation

Fixed tilt arrays are defined by two static angles – the array tilt angle ($\theta_T$) and the array azimuth angle ($\theta_{AA}$). Tilt angle is the angle between the module and horizontal surfaces. The array azimuth angle is the compass direction in degrees of the normal vector pointing away from the module surface and projected to the horizontal plane. For example, a south-facing array would have an array azimuth of 180°, while a west-facing array would have an array azimuth of 270°. Figure 4.3 shows an illustration of these angle conventions.

Choosing an array orientation that will maximize the amount of energy produced is not a simple task and often involves running a PV performance model iteratively to see which pair of tilt and azimuth angles result in the maximum possible energy for the site and conditions. In the absence of clouds and other obstructions of the Sun (e.g. mountains, buildings, and snow), arrays facing toward the equator (south in the northern hemisphere) and tilted to an angle equal to the latitude of the site will maximize the total insolation on the array. However, all sites experience at least some cloud and site effects, and the true optimal angles are usually different than the ideal case. For example, many coastal sites experience fog in the morning and have a greater likelihood of having clear sky in the afternoon. In such a case, it would probably make more sense to orient the array slightly west of the theoretical

**Figure 4.3** Array tilt and azimuth angle examples. Photo credit: Sandia National Laboratories.

optimum. For sites with frequent overcast conditions, it may make sense to decrease the tilt angle to increase the view angle of the sky, which is much brighter during overcast periods than the ground, which is usually darker. However, if the site experiences snowfall, steeper tilt angles may be justified to enhance snow shedding from the array.

### 4.3.2 Tracking Schemes and Algorithms

Tracking systems move the PV array with the Sun to maximize the POA irradiance on the system. There are two main types of tracking, single- and dual-axis tracking. Single-axis tracking works by the rotation of the array on a single axis, as the name suggests. It reduces the angle of incidence of the sunlight on the array but does not keep the array normal to the Sun. Dual-axis tracking allows rotation around two axes that are usually orthogonal and allows the array to be held normal to the Sun or in any desired orientation (Figures 4.4 and 4.5).

**Figure 4.4** Two examples of single-axis trackers. On the left is a horizontal, single-axis tracker. On the right is a tilted single-axis tracker.

**Figure 4.5** Example dual-axis tracker.

Backtracking avoids tracker-to-tracker shading when Sun is low in sky

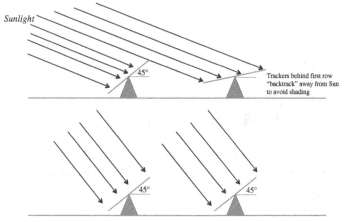

*Sunlight*

45°

Trackers behind first row
"backtrack" away from Sun
to avoid shading

45°     45°

When Sun elevation is greater, trackers can be at equal tilt angles

**Figure 4.6**  Backtracking concept.

### 4.3.3  Backtracking

Some trackers include advanced control algorithms that allow them to "backtrack" at times when the solar elevation is low enough to cast shadows on adjacent rows. The illustration below shows a typical example. Here, the front row is at a tilt angle of 45°, which is a typical maximum value for single-axis trackers. In the case of an infinite single row, the tracker would stay at 45° until the solar elevation increased enough for the movement of the tracker to result in a lower angle of incidence. However, when there are many rows of trackers, rows behind the front row can be told to "backtrack" away from the Sun and move to a lower tilt angle to avoid row-to-row shading. Despite resulting in a higher angle of incidence, this configuration produces more energy than one with row–row shading (Figure 4.6).

## 4.4  Plane-of-Array Beam Irradiance

The simplest of these factors to calculate is the beam irradiance. This is defined as the portion of the DNI that hits the plane of the array. Since the DNI only consists of light coming directly from the Sun without scattering, the calculation of $E_b$ is simply a function of the angle of incidence of this sunlight on the surface of the array ($\theta_{AOI}$):

$$E_b = DNI \cos \theta_{AOI} \tag{4.4}$$

## 4.5  Angle of Incidence

The angle of incidence ($\theta_{AOI}$) can be calculated as:

$$\theta_{AOI} = \cos^{-1}[\cos(\theta_z)\cos(\theta_T) + \sin(\theta_z)\sin(\theta_T)\cos(\theta_A - \theta_{AA})] \tag{4.5}$$

where:

- $\theta_z$ is the solar zenith angle
- $\theta_A$ is the solar azimuth angle
- $\theta_T$ is the array tile angle
- $\theta_{AA}$ is the array azimuth angle

## 4.6 Plane-of-Array Ground Reflected Irradiance

The ground reflected irradiance on the plane of the array ($E_g$) contributes to the diffuse irradiance in the array field of view and is a function of GHI, albedo, and the ground VAF:

$$E_g = \text{GHI} \times \text{albedo} \times \text{VAF}_g \tag{4.6}$$

## 4.7 Albedo

The albedo of a surface quantifies the amount of light that is reflected from it. When working with broadband measurements of light, as is common with PV system modeling, albedo can be expressed as a scalar value. In this case, the ground albedo is defined as the fraction of the GHI that is reflected off the ground surface and is derived from measurements using two horizontal pyranometers, one facing up and the other facing down as:

$$\text{Albedo} = \frac{\text{GRI}}{\text{GHI}} \tag{4.7}$$

where GRI is the ground-reflected irradiance on a horizontal plane.

Figure 4.7 shows an example calculation of albedo performed for about six days of data collected over a compacted gravel surface. Fitting a line to the (GRI, GHI) data provides a good estimate of the broadband albedo, 0.23 in this case.

For a more detailed treatment, one must recognize that the albedo is spectrally dependent. That is, the amount of light that reflects off a surface depends on the wavelength of the

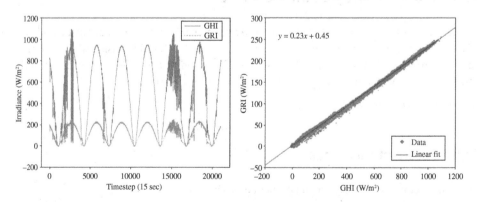

**Figure 4.7** Example broadband albedo measurement data from a uniform compacted gravel surface. The slope of the best fit line indicates that the albedo is 0.23 for this surface.

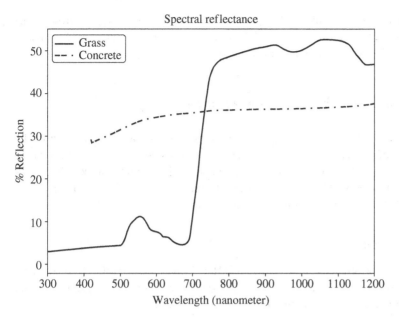

**Figure 4.8** Spectral reflectance data from grass and concrete. Data is from the ASTER Spectral Library. Source: Adapted from Baldridge et al. (2009).

light. This should be no surprise since this effect explains why different materials come in different colors. A "green" material reflects green light and absorbs other colors. For most PV cells, which can use wavelengths from 350–1200 nm (visible and near IR), the spectral reflectance in this range can influence the output of PV systems. Figure 4.8 compares the spectral reflectivity of grass and concrete in the range of typical crystalline silicon (c-Si) PV cell spectral response. Notice that the grass shows a slight peak at about 530 nm. This is the wavelength of green and explains why grass appears green. What is perhaps more significant for PV is the much larger reflectance above 700 nm. This is in the range of near IR and is within the spectral response of most PV cell materials. In contrast, concrete has a flatter reflectance that does not vary as much with wavelength.

In order to calculate the significance of these differences, Brennan et al. (2014) have combined the solar spectrum with the spectral reflectance data for different surface materials and the spectral response of various PV cell types to calculate an effective albedo for different surface/PV cell type combinations. They define effective albedo as:

$$\text{albedo}_{SR} = \frac{\int A(\lambda)\text{SR}(\lambda)G(\lambda)d\lambda}{\int \text{SR}(\lambda)G(\lambda)d\lambda} \tag{4.8}$$

Where:

- $\lambda$ is the wavelength (nm)
- $A(\lambda)$ is the spectral reflectance at wavelength, $\lambda$
- $\text{SR}(\lambda)$ is the spectral response of the PV cell material at wavelength, $\lambda$
- $G(\lambda)$ is the spectral irradiance at wavelength, $\lambda$

Several researchers have made similar measurements over a wide range of ground surfaces and have summarized the resulting albedo values in tables. Coakley (2002) provides a good summary of albedo data for natural surfaces.

## 4.8 Plane-of Array Sky Diffuse Irradiance

The diffuse irradiance from the sky ($E_{sd}$) can be influenced by many factors such as the optical properties of the atmosphere, position of the Sun, presence, and type of clouds. There exist a number of models for estimating this quantity. In this section, we will describe a selection of these models that increase in complexity and hopefully accuracy. Loutzenhiser et al. (2007) provide a nice review of more of these models.

### 4.8.1 Isotropic Model

The simplest POA sky diffuse model assumes that the diffuse light from the sky is isotropic (equal from all directions). While this is rarely a good assumption, the model is a good introduction to the topic because of its simplicity.

$$E_{sd} = \text{DHI} \times \text{VAF}_s \qquad (4.9)$$

The only array parameter that influences this quantity is the tilt angle of the array.

### 4.8.2 Circumsolar Brightening (e.g. Hay and Davies Model)

An important effect of small particles (e.g. 0.1–1 μm) in the atmosphere is the scattering of sunlight. Scattering causes a large increase in the diffuse light on the Earth's surface. Most scattering causes only small angular deviations of incoming light. Therefore, when direct irradiance from the Sun is forward scattered for the first time, it creates an effect known as circumsolar brightening (Unsworth and Monteith 1972). Much of the light is repeatedly scattered, resulting in a wide distribution of angles, but there is a concentration of diffuse light around the Sun.

The Hay and Davies (1980) model for diffuse sky irradiance tries to account for circumsolar brightening by adding a new term to the isotropic model.

$$E_s = \text{DHI} \times \left[ \frac{\text{DNI}}{E_a} \cos(\theta_{AOI}) + \left( 1 - \frac{\text{DNI}}{E_a} \right) \text{VAF}_s \right] \qquad (4.10)$$

where $E_a$ is extraterrestrial radiation.

For example, assuming that DHI = 100 W/m², DNI = 850 W/m², and $E_a$ = 1360 W/m², Figure 4.9 shows the effect of angle of incidence on diffuse sky irradiance predicted by this model for three different array tilt angles. The plot shows an increase in sky diffuse on the array of about 17% from an angle of incidence of 60° to 1° due to circumsolar brightening. Sky diffuse irradiance also increases as tilt angle decreases due to the increase in the view angle fraction for low tilt angles.

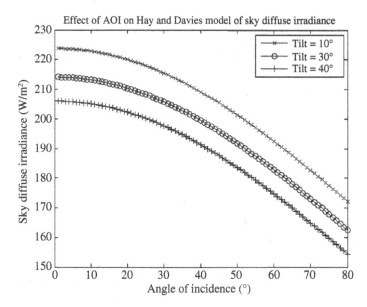

**Figure 4.9** Effect of angle of incidence and tilt angle on the sky diffuse irradiance predicted by the Hay and Davies diffuse irradiance model.

### 4.8.3 Horizon Brightening (e.g. Reindl Model)

In addition to the circumsolar brightening, scattering of light in the atmosphere can cause a slight brightening of the horizon, especially for clear skies. This effect is included in the next two models that are discussed. The first model was developed by Reindl et al. (1990) which adds an additional factor to the Hay and Davies model to represent this effect as a function of zenith angle ($\theta_Z$).

$$E_s = \text{DHI} \times \left[ \frac{\text{DNI}}{E_a} \cos(\theta_{\text{AOI}}) + \left( 1 - \frac{\text{DNI}}{E_a} \right) \text{VAF}_s \left( 1 + \sqrt{\frac{\text{DNI} \times \cos(\theta_Z)}{\text{GHI}}} \sin^3 \left( \frac{\theta_T}{2} \right) \right) \right]$$

$$(4.11)$$

This model differs from the Hay and Davies model above as the array tilt angle increases and the effect of the horizon brightening increases. For example, for the case where DHI = 100 W/m², DNI = 850 W/m², GHI = 950 W/m², and $E_a$ = 1360 W/m², Figure 4.10 compares the two models as a function of tilt angle. At low tilt angles, the models are nearly identical, but as tilt angle increases and the horizon fills more of the VAF, the Reindl model predicts higher sky diffuse irradiance values (about 5% increase at a 50° tilt angle).

### 4.8.4 Empirical Models (e.g. Perez Model)

While the sky diffuse models presented up to this point separated the isotropic, circumsolar, and horizon components explicitly, Perez et al. (1987, 1988, 1990) developed a more complex

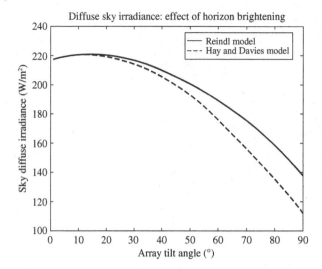

**Figure 4.10** Comparison between the Hay and Davies and Reindl models of diffuse sky irradiance as a function of tilt angle.

model that relies on a set of empirical coefficients derived from regional measurements to fit each of these components. The basic form of the model is:

$$E_s = DHI \times \left[ ((1 - F_1)VAF_s) + F_1\left(\frac{a}{b}\right) + F_2 \sin(\theta_T) \right] \tag{4.12}$$

where:

- $F_1$ and $F_2$ are empirically fitted functions that describe circumsolar and horizon brightening, respectively.
- $a = \max(0, \cos(\theta_{AOI}))$
- $b = \max(\cos(85°), \cos(\theta_Z))$

$$F_1 = \max\left[ 0, \left( f_{11} + f_{12}\Delta + \frac{\pi\theta_Z}{180°}f_{13} \right) \right] \tag{4.13}$$

$$F_2 = f_{21} + f_{22}\Delta + \frac{\pi\theta_Z}{180°}f_{23} \tag{4.14}$$

The $f$ coefficients are defined for specific bins of clearness ($\varepsilon$), which are defined as:

$$\varepsilon = \frac{(DHI + DNI)/DHI + \kappa\theta_Z{}^3}{1 + \kappa\theta_Z{}^3} \tag{4.15}$$

where $\kappa$ is a constant equal to 1.041 for angles in radians or $5.535 \times 10^{-6}$ for angles in degrees.

$$\Delta = \frac{DHI \times AM_a}{E_a} \tag{4.16}$$

where:

- $AM_a$ is the absolute air mass.
- $E_a$ is extraterrestrial radiation.

To implement the model tables of $f$ coefficients and clearness ($\varepsilon$) bins are needed. Perez has compiled several versions of these parameters, see Tables 4.1 and 4.2 (Perez et al. 1987, 1988, 1990).

Since this is an empirical, fitted model, the best coefficients to use are those specifically fit to data measured on site. However, it is rare that such data exists and, therefore, Perez recommends using data published in Perez et al. (1990) and presented in Tables 4.1 and 4.2, which is a compilation of results using high-quality data from several regions. Figure 4.10 shows the effect of using different site-specific model parameters that were published

**Table 4.1** Perez diffuse radiation model clearness bins.

| Clearness ($\varepsilon$) bin | Lower bound | Upper bound |
| --- | --- | --- |
| 1 Overcast | 1 | 1.065 |
| 2 | 1.065 | 1.230 |
| 3 | 1.230 | 1.500 |
| 4 | 1.500 | 1.950 |
| 5 | 1.950 | 2.800 |
| 6 | 2.800 | 4.500 |
| 7 | 4.500 | 6.200 |
| 8 Clear | 6.200 | — |

**Table 4.2** Perez diffuse radiation model $f$ coefficients.

| $\varepsilon$ bin | $f_{11}$ | $f_{12}$ | $f_{13}$ | $f_{21}$ | $f_{22}$ | $f_{23}$ |
| --- | --- | --- | --- | --- | --- | --- |
| 1 Overcast | −0.008 | 0.588 | −0.062 | −0.06 | 0.072 | −0.022 |
| 2 | 0.13 | 0.683 | −0.151 | −0.019 | 0.066 | −0.029 |
| 3 | 0.33 | 0.487 | −0.221 | 0.055 | −0.064 | −0.026 |
| 4 | 0.568 | 0.187 | −0.295 | 0.109 | −0.152 | −0.014 |
| 5 | 0.873 | −0.392 | −0.362 | 0.226 | −0.462 | 0.001 |
| 6 | 1.132 | −1.237 | −0.412 | 0.288 | −0.823 | 0.056 |
| 7 | 1.06 | −1.6 | −0.359 | 0.264 | −1.127 | 0.131 |
| 8 Clear | 0.678 | −0.327 | −0.25 | 0.156 | −1.377 | 0.251 |

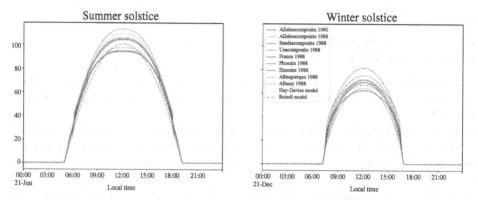

**Figure 4.11** Modeled sky diffuse plane-of-array irradiance for clear sky conditions in Albuquerque, New Mexico, for a south-facing, 30-degree tilted plane. Models include nine parameter sets published for the Perez model as well as results of the Hay–Davies and Reindl models. Calculation was done with pvlib-python.

by Perez et al. (1988) for various locations around the world as well as the Hay–Davies and Reindl models. The example is for two clear days in Albuquerque for a 30° tilted south-facing plane. There is about an 18% variation between results near midday. Note that the Hay–Davies and Reindl models are nearly identical for this case and that their relative position to other models changes with season (Figure 4.11).

## 4.9 Conclusion

POA irradiance is the irradiance that falls on a plane that is coincident with the orientation of a photovoltaic array. Transposition models are used to estimate the irradiance that falls on

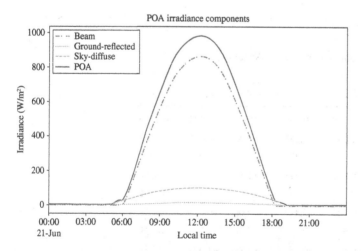

**Figure 4.12** Example of the relative magnitude of the components of POA irradiance calculated for clear sky conditions on the summer solstice in Albuquerque, New Mexico, for a south-facing, 30-degree tilted plane. Calculation was done with pvlib-python.

a tilted plane from the standard quantities of GHI, DNI, and DHI. POA irradiance comprises three irradiance components: beam, ground-reflected, and sky diffuse. The beam irradiance depends on the DNI and angle of incidence. Ground-reflected irradiance depends on the tilt angle of the array, angle of incidence, GHI, and albedo of the ground surface. Sky diffuse depends on the tilt angle of the array and the complex distribution of irradiance from the sky dome (e.g. circumsolar and horizon brightening as well as clouds). There are several types of models to represent this complexity. Overall, these models result in a significant amount of uncertainty in predicted sky diffuse irradiance. However, in sunny locations, these differences are less important since most of the energy reaching the array is from the beam irradiance (see example in Figure 4.12).

## Author Biographies

**Joshua S. Stein** is a Senior Scientist at Sandia National Laboratories where he leads research and development projects in the areas of photovoltaic module and system performance and reliability, specializing in modeling and analysis of new technologies. He is the founder of the PV Performance Modeling Collaborative (PVPMC) and represents the United States in the International Energy Agency PVPS Task 13 on PV Performance and Reliability.

**Clifford W. Hansen** is a Distinguished Member of the Technical Staff at Sandia National Laboratories. His research focuses on improving accuracy and confidence in PV system performance modeling, and methods for analysis of system monitoring data. He co-maintains the python libraries pvlib-python for modeling and pvanalytics for system analysis, and co-chairs the Orange Button technical working group to promote data exchange interoperability between solar system-related software.

## References

Baldridge, A.M., Hook, S.J., Grove, C.I., and Rivera, G. (2009). The ASTER spectral library version 2.0. *Remote Sensing of Environment* 113: 711–715.

Brennan, M.P., Abramase, A.L., Andrews, R.W., and Pearce, J.M. (2014). Effects of spectral albedo on solar photovoltaic devices. *Solar Energy Materials and Solar Cells* 124: 111–116. https://doi.org/10.1016/j.solmat.2014.01.046.

Coakley, J.A. Jr., (2002). Reflectance and albedo, surface. In: *Encyclopedia of Atmospheric Sciences* (ed. J.R. Holton, J.A. Curry, and J.A. Pyle), 1914–1923. Academic Press https://doi.org/10.1016/B0-12-227090-8/00069-5.

Hay, J.E. and Davies, J.A. (1980). Calculations of the solar radiation incident on an inclined surface. In: *Proc. of First Canadian Solar Radiation Data Workshop*, vol. 59 (ed. J.E. Hay and T.K. Won). Canada: Ministry of Supply and Services.

Holmgren, W.F., Hansen, C.W., and Mikofski, M.A. (2018). pvlib python: a python package for modeling solar energy systems. *Journal of Open Source Software* 3 (29): 884. 1 https://doi.org/10.21105/joss.00884.

Loutzenhiser, P.G., Manz, H., Felsmann, C. et al. (2007). Empirical validation of models to compute solar irradiance on inclined surfaces for building energy simulation. *Solar Energy* 81: 254–267.

Perez, R., Seals, R., Ineichen, P. et al. (1987). A new simplified version of the Perez diffuse irradiance model for tilted surfaces. *Solar Energy* 39 (3): 221–232.

Perez, R., Stewart, R., Seals, R., and Guertin, T. (1988). *The Development and Verification of the Perez Diffuse Radiation Model*. Sandia Report. Albuquerque, NM: Sandia National Laboratories.

Perez, R., Ineichen, P., and Seals, R. (1990). Modeling daylight availability and irradiance components from direct and global irradiance. *Solar Energy* 44 (5): 271–289.

Reindl, D.T., Beckman, W.A., and Duffie, J.A. (1990). Diffuse fraction correlations. *Solar Energy* 45 (1): 1–7.

Unsworth, M.H. and Monteith, J.L. (1972). Aerosol and solar radiation in Britain. *Quarterly Journal of the Royal Meteorological Society* 98: 778–797. https://doi.org/10.1002/qj .49709841806.

**Part Three**

**Crystalline Silicon Technologies**

# 5

# Crystalline Silicon Ingot Pulling and Wafering Technology

*Nannan Fu and Jiarui Fan*

Longi, R&D Department, Wafer Business Unit, Xi'An, Shaanxi, China

## 5.1 Origin of Czochralski-Growth Technology

The Czochralski (Cz) crystal growth process was originally developed by Jan Czochralski in 1918 (Czochralski 1918). At first, the polysilicon is put into a high-purity quartz crucible and heated by the graphite thermal field until the silicon melts. The temperature of the liquid surface, above 1410 °C, is stabilized at the critical point of crystallization by controlling the thermal field. At that point, a mono-crystalline seed with the desired orientation is lowered toward the liquid level. When the tip of the seed starts to melt, it is put in contact with the silicon in the crucible. The crucible and the seed are turning in opposite direction and the seed is slowly moved up as the ingot forms. This starts the process of ingot pulling. Ingot pulling can be divided into five steps: seeding, necking, crown, bodying, and tail (Friedrich et al. 2015). The pulling furnace system consists of six parts, as shown in Figure 5.1a–f, and includes a water cooling system (Fig. 5.1a), vacuum system (Fig. 5.1b), argon system (Fig. 5.1c), electrical equipment (Fig. 5.1d), furnace body (Fig. 5.1e), and a thermal system (hot zone) (Fig. 5.1f) which are briefly described in the following:

(1) Water cooling system (Fig. 5.1a)

The production of monocrystalline silicon using a Cz furnace operates at high temperatures. All parts of the furnace must be water cooled. The water cooling system includes: water pool, circulating water pump, main pipe, cooling tower, furnace water separator, inlet pipe, outlet pipe, etc.

(2) Vacuum system (Fig. 5.1b)

The vacuum system allows the furnace to operate at a desired pressure below atmosphere. The vacuum system includes: vacuum pump, electromagnetic valve, vacuum pipeline, ball valve, vacuum gauge, etc. A Cz single crystal furnaces generally use oil-sealed rotary vane mechanical vacuum pumps.

(3) Argon system (Fig. 5.1c)

When preparing monocrystalline silicon by the Cz method, high-purity argon (≥99.999%) is generally used as the shielding gas. The argon system generally includes:

*Photovoltaic Solar Energy: From Fundamentals to Applications, Volume 2*, First Edition.
Edited by Wilfried van Sark, Bram Hoex, Angèle Reinders, Pierre Verlinden, and Nicholas J. Ekins-Daukes.
© 2024 John Wiley & Sons Ltd. Published 2024 by John Wiley & Sons Ltd.
Companion website: www.wiley.com/go/PVsolarenergy

**Figure 5.1a** Colling tower.

**Figure 5.1b** Vacuum pump.

**Figure 5.1c** Argon tank.

**Figure 5.1d** Power supply cabinet (left) and control cabinet (right).

**Figure 5.1e** Furnace body.

**Figure 5.1f** Schematic structure of the thermal system. Source: Adapted from Zhang et al. (2021).

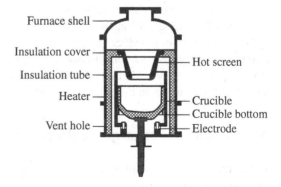

liquid argon tank, vaporizer, buffer tank, main pipeline (pipeline), pressure stabilizing valve, argon electromagnetic valve, etc.
(4) Electrical equipment (Fig. 5.1d)
   The electrical part consists of a transformer room, power distribution room, power supply cabinet, and control cabinet. Electricity enters the substation and is transmitted to the power distribution room, which is distributed to each furnace's power cabinets. Through it, the current is transmitted to the control cabinet and heater, which controls the operation of the entire furnace, including pumps, valves, heaters, rotation motors, pulling, etc.
(5) Furnace body (Fig. 5.1e)
   Single crystal silicon production device, which mainly includes lower shaft (crucible drive), furnace barrel, furnace cover, valve body, auxiliary chamber, and upper shaft.
(6) Thermal system (hot zone) (Fig. 5.1f)
   Melt and heat the silicon material.

## 5.2 Technical Principles and Application

Figure 5.2 gives information about the steps of the Cz growth process. After melting, a seed is lowered to the silicon melt, leading to non-spontaneous nucleation. The seed is monocrystalline and it provides an initial crystal nucleus and enables to reduce the energy required for monosilicon nucleation (Ceccaroli et al. 2016; Eranna 2015). Before growing the crystal to the desired diameter, it is first reduced to the smallest diameter capable of carrying the total weight of the future ingot (up to 250 kg). The purpose of this step, called "necking," is to reduce the density of crystallographic defects. Crowning is a process of lateral growth, and the diameter of the silicon ingot will continuously increase in this step. The following step is ingoting the body after the diameter reaches the required size. The final step is tailing to end the whole process.

For industrial production, the ingot grows in a low-pressure argon atmosphere and requires a very stable environment temperature (Eranna 2015). On this basis, to guarantee that the seed crystal is the only non-spontaneous nucleation in the molten silicon, the overall growth environment should be kept as clean as possible (Eranna 2015). Figure 5.3 shows a range of additional processes that are required on top of the process flow from Figure 5.2, to prevent spontaneous nucleation.

The ingot growth process only takes 60–70% of the entire ingot pulling time, all the other steps, mostly melting, cooling, shutdown and cleaning, take more than 30% of the entire cycle. It is important to reduce the time of the non-silicon-growth process to increase the output. Currently, two technologies have emerged: Recharge-Cz (RCz) and Continuous-Cz (CCz) ingot growth. We will describe these two technologies in the following two sections as well as their impact on n-type ingot growth.

## 5.3 Status of Recharge-Cz (RCz) Ingot Pulling Technology

In the RCz process (Figure 5.4), after finishing a single ingot growth, the conventional method of cooling the entire furnace, crucible, and ingot for pot scrap removal (residual

**Figure 5.2** Steps of Cz crystal growth process.

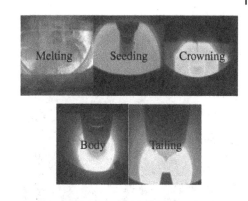

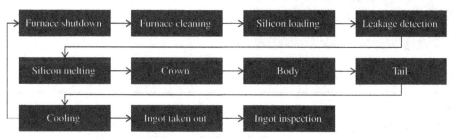

**Figure 5.3** Full process flow for Cz growth.

silicon in the crucible), furnace cleaning, and reloading is circumvented (Figures 5.4–5.5). Instead, the furnace is maintained at the silicon melting temperature, utilizing a silicon feeder apparatus (Figure 5.4) – commonly a quartz tube containing polysilicon fragments – to replenish the crucible. In the first step, the furnace with the crucible still at high temperature is isolated from the pull chamber. The grown monocrystalline ingot is then removed by rotating the pull chamber, and the feeder is inserted into the pull chamber. Finally, the pull chamber is rotated to allow the silicon feeder to discharge and load the polysilicon chunks in the hot crucible. After that, following the same steps in reverse, the feeder is removed from the pull chamber, the same way as the grown ingot was removed. The RCz process reduces idle production time (Eranna 2015; Fickett and Mihalik 2001; Mosel et al. 2016; Fiegl 1983) by avoiding furnace cooling, shutdown, disassembly, cleaning and leakage detection for every single ingot pull. Therefore, the utilization degree increases from 70% to more than 90% for one growth cycle according to RCz technology, and the time for each step can be found in Table 5.1 (Figure 5.5).

In addition to the obvious improvement of the production efficiency, RCz also allows for a significant reduction in the cost of non-silicon material. For example, the lifespan of the quartz crucible and the hot zone are improved. Owing to the decreased frequency of furnace disassembly and assembly, the purity of the crystal puller is enhanced, concurrently reducing the instances of hot zone exposure to the atmosphere. In the RCz method, the hot zone is much less often subjected to hot-cold cycles, thermal expansion stresses, and oxidation, which results in an improved lifetime.

The realization of RCz technology involved the development of an auxiliary chamber, silicon feeder, long-life vacuum system, and long-life crucible. At present, the continuous

**Figure 5.4** RCz feeder.

**Table 5.1** Time for each RCz step.

| Step | Time | Comment |
| --- | --- | --- |
| Furnace cleaning | 1 h | Clean the inner/outer parts of furnace |
| Silicon loading | 1 h | — |
| Furnace closing | | — |
| Air evacuating | 20 min | — |
| Leakage detection | 5 min | — |
| Pressuring | 1 min | — |
| Silicon melting | 5.5 h | Melt around 370 kg of silicon at a time |
| Temperature adjusting | 2.5 h | |
| Seeding | 30 min | |
| Necking growth | 6 min | |
| Crowning | 2 h | The crowning and body transitional stage |
| Body | 45 h | 4000 mm length |
| Tailing | 1.5 h | |
| Ingot taken out | 0.5 h | |
| **Recharge** | 0.5 h | After recharge, return to step 8 "Temperature adjusting." Skip this step after the 6th ingot |
| Furnace shutdown | 6 h | Occurs one time only after sequentially producing six ingots |

pulling time is more than 400 hours (around 3 tons ingot). The future development of RCz technology focuses on a larger hot zone, lower silicon ingot oxygen density, and higher pulling speed.

Similar ingot outputs are currently achieved for gallium (Ga, p-type) doped and phosphorus (P, n-type) ingots. The yields are, however, restricted by two different factors.

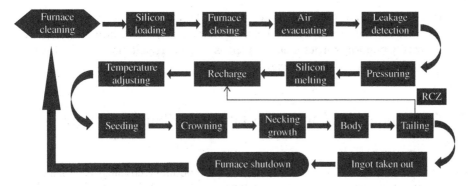

**Figure 5.5** Full ingot recharge process.

The production rate and production yield of p-type ingots is restricted by resistivity requirements while the output of n-type ingots is limited by bulk minority carrier lifetime requirements. For n-type ingots, the principal causes of this limitation are widely recognized as the ring-type defects related to interstitial oxygen concentration at the seed end and the concentration of transition metal impurities accumulation at the tail end. Figure 5.6 shows the minority carrier lifetime in an n-type ingot after up to five recharge cycles. The minority lifetime degrades after subsequent ingoting. In addition, the lifetime

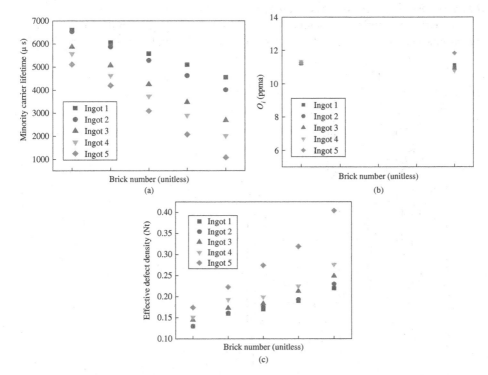

**Figure 5.6** (a) Minority carrier lifetime, (b) interstitial oxygen concentration, and (c) effective defect density in an n-type ingot after five recharges.

decreases along the ingot, even at a faster rate for the 5th ingot compared to the first ingot. Figure 5.6 also shows the effective defect (impurities that affect the quality of silicon wafers, especially minority carrier lifetime, such as oxygen precipitation, Fe, Cu, Co, etc.) density along the ingot (Davis et al. 1980).

## 5.4 Development of Continuous Crystal Pulling Technology (CCz)

An alternative method to RCz is continuous Cz crystal pulling technology (CCz), another instance of the Cz-growth technology (Fiegl 1983). CCz technology was first developed in Japan and the United States. An example of hot zone composition is shown in Figure 5.7.

Compared with RCz, CCz is a newer technology and has been put into production on a small scale. The CCz technology utilizes a new design for crucibles with inside quartz walls, called double crucible, to separate the pulling and feeding zones of the crucible and to allow a constant replenishment of the crucible with polysilicon chunks while the ingot growth is in process. In addition to the double crucible design, the key technological innovations that allowed the CCz technology to become a reality are (i) the development of an innovative feeding device with a doping function that ensures a controllable feeding rate and (ii) the improvement of the control system that becomes quite complex to manage at the same time the pulling process and the hot zone while feeding polysilicon.

Because the segregation coefficient of doping and other impurities at the silicon/molten silicon interface is usually smaller than 1, or even much smaller than one for metallic

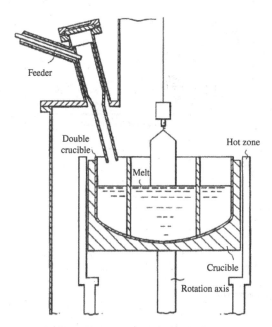

**Figure 5.7** Schematic of CCz pulling technology. Source: Naoki et al. (1997)/US Patent/Public Domain.

**Table 5.2** Equilibrium segregation coefficients
of the main dopants at the Si/molten Si interface.

| Element | $k_0$ |
|---|---|
| Boron | 0.8 |
| Gallium | 0.008 |
| Aluminum | 0.002 |
| Indium | 0.0004 |
| Phosphorus | 0.35 |
| Arsenic | 0.3 |
| Antimony | 0.023 |

impurities, the impurities preferentially stay in the liquid phase of silicon (Davis et al. 1980). As a beneficial consequence, the crystallization process also serves as a purification process, which results in an accumulation of metallic impurities in the molten silicon and eventually in the tail end of the ingot. Fortunately for p-type ingots with boron doping, the segregation coefficient of boron is only slightly smaller than 1 (see Table 5.2), resulting in a quite uniform boron doping throughout the length of the silicon Cz ingot. This is unfortunately not the case for phosphorus or gallium doping, which both result in a significant doping gradient from head to tail end of the ingot. In general, in the traditional Cz technology, technologists face the problem of increasing impurities at the ingot tail. It leads to a wide resistivity range throughout the produced ingot. The main dopant for silicon in the PV industry is currently phosphorus for n-type, and gallium for p-type ingots. Note that boron used to be the mainstream p-type dopant, but was replaced by gallium around 2020 to circumvent the light-induced degradation (LID) linked to the boron-oxygen defect in crystalline silicon (Fischer and Pschunder 1973).

Table 5.2 shows the equilibrium segregation coefficients of the main dopants in Si. We can find that both gallium and phosphorus have lower segregation coefficients compared to boron. So the distribution of gallium and phosphorus tends to be significantly less uniform than boron in a Cz silicon ingot. For best manufacturing control and best reproducibility, the resistivity range of silicon wafer needs to be as narrow as possible, although with recent improvements in minority carrier lifetime, the acceptable resistivity range is nowadays larger than before. This means that using conventional Cz technology to pull gallium or phosphorus-doped ingots leads to shorter ingots to avoid a large variation in impurity density in the crystal as the concentration in the liquid phase increases during the pulling process. Shorter ingot length means shorter ingot growth period, and longer time spent in feeding, melting, bodying, tailings treatment, and other non-growth stages, all of which increase pulling cost. CCz technology significantly reduces the time of non-growth stages and can narrow the resistivity range through continuous feeding and pulling to meet the resistivity requirements. However, the commercial CCz technology is still not mature for the following reasons: (i) Currently, the feed inlet of CCz is relatively small, and can feed only small chunks of silicon. Large chunks can easily get stuck. There is a significant risk of oxidation of the silicon chunks that could affect the crystallization process. (ii) CCz uses a double-crucible design, which results in more contact between the molten silicon and the crucible, leading to the introduction of more contamination, including oxygen. (iii) The quality of CCz are currently relatively low due to the instability of the pulling process and the declining minority carrier lifetime caused by more impurities.

## 5.5 N-Type Mono-Wafer Slicing Technology

Cutting ingots into thin wafers is an important step before the photovoltaic cell manufacturing process. Diamond wire slicing has recently become the dominant slicing technology in the photovoltaic industry, rapidly displacing the previous wire-saw slicing technology using a slurry of silicon carbide (SiC) in a solution containing polyethylene glycol (PEG) (Möller 2006; Enomoto et al. 1999). Diamond wire sawing (DWS) has brought several significant improvements in slicing efficiency and quality, and greatly reduced production costs. As an additional benefit of this technology change, wafers can be cut thinner, which is particularly important for n-type technology that often requires thinner wafers (Figure 5.8).

In typical wire-saw technology, a thin steel wire is run through a supply spool and wound on rollers multiple times at a constant pitch (~200 μm), and eventually wound up on a take-up spool at the end. The wire is kept at a specified tension (3.6–4 N) and moved at high speeds (30–40 m/s). Silicon bricks, glued on thick glass plate (~10 mm), are fed through the moving wire web. Multiple wafers are sliced at the same time. In the previously used slurry wire-saw process, the cutting fluid is a slurry containing SiC in PEG (Möller 2006). The main difference with the DWS process is that the abrasive grits (diamond) are attached to the steel wire, and the PEG slurry is replaced by a water-based cutting fluid. The diamond abrasive particles are fixed on the steel wire by bonding with an electroplated nickel coating for high-speed slicing (Enomoto et al. 1999). The diamond wire is usually short and does not allow to slice a silicon brick in one run. The wire speed is reversed back and forth until the entire brick is sliced. There are two types of diamond wire: resin wire and electroplated wire. Because of its better slicing ability, DWS has completely replaced slurry slicing as the mainstream slicing method in the photovoltaic industry. DWS presents the following advantages (Wu 2016):

(1) The diamond has higher hardness and wear resistance, greatly extending the life of the cutting wire.
(2) Diamond wires can be made of much smaller diameter, so thinner silicon wafers can be cut and silicon kerf loss caused by cutting can be significantly reduced. The diameter of a typical diamond wire is currently less than 50 μm, compared to about 200 μm for the steel wire for slurry wire saw, which has significantly reduced the quantity of silicon feedstock per MW to less than 4 tons per MW. Production with diamond wires smaller than 30 μm is currently being trialed.
(3) The large contact area between the diamond and wafer leads to more grinding and improves grinding efficiency.

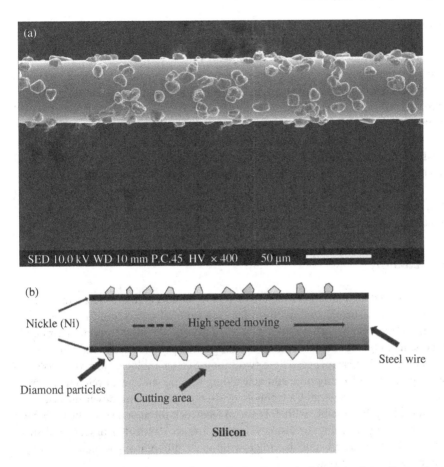

**Figure 5.8** (a) Scanning electron micrograph of a diamond wire. (b) A schematic drawing of a wire-saw.

From a manufacturing point of view, these advantages of diamond wire slicing are three aspects: high cutting efficiency, low cost per wafer and more controllable quality, as explained below (Enomoto et al. 1999; Wu 2016):

(1) High slicing efficiency: improved by a factor of 3–4 due to reduced slicing time, reduced cost, and improved manufacturing capacity per tool
(2) Low cost per piece: faster slicing speed and more durable cutting wire make the cutting cost of diamond wire only half that of slurry slicing
(3) Better quality control:
    a. From the perspective of quality control, for slurry slicing, three different suppliers are providing the key consumables: slurry (PEG), steel wire, and abrasive slicing agent (SiC). Diamond wire slicing, by comparison, needs only one supplier since the abrasive agent and wire are integrated and the cutting fluid is simply water. The potential fluctuation in manufacturing quality caused by consumable materials from the suppliers is reduced significantly.

b. Wire breakage during slicing is an important factor affecting yield. A diamond wire, which requires multiple electroplating, polishing, and cleaning, has more strict quality control requirements than the old standard wire.

c. There are multiple inspection processes for checking the quality of finished products in the production process of the diamond wire, including multichannel diamond wire pull test machine (pull strength), and inspection of the electroplated surface.

(4) Environmentally friendly: diamond wire slicing does not use PEG or SiC particles. The cutting fluid (water) and silicon powder mixture contains less metallic impurities due to the diamond wire's much lower contact pressure on the silicon surface, potentially more conducive to the subsequent recovery and reuse of silicon powder for reuse in silicon feedstock production. In addition, compared to slurry slicing, diamond wire cutting significantly reduces silicon wafer damage and increases silicon wafer yield by approximately 15%, and reduces silicon waste.

## 5.6 Conclusion

This chapter reviewed the manufacturing technology of monocrystalline silicon ingots (Cz technology) and wafer slicing. Derived from the conventional Cz crystallization process, two innovative Cz technologies have been developed to solve the most important problem of Cz process, namely, the fact that actual crystal growth process represents only about 60% of the entire pulling sequence. These two innovative technologies aim primarily at reducing the polysilicon feeding time and at avoiding the cool and heat cycle for the crucible and hot zone. The Recharge-Cz technology, RCz, avoids cooling down the crucible and allows reloading the crucible with polysilicon feedstock through a specially designed feeder without cooling, opening, or dismantling the furnace. With RCz process and specially designed RCz puller, it is possible to grow up to six different ingots before cooling, opening, and cleaning the furnace.

Continuous-Cz technology, CCz, is another innovative Cz technology that allows continuous feeding through a double crucible design and can effectively narrow the resistivity range of ingot, a specific problem of n-type ingots and p-type ingots with gallium doping. This technology is still in development or experimental production.

DWS technology has dramatically changed silicon ingots' slicing process and rapidly displaced the old slurry wire saw process. Compared to traditional slurry slicing, the diamond wire technology presents significant advantages in improving slicing efficiency and quality control and reducing slicing cost.

## Abbreviations and acronyms

Cz     Czochralski
CCz   continuous Czochralski
DWS  diamond wire saw
LID    light-induced degradation
PEG   polyethylene glycol
RCz   recharge Czochralski

## Author Biographies

**Nannan Fu** is a Senior Staff Engineer at Longi in the R&D department. He has more than eight years of work experience in the field of photovoltaic technology development. He has led several engineering projects or scientific research related to Cz ingot technology and monocrystalline silicon wafers. His focus is on minority carrier life enhancement, impurity reduction, tool development research, and other related fields.

**Jiarui Fan** is a Staff Engineer at Longi in the R&D department. He is an assistant and project team members of Nannan Fu.

## References

Ceccaroli, B., Ovrelid, E., and Pizzini, S. (2016). *Solar Silicon Processes: Technologies, Challenges, and Opportunities*. Boca Raton, FL, USA: CRC Press.

Czochralski, J. (1918). A new method for the measurement of the crystallization rate of metals. *Zeitschrift für Physikalische Chemie* 92: 219–221.

Davis, J.R., Rohatgi, A., Hopkins, R.H. et al. (1980). Impurities in silicon solar cells. *IEEE Transactions on Electron Devices* 27 (4): 677–687.

Enomoto, T., Shimazaki, Y., Tani, Y. et al. (1999). Development of a resinoid diamond wire containing metal powder for slicing a slicing ingot. *CIRP Annals* 48 (1): 273–276.

Eranna, G. (2015). *Crystal Growth and Evaluation of Silicon for VLSI and ULSI*. Boca Raton, FL, USA: CRC Press.

Fickett, B. and Mihalik, G. (2001). Multiple batch recharging for industrial CZ silicon growth. *Journal of Crystal Growth* 225 (2–4): 580–585.

Fiegl, G. (1983). Recent advances and future directions in Cz-silicon crystal growth technology. *Solid State Technology* 26 (8): 121–131.

Fischer, H. and Pschunder, W. (1973). *Investigation of Photon and Thermal Induced Changes in Silicon Solar Cells*, 404–411. New York: IEEE.

Friedrich, J., von Ammon, W., and Müller, G. (2015). Czochralski growth of silicon crystals. In: *Handbook of Crystal Growth*, 45–104. Elsevier.

Möller, H.J. (2006). Wafering of silicon crystals. *Physica Status Solidi (A)* 203 (4): 659–669.

Mosel, F., Denisov, A.V., Klipp, B. et al. (2016). Cost effective growth of silicon mono ingots by the application of a mobile recharge system in CZ-puller. Paper Presented at the *32nd European PV Solar Energy Conference and Exhibition*, Munich, Germany.

Naoki, N., Isamu, H., and Michiaki, O. (1997). Method of manufacturing silicon moncrystal using continuous czochralski method. US: EP0792952A1.

Wu, H. (2016). Wire sawing technology: a state-of-the-art review. *Precision Engineering* 43: 1–9.

Zhang, X., Gao, D., Wang, S. et al. (2021). Design of followed-up heater for Czochralski single crystal furnace. *Journal of Synthetic Crystals* 50 (2): 361–367.

# 6

# Tunnel Oxide Passivated Contact (TOPCon) Solar Cells

*Bishal Kafle and Armin Richter*

*Fraunhofer ISE, Freiburg im Breisgau, Germany*

## 6.1 Introduction

An ideal contact to a solar cell would be designed to facilitate efficient transport of a preferred (majority) type of charge carriers while suppressing the recombination of the opposite type. Conventionally, for instance, in the state-of-the-art $p$-PERC (passivated emitter and rear cell), this is achieved by applying localized heavily doped regions ($n^{++}$or $p^{++}$) beneath the electron and hole contacts (respectively). This solution, however, has fundamental limitations due to enhanced Auger recombination in highly doped areas and high recombination at the metal-silicon interface. One way of overcoming these limitations is to form a so-called carrier-selective passivating contact, aiming to decouple the recombination and transport properties of the contact that are typically represented by the recombination current density ($J_0$) and contact resistivity ($\rho_C$), respectively. The theory of passivating contacts was introduced in the previous edition of the book by Martin Hermle (Reinders et al. 2017). Aligning to the nomenclature used in the previous edition, the term "contact(s)" here refers to metal/semiconductor interface, including any diffused/induced regions in silicon that are primarily responsible for the collection of charge carriers. In brief, following are the main requirements of an ideal passivating contact – (i) minimizing the minority carrier recombination at contact areas to allow a high internal voltage or, in other words, a high implied open-circuit voltage ($iV_{OC}$) of the solar cell, (ii) facilitating a low resistance to majority transport enabling high external voltage ($V_{OC}$) measured at cell contacts.

These requirements are fulfilled by first maintaining the carrier-selective nature of the contact where majority carriers have significantly higher conductivity than minority carriers, and second by essentially decoupling the direct metal contact from the Si absorber to minimize recombination. One successful method is to form a heterojunction featuring a stack of suitably doped and intrinsic amorphous silicon layers (a-Si:H($n/p$)/a-Si:H($i$)) on $c$-Si absorber. This concept, with predominantly low-temperature process steps, is discussed in Chapter 7. In this chapter, we focus on another prominent technology widely known as tunnel oxide passivated contact (TOPCon) (Feldmann et al. 2014b) that combines the advantages of heterojunctions with high-temperature processing capability.

*Photovoltaic Solar Energy: From Fundamentals to Applications, Volume 2*, First Edition.
Edited by Wilfried van Sark, Bram Hoex, Angèle Reinders, Pierre Verlinden, and Nicholas J. Ekins-Daukes.
© 2024 John Wiley & Sons Ltd. Published 2024 by John Wiley & Sons Ltd.
Companion website: www.wiley.com/go/PVsolarenergy

## 6.2 Concept

Inspired by the early research in bipolar junction transistor (BJT) emitters, the concept was first applied in the 1980s in a working solar cell with heavily doped polysilicon emitters (Lindholm et al. 1985; Green and Blakers 1983; Tarr 1985). The potential of a passivating contact in solar cells was underlined by reaching high $V_{OC}$ (720 mV) (Yablonovitch et al. 1985), low $J_0$ (<20 fA/cm$^2$), and $\rho_C$ (<0.1 m$\Omega$ cm$^2$) values (Gan and Swanson 1990) using a so-called "semi-insulating polycrystalline-silicon" (SIPOS) junction on $c$-Si. Interestingly, the concept of passivating contacts did not get the deserved attention in the research community for a long time, although some solar cell companies leveraged it to fabricate solar cells with high $V_{OC}$ (>710 mV) and efficiencies (Cousins et al. 2010; Schultz-Wittmann et al. 2016). This changed after the re-discovery of "TOPCon" contacts by Feldmann et al. in 2015, who replaced the rear-side point contacts of a $n$-type cell with a stack of thin SiO$_x$ and doped semicrystalline silicon-containing layer in a hybrid cell structure leading to 23.0% conversion efficiency (Feldmann et al. 2014a). Similar concepts were also developed elsewhere and named, for instance, POLO (polysilicon on oxide) (Römer et al. 2014).

TOPCon consists of an ultra-thin wide bandgap dielectric layer sandwiched between the silicon absorber and a doped polycrystalline silicon or polysilicon (poly-Si) layer. Figure 6.1a schematically shows the band diagram of $n$-TOPCon on $c$-Si($n$) absorber. Here, the heavily phosphorous doped poly-Si($n$) layer induces a band bending in $c$-Si($n$) absorber. Additionally, the thin oxide could have a slightly asymmetric tunneling probability for electrons and holes due to differences in energy barrier heights ($\Delta\Phi_e < \Delta\Phi_h$) (Post et al. 1992). This provides carrier selectivity by preventing the migration of minority carriers (holes) from $c$-Si to the poly-Si($n$) layer. The carrier selectivity is also enhanced by a shallow high–low junction ($n^+/n$) in $c$-Si($n$) that is typically formed during the high-temperature step required to perform doping and crystallization of the poly-Si layer. In Figure 6.1b, a transmission electron microscope (TEM) cross-section image of the TOPCon contact shows a uniform oxide layer essential to reducing the defect density ($D_{it}$) at $c$-Si/SiO$_x$ interface. Hence, the $n$-TOPCon acts here as a passivating electron-selective contact.

The mechanism of the majority carrier transport through oxide depends largely upon two factors: (i) thickness and stoichiometry of the oxide layer, and (ii) the annealing

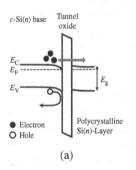

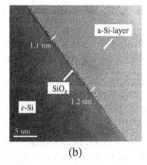

(a)  (b)

**Figure 6.1** (a) Schematic band diagram of $n$-TOPCon on $c$-Si($n$) (Source: adapted from (Glunz et al. 2017)), and (b) TEM cross-section image showing the $c$-Si/SiO$_x$/poly-Si passivating contact scheme (Moldovan et al. 2015 with permission from Elsevier).

temperature that is required to transform the deposited silicon layer of predominantly amorphous nature into polycrystalline silicon (poly-Si) layer with mixed fractions of amorphous and crystalline phases. For a sufficiently thin and stoichiometric oxide, the tunneling mechanism is reported to dominate the majority of carrier transport after performing annealing at lower temperatures. However, the tunneling probability exponentially decreases with the thickness of the oxide layer (de Graaff and de Groot 1979). This implies that if thick oxides are used as barriers, the thermal budget during annealing must be deliberately kept high enough ($T_a > 1000\,°C$) to break the oxide and form local direct pathways for current flow. For thick oxides, the formation of pinholes does not necessarily impact the interface passivation negatively (Peibst et al. 2016). Hence, the dominant charge carrier transport mechanism (i.e. tunneling or pinhole-mediated transport) depends strongly on the sample preparation and, therefore, cannot be generalized (Glunz and Feldmann 2018).

## 6.3 Both Side Contacted Cells with TOPCon

### 6.3.1 Cell Architectures

To date, polysilicon-based passivating contacts of either polarity are used in several cell architectures, most notably as: (i) rear (and optionally front) contact of a cell featuring a diffused front junction (FJ) commonly referred to as TOPCon cell (Feldmann et al. 2014a), (ii) rear emitter in a back junction configuration (POLO BJ/TOPCoRE) (Richter et al. 2021; Min et al. 2020), and (iii) both contacts of an interdigitated back contact (IBC) solar cell (Haase et al. 2018). Efficiency in the range of 26% and above were reported with such TOPCon solar cells, which demonstrates impressively the potential of this technology. For instance, with small-area lab-type cells, conversion efficiencies of 25.8% (Richter et al. 2021), 26.0% (Richter et al. 2021), and 26.1% (Haase et al. 2018), were achieved with TOPCon, TOPCoRE and IBC cells, respectively. The respective cell architectures are schematically shown in Figure 6.2. Please note that the polarity of the wafer substrate is shown based on the best cell fabricated for each scheme.

Among the abovementioned cell architectures, the IBC concept (Figure 6.2c), where both contact polarities are located at the rear of the cell, offers the highest potential in terms of monofacial conversion efficiency. However, it is also the most complex architecture for transfer to mass production, resulting in higher cost, due to several structuring steps required during fabrication. While efforts toward a leaner process flow for the IBC concept are currently underway (Hasse et al. 2020), both side-contacted cells also have an additional advantage of high bifacial power gain that is especially relevant for large-scale power plants. In the following, we will focus on cell architectures with both-sided contacts that feature a full-area TOPCon stack at the rear side. In general, $n$-TOPCon layers are preferred as rear passivating contacts in comparison to $p$-TOPCon layers. This is mainly due to easier process development, higher passivation quality on rough/textured morphology (low $J_0$) (Feldmann et al. 2017b), and a more straightforward metallization of $n$-TOPCon layers (low $\rho_C$ and $J_{0,met}$) using commercially available Ag-pastes in the screen printing process (Mack et al. 2021). Nevertheless, future improvements in the passivation properties of $p$-TOPCon could allow the use of the opposite wafer polarity as represented in Figure 6.2b,c. We

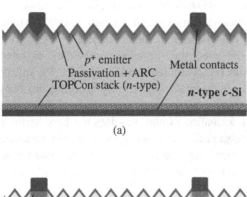

(a)

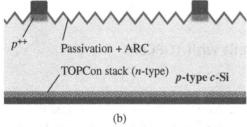

(b)

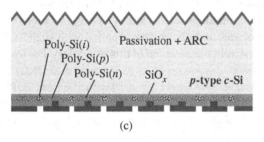

(c)

**Figure 6.2** Schematic cross-section of various cell concepts utilizing TOPCon-like passivating contacts. (a) TOPCon solar cell (with diffused $p^+$ FJ); (b) TOPCoRE solar cell (with TOPCon BJ); (c) IBC solar cell with poly junctions

will shortly review the existing challenges of $p$-TOPCon development in Section 6.4.1. As poly-Si has a similar band gap as $c$-Si, it is preferably placed at the rear contact to limit the parasitic absorption of incident light with approximated $J_{SC}$ loss of up to 0.5 mA/cm$^2$ per 10 nm of phosphorous doped poly-Si (Feldmann et al. 2017b).

The TOPCon cell (Figure 6.2a) is a FJ cell based on $n$-type $c$-Si substrate with boron ($p^+$) emitter on the textured (front) side. The front side is passivated by a dielectric layer stack, typically AlO$_x$/SiN$_x$. The rear with textured/semi-polished/polished surface features $n$-TOPCon as electron selective passivating contact. Optionally, $n$-TOPCon is stacked with a hydrogenated dielectric layer like silicon nitride (a-SiN$_x$:H) for passivating defects in poly-Si and at the SiO$_x$/$c$-Si interfaces. Although $n$-TOPCon layer results in a low total saturation current density ($J_{0,\text{rear}} < 7$ fA/cm$^2$) of the rear side (Chen et al. 2020), the device performance is limited by the recombination losses at the front side, i.e. at the surface, $p^+$ emitter bulk and metal contacts (Altermatt et al. 2021). To some extent, the recombination at the front could be minimized by introducing either a $p^{++}$ selective emitter (Richter et al. 2021) or a $p$-TOPCon stack locally under the front contact (Singh et al. 2020), which, however, increases the complexity in solar cell processing. The FJ TOPCon cell architecture is used to reach up to 25.8% conversion efficiency and a high $V_{OC}$ of 724 mV in small area devices with evaporated front and rear metal contacts (Richter et al. 2021). The industrial version of this concept is discussed in Section 6.3.2.

For the back-junction (BJ) cell concept (Figure 6.2b), the $n$-TOPCon stack acts as a rear emitter on a $p$-type Si substrate. The main advantage of this configuration is that the whole $c$-Si substrate contributes to the charge carrier transport toward the local front side contacts that makes a full-area highly conductive layer at the front surface obsolete, such as the full-area B-diffusion in the case of the $n$-TOPCon cell. Therefore, the front surface recombination can be strongly reduced. However, a localized $p^{++}$ region is placed under the contact to limit the recombination and allow contact formation. Since the minority carriers (here electrons) are collected at the rear junction, high diffusion lengths, i.e. a high-quality base material is a primary requirement for the BJ concept. Additionally, omission of a conductive front surface ($p^+$ emitter for $p$-BJ) is found not to limit the majority transport to the front surface. Moreover, a $p$-BJ design is found to yield higher performance in the practical doping levels ($\rho_b \sim 2\,\Omega$ cm) than similar concept of opposite polarity ($n$-BJ) due to a better balance of electron and hole transport losses in the former (Richter et al. 2021). The cell architecture showed conversion efficiencies of up to 26.0% in small area devices with evaporated contacts, with a clear advantage in FF (more than 1% absolute) and $V_{OC}$ (8 mV) to the $n$-FJ TOPCon cell (Richter et al. 2021). Using POLO junctions, the industrial version of this cell architecture featuring screen-printed contacts reached a very high $V_{OC}$ of 716 mV and a promising efficiency of 22.6% (Min et al. 2021). Transfer of this process in mass production is expected to meet a challenge in solving light and elevated temperature-induced degradation (LeTID) (see Chapter 9), which is more critical to the industrial standard $p$-type Cz-Si in comparison to $n$-type Cz-Si.

### 6.3.2 Industrial TOPCon (i-TOPCon)

Currently, the industrial version of the FJ TOPCon cells on $n$-type $c$-Si, so-called industrial TOPCon (i-TOPCon) cell (Chen et al. 2020; Feldmann et al. 2020) is widely seen as the potential evolutionary upgrade to the incumbent $p$-PERC cells. The i-TOPCon cell design envisions a process route that benefits from the processing similar to the PERC cell, thus requiring the integration of only a few additional process steps in the cell process chain. The cell architecture is reported to yield high efficiencies of > 24.0% in pilot line (Fertig et al. 2022) and volume production (Mark Osborne 2021) of leading cell manufacturers, with record efficiency claims of > 26.0% (Press Release – JinkoSolar Holding Co. Ltd., 2023; Press Release – Jolywood, 2023).

Figure 6.3a summarizes the typical process steps used in a i-TOPCon cell. Please note that the process flow is nonexhaustive and various process routes and a wide range of technology options for the TOPCon concept are currently in consideration by the PV industry in terms of their technological and economic viability. The process flow is primarily dictated by the

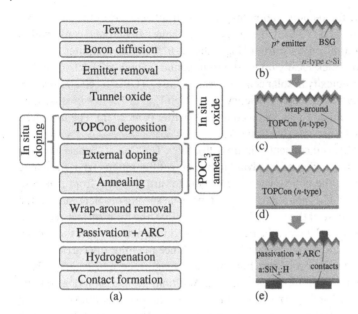

**Figure 6.3** (a) Exemplary process flow for i-TOPCon cell with the potential combination of two processes in a single step; (b–e) schematic cross-sections of cell precursors after some crucial process steps in a typical TOPCon processing: (b) after texturing, boron emitter diffusion and single-sided emitter removal, (c) after TOPCon deposition, doping and annealing, (d) after wraparound removal of poly-Si($n$) at the front and the edges, (e) after dielectric deposition and contact formation

choice of the deposition technology to form TOPCon layers and whether the layers are in situ doped or require an external doping. If technologically feasible, more process steps are combined in a single tool to ensure a lean process flow.

Typically, boron doping is performed using a tube diffusion process to form $p^+$ emitters on a textured $n$-type $c$-Si substrate. This is followed by an inline wet-chemical process for single-sided removal of rear-side emitter. During emitter removal, the borosilicate glass (BSG) layer is typically kept intact at the front to act as a barrier against wet or dry chemicals, which are used later during wraparound removal of the parasitic poly layer at the front side, and the wafer edges. The doping of the TOPCon layers is performed either in the same step (in situ), or in a successive step (ex situ). If an intrinsic amorphous silicon (a-Si) layer is deposited, external doping is performed typically in a POCl$_3$-based tube furnace that simultaneously crystallizes the a-Si to form a poly-Si layer. For in situ doped layers, crystallization is achieved by a thermal annealing step. If single-sided poly-Si deposition is not warranted, a wraparound removal process is required to remove unwanted poly-Si on the textured side before depositing passivation layers. The wraparound removal is typically performed using wet-chemical processes (Glunz et al. 2023), although, recently, a dry alternative is also reported (Kafle et al. 2021b). This is followed by a hydrogenation step aiming to improve the passivation property of the TOPCon layer. In an industrial scenario, this is typically performed by depositing hydrogen-rich dielectrics, for example, an amorphous silicon nitride layer (a-SiN$_x$:H), which acts as an efficient hydrogen source during the contact firing step. Front and rear metallization is typically achieved by using screen-printed Ag-based paste for the rear (n-type contact) and Al-Ag paste for the front (*p*-type contact), followed by a fast-firing process to form external contacts. In the next subsections, we will briefly look into the formation of TOPCon layers and the contact formation in i-TOPCon cells.

### 6.3.2.1 Formation of Interfacial Oxide

The thin interfacial oxide layer (1–3 nm) not only provides chemical passivation of the *c*-Si surface by decoupling it from the poly-Si layer, but also ensures the majority transport from *c*-Si to poly-Si. Additionally, it also works as a barrier to excess diffusion of dopants into the *c*-Si substrate during dopant activation of poly-Si. The stoichiometry and thickness of the oxide layer mainly determine the thermal stability of the layer (Moldovan et al. 2015; Polzin et al. 2019), which is a crucial aspect considering the current PV manufacturing process route that features several high-temperature steps. The oxides are formed either by wet-chemical or dry processes. Among wet-chemical methods, DI-O$_3$ oxide is shown to have a very good thermal stability (Moldovan et al. 2015). Among dry methods, thermal oxidation at temperatures close to 600–700 °C is the current mainstream process as thermally stable oxides can be formed in situ before depositing LPCVD layers in the same tool (Naber et al. 2019). Other appealing candidates are ultraviolet (UV) light-assisted O$_3$-based oxidation (Moldovan et al. 2015), and plasma-based oxidation using oxygen or nitrous oxide in a PECVD tool (Huang et al. 2020; Glunz et al. 2023), the latter allowing simultaneous deposition of doped a-Si layers in the same tool. All of the above oxidation methods could result in low $J_0$ values ($J_0$ <10 fA/cm$^2$) on both *n* and *p*-type poly-Si/SiO$_x$ junctions (Yan et al. 2021), although thermal oxides show the best thermal stability for a wider range of annealing temperatures (Polzin et al. 2019). In PV production, their choice largely depends upon the used a-Si/poly-Si deposition technology that can integrate the oxidation step in a single tool.

### 6.3.2.2 Deposition and Doping of Polysilicon

Except for LPCVD, which allows for the deposition of polysilicon, in all the other techniques of deposition, a-Si layers are first deposited and then subjected to a high-temperature annealing step to form poly-Si layers. Depending upon the deposition technology, poly-Si doping is performed either during the deposition process (in situ), or in a subsequent process like thermal diffusion or ion implantation (ex situ). During the dopant activation step, the dopants aggregate at the SiO$_x$/*c*-Si interface, with a decreasing dopant concentration toward the wafer bulk (Kale et al. 2018). In particular, boron could readily diffuse into the tunneling SiO$_x$ layer that might be responsible for generally higher $J_0$ of *p*-TOPCon layers (Kale et al. 2018; Feldmann et al. 2019). The doping profile in *c*-Si should be carefully optimized by considering the trade-off between Auger recombination and majority carrier transport. A high dopant incorporation also leads to oxide break-up and increases $J_0$ values (Yan et al. 2021). In addition, the layer properties, for example, crystallinity and doping profile, could influence the extent of H diffusion into the SiO$_x$/*c*-Si interface during the hydrogenation/firing step (Kang et al. 2022).

Importantly, the choice of the a-Si or poly-Si deposition technology dictates almost all the other cell processing steps before metallization, especially based on whether it allows an in situ doping and a true single-sided deposition. Although several deposition methods are under investigation, most notable are chemical vapor deposition (CVD) and physical vapor deposition (PVD). CVD is either performed at low pressure (LPCVD), using plasma (PECVD), or atmospheric pressure (APCVD) using Silane ($SiH_4$) as silicon precursor and optionally Phosphine ($PH_3$) or Diborane ($B_2H_6$) as dopant gases. LPCVD is the current state-of-the-art process used in the industry to form pinhole-free poly-Si layers with good thickness distribution in a batch-type process performed at temperatures close to 600 °C. Some major limitations of LPCVD are: (i) almost 30–40% lower deposition rate ($n$-doped) and challenging homogeneity if in situ doping is performed, (ii) significant wraparound of poly-Si. Among alternative CVD methods, the most notable is PECVD which offers significantly higher deposition rates for in situ doped a-Si layers in comparison to LPCVD, thereby promising cost-effectiveness (Kafle et al. 2021a). The challenges of PECVD are to avoid blistering in thick layers ($d > 100$ nm) due to an inherently high H concentration and to offer a true single-sidedness (Grübel et al. 2021; Glunz et al. 2023). Meanwhile, excellent efficiencies exceeding 22.5% are reached using both inline (Nandakumar et al. 2019) and batch (Feldmann et al. 2020) PECVD tools. PVD can deposit high-quality a-Si layers using a silicon target, with the major benefit of providing true single-sidedness (Yan et al. 2018).

### 6.3.2.3 Hydrogenation and Contact Formation

A subsequent hydrogenation step after the crystallization of poly-Si is important to further lower the $J_0$ value of the TOPCon contact on $c$-Si. Although the exact mechanism is not yet understood, studies point to the significance of atomic H species in passivating the electrically active defects at $SiO_x/c$-Si interface. The H species can be either introduced using remote hydrogen plasma treatment or by using hydrogenated dielectric layers like $AlO_x$ or $SiN_x$ (Truong et al. 2020) and their combinations (van de Loo et al. 2020; Polzin et al. 2021). Typically, a thermal step is required for diffusion of H species from the dielectrics toward the $c$-Si/$SiO_x$ interface. The release of atomic H from the dielectric layer could depend upon the layer stoichiometry defined by the deposition conditions (Steinhauser et al. 2020a), and H effusion profile of dielectrics at different annealing/firing temperatures (Dingemans et al. 2010). Although H species are known to play an essential role in interface passivation, recent studies show that the thermal stability deteriorates in the presence of excess hydrogen around the interfacial oxide (Kang et al. 2021). For i-TOPCon cells, the hydrogenation step is typically combined with the fast-firing step that is typically used to form screen-printed contacts. The firing temperature strongly impacts the surface passivation of poly-Si/$SiO_x$ junctions by determining the hydrogen diffusion from the dielectric capping into the $c$-Si/$SiO_x$ interface.

## 6.4 Challenges

### 6.4.1 Technological Challenges

One of the major challenges of poly/$SiO_x$ junctions is to minimize the passivation degradation during the contact formation process. The typical firing profiles with peak temperatures in the range of 750–850 °C, designed to reach low contact resistivities, could lead to oxide breakage or spiking of metal to $c$-Si. The latter significantly increases the recombination under metalized area ($J_{0,met}$) and thus $V_{OC}$, and challenges the very essence of true passivating contact. This trade-off between FF and $V_{OC}$ could be partly addressed by the deposition of a thicker polysilicon layer ($d > 150\,\mu m$). However, it increases the process costs (Kafle et al. 2021a) as well as $J_{SC}$ losses due to the free carrier absorption (FCA) in poly-Si layers (Feldmann et al. 2017a). As commercially available pastes already allow low $J_{0,met} < 35\,fA/cm^2$ on 200 nm thick poly-Si (Padhamnath et al. 2020), further advances in paste development and more understanding of contact formation are expected to result in lower $J_{0,met}$ for thinner poly-Si layers in the near future.

Another industrially relevant approach is directly replacing the $AlO_x/SiN_x$ stack at the rear side of a $p$-PERC cell with $p^+$ TOPCon layers. However, an inferior passivation quality of $p$-TOPCon layers, especially on rougher surfaces, currently hinders their application as passivating contact. Meanwhile, recent studies show the potential of decent passivation ($J_{0,pass} < 10\,fA/cm^2$) and contact formation ($J_{0,met} < 100\,fA/cm^2$) for thick $p$-TOPCon ($d > 200\,nm$) after processing at industrially relevant peak firing temperatures (Mack et al. 2021) with efficiencies close to 22% (Choi et al. 2021). Nevertheless, it should be noted that placing TOPCon layers at the rear in front of pn-junction configuration does not solve the major limitation of a PERC cell. For instance, the total recombination of the i-TOPCon device ($J_{0,tot}$) in a $n$-TOPCon FJ cell is still limited by its front side. Apart from the BJ-approach discussed earlier (Figure 6.2b), another approach to reduce $J_{0,front}$ is to place a $p$-TOPCon layer locally under the front contacts. This concept was applied on plated contacts and showed promising front passivation of $p$-TOPCon on boron emitter ($J_{0,front} = 30\,fA/cm^2$) and excellent contact formation ($\rho_C < 1\,m\Omega\,cm$) (Singh et al. 2020). The drawback is a nonperfect alignment of TOPCon stripes with the metal grid in the front that could lead to parasitic absorption in poly-Si, and restrictions in the number of fingers in the front metallization limiting the current transport (Altermatt et al. 2021). Alternatively, a local doping of poly-Si layers can be achieved using inkjet-printed doping pastes (Kiaee et al. 2022). Another solution is to apply a full area stack of transparent conductive oxide (TCO) and thin oxides ($AlO_x$ and/or $SiO_x$) as passivating front contact (van de Loo et al. 2019; Macco et al. 2021). However, a trade-off between optical transparency in N(IR) region and conductivity remains a major obstacle for using TCO as a front contact (Macco et al. 2021). Additionally, the integration of screen printing in this process sequence needs to be demonstrated as a prerequisite for its applicability in current industrial manufacturing.

## 6.4.2 Alternative to Ag Contacts

Current state-of-the-art metallization with screen printing requires Ag-based pastes applied on both sides of the i-TOPCon solar cell, which is about 1.5 times the required Ag consumption per cell in comparison to $p$-PERC. Although Ag consumption per cell is forecasted to reduce by up to 40% in the next decade (ITRPV 2021) by further pushing the limits of screen printing, it is still likely to remain one of the major cost drivers in cell production. In fact, Ag consumption is forecasted to be a major bottleneck for terra-watt PV production due to sustainability and supply constraints (Verlinden 2020; Goldschmidt et al. 2021). Plating is an alternative metallization scheme to replace Ag with cheaper and more abundant metals like copper. Additional advantage of plating is that a low contact resistivity ($\rho_C < 1\,m\Omega\,cm^2$) can be reached with thinner poly-Si layers ($d < 80\,nm$) (Steinhauser et al. 2020b) and narrow finger widths ($w < 25\,\mu m$). Recently, both front and rear contacts of a 23.8% efficient i-TOPCon cell were formed by combining Ni-Cu plating with laser patterning (Grübel et al. 2022). A reduction in poly-Si thickness and Ag consumption is likely to drive further reduction in the cost of ownership (COO) of i-TOPCon cell manufacturing (Kluska et al. 2020).

### 6.4.3 Economic Competitiveness to *p*-PERC Cells

Last, but not least, the TOPCon concept requires to be economically competitive with the existing PERC cells in terms of COO and levelized cost of electricity (LCOE). Recent studies suggest that an efficiency gain higher than $0.5\%_{abs.}$ to *p*-PERC at cell level already allows cost-effective, high-volume manufacturing of TOPCon cells (Kafle et al. 2021a). Nevertheless, further simplification and standardization of process routes are required to allow stable production with equivalent uptime and utilization rates to current PERC cell manufacturing lines. However, given the tremendous progress in technology and efficiency in particular in the past two years, demonstrated with champion efficiencies in the range of 26% and above even from leading cell manufactures, TOPCon seems to out-perform PERC and is about to become the new standard in silicon solar cell mass production.

## 6.5 Conclusion

Application of poly-Si/SiO$_x$ junction on the *c*-Si allows excellent passivation of contact areas while providing low resistance to the majority transport. These properties are exploited to reach excellent efficiencies of around 26.0% in laboratory and >25.0% on industrial scale. Theoretically, TOPCon-like junctions are applicable to a wide range of cell architectures featuring both *p* and *n*-type *c*-Si substrates. However, a generally inferior passivation quality of *p*-TOPCon currently hinders its application as a full area hole contact. Hence, TOPCon architecture featuring "*n*-TOPCon" at rear side is currently favored for industrialization and is widely seen as the potential evolutionary cell technology upgrade to the incumbent *p*-PERC.

## List of symbols

| | |
|---|---|
| $E_C, E_V,$ | Conduction/valence band edge (eV) |
| $E_g/E_F$ | Band gap/Fermi level energy (eV) |
| $V_{OC}$ | Open circuit voltage (V) |
| FF | Fill Factor |
| $J_{SC}$ | Short-circuit current density (A cm$^{-2}$) |
| $J_0$ | Saturation current density or recombination prefactor (A cm$^{-2}$) |
| $J_{0,front}/J_{0,rear}$ | Saturation current density of front/rear-side (A cm$^{-2}$) |
| $J_{0,tot}$ | Total recombination current density (A cm$^{-2}$) |
| $J_{0,pass}/J_{0,met}$ | Recombination current density of passivated/metalized area (A cm$^{-2}$) |
| $\rho_C$ | Specific contact resistivity ($\Omega$ cm$^2$) |
| $D_{it}$ | Interface trapped charge density (cm$^{-2}$eV$^{-1}$) |
| $\Delta\Phi_e/\Delta\Phi_h$ | Barrier heights for electrons and holes (eV) |

## Author Biographies

**Bishal Kafle** worked as a Scientist at the Fraunhofer Institute for Solar Energy Systems ISE, Germany, where he focused on integration of industrial relevant processes into novel cell architectures such as passivated contact solar cells, and techno-economic assessment of various cell architectures in terms of cell/module/system costs and levelized cost of electricity (LCOE). He is currently working as Process Integration Team Lead at Black Semiconductor GmbH, integrating graphene-based photonics platform on existing electronics technology. He received his MSc and PhD degree in Microsystems engineering from University of Freiburg, Germany in 2011 and 2017, respectively.

**Armin Richter** is Senior Scientist at the Fraunhofer Institute for Solar Energy Systems ISE, Germany. His research interests include atomic layer deposition (ALD) of functional thin films (e.g. surface passivation, electron/hole transport layers for perovskite solar cells or TCOs), the in depth characterization of dielectric surface passivation as well as the development of high efficiency silicon solar cells along the whole process chain including 3d device simulations. He received his PhD in physics in 2014 with work on n-type silicon solar cells and $Al_2O_3$ based silicon surface passivation.

## References

Altermatt, P.P., Guanchao, X., Xueling, Z. et al. (2021). From upscaling PERC to the next technology cycle: transparent passivating contacts may merge n- and p-type cell from upscaling PERC to the next technology cycle: transparent passivating contacts may merge n- and p-type cell. In: *Proceedings of the 38th EUPVSEC*, 100–106.

Chen, D., Chen, Y., Wang, Z. et al. (2020). 24.58% total area efficiency of screen-printed, large area industrial silicon solar cells with the tunnel oxide passivated contacts (i-TOPCon) design. *Solar Energy Materials and Solar Cells* 206: 110258.

Choi, W.-J., Madani, K., Huang, Y.-Y. et al. (2021). Optimization of in-situ and ex-situ doped p+ passivating contact for high efficiency p-TOPCon solar cell application. In: *2021 IEEE PVSC*, 1907–1912.

Cousins, P.J., Smith, D.D., Luan, H.-C. et al. (2010). Generation 3: improved performance at lower cost. In: *Proceedings of the 35th IEEE Photovoltaic Specialists Conference*, 275–278.

Dingemans, G., Beyer, W., van de Sanden, M.C.M., and Kessels, W.M.M. (2010). Hydrogen induced passivation of Si interfaces by Al2O3 films and SiO2/Al2O3 stacks. *Applied Physics Letters* 97 (15): 152106.

Feldmann, F. (2015). Carrier-selective contacts for high-efficiency Si solar cells. Dissertation. In: . Freiburg: Albert-Ludwigs-Universität Freiburg.

Feldmann, F., Bivour, M., Reichel, C. et al. (2014a). Passivated rear contacts for high-efficiency n-type Si solar cells providing high interface passivation quality and excellent transport characteristics. *Solar Energy Materials and Solar Cells* 120: 270–274.

Feldmann, F., Bivour, M., Reichel, C. et al. (2014b). Tunnel oxide passivated contacts as an alternative to partial rear contacts. *Solar Energy Materials and Solar Cells* 131: 46–50.

Feldmann, F., Nicolai, M., Müller, R. et al. (2017a). Optical and electrical characterization of poly-Si/SiOx contacts and their implications on solar cell design. *Energy Procedia* 124: 31–37.

Feldmann, F., Reichel, C., Müller, R., and Hermle, M. (2017b). The application of poly-Si/SiOx contacts as passivated top/rear contacts in Si solar cells. *Solar Energy Materials and Solar Cells* 159: 265–271.

Feldmann, F., Schön, J., Niess, J. et al. (2019). Studying dopant diffusion from poly-Si passivating contacts. *Solar Energy Materials and Solar Cells* 200: 109978.

Feldmann, F., Steinhauser, B., Pernau, T. et al. (2020). Industrial TOPCon solar cells realized by a PECVD tube process. In: *Proceedings of the 37th EUPVSEC* Online, 120.

Fertig, F., Kloter, B., Hoger, I. et al. (2022). Q CELLS > 24% silicon solar cells with mass-production processes. *IEEE Journal of Photovoltaics* 12 (1): 22–25.

Gan, J.Y. and Swanson, R.M. (1990). Polysilicon emitters for silicon concentrator solar cells. In: *Proceedings of the IEEE Conference on Photovoltaic Specialists*, vol. vol. 1, 245–250.

Glunz, S.W. and Feldmann, F. (2018). SiO2 surface passivation layers – a key technology for silicon solar cells. *Solar Energy Materials and Solar Cells* 185: 260–269.

Glunz, S.W., Bivour, M., Messmer, C. et al. (2017). Passivating and carrier-selective contacts – basic requirements and implementation. In: *Proceedings of the IEEE 44th Photovoltaic Specialist Conference (PVSC)*, 2064–2069.

Glunz, S.W., Steinhauser, B., Polzin, J. et al. (2023). Silicon-based passivating contacts: the TOPCon route. *Progress in Photovoltaics: Research and Applications* 31 (4): 341–359.

Goldschmidt, J.C., Wagner, L., Pietzcker, R., and Friedrich, L. (2021). Technological learning for resource efficient terawatt scale photovoltaics. *Energy & Environmental Science* 14 (10): 5147–5160.

de Graaff, H.C. and de Groot, J.G. (1979). The SIS tunnel emitter: a theory for emitters with thin interface layers. *IEEE Transactions on Electron Devices* 26 (11): 1771–1776.

Green, M.A. and Blakers, A.W. (1983). Advantages of metal-insulator-semiconductor structures for silicon solar cells. *Solar Cells* 8 (1): 3–16.

Grübel, B., Nagel, H., Steinhauser, B. et al. (2021). Influence of plasma-enhanced chemical vapor deposition poly-Si layer thickness on the wrap-around and the quantum efficiency of bifacial n-TOPCon (tunnel oxide passivated contact) solar cells. *Physica Status Solidi (A)* 218 (16): 2100156.

Grübel, B., Cimiotti, G., Schmiga, C. et al. (2022). Progress of plated metallization for industrial bifacial TOPCon silicon solar cells. *Progress in Photovoltaics* 30 (6): 615–621.

Haase, F., Hollemann, C., Schäfer, S. et al. (2018). Laser contact openings for local poly-Si-metal contacts enabling 26.1%-efficient POLO-IBC solar cells. *Solar Energy Materials and Solar Cells* 186: 184–193.

Hasse, F., Min, B., Hollemann, C. et al. (2020). Fully screen-printed silicon solar cells with local Al-BSF base contacts and a Voc of 711 mV. In: *Presented at 37th EUPVSEC*. Online.

Huang, Y., Liao, M., Wang, Z. et al. (2020). Ultrathin silicon oxide prepared by in-line plasma-assisted N2O oxidation (PANO) and the application for n-type polysilicon passivated contact. *Solar Energy Materials and Solar Cells* 208: 110389.

ITRPV (2021). *International Technology Roadmap for Photovoltaic*. VDMA e. V. Photovoltaic Equipment Retrieved December 07, 2021.

JinkoSolar Holding Co. Ltd. (2023). JinkoSolar's high-efficiency N-type monocrystalline silicon solar cell sets new record with maximum conversion efficiency of 26.89%. Available at:

https://ir.jinkosolar.com/news-releases/news-release-details/jinkosolars-high-efficiency-n-type-monocrystalline-silicon-3.

Kafle, B., Goraya, B.S., Mack, S. et al. (2021a). TOPCon – technology options for cost efficient industrial manufacturing. *Solar Energy Materials and Solar Cells* 227: 111100.

Kafle, B., Mack, S., Teßmann, C. et al. (2021b). Atmospheric pressure dry etching of polysilicon layers for highly reverse bias-stable TOPCon solar cells. *Solar RRL* 2021: 2100481.

Kale, A.S., Nemeth, W., Harvey, S.P. et al. (2018). Effect of silicon oxide thickness on polysilicon based passivated contacts for high-efficiency crystalline silicon solar cells. *Solar Energy Materials and Solar Cells* 185: 270–276.

Kang, D., Sio, H.C., Stuckelberger, J. et al. (2021). Optimum hydrogen injection in phosphorus-doped polysilicon passivating contacts. *ACS Applied Materials & Interfaces* 13 (46): 55164–55171.

Kang, D., Sio, H.C., Yan, D. et al. (2022). Firing stability of phosphorus-doped polysilicon passivating contacts: factors affecting the degradation behavior. *Solar Energy Materials and Solar Cells* 234: 111407.

Kiaee, Z., Fellmeth, T., Steinhauser, B. et al. (2022). TOPCon silicon solar cells with selectively doped PECVD layers realized by inkjet-printing of phosphorus dopant sources. *IEEE Journal of Photovoltaics* 12 (1): 31–37.

Kluska, S., Hatt, T., Grübel, B. et al. (2020). Plating for passivated-contact solar cells. *Photovoltaics International* 44.

Lindholm, F.A., Neugroschel, A., Arienzo, M., and Iles, P.A. (1985). Heavily doped polysilicon-contact solar cells. *IEEE Electron Device Letters* 6 (7): 363–365.

van de Loo, B.W.H., Macco, B., Melskens, J. et al. (2019). Silicon surface passivation by transparent conductive zinc oxide. *Journal of Applied Physics* 125 (10): 105305.

van de Loo, B.W., Macco, B., Schnabel, M. et al. (2020). On the hydrogenation of poly-Si passivating contacts by Al2O3 and SiN thin films. *Solar Energy Materials and Solar Cells* 215: 110592.

Macco, B., van de Loo, B.W., Dielen, M. et al. (2021). Atomic-layer-deposited Al-doped zinc oxide as a passivating conductive contacting layer for n+-doped surfaces in silicon solar cells. *Solar Energy Materials and Solar Cells* 233: 111386.

Mack, S., Herrman, D., Lenes, M. et al. (2021). Progress in p-type tunnel oxide-passivated contact solar cells with screen-printed contacts. *Solar RRL* 5: 2100152.

Mark Osborne, P. T. (2021). Jolywood touts 24.5% TOPCon cell efficiency for volume manufacturing. Available at: https://www.pv-tech.org/jolywood-touts-24-5-topcon-cell-efficiency-for-volume-manufacturing/. [].

Min, B., Wehmeier, N., Brendemuehl, T. et al. (2020). A 22.3% efficient p-type back junction solar cell with an Al-printed front-side grid and a passivating n + -type polysilicon on oxide contact at the rear side. *Solar RRL* 4 (12): 2000435.

Min, B., Wehmeier, N., Brendemuehl, T. et al. (2021). 716 mV open-circuit voltage with fully screen-printed p-type back junction solar cells featuring an aluminum front grid and a passivating polysilicon on oxide contact at the rear side. *Solar RRL* 5 (1): 2000703.

Moldovan, A., Feldmann, F., Zimmer, M. et al. (2015). Tunnel oxide passivated carrier-selective contacts based on ultra-thin SiO2 layers. *Solar Energy Materials and Solar Cells* 142: 123–127.

Naber, R., van de Loo, B.W.H., and Luchies, J. (2019). LPCVD in-situ n-type doped poly-silicon process throughput optimization and implementation into an industrial solar cell process flow. In: *Proceedings of the 34th Eur. Photovolt. Sol. Energy Conf*, 53.

Nandakumar, N., Rodriguez, J., Kluge, T. et al. (2019). Approaching 23% with large-area monoPoly cells using screen-printed and fired rear passivating contacts fabricated by inline PECVD. *Progress in Photovoltaics: Research and Applications* 27: 107–112.

Padhamnath, P., Buatis, J.K., Khanna, A. et al. (2020). Characterization of screen printed and fire-through contacts on LPCVD based passivating contacts in monoPoly™ solar cells. *Solar Energy* 202: 73–79.

Peibst, R., Römer, U., Larionova, Y. et al. (2016). Working principle of carrier selective poly-Si/c-Si junctions: is tunnelling the whole story? *Solar Energy Materials and Solar Cells* 158: 60–67.

Polzin, J.-I., Feldmann, F., Steinhauser, B. et al. (2019). Study on the interfacial oxide in passivating contacts. *AIP Conference Proceedings* 2147 (040016): 1–7.

Polzin, J.-I., Hammann, B., Niewelt, T. et al. (2021). Thermal activation of hydrogen for defect passivation in poly-Si based passivating contacts. *Solar Energy Materials and Solar Cells* 230: 111267.

Post, I., Ashburn, P., and Wolstenholme, G.R. (1992). Polysilicon emitters for bipolar transistors: a review and re-evaluation of theory and experiment. *IEEE Transactions on Electron Devices* 39 (7): 1717–1731.

Press Release – Jolywood (Taizhou). Solar Technology Co., Ltd. (2023). Available at: http:// jolywood.cn/NewsDetails-436-1442.html [].

Reinders, A., Verlinden, P., van Sark, W., and Freundlich, A. (ed.) (2017). *Photovoltaic Solar Energy: From Fundamentals to Applications*. Chichester West Sussex United Kingdom, Hoboken NJ: John Wiley & Sons.

Richter, A., Müller, R., Benick, J. et al. (2021). Design rules for high-efficiency both-sides-contacted silicon solar cells with balanced charge carrier transport and recombination losses. *Nature Energy* 6 (4): 429–438.

Römer, U., Peibst, R., Ohrdes, T. et al. (2014). Recombination behavior and contact resistance of n+ and p+ poly-crystalline Si/mono-crystalline Si junctions. *Solar Energy Materials and Solar Cells* 131: 85–91.

Schultz-Wittmann, O., Turner, A., Eggleston, B. et al. (2016). High volume manufacturing of high efficiency crystalline silicon solar cells with shielded metal contacts. In: *Proceedings of the 32nd EUPVSEC, Munich, Germany*, 456–459.

Singh, S., Choulat, P., Duerinckx, F. et al. (2020). Development of 2-sided polysilicon passivating contacts for co-plated bifacial n-PERT cells. In: *Proceedings of the 47th IEEE PVSC*, 449–452.

Steinhauser, B., Feldmann, F., Ourinson, D. et al. (2020a). On the influence of the SiNx composition on the firing stability of poly-Si/SiNx stacks. *Physica Status Solidi (A)* 217 (21): 2000333.

Steinhauser, B., Grübel, B., Nold, S. et al. (2020b). Plating on TOPCon as a way to reduce the fabrication costs of I-TOPCon solar cells. In: *Proceedings of the 37th EUPVSEC*, 179–183.

Tarr, N.G. (1985). A polysilicon emitter solar cell. *IEEE Electron Device Letters* 6 (12): 655–658.

Truong, T.N., Yan, D., Chen, W. et al. (2020). Hydrogenation mechanisms of poly-Si/SiOx passivating contacts by different capping layers. *Solar RRL* 4 (3): 2070033.

Verlinden, P.J. (2020). Future challenges for photovoltaic manufacturing at the terawatt level. *Journal of Renewable and Sustainable Energy* 12 (5): 53505.

Yablonovitch, E., Gmitter, T., Swanson, R.M., and Kwark, Y.H. (1985). A 720 mV open circuit voltage SiOx:c-Si:SiOx double heterostructure solar cell. *Applied Physics Letters* 47 (11): 1211–1213.

Yan, D., Cuevas, A., Phang, S.P. et al. (2018). 23% efficient p-type crystalline silicon solar cells with hole-selective passivating contacts based on physical vapor deposition of doped silicon films. *Applied Physics Letters* 113 (6): 61603.

Yan, D., Cuevas, A., Michel, J.I. et al. (2021). Polysilicon passivated junctions: the next technology for silicon solar cells? *Joule* 5: 1–18.

# 7

# Heterojunction Silicon Solar Cells: Recent Developments

*Mathieu Boccard[1] and Jun Zhao[2]*

[1]*École Polytechnique Fédérale de Lausanne (EPFL), Institute of Electrical and Micro Engineering (IEM), Photovoltaics and Thin Film Electronics Laboratory (PV-LAB), Neuchâtel, Switzerland*
[2]*PV-Center, Swiss Center for Electronic and Microengineering (CSEM), Neuchâtel, Switzerland*

## 7.1 Introduction

Since the first volume of "Photovoltaics from Fundamentals to Applications" was published, the field of silicon heterojunction technology (HJT) has massively evolved. Efficiency values reached by research institutes and industries have steadily increased. The absolute world record efficiency for silicon solar cells is now held by an HJT device using a fully rear-contacted structure, and HJT also demonstrated the highest full-wafer efficiency. Several production lines are now running, and multi-gigawatt expansions or new capacities are announced. This progress was neither triggered by a revolution in the fundamental understanding of these devices nor by drastic changes in the materials or deposition techniques used, but rather by systematic improvement of all elements involved in HJT fabrication. Here we will review the recent research and industry developments which have enabled this technology to reach unprecedented performance and discuss challenges and opportunities for its future industrial expansion.

## 7.2 Material Evolutions

### 7.2.1 Thin Silicon Layers

Silicon heterojunction devices rely on the use of thin-film silicon coatings on either side of the wafer to provide surface passivation (using an intrinsic layer) and charge carrier-selectivity (using p- or n-type layers, respectively doped with boron or phosphorus). Recent evolution lies in the use of elaborate stacks for the intrinsic layer and doped layers, following the realization that different properties are required at each interface. From the initial layout shown in Figure 7.1a, more complex structures are used nowadays as shown in Figure 7.1b.

Concerning the intrinsic layer, advanced stacks have replaced single layers. The first few nanometers should ensure a sharp transition between the c-Si and the a-Si:H layer

*Photovoltaic Solar Energy: From Fundamentals to Applications, Volume 2*, First Edition.
Edited by Wilfried van Sark, Bram Hoex, Angèle Reinders, Pierre Verlinden, and Nicholas J. Ekins-Daukes.
© 2024 John Wiley & Sons Ltd. Published 2024 by John Wiley & Sons Ltd.
Companion website: www.wiley.com/go/PVsolarenergy

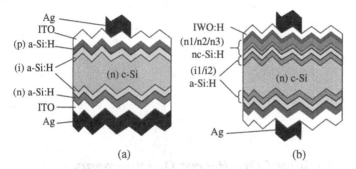

**Figure 7.1** Schematic cross section of silicon heterojunction devices: (a) simplest architecture historically considered and (b) one example of the advanced architectures considered (at least in research and development) in 2021.

(without any damage to the c-Si, nor any epitaxial incubation layer), which is generally obtained for layers presenting a high microstructure ratio, which corresponds to a high hydrogen content (Descoeudres et al. 2017; Sai et al. 2018). Then, to minimize series resistance and ensure high passivation, a low microstructure ratio (lower hydrogen content) is necessary for the bulk of the layer. Sequences of deposition/$H_2$-plasma or multilayer stacks can enable these two properties (Sai et al. 2018; Ruan et al. 2019; Ru et al. 2020).

Nanocrystalline silicon (nc-Si:H) and silicon oxide (nc-SiO$_x$:H) have recently been adopted in the doped layers due to improved doping efficiency, better transparency, and lower contact resistance for charge carrier extraction from the wafer to the transparent electrode. These layers were initially developed for use in thin-film silicon solar cells; their deposition process thus needed some adaptation to combine excellent passivation, conductivity, and transparency. These adaptations typically included the use of multilayer stacks, and/or dedicated plasma treatments prior to growth (Mazzarella et al. 2015; Boccard et al. 1999; Peng et al. 2019). After years of incubation in academic research laboratories, this architecture has recently been adopted by the research and development units of several industrial actors. Therefore, there is strong industrial demand for processing equipment that enables such processes with excellent uniformity and high throughput in order for this architecture to become standard in production in the near future.

### 7.2.2 Transparent Electrodes

Beyond traditional indium tin oxide (ITO), multiple higher-mobility indium-based transparent conductive oxides (TCO) have been employed successfully in HJT cells. These follow the initial works from Koida and Kondo, who demonstrated that hydrogen-doped indium oxide can be grown amorphous and solid-phase crystallized into large crystalline grains (several hundreds of nanometers of typical lateral size for a thickness below 100 nm) (Koida et al. 2007). This yields very large mobility values (typically around 100 cm$^2$/Vs), which in turn enables a high lateral conductivity while minimizing the parasitic free carrier absorption (FCA) of infrared light. Although no dopant other than hydrogen is mandatory, several dopant atoms (Mo, Ce, W, Zr) were shown to enable similar performance, with eventually a higher doping density, and perhaps with improved properties such as device stability, ease of processing, contact resistance to the screen-printed silver electrode. Also, similar to the thin silicon layers, multilayer stacks are increasingly used to enable independent tuning of the properties of the TCO layer to the requirements of each interface and the bulk properties. Relatively high doping at the Si-TCO interface is typically beneficial for reaching low contact resistances but is detrimental to transparency in the infrared (Procel et al. 2020). Also, a low contact resistance at the TCO-Ag interface is desirable, which might require a dedicated capping layer, e.g. to prevent oxidation (Barraud et al. 2013).

Different strategies to reduce the usage of indium in HJT are investigated, which include thinning the doped indium-oxide layer down, or replacing it with another metal-oxide, typically zinc oxide (usually aluminum-doped). The latter was demonstrated to yield similar performances as ITO when used at the rear side (Senaud et al. 2019) and enable over 23% of devices when used on each side (Morales-Vilches et al. 2019). On the front side, reducing the thickness of indium-based TCO and complementing it with a dielectric has also brought optical advantages. The dielectric layer can either be silicon nitride which has a similar refractive index as the TCO ($\sim$2) and completes the single-layer antireflection coating (ARC) (Bätzner et al. 2019), or silicon oxide, which provides a very good dual-layer ARC between the silicon and air (resulting in high current densities at the solar-cell level, which however, does not directly translate into high current densities at the module level) (Herasimenka et al. 2016).

### 7.2.3 Alternative Materials

A surge of interest occurred since 2014 towards the use of metal compounds to replace doped amorphous silicon in the carrier-selective contacts of HJT devices. This was motivated mainly by the large parasitic absorption in this layer on the light-incoming side (Holman et al. 2012; Battaglia et al. 2014). Several remarkable efficiency devices were produced using metal oxide contacts (typically using $MoO_{3-x}$ or $V_2O_{5-x}$ for hole-selectivity and ZnO or $TiO_{2-x}$ for electron-selectivity) or alkali/alkaline-earth metal halide (typically $LiF_{1-x}$ or $MgF_{2-x}$). Nevertheless, the performance of all devices employing metal compound contacts remains below the one using traditional doped-silicon contacts. Best performances top just above 23% and the processes required remain similarly complex as traditional contacting approaches (Dréon et al. 2020; Bullock et al. 2019). These alternative approaches thus remain purely academic so far, with limited foreseen applicability for production. Their study nevertheless contributed to the broad understanding of contacts for silicon solar cells, and the ease of processing of their simplest implementation (only requiring an evaporation or solution-processing tool) made silicon photovoltaics more broadly accessible to laboratories with limited infrastructures. However, as detailed in this chapter, the significant progress demonstrated in the last few years with thin silicon-based contacts places a very high-efficiency benchmark for alternative materials, which questions the benefits of these concepts. Similar approaches were noteworthily investigated half a century ago, before fading out with the rise of the dopant diffusion approaches (Singh et al. 1981).

## 7.3 Device Processing

### 7.3.1 Contacting Strategies

Two-side contacted, rear-junction devices are currently the most popular strategy as shown in Figure 7.1, although high-efficiency values were also demonstrated in front-junction configuration. The main advantage of the rear-junction architecture is that electrical charges can be transported laterally in the wafer, to relax the need for a highly conductive front TCO. This advantage stems from the combination of a higher electron than hole mobility in c-Si, and from the lower contact resistance typically reached on the electron-contact side than hole-contact side (Luderer et al. 2020). Wafer type does not matter due to the long minority carrier lifetimes in relevant HJT devices, leading to a sufficiently high density of photogenerated carriers in most relevant situations.

The opportunity to remove the transparent electrodes has also been explored in recent years. In that case, since a full-area metal contact leads to strong plasmonic absorption (Holman et al. 2013), lateral charge transport has to occur via the wafer itself. Then, the reduction of the contact area from full area (in case of TCO usage) to a few percent requires proportionally lower contact resistance values to maintain similar series resistance losses. However, mastering the direct contact between metal and doped thin-film silicon layers is not straightforward. Annealing can cause metal interdiffusion in Si, which improves contact resistance, but degrades surface passivation (Bryan et al. 2020). Combining dedicated interface treatments with adequate metal stacks, close to 22% efficiency was nevertheless demonstrated with a TCO-free HJT device, validating this concept (Li et al. 2021).

In parallel to these approaches, the back-contacted architecture for which contacts of both polarities are placed on the rear side yielded the most efficient HJT devices until 2023. Beyond removing shadowing from the metal grid, this also relaxes the optical constraint on the thin silicon contact layers. This strategy enabled the first demonstration of a silicon solar cell with an efficiency of over 25% (Masuko et al. 2014), then over 26% (Yoshikawa et al. 2017), and held the world record efficiency until 2023. A selection of remarkable efficiency values reached on research-scale devices is shown in Table 7.1, evidencing the superiority of back-contacted approaches. The drawback of this approach is the higher complexity of cell manufacturing (which involves patterning of the rear contacts) and module interconnection. Disruptive techniques exploiting the properties of doped thin silicon layers nevertheless emerge (Tomasi et al. 2017; Lachenal et al. 2019).

### 7.3.2 Wafer Type

The wafer type of choice has been n-type since the early days of HJT, and all record efficiency devices and production lines use n-type wafers. Nevertheless, interest spiked for the use of p-type wafers, which are slightly less expensive than n-type. Slightly lower performance is reported when using p-type wafers compared to n-type, which is attributed to the lower minority carrier lifetime in these wafers. Several strategies are proposed to improve performance when using p-type wafers, although the economic viability of using p-type wafers is questionable if performance is even so slightly lower than using n-type (Chen et al. 2019).

**Table 7.1** Remarkable efficiencies reported by research laboratories for HJT devices in designated areas ("BC" denotes back-contacted devices and "FJ" refers to front-junction devices).

| Company/Institute | Efficiency (%) | $V_{oc}$ (mV) | $J_{sc}$ (mA/cm$^2$) | FF (%) | Cell area (cm$^2$) | Year |
|---|---|---|---|---|---|---|
| Kaneka (BC) | 26.7 | 738 | 42.7 | 84.9 | 79 | 2017 |
| Panasonic (BC) | 25.6 | 740 | 41.8 | 82.7 | 144 | 2014 |
| **Meyer Burger (BC)** | **25.4** | **745** | **41.5** | **82.6** | **200** | **2021** |
| Kaneka (FJ) | 25.1 | 738 | 40.8 | 83.5 | 152 | 2015 |
| CSEM (BC) | 25.0 | 736 | 41.5 | 81.9 | 25.0 | 2019 |
| Panasonic (FJ) | 24.7 | 750 | 39.5 | 83.2 | 102 | 2012 |
| **HZB/PVcomB** | **24.6** | **747** | **39.8** | **82.8** | **3.9** | **2021** |
| **EPFL/CSEM (FJ)** | **24.4** | **731** | **40.7** | **82.1** | **4.0** | **2021** |
| CSEM (p-type wafer) | 23.7 | 723 | 40.7 | 80.8 | 4.0 | 2020 |

### 7.3.3  Post-processing Improvements

Multiple institutions reported post-fabrication light soaking to improve the efficiency of HJT devices. The physical origin of this effect is not accurately understood yet but is related to passivation improvements. The presence of doped layers appears to be necessary to observe this effect (Kobayashi et al. 2016). The light-induced performance improvement is suppressed when short-circuiting the illuminated devices and reproduced in the dark by injecting forward current instead of illuminating. This indicates that recombination from excited charge carriers is responsible for this effect and not light itself. A degradation of poorly performing HJT devices under prolonged illumination was also reported, which was shown to originate from UV illumination of suboptimal p-type a-Si:H layers. Such degradation was not observed when using thick enough p-type layers or a rear-junction architecture for which UV light does not reach the p-type layer (Cattin et al. 2021). Light soaking at temperatures higher than 50 °C can also lead to degradation, at times followed by recovery (Madumelu et al. 2020).

## 7.4  Revolution in Industry

The industrial landscape around HJT has evolved rapidly in the past decade following the expiry of a key patent held by Panasonic Corporation (formerly Sanyo). Beyond being a topic of interest for academic institutions, several companies have active research and development teams investigating HJT both at the cell and module level.

### 7.4.1  Research and Development

Many companies reached remarkable efficiency values in the year 2021 over the full area of large wafers, as shown in Table 7.2. The 26.8% efficiency value reached by LONGi is

**Table 7.2** Remarkable efficiencies reported by companies on full-wafer HJT devices.

| Company/Institute | Efficiency (%) | $V_{oc}$ (mV) | $J_{sc}$ (mA/cm$^2$) | FF (%) | Cell area (cm$^2$) | Year |
|---|---|---|---|---|---|---|
| **LONGi Solar** | **26.8** | **751** | **41.5** | **86.1** | **274** | **2022** |
| **LONGi Solar (p-type)** | **26.6** | **751** | **41.3** | **85.6** | **274** | **2022** |
| **Sundrive & Maxwell** | **26.4** | **750** | **40.5** | **86.3** | **274** | **2022** |
| **LONGi Solar** | **26.3** | **750** | **40.5** | **86.6** | **274** | **2021** |
| **LONGi Solar (in free)** | **26.1** | **750** | **40.5** | **85.9** | **274** | **2022** |
| **Sundrive & Maxwell** | **26.1** | **747** | **40.7** | **85.7** | **274** | **2022** |
| **LONGi Solar** | **25.8** | **750** | **40.2** | **85.6** | **274** | **2021** |
| **Sundrive & Maxwell (in free)** | **25.5** | **748** | **40.6** | **83.8** | **274** | **2022** |
| **Sundrive & Maxwell** | **25.5** | **746** | **40.2** | **85.1** | **274** | **2021** |
| **HuaSun & Maxwell** | **25.3** | **746** | **40.0** | **84.6** | **274** | **2021** |
| **LONGi Solar** | **25.3** | **749** | **39.5** | **85.5** | **244** | **2021** |
| **HuaSun & Maxwell** | **25.2** | **746** | **39.8** | **85.0** | **274** | **2021** |
| **GS Solar** | **25.2** | **747** | **39.3** | **85.8** | **252** | **2021** |
| **SIMIT + Tongwei** | **25.2** | **749** | **39.4** | **85.4** | **244** | **2021** |
| Hanergy | 25.1 | 747 | 39.6 | 85.0 | 244 | 2019 |
| **Maxwell** | **25.1** | **746** | **39.6** | **84.8** | **274** | **2021** |
| ENEL/CEA-ines | 24.6 | 742 | 39.8 | 83.4 | 244 | 2020 |
| Kaneka | 24.5 | 741 | 40.1 | 82.5 | 239 | 2015 |
| **Jülich** | **24.5** | **741.8** | **39.5** | **83.6** | **244** | **2021** |
| CIC | 24.1 | 745 | 38.8 | 83.2 | 243 | 2016 |
| ZhongWei | 24.1 | 745 | 39.1 | 84.3 | 244 | 2020 |
| Meyer Burger | 24.0 | 739 | 39.3 | 82.7 | 244 | 2018 |
| CSEM | 24.0 | 740 | 39.5 | 82.1 | 244 | 2020 |
| CSEM (p-type) | 23.4 | 741 | 39.4 | 80.0 | 244 | 2020 |

Beyond solar cells, most big actors in the silicon photovoltaics industry announced high-power HJT modules in 2021 with efficiencies well over 21%. Trina, Tongwei, and Akcome demonstrated over 700 W modules using either half-cut wafers or shingled interconnection, corresponding to efficiency values above 22.5%.

the highest performance for a silicon solar cell without concentration. This undoubtedly results from a thorough optimization of all device elements, evidencing the strong company interest in this technology. Particularly noticeable are the fill-factor results, which regularly overpasses 85% and even reach 86% for the very best device. Beyond simply evidencing a low series resistance, the clear correlation with $V_{oc}$ shown in Figure 7.2 indicates that excellent passivation is a key element in reaching high FF values. This requires finely tuned processes, likely involving multilayer stacks, for all the elements of the passivating contact (intrinsic and doped thin silicon layers and TCO film).

**Figure 7.2** Relationship between the fill factor and open-circuit voltage of devices reported in Table 7.1.

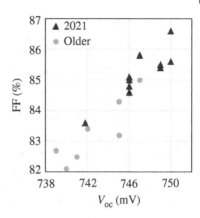

## 7.4.2 Status of Production

In 2021, an estimated 1 GW of HJT modules was shipped from Chinese manufacturers and an additional 1 GW from non-Chinese ones. This value represents around 1% of the total module shipment. There is, however, a strong momentum for increasing production capacity, which reached 5–10 GW at the end of 2021, corresponding to ~5% of the total production capacity. This is expected to be doubled by the end of 2022 following concrete announcements from companies already active in this technology as well as newcomers. The higher efficiency offered by HJT compared to alternative technologies is a strong positive aspect of the technology, as well as its natural bifaciality and higher performance at high temperatures, resulting in improved energy yield compared to mainstream products. Contrarily, pressing challenges for HJT mass production include higher production costs (both in terms of capital and operational expenditures). Historical industrial players (notably Panasonic, which brought the technology to the market over the past 30 years) are being replaced by newcomers and several production equipment manufacturers are offering high-throughput equipment. Standard size modules with over 22% efficiency are already available on the market, which are amongst the highest available efficiencies. With the unremitting efforts of some Chinese equipment manufacturers, the CAPEX of a 1 GW turnkey HJT line is reduced from 81 million USD in 2019 to 58 million USD in 2022. But compared with a PERC turnkey production line with the same 1 GW scale and worth 14.5 million USD, HJT still has high potential to reduce its CAPEX. Challenges and opportunities for large-scale development.

The rapid growth of the PV industry brings up challenges and opportunities for all technologies. Thanks to the newly developed nano-crystalline doped silicon layer, a significant increase of fill factor has been achieved in 2022 by most of the record cells, and this new concept has been adapted in most new mass production PECVD tools as standard configuration. Silver consumption is common to all technologies, yet more pressing for HJT which has a historically higher usage. This stems from the use of silver on both sides (compared to the use of aluminum on the rear side for the mainstream PERC technology), and the use of low-curing temperature paste, which has a lower conductivity than high-curing-temperature paste. A drastic reduction in Ag usage is necessary for HJT to reach the TW production scale. The HJT architecture, however, can tolerate two alternative copper-based metallization schemes (copper plating and printing of paste composed of silver-coated copper particles) readily implementable due to the use of low temperatures and a TCO. A 50–60% silver content in the silver-coated cupper pastes has been tested partially in mass production and reached a minimized efficiency drop to standard silver paste. The modules with such pastes have passed the IEC reliability tests. Further reduction of the silver content to 30% is evaluated on R&D scale. Then, the use of indium for the TCO is another roadblock for TW-scale deployment and a source of cost volatility. Solutions also exist here with alternative oxides and/or drastic thickness reduction. Then, since HJT relies on thin-film coatings, increasing the wafer size – which is currently occurring in mainstream production to improve line throughput – does not bring a significant advantage in terms of cell processing. Also, the very long minority carrier lifetimes in HJT devices bring additional losses when cutting the wafer in half (which is again a strong trend in current module production to minimize resistive losses in the module interconnection). Therefore, since 2022, most of the HJT mass production in China started producing half as-cut wafers from the beginning of the cell processing to minimize the cutting losses on finished cell level. Conversely, one strong advantage of HJT over competing technologies is the ability to use thinner wafers, stemming from the symmetry of the contacts of both polarities and the use of low temperatures, currently wafer thickness below 130 µm becomes mostly standard in HJT mass production. Together with an improved module efficiency and system energy yield, this provides an opportunity to reduce the $CO_2$ emissions needed to generate PV electricity (in $g_{CO2}$/kWh) since monocrystalline silicon production remains a large contributor to the carbon footprint of c-Si-based PV electricity. There are, therefore, opportunities for HJT to gain market shares in the coming decade, which will depend on the evolution of the technologies currently in production and on the market pressure.

## 7.5 Conclusions

The silicon HJT field has been continuously evolving for the past few years. Systematic optimization of each material involved in these devices enabled to mitigate all limitations impeding their performance. Furthermore, the industrial landscape is diversifying, with around a dozen companies being in production. This technology remains an outsider with a competitive edge in terms of performance but slightly higher production costs than mainstream technology. Its market share is nevertheless progressing with challenges in manufacturing still to be fully solved (notably the usage of silver and indium, production cost) to continue expanding. Beyond the current market, silicon heterojunctions have been used extensively as bottom sub-cells in two-junction ("tandem") solar-cell devices and in particular with top sub-cells using a perovskite absorber. Silicon HJT is therefore well established as worthy of consideration, and it leads the race in several key areas. It thus holds a place of choice in the intense competition occurring nowadays for market supremacy in the photovoltaics world.

# References

Barraud, L., Holman, Z.C., Badel, N. et al. (2013). Hydrogen-doped indium oxide/indium tin oxide bilayers for high-efficiency silicon heterojunction solar cells. *Solar Energy Materials & Solar Cells* 115: 151–156.

Battaglia, C., Yin, X., Zheng, M. et al. (2014). Hole Selective $MoO_x$ contact for silicon solar cells. *Nano Letters* 14 (2): 967–971.

Bätzner, D.L., Papet, P., Legradic, B. et al. (2019). Alleviating performance and cost constraints in silicon heterojunction cells with HJT 2.0. In: *2019 IEEE 46th Photovoltaic Specialists Conference (PVSC)*, 1471–1474. IEEE.

Boccard, M., Monnard, R., Antognini, L., and Ballif, C. (2018, 1999). Silicon oxide treatment to promote crystallinity of p-type microcrystalline layers for silicon heterojunction solar cells. *AIP Conference Proceedings* 1: 40003.

Bryan, J.L., Carpenter, J.V. III, Yu, Z.J. et al. (2020). Aluminum–silicon interdiffusion in silicon heterojunction solar cells with a-Si:H(i)/a-Si:H(n/p)/Al rear contacts. *Journal of Physics D: Applied Physics* 54 (13): 134002.

Bullock, J., Wan, Y., Hettick, M. et al. (2019). Dopant-free partial rear contacts enabling 23% silicon solar cells. *Advanced Energy Materials* 9 (9): 1803367.

Cattin, J., Senaud, L.-L., Haschke, J. et al. (2021). Influence of light soaking on silicon heterojunction solar cells with various architectures. *IEEE Journal of Photovoltaics* 11 (3): 575–583.

Chen, D., Kim, M., Shi, J. et al. (2019). Defect engineering of p-type silicon heterojunction solar cells fabricated using commercial-grade low-lifetime silicon wafers. *Progress in Photovoltaics: Research and Applications* (November): 1–15.

Descoeudres, A., Allebe, C., Badel, N. et al. (2017). Advanced silicon thin films for high-efficiency silicon heterojunction-based solar cells. In: *2017 IEEE 44th Photovoltaic Specialist Conference (PVSC)*, 50–55. IEEE.

Dréon, J., Jeangros, Q., Cattin, J. et al. (2020). 23.5%-efficient silicon heterojunction silicon solar cell using molybdenum oxide as hole-selective contact. *Nano Energy* 70: 104495.

Herasimenka, S.Y., Dauksher, W.J., Boccard, M., and Bowden, S. (2016). ITO/SiOx:H stacks for silicon heterojunction solar cells. *Solar Energy Materials & Solar Cells* 158: 98–101.

Holman, Z.C., Descoeudres, A., Barraud, L. et al. (2012). Current losses at the front of silicon heterounction solar cells. *IEEE Journal of Photovoltaics* 2 (1): 7–15.

Holman, Z.C., De Wolf, S., and Ballif, C. (2013). Improving metal reflectors by suppressing surface plasmon polaritons: a priori calculation of the internal reflectance of a solar cell. *Light: Science & Applications* 2 (10): e106–e106.

Kobayashi, E., De Wolf, S., Levrat, J. et al. (2016). Light-induced performance increase of silicon heterojunction solar cells. *Applied Physics Letters* 109 (15): 1–6.

Koida, T., Fujiwara, H., and Kondo, M. (2007). Hydrogen-doped in2O3 as high-mobility transparent conductive oxide. *Japanese Journal of Applied Physics, Part 2: Letters* 46 (25–28): 685–687.

Lachenal, D., Papet, P., Legradic, B. et al. (2019). Optimization of tunnel-junction IBC solar cells based on a series resistance model. *Solar Energy Materials & Solar Cells* 200: 110036.

Li, S., Pomaska, M., Lambertz, A. et al. (2021). Transparent-conductive-oxide-free front contacts for high-efficiency silicon heterojunction solar cells. *Joule* 5 (6): 1535–1547.

Luderer, C., Messmer, C., Hermle, M., and Bivour, M. (2020). Transport losses at the TCO/a-Si:H/c-Si heterojunction: influence of different layers and annealing. *IEEE Journal of Photovoltaics* 10 (4): 952–958.

Madumelu, C., Wright, B., Soeriyadi, A. et al. (2020). Investigation of light-induced degradation in N-Type silicon heterojunction solar cells during illuminated annealing at elevated temperatures. *Solar Energy Materials & Solar Cells* 218: 110752.

Masuko, K., Shigematsu, M., Hashiguchi, T. et al. (2014). Achievement of more than 25% conversion efficiency with crystalline silicon heterojunction solar cell. *IEEE Journal of Photovoltaics* 4 (6): 1433–1435.

Mazzarella, L., Kirner, S., Gabriel, O. et al. (2015). nanocrystalline silicon oxide emitters for silicon hetero junction solar cells. *Energy Procedia* 77: 304–310.

Morales-Vilches, A.B., Cruz, A., Pingel, S. et al. (2019). ITO-free silicon heterojunction solar cells with ZnO:Al/SiO$_2$ front electrodes reaching a conversion efficiency of 23%. *IEEE Journal of Photovoltaics* 9 (1): 34–39.

Peng, C., Lei, C., Ruan, T. et al. (2019). High phosphorus-doped seed layer in microcrystalline silicon oxide front contact layers for silicon heterojunction solar cells. In: *2019 IEEE 46th Photovoltaic Specialists Conference (PVSC)*, 2550–2553. IEEE.

Procel, P., Xu, H., Saez, A. et al. (2020). The role of heterointerfaces and subgap energy states on transport mechanisms in silicon heterojunction solar cells. *Progress in Photovoltaics: Research and Applications* 28 (9): 935–945.

Ru, X., Qu, M., Wang, J. et al. (2020). 25.11% efficiency silicon heterojunction solar cell with low deposition rate intrinsic amorphous silicon buffer layers. *Solar Energy Materials & Solar Cells* 215: 110643.

Ruan, T., Qu, M., Wang, J. et al. (2019). Effect of deposition temperature of a-Si:H layer on the performance of silicon heterojunction solar cell. *Journal of Materials Science: Materials in Electronics* 30 (14): 13330–13335.

Sai, H., Chen, P.W., Hsu, H.J. et al. (2018). Impact of intrinsic amorphous silicon bilayers in silicon heterojunction solar cells. *Journal of Applied Physics* 124: 103102.

Senaud, L.-L., Christmann, G., Descoeudres, A. et al. (2019). Aluminium-doped zinc oxide rear reflectors for high-efficiency silicon heterojunction solar cells. *IEEE Journal of Photovoltaics* 9 (5): 1217–1224.

Singh, R., Green, M.A., and Rajkanan, K. (1981). Review of conductor-insulator-semiconductor (CIS) solar cells. *Solar Cells* 3 (2): 95–148.

Tomasi, A., Paviet-Salomon, B., Jeangros, Q. et al. (2017). Simple processing of back-contacted silicon heterojunction solar cells using selective-area crystalline growth. *Nature Energy* 2 (5): 17062.

Yoshikawa, K., Kawasaki, H., Yoshida, W. et al. (2017). Silicon heterojunction solar cell with interdigitated back contacts for a photoconversion efficiency over 26%. *Nature Energy* 2 (5): 17032.

# 8

## Update on Non-silicon-based Low-Temperature Passivating Contacts for Silicon Solar Cells

*Md. Anower Hossain and Bram Hoex*

*School of Photovoltaic and Renewable Energy Engineering, University of New South Wales, Sydney, NSW, Australia*

## 8.1 Introduction

In order to increase the crystalline silicon (c-Si) solar cell efficiency, further reducing the electrical and optical losses at the surfaces and, specifically, the contact between the semiconductor and the metallization is necessary. Diffused contact-based c-Si solar cells have reached their efficiency limit, and alternative solutions are now required. As a result, carrier selective or passivating contacts have recently attracted significant interest in the photovoltaic scientific community (Allen et al. 2019). This chapter builds upon the chapter in the first volume (Reinders et al. 2017) and discusses recent progress in low-temperature deposited transition metal oxides (TMOs) passivating contact materials for c-Si solar cells.

Figure 8.1 shows a bulk semiconductor, in this case, c-Si, where light is absorbed and creates electron–hole pairs. These electron–hole pairs need to be separated, which is done by using interfaces that are either selective for holes or electrons. By design, these passivating or carrier-selective contacts exhibit extremely low recombination currents for one type of carrier (i.e. interface passivation) while having a low resistivity for the opposite polarity. An elegant way to look at this was recently proposed by Brendel and Peibst (2016), based on the theoretical framework proposed by Würfel et al. (2015), which attributes carrier selectivity to the asymmetry in the electron and hole conduction. Brendel et al. defined the electron resistance $\rho_n$ and hole resistance $\rho_p$ as follows (in this case for a hole-selective contact):

$$\rho_n = \left(\frac{dJ_n}{dV}\right)^{-1} = \frac{v_{th}}{f_c j_c} \tag{8.1}$$

$$\rho_p = \left(\frac{dJ_p}{dV}\right)^{-1} = \frac{\rho_c}{f_c} \tag{8.2}$$

where $J_{n,p}$ is the electron/hole current density, $V$ is the voltage, $v_{th}$ is the thermal voltage, $f_c$ is the contact fraction, and $j_c$ is the minority carrier combination current (all at $V = 0$). When dividing the electron and hole resistance, we can define the selectivity $S_{10}$:

$$S_{10} = \log_{10}\left(\frac{\rho_n}{\rho_p}\right) = \log_{10}\left(\frac{v_{th}}{j_c \rho_c}\right) \tag{8.3}$$

*Photovoltaic Solar Energy: From Fundamentals to Applications, Volume 2*, First Edition.
Edited by Wilfried van Sark, Bram Hoex, Angèle Reinders, Pierre Verlinden, and Nicholas J. Ekins-Daukes.
© 2024 John Wiley & Sons Ltd. Published 2024 by John Wiley & Sons Ltd.
Companion website: www.wiley.com/go/PVsolarenergy

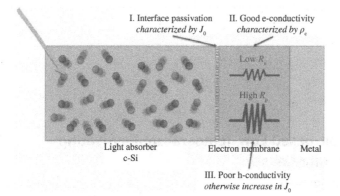

I. Interface passivation
*characterized by $J_0$*

II. Good e-conductivity
*characterized by $\rho_c$*

Low $R_s$

High $R_p$

Light absorber
c-Si

Electron membrane

Metal

III. Poor h-conductivity
*otherwise increase in $J_0$*

**Figure 8.1** Schematic representation of a passivating electron contact (Zhou et al. 2022/with permission of Elsevier).

where $\log_{10}$ is the logarithm with base 10. This selectivity is a very useful parameter to consider as it is an excellent figure of merit to estimate the efficiency potential of a particular contact. This is shown schematically in Figure 8.2, where we plot the upper limit efficiency as a function of the $j_c$ and $\rho_c$ of a contact. It should be noted that you can move along the iso-selectivity lines (dotted lines in Figure 8.2) by changing the contact fraction of the

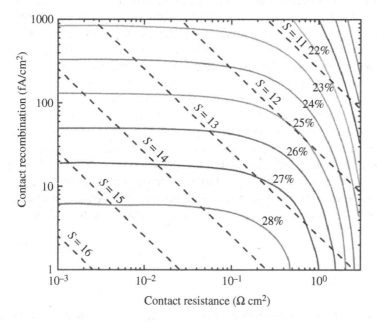

**Figure 8.2** Upper limit efficiency of a silicon solar cell as a function of the minority carrier recombination current $j_c$ and majority carrier contact resistance $\rho_c$. The dashed lines indicate iso-selectivity lines. Please note that according to the definitions in Eqs. (8.1)–(8.3), we can move along the iso-selectivity line by changing the contact fraction $f_c$. Hence, this allows for optimizing the efficiency ceiling for a certain contact system when the $j_c$ and $\rho_c$ values are known. This calculation was adapted from Brendel and Peibst (2016) and a short-circuit density of 43.6 mA/cm² and assumes that only that contact limits the solar cell efficiency.

contact. This is, for example, already exploited in the passivated emitter and rear cell (PERC) solar cell where the optimal rear contact fraction (with a selectivity of 13) is around 1%, thereby limiting the solar cell efficiency to 27.2% (Brendel and Peibst 2016). Unfortunately, the selectivity of the phosphorus-diffused electron contact in a PERC is only ~12 (Brendel and Peibst 2016), and this limits the PERC solar cell to roughly 25%, in excellent agreement with the 25% champion result achieved by UNSW in 1999 (Zhao et al. 1999).

The most widely investigated and popular passivating contacts are silicon heterojunction (SHJ, Chapter 3.3) and doped polysilicon (poly-Si, Chapter 6). However, these contacts are not yet perfect; most notably, they have significant parasitic absorption and Auger recombination (Baker-Finch et al. 2014). To overcome these downsides, wide bandgap TMOs have been explored (Figure 8.3). The materials exhibit high transparency extending well into the visible spectrum, which significantly reduces or potentially eliminates optical losses. Additionally, these materials mitigate interface carrier recombination and exhibit no Auger recombination in comparison to doped silicon contacts. Furthermore, they can be fabricated via a low-temperature deposition technique, making them suitable for various applications. These benefits of TMOs are used to get more carrier generation in c-Si provided that they are used as front electron or hole passivating contacts, which also simplifies the device fabrication.

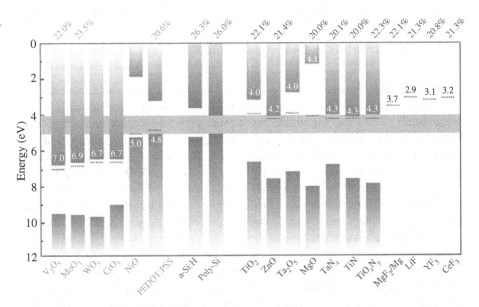

**Figure 8.3** Shows band alignment between c-Si and different passivating contact materials. Materials that can be used for passivating electron or hole contacts are labeled orange and green, respectively, and approximate Fermi-level positions are shown as red dashed lines. The band edges of c-Si (gray bar) and the band position of a-Si:H and poly-Si are shown for reference. The power conversion efficiencies (PCEs) of c-Si solar cells, featuring a full-area passivating contact, are listed at the top, and the new world record for SHJ solar cell efficiency is 26.81% by LONGi Solar (LONGi 2022). Note that the best PCEs achieved for both side-contacted c-Si solar cells (≥20%) are shown, except for the MgF$_2$/Mg/Al contact, which features an interdigitated back contact (IBC) solar cell design (K. Gao et al. 2022/John Wiley & Sons).

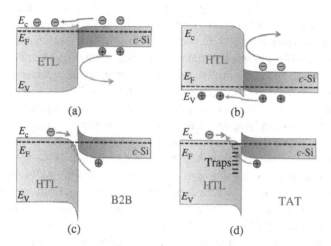

**Figure 8.4** (a) Shows band alignment between *n*-type electron transport layer (ETL) material and c-Si, where low $\Delta E_C$ facilitates electron transport. In contrast, large $\Delta E_V$ blocks holes. (b) Band alignment of *p*-type HTL material with c-Si, where low $\Delta E_V$ facilitates hole transport while large $\Delta E_C$ blocks holes. (c) The band alignment between *n*-type high $W_F$ TMOs and c-Si where the hole transport is enabled by B2B tunneling or (d) trap-assisted tunneling (Wang et al. 2023). It should be noted that an interface passivation layer between the contact, such as silicon oxide or amorphous silicon, is typically added to reduce the interface recombination current.

The working principle of TMOs-based passivating electron and hole contacts is shown schematically in Figure 8.4.

- The first method is, for example, an n-type wide bandgap semiconductor material with a good alignment with the c-Si conduction band (CB) and a large offset with the c-Si valence band (VB) at the c-Si interface, making sure a small conduction band offset ($\Delta E_C$) and large valence band offset ($\Delta E_V$) with c-Si (Figure 8.4a) for easy transport of electrons from c-Si and block holes (Wang et al. 2023). The electron-selective contact materials typically have low work functions ($W_F$), such as $TiO_{2-x}$, $Nb_2O_{5-x}$, and $MgO_x$. Other low $W_F$ ultrathin metal fluoride materials like $MgF_x$, $LiF_x$ are also proven to be suitable electron-selective contact materials. They can significantly lower the effective $W_F$ of the metal electrode contact stack. For hole transport layer (HTL), a large $\Delta E_C$ and small $\Delta E_V$ facilitate the transport of holes from c-Si to the contacts (Figure 8.4b).
- In the second method, one can form an electron (hole) contact by using a material with an extremely low (high) $W_F$ (with reference to silicon) where the carrier transport mechanisms relied on band-to-band tunneling (B2B) (Figure 8.4c) or trap-assisted tunneling (TAT) (Figure 8.4d). The band-alignment between the two materials results in band-bending in c-Si, and this results in an asymmetry in conduction in the c-Si, in a similar way to conventional diffusion-based silicon contact. TMOs like $MoO_{3-x}$, $WO_{3-x}$, $V_2O_{5-x}$, $NiO_x$, $CuO_{2-x}$, and $CrO_x$ are promising candidates for passivating hole contacts. Among these, the p-type $NiO_x$, and $CuO_{2-x}$ have low $\Delta E_V$ with c-Si.

However, high interfacial trap density resulted in poor solar cell performance. A unique feature is that all these materials can be deposited at low temperatures, which enables the fabrication of dopant-free c-Si solar cells using only low-temperature processes (Wan

**Table 8.1** Efficiency of low-temperature passivation electron contacts-based c-Si solar cells.

| Passivation material | Contact structure | $J_{sc}$ (mA/cm$^2$) | $V_{oc}$ (V) | FF (%) | Efficiency (%) | Refs. |
|---|---|---|---|---|---|---|
| $TiO_{2-x}$ | n-Si/SiO$_{2-x}$/TiO$_{2-x}$/Al/Ag | 39.8 | 0.674 | 82.5 | 22.1 | Yang et al. (2017) |
| $TiO_{2-x}$/LiF | n-Si/TiO$_{2-x}$/LiF$_x$/Al | 41.5 | 0.695 | 80.0 | 23.1 | Bullock et al. (2019) |
| $TiO_xN_y$ | n-Si/a-Si:HTiO$_x$N$_y$/Al/Ag | 39.5 | 0.698 | 80.8 | 22.3 | Yang et al. (2020a) |
| $Ta_2O_{5-x}$ | n-Si/Ta$_2$O$_{5-x}$/Mg/Ag | 37.8 | 0.638 | 79.3 | 19.1 | Wan et al. (2018) |
| $ZnO_x$ | n-Si/a-Si:H/ZnO$_x$/LiF$_x$/Al | 39.2 | 0.706 | 77.3 | 21.4 | Zhong et al. (2020) |
| $SnO_{2-x}$ | i-a-Si:H/SiO$_{2-x}$/SnO$_{2-x}$/Mg | 37.45 | 0.685 | 71.52 | 18.35 | Liu et al. (2019) |
| $MgO_x$ | n-Si/MgO$_x$/Al | 39.5 | 0.628 | 80.6 | 20.0 | Wan et al. (2017) |
| $MgF_2$ | n-Si/a-Si:H/MgF$_x$ | 41.5 | 0.718 | 74.2 | 22.1 | Wu et al. (2020) |
| $TaN_x$ | n-Si/TaN$_x$/Al | 38.8 | 0.632 | 81.8 | 20.1 | Yang et al. (2018) |
| $TiN_x$ | n-Si/SiO$_{2-x}$/TiN$_x$ | 37.9 | 0.644 | 81.9 | 20.0 | Yang et al. (2019) |
| $YF_3$ | n-Si/YF$_3$/Al/Ag | 39.8 | 0.645 | 80.8 | 20.8 | Chen et al. (2021) |
| $YbSi_x$ | i-a-Si:H/YbSi$_x$ | 33.7 | 0.654 | 76.9 | 17.0 | Cho et al. (2020) |
| $CeF_3$ | n-Si/SiO$_{2-x}$/CeF$_{3-x}$ | 41.6 | 0.646 | 79.13 | 21.27 | Wang et al. (2021) |
| $GdF_3$ | n-Si/GdF$_3$/Al | 40.97 | 0.652 | 77.5 | 20.71 | Chen et al. (2022) |

et al. 2017). However, TMOs are prone to structural change in thermal annealing, which varies for different TMOs. For example, $NiO_x$ is stable up to 400 °C, and $TiO_{2-x}$ is stable up to 250 °C. Doping typically improves their thermal stability.

Table 8.1 summarizes passivating electron contact materials that have been investigated to date, showing that passivating electron contacts have been developed, resulting in solar cell efficiencies of over 20% (Bullock et al. 2019), suggesting their efficiency potential. The champion efficiency of 23.1% (Bullock et al. 2019) and 23.83% (Cao et al. 2022) of the TMOs-based electron and hole passivating contacts, respectively, is significantly lower than the current record efficiency of 26.81% (LONGi 2022) for SHJ device, suggesting that significant effort is needed to make this low-temperature processing technology a competitive c-Si solar cell technology. It is worth mentioning that typically TMOs possess a strong asymmetry in carrier conductivity but are not good in silicon surface passivation, and efficient devices rely on a thin $SiO_x$ or an a-Si:H interface passivation layer between the contact to achieve a low $j_c$. Therefore, improving both the contact and passivation properties and their thermal stability is crucial for the long-term sustainability of these contacts. The next sections will summarize the most notable contributions for passivating electron and hole contacts.

## 8.2 TMOs as Passivating Electron Contacts

TMOs are attractive materials to be used in passivating contacts because of their versatile electronic properties, which can be varied by changing the transition-metal (TM)-to-oxygen

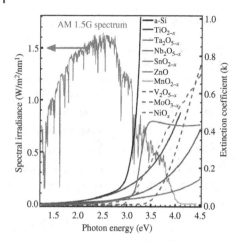

**Figure 8.5** The wavelength-dependent extinction coefficient of various TMOs compared to a-Si counterpart. The energy-dependent spectral energy of the AM 1.5G spectrum is added to illustrate that the extinction coefficients of the TMOs are lower in parts of the solar spectrum with significant irradiance.

ratio. The physical properties of TMOs originate from the partially filled d-orbitals of the TM in TMOs (Tokura and Nagaosa 2000). While stoichiometric TMOs are usually insulators, their non-stoichiometry along with metal or oxygen vacancies make them p- or n-type semiconductors. The oxygen vacancies tend to fill the d-bands of TM, which moves the Fermi level toward the CB and makes the semiconductor n-type. The $W_F$ of TMOs results in an upward or downward band bending and dictates the passivation electron and hole contact properties. The strongly correlated nature of these d-electrons results in narrow electronic d-bands within the bandgap. These bandgap states close to the CB and VB make the TMO *n*- or *p*-type semiconductors, respectively, and determine material properties, such as carrier concentration, bandgap, and optical properties. Some TMOs, such as $TiO_{2-x}$ (Liu et al. 2021; Yang et al. 2016), $Ta_2O_{5-x}$ (Wan et al. 2018), $Nb_2O_{5-x}$ (Macco et al. 2018), ZnO (Ding et al. 2018; Lin et al. 2020), and $Sc_2O_{3-x}$ (Quan et al. 2018), and non-TMOs such as $MgO_x$ (Wan et al. 2017), BaO (Kim et al. 2020), and $SnO_{2-x}$ (M. Liu et al. 2019) are used as electron-selective contacts. For example, dipoles at the c-Si/TMOs interface originating from shallow oxygen vacancies in TMOs play a key role in defining the $W_F$ and band alignment with c-Si (Gerling et al. 2017; Mazzarella et al. 2021; Meyer et al. 2012b). The parasitic absorption can be mitigated by exploiting wide bandgap TMOs and shows better optical transparency for photon energies of >3 eV (Figure 8.5) (Gupta et al. 2022; Liao et al. 2014; Liu et al. 2021; Zhang et al. 2018). TMOs have significantly lower absorption in the short-wavelength region of the solar spectrum compared to a-Si, thus allowing for higher photogenerated current densities.

### 8.2.1 $TiO_{2-x}$ as Passivating Electron Contact

As stoichiometry and defects are crucial to change the electronic properties of the electron-selective TMOs, their synthesis and post-treatment can affect their electronic properties, especially when they are synthesized in an oxidizing or reducing environment. $TiO_{2-x}$ is currently the most studied passivating electron contact material for c-Si devices

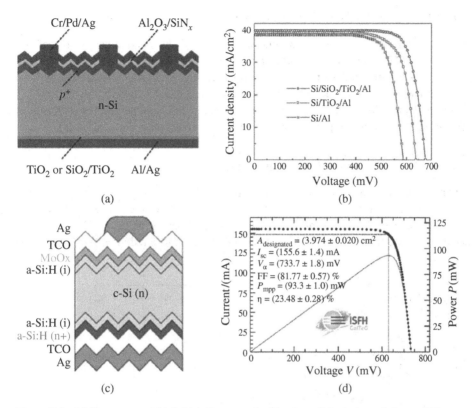

**Figure 8.6** (a) Structure and (b) light $J–V$ curves of $n$-Si solar cell featuring a full-area $TiO_2$ or $SiO_2/TiO_2$ stack contact. (c) Device cross-section and (d) certified light $J–V$ characteristics of the SHJ solar cell employing $MoO_{3−x}$ as HTL (Dréon et al. 2020; Wang et al. 2023; Yang et al. 2016).

(Bullock et al. 2019). It has shown excellent surface passivation of c-Si and is thermally stable up to 575 °C. $TiO_{2−x}$ works according to the mechanism shown in Figure 8.4a, the selectivity originating from a low $\Delta E_C$ of ~0.05 eV and high $\Delta E_V$ of ~2 eV with c-Si (Jhaveri et al. 2014; Yang et al. 2016), ensuring easier pathways for electrons, and a barrier for the hole extraction, respectively (Yang et al. 2017). In addition, high-quality surface passivation was achieved from atomic layer deposition (ALD)-grown $TiO_{2−x}$ films (Liao et al. 2014, 2015). The $TiO_{2−x}$-based champion c-Si solar cells were fabricated using ALD-grown $TiO_{2−x}$ as a passivating electron contact layer with a record efficiency of 23.1% (Bullock et al. 2019).

The efficiency for $TiO_{2−x}$ contacts-based c-Si solar cells increased rapidly after the first demonstration, followed by understanding mechanisms for contact selectivity and passivation. An efficiency of 22.6% was demonstrated using a thermally grown ultrathin $SiO_2$ interface passivation layer which provided surface passivation and improved the $V_{oc}$ (Figure 8.6a,b) (Yang et al. 2016, 2017). The use of an Al capping layer favored the formation of oxygen-deficient $TiO_{2−x}$ and significantly improved the contact and the passivation properties. In addition, the contact became much more stable at a temperature of 575 °C (Yang et al. 2017), suggesting its superior thermal stability over the a-Si:H-based contacts. The use of low $W_F$ metals, such as Al, Ca, and Yb, resulted in a very low $\rho_c$ of ~5 m$\Omega$ cm$^2$ (Allen et al. 2017).

### 8.2.2 Nb$_2$O$_{5-x}$ as Passivating Electron Contact

The band energetics of Nb$_2$O$_{5-x}$ is similar to TiO$_{2-x}$ and shows good passivation of c-Si (Macco et al. 2018). Like TiO$_{2-x}$, an ultrathin chemically grown SiO$_{2-x}$ interface layer and Al capping layer help achieve the lowest contact resistivity value of 70 mΩ cm$^2$ for 1 nm Nb$_2$O$_{5-x}$ thick films. A strong negative fixed charge $Q_f$ of 1.6–1.9 × 10$^{12}$ cm$^{-2}$ in the Nb$_2$O$_{5-x}$ film assists in providing excellent passivation and an upward band bending of ~850 mV on 3 Ω cm n-type c-Si, suggesting its potential as n-type contact (Messmer et al. 2018).

### 8.2.3 Ta$_2$O$_{5-x}$ as Passivating Electron Contact

ALD-grown Ta$_2$O$_{5-x}$ has also shown electron selectivity and good surface passivation due to a small $\Delta E_C$ and large $\Delta E_V$, which resulted in 19.1% device efficiency (Wan et al. 2018). In the case of Ta$_2$O$_{5-x}$ films, it was found that hydrogenation provided by a capping plasma-enhanced chemical vapor deposited (PECVD) SiN$_x$ film reduced its $W_F$ from 3.56

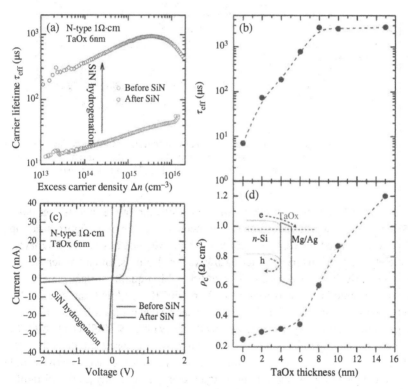

**Figure 8.7** Carrier selectivity characterizations of Ta$_2$O$_{5-x}$ passivated electron hetero-contacts to *n*-type c-Si. Panel (a) presents the effective carrier lifetime $\tau_{eff}$ versus excess carrier density $\Delta n$ for c-Si passivated with Ta$_2$O$_{5-x}$ films before and after SiN$_x$ hydrogenation. Panel (b) shows the effect of hydrogenated Ta$_2$O$_{5-x}$ film thickness on $\tau_{eff}$. Panel (c) illustrates representative *I–V* measurements of Ta$_2$O$_{5-x}$ samples before and after SiN$_x$ hydrogenation. Panel (d) shows the effect of hydrogenated Ta$_2$O$_{5-x}$ film thickness on the contact resistivity $\rho_c$. The inset in Panel d shows a schematic illustration of the band diagram with hydrogenated Ta$_2$O$_{5-x}$ (Wan et al. 2018).

to 3.27 eV, while not changing its band energetics, which increased band bending in the c-Si and, consequently, the contact resistance for electrons as shown in Figure 8.7 (Wan et al. 2018). A low $W_F$ metal (Mg in this case) was used to increase electron transport, and $SiN_x$ hydrogenation treatment of ~6 nm thick $Ta_2O_{5-x}$ film resulted in a low contact resistivity of ~0.35 $\Omega\,cm^2$ to the $n$-type c-Si, which is sufficiently low for efficient use as a hetero-contact for solar cells. The $\tau_{eff}$ and $\rho_c$ increase with the $Ta_2O_{5-x}$ film thickness, suggesting a trade-off between contact resistivity and surface passivation on c-Si.

### 8.2.4 $Sc_2O_{3-x}$ as Passivating Electron Contact

$Sc_2O_{3-x}$ formed from an interface redox reaction between $SiO_{2-x}$, and Sc was found to modify interfacial $W_F$; thus, it exhibited electron selectivity and full area surface passivation to c-Si with the lowest contact resistivity of 20 $m\Omega\,cm^2$ and $J_c$ of 61 $fA\,cm^{-2}$, respectively, with a device efficiency of 15.1% (Quan et al. 2018).

### 8.2.5 ZnO as Passivating Electron Contact

ZnO with varying dopant concentrations and improved electrical band structure properties have shown improvement in contact properties with c-Si, thus, better solar cell performance (Lin et al. 2020; Panigrahi et al. 2017). A 75 nm thick ZnO film layer deposited on a-Si:H and capped with $LiF_x$/Al showed a contact resistance of 136 $m\Omega\,cm^2$ and low $j_c$ of 3.5 $fA/cm^2$, confirming improvement of both the electron selectivity and reduced optical loss from parasitic absorption, which yielded 21.4% efficiency device with $MoO_{3-x}$ hole contact (Zhong et al. 2020).

## 8.3 Non-TMOs as Electron-Selective Contacts

### 8.3.1 $MgO_x$ as Passivating Electron Contact

$MgO_x$ is a wide bandgap material and insulator. However, the insertion of 1 nm $MgO_x$ film dramatically improved the contact resistivity to 17.5 $\pm$ 2 $m\Omega\,cm^2$ with n-type c-Si/Al. This high electron transport is attributed to the high carrier concentration and conductivity of $MgO_x$ bulk film and passivation of the interface gap states (Wan et al. 2017). A suitable $W_F$ of 4.2 eV for $MgO_x$ makes it a suitable electron contact material with c-Si (Yu et al. 2018). ALD $MgO_x$ deposited at ≤200 °C on c-Si and an a-Si:H interface passivation layer initially resulted in a reduction in passivation provided by the a-Si:H layer. However, post-deposition annealing for five minutes at 200–250 °C resulted in a recovery of the surface passivation and the formation of ohmic contact with $MgO_x$/i-aSi:H/c-Si and $MgO_x$/c-Si stacks (Chistiakova et al. 2020). A 3 nm thick $MgO_x$ film showed a contact resistivity of 180 $m\Omega\,cm^2$, and 950 $m\Omega\,cm^2$ in the $MgO_x$/c-Si and $MgO_x$/i-aSi:H/c-Si stacks, respectively. The use of the $MgO_x$ layer significantly improved n-type contact quality due to the reduction of Schottky barrier height originating from the $MgO_x$-induced reduction of Al; however, the device efficiency remains still in the range of 14.2–20.0% (Chistiakova et al. 2020; Wan et al. 2017; Yu et al. 2018, 2019).

### 8.3.2 SnO$_{2-x}$ as Passivating Electron Contact

Liu et al. demonstrated that SnO$_{2-x}$ prepared by sol–gel could be used as an electron-selective contact structure. However, it does not yield any significant surface passivation, therefore, a passivating interlayer (a-Si:H and/or SiO$_{2-x}$) is used for contacting (Liu et al. 2019). There was an explicit dependency of device performance on the $W_F$ of the contacting metal, where Mg worked best, followed by Al. In contrast, poor-performing solar cells were obtained for the higher WF metals Ni, Ag, and Au. Excellent contact properties with a $j_c$ of 2.4 fA/cm$^2$ and a $\rho_c$ of 35 mΩ cm$^2$ were obtained from test structures, although this was not reflected by the device characteristics, with the $V_{oc}$ and FF being 695 mV and 71.1%, respectively. Incorporating BaO$_x$ film between c-Si and Al significantly improved rear contact properties, solar cell parameters, and an efficiency of 17.25% (Kim et al. 2020).

### 8.3.3 Other Metal Compounds as Electron-Selective Contacts

Low $W_F$ materials, TiN$_x$ (Yang et al. 2019) TaN$_x$ (Yang et al. 2018), MgF$_x$ (Wu et al. 2020), and rare earth metal compounds (Table 8.1) YF$_3$ (Chen et al. 2021), GdF$_3$ (Chen et al. 2022), CeF$_3$ (Wang et al. 2021), and CsCO$_3$, have been explored as passivating electron contact. Sputtered TiN$_x$ film having high conductivity has been demonstrated as an effective electron selective passivation and stable contacts layer with SiO$_{2-x}$/TiN$_x$, which yielded an efficiency of 20.0% with n-type c-Si (Yang et al. 2019). A contact resistivity as low as 2.0 mΩ cm$^2$ was obtained for the $n$-Si/TiN$_x$/metal table contact due to the defect bands near the Fermi-level and rational energy band alignment (Yu et al. 2020). A low contact recombination and contact resistivity can be maintained at the device level as it can resist intermixing to form a solid solution of TiN$_x$ and Al or Si, making it stable contact material. A solid solution of TiN$_x$ and TiO$_{2-x}$ called titanium oxynitride (TiO$_x$N$_y$) resulted in an efficiency of 22.3% (Yang et al. 2020a).

TaN$_x$ interfacial layers at c-Si/Al contact improved passivation properties on c-Si, and reduced electron contact resistivity at the rear surface, consequently improving the efficiency of c-Si solar cells to 20.1% (Yang et al. 2018). A transparent and conductive MgF$_x$O$_y$ layer with LiF$_x$ showed a very low contact resistance of ~2.0 mΩ cm$^2$ and high lateral conductivity of ~2978.4 S/cm and 21.3% efficiency device was reported (Li et al. 2022). YbSi$_x$ resulted from Yb deposited onto i-a-Si:H and annealing at 200 °C yielded a contact resistivity of 100 mΩ cm$^2$ and $j_c$ of 5 fA/cm$^2$ and device efficiency of 17.0% (Cho et al. 2020) have been used as electron-selective contact. CeF$_3$ films helped alleviate the Fermi-level pinning at the c-Si/Al interface and obtained a low contact resistivity of 10.96 mΩ cm$^2$ for n-type PERC solar cells, which led to device efficiency of 21.27% on n-type c-Si (Wang et al. 2021).

## 8.4 TMOs as Passivating Hole Contacts for c-Si Solar Cells

Passivating hole contacts are notoriously poor compared to their electron counterparts because of their intrinsic poor hole conductivity, thus resulting in mediocre solar cell efficiency. A few n-type TMOs with high $W_F$, such as MoO$_{3-x}$ (Bullock et al. 2014; Dréon et al. 2020; Cao et al. 2020; Gao et al. 2018), WO$_{3-x}$ (Bivour et al. 2015), doped NiO (Hossain

**Figure 8.8** (a) Dark *I-V* curves for p-Si/Zn$_x$Ni$_{1-x}$O hole contact, the inset shows the schematic of the testing structure, and (b) extracted contact resistivity of the three Zn$_x$Ni$_{1-x}$O films deposited on *p*-Si before (RT: room temperature) and after rapid thermal annealing (RTA) at 200–500 °C (Tian Zhang et al. 2018/American Institute of Physics).

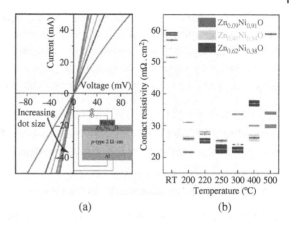

(a)          (b)

et al. 2021; Zhang et al. 2019, 2018), and V$_2$O$_{5-x}$ (Bivour et al. 2015), work as passivating hole contact and have good transparency to sunlight for better light management in solar cells. These TMOs help induce strong upward band bending within c-Si in contact with c-Si, facilitating the hole collection at the interface through TAT or B2B methods. As shown in Figure 8.8, Zn-incorporated NiO (Zn$_x$Ni$_{1-x}$O) makes excellent Ohmic contact with c-Si (Zhang et al. 2018).

### 8.4.1 MoO$_{3-x}$ as Passivating Hole Contact

Among various passivating hole contact TMOs, an optimization step both for MoO$_{3-x}$ thickness and device fabrication steps led to an improved device efficiency of 23.5% when p-type a-Si:H was replaced by MoO$_{3-x}$ in an SHJ device (Figure 8.6c,d) (Dréon et al. 2020). As shown in Figure 8.9, external quantum efficiency (EQE) and reflectance measurements show that an optimal thickness of a-Si:H and MoO$_{3-x}$ are crucial to reduce sub-bandgap parasitic absorption (Dréon et al. 2020). Overall, reducing the thickness of i-a-Si:H and MoO$_{3-x}$ layers to 6 and 4 nm, respectively, sub-bandgap states and parasitic absorption were

**Figure 8.9** EQE and reflectance of (a) devices with 4 nm MoO$_{3-x}$ and variable (i)a-Si:H thickness; (b) devices with 6 nm (i)a-Si:H and a variable MoO$_{3-x}$ thickness (Dréon et al. 2020/Elsevier).

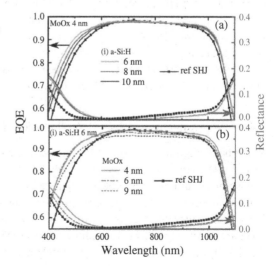

mitigated to enhance charge carrier transport properties while maintaining hole-selectivity and passivation properties which led to 23.5% solar cells (Figure 8.8; Dréon et al. 2020). The most successful hole-selective contact to date yielded a champion device efficiency of 23.83% using a precisely controlled oxygen content for 1.7 nm $MoO_{3-x}$ layer. A barrier to oxygen diffusion or reaction by plasma treatment resulted in optimal electrical properties for $MoO_{3-x}$ layer, which further improved the interface properties with s-Si:H. The use of an ultrathin $MoO_{3-x}$ and controlled interface resulted in $J_{sc}$ well above 40 mA/cm$^2$ and contributed to the champion 23.83% device. However, thermally evaporated $MoO_{3-x}$ suffers from unstable hole selectivity due to the hydrogen-induced reduction of Mo into $MoO_{3-x}$.

### 8.4.2 $V_2O_{5-x}$ as Passivating Hole Contact

ALD-based $V_2O_{5-x}$ films have proven to be an effective hole-selective contact on c-Si which simultaneously provided good passivation of 40 fA/cm$^2$ and hole contact of 95 m$\Omega$ cm$^2$ (Yang et al. 2020b). Thermally evaporated $V_2O_{5-x}$ as front hole contact and UV/$O_3$-treated ultrathin $SiO_{2-x}$ passivation layer resulted in device efficiency of 21.01% (Du et al. 2021). The excellent passivating hole contact properties of $V_2O_{5-x}$ resulted in a device efficiency of 21.6% with a full area $V_2O_{5-x}$ contact, showing its potential as a stable passivating hole contact for c-Si solar cells (Yang et al. 2020b). The highest solar cell efficiency of 22.03% was achieved on p-Si with a full-area $V_2O_X/SiO_{2-x}/NiO_X$ stack as passivation hole contacts with reduced contact resistivity (Du et al. 2022).

Low-temperature ALD and sputtered $MoO_{3-x}$ films have been studied, however, their $W_F$ was found to be lower due to induced gap states, resulting in a decreased performance as passivating hole contact (Liu et al. 2020).

### 8.4.3 $WO_{3-x}$ as Passivating Hole Contact

$WO_{3-x}$ also provides a device efficiency of 16.6% in c-Si solar cells, where it was deposited by sputtering or thermal evaporation approaches. Moderate surface passivation was observed once $WO_x$ is directly placed on the c-Si, which yielded 17.7% device efficiency for $WO_{3-x}$ (Bivour et al. 2015). It is worth mentioning that $MoO_{3-x}$, $V_2O_{5-x}$, and $WO_{3-x}$ alone provide limited passivation to the c-Si surfaces. For this reason, typically TMOs-based c-Si solar cells performed poorly, for example, 17.7% for $WO_{3-x}$ (Bivour et al. 2015) 16.7% for $MoO_{3-x}$ (Gao et al. 2018), and 19.0% for $V_2O_{5-x}$. To solve these issues, an additional interface passivation layer, such i-a-Si:H, $SiO_2$:H, or $Al_2O_3$, are used.

### 8.4.4 Non-conventional $TiO_xSi_y$/c-Si Stack as Passivating Hole Contact

While band alignment is crucial for passivating contacts, the fixed charge can also affect carrier selectivity. For example, although $TiO_{2-x}$ intrinsically is an n-type, under certain deposition conditions, a significant negative fixed charge density originated from a change in the composition of $TiO_{2-x}$/c-Si leading to a band bending in n-Si resulting in an induced junction that attracts holes toward the c-Si/$TiO_{2-x}$ interface (Matsui et al. 2020). Therefore, holes are selectively collected at indium tin oxide (ITO) via the defect states in $TiO_{2-x}$, interfacial $TiO_xSi_y$, and $SiO_{2-x}$ layers, ultimately determining the hole selectivity and surface

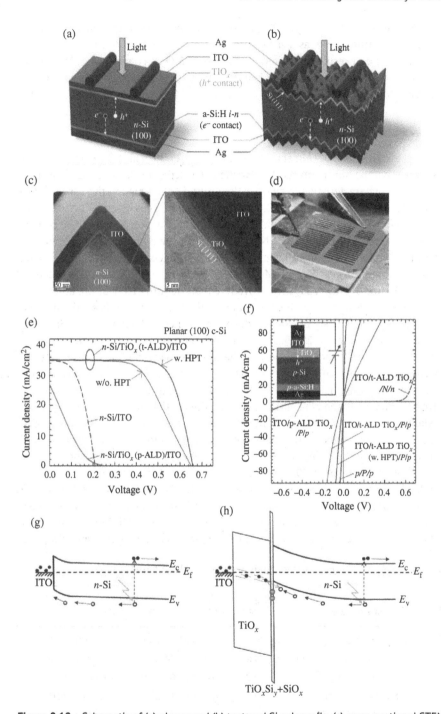

**Figure 8.10** Schematic of (a) planar and (b) textured Si solar cells, (c) cross-sectional STEM images of ITO/t-ALD TiO$_x$ layer on textured Si, (d) a digital photo of a solar cell, (e) $J$–$V$ characteristics, (f) contact resistivity measurements, (g) schematic energy band diagrams of the ITO/n-Si, and (h) ITO/t-ALD TiO$_x$/n-Si contact systems. Source: Reproduced with permission (Matsui et al. 2020/American Chemical Society).

passivation (Matsui et al. 2020). Figure 8.10a–f shows a proof-of-concept c-Si device where a 5 nm ALD $TiO_{2-x}$ film works as a passivating hole contact. A post-annealing of ALD deposited $TiO_{2-x}$ on c-Si resulted in interfacial layers of $SiO_{2-x}$:H and $TiO_xSi_y$:H and the fixed charge density, which induced band bending in $n$-Si (Figure 8.10g,h). As a result, the junction attracts holes toward the $Si/TiO_{2-x}$ interface, and holes are collected at contact.

### 8.4.5 Other Passivating Hole Contacts

$NiO_x$ (Zhang et al. 2018), $Cr_2O_3$ (Lin et al. 2018), and $Cu_{2-x}O$ (Li et al. 2021b) are promising materials for passivating hole contacts. Among these, Cr-deficient $Cr_2O_3$ is the most stable and showed p-type characteristic type with a $W_F$ around 5.0 eV among the various oxide phases of $Cr_2O_3$, $CrO_2$, $CrO_3$, $Cr_2O$, $CrO$, $Cr_3O_4$, and $Cr_8O_{11}$ (Greiner and Lu 2013; Qin et al. 2011). NiO forms a near-perfect band alignment for extraction of holes from c-Si due to low $\Delta E_V$. However, both $Cu_{2-x}O$ and $NiO_x$ show relatively high contact resistivity and poor surface passivation due to insufficient concentrations of copper and nickel vacancies, respectively, and many surface defects (Meyer et al. 2012b; Xue et al. 2018). The electrical and optical properties of $NiO_x$ can be improved by incorporating Al, Cu, Zn, and Al and obtaining low contact resistivity with c-Si (Zhang et al. 2018). Among these, the incorporation of Zn and Al into NiO significantly reduced the contact resistivity with c-Si (Zhang et al. 2019, 2018; Hossain et al. 2021), suggesting an improvement of the conductivity of $NiO_x$ and increase of $\Delta E_V$ with c-Si. Despite the improvement in contact resistivity, the best devices based on thermally evaporated $NiO_x$ on a $SiO_{2-x}$ interface passivation layer only resulted in a modest efficiency of only 15.2% (Nayak et al. 2019). $NiO_x$ is successfully used in dye-sensitized (Huang et al. 2011), quantum dots (Safari-Alamuti et al. 2013), perovskite (Islam et al. 2017), and organic (Tran et al. 2021) solar cells. Table 8.2 summarizes passivating hole contact materials that have been investigated to date, showing that passivating hole contacts have been developed that have resulted in solar cell efficiencies of over 20% (Cao et al. 2022; Dréon et al. 2020) suggesting their efficiency potential.

Two stable phases of copper oxides are cupric oxide (CuO) and cuprous oxide ($Cu_{2-x}O$) (Ravindra et al. 2017), which have different optical and electrical characteristics (Meyer et al. 2012a; Zhang et al. 2016). These are both low-cost, environmentally benign, and can be processed at low-temperature which shows intrinsic p-type conductivity originating from

**Table 8.2** Champion efficiency of low-temperature passivated hole-based c-Si solar cells.

| Passivation material | Contact structure | $J_{SC}$ (mA/cm²) | $V_{OC}$ (V) | FF (%) | Efficiency (%) | Refs. |
|---|---|---|---|---|---|---|
| $MoO_{3-x}$ | n-Si/a-Si:H/$MoO_{3-x}$/TCO/Ag | 39.2 | 0.734 | 81.8 | 23.5 | Dréon et al. (2020) |
| $MoO_{3-x}/NiO_x$ | p-Si/$SiO_{2-x}$/$MoO_{3-x}$/$NiO_x$/Ag | 39.8 | 0.650 | 83.34 | 21.60 | Li et al. (2021a) |
| $V_2O_{5-x}$ | Si/ $V_2O_{5-x}$/$SiO_{2-x}$/$NiO_x$ | 40.24 | 0.659 | 83.07 | 22.03 | Du et al. (2022) |
| $CrO_x$ | p-Si/$CrO_x$/Ag/$CrO_x$ | 39.8 | 0.638 | 80.1 | 20.3 | Lin et al. (2018) |
| $NiO_x$ | n-Si/$SiO_{2-x}$/$NiO_x$/Au | 36.9 | 0.580 | 71.06 | 15.2 | Nayak et al. (2019) |
| $Cu_{2-x}O$ | Si/$SiO_{2-x}$/$Cu_{2-x}O$/Au | 13.2 | 0.528 | 48.6 | 3.39 | Ravindra et al. (2017) |
| $Cu_{2-x}O$ | p-Si/$Al_2O_3$/$Cu_{2-x}O$/Au/Ag | 38.38 | 0.622 | 82.57 | 19.71 | Li et al. (2021b) |

**Figure 8.11** Interface recombination current density for minority carriers as a function of the contact resistance for majority carriers for various TMOs-based passivating electron (solid markers) and hole contacts (hollow markers) with c-Si (Allen et al. 2017; Cao et al. 2020; Du et al. 2022; Gao et al. 2022; Macco et al. 2018; Yang et al. 2018, 2019).

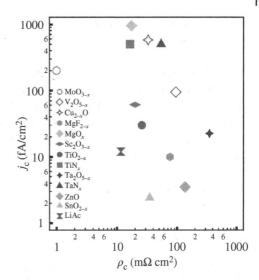

Cu cation deficiency and have low bandgap values of 1.2–1.8 and 2.1–2.2 eV, and $W_F$ of 5.9 and 5 eV, respectively (Meyer et al. 2012a). These low bandgaps result in a high parasitic absorption if used as front contacts. $Cu_{2-x}O$ has a negligible VB offset and 0.9 eV CB offset with c-Si, which allows hole transport but blocks electron transport (Ravindra et al. 2017). A low contact resistivity of ~11 mΩ cm² has been established; however, only modest device efficiencies of 1.2–3.39% were reported due to a high interface recombination current. Strategies, such as N-doped $Cu_{2-x}O$ combined with 1.2 nm $SiO_{2-x}$ tunneling layer, improve the $V_{oc}$ of solar cells to 0.528 V; however, it still suffers from low device efficiency (Masudy-Panah et al. 2014; Ravindra et al. 2017).

Figure 8.11 shows the plot of $j_c$ vs $\rho_c$ for the electron and hole-selectivity of the TMOs in c-Si solar cells. Among the hole contacts, the $MoO_{3-x}$ showed the lowest $\rho_c$ of 1 mΩ cm², and a relatively high $j_c$ of 200 fA/cm² which resulted in the champion device efficiency of 23.5% for the device (Dréon et al. 2020). The $V_2O_{5-x}$ hole contact with inferior $\rho_c$ of 95 mΩ cm² and better $j_c$ of 95 fA/cm² showed decent device efficiency of 22.03% (Du et al. 2022). For the electron-contacts, $TiO_2$ performed the best due to decent values of $\rho_c$ and $j_c$ of 26 mΩ cm² and 30 fA/cm², thus, a champion efficiency of 23.1% (Bullock et al. 2019). These intuitive key parameters suggest why the best-performing hole-selective and electron-selective c-Si solar cells were fabricated using the $MoO_{3-x}$ and $TiO_{2-x}$ carrier-selective contacts, respectively. Nevertheless, the rest of the TMOs also show reasonable $\rho_c$ and $j_c$ values which still allow obtaining high efficiency c-Si solar cells.

## 8.5 Summary

Significant work has been done on low-temperature synthesized dopant-free TMOs-based passivation contact materials for c-Si solar cells. Although the carrier-selectivity of these TMOs is still inferior in performance compared to the doped a-Si:H and poly-Si contacts, excellent results have been obtained for the widely tunable passivating electron and hole contacts properties and can be further improved by choice of contact materials. These types of contacts could be appealing for future silicon-based tandem solar cells. The use of $TiO_{2-x}$ and $MoO_{3-x}$ contacts as electron and hole-selective contacts for c-Si solar cells shows their best device performance at lower cost and the potential to compete with mainstream commercial c-Si solar cells. However, significant work is required to further improve the properties of TMOs and the c-Si/TMO interface for better carrier transport across the interface. To optimize the performance of TMO-based c-Si solar cells, it is crucial to systematically fine-tune key parameters, including the thickness and defect density of TMOs. By enhancing the electronic and optical properties and improving carrier-selective passivation at the TMO-c-Si interface, the charge collection efficiency in these solar cells can be significantly increased.

## Author Biographies

**Professor Bram Hoex** completed both an MSc and PhD degree from Eindhoven University of Technology in the Netherlands in 2003 and 2008, respectively. From 2008 to 2015, he worked at the Solar Energy Research Institute of Singapore (SERIS) at the National University of Singapore (NUS) as a Group Leader and from 2012 also as Director. In 2015, he joined UNSW, where he is currently a Deputy Head at UNSW's School of Photovoltaic and Renewable Energy Engineering. His research focuses on developing and applying nanoscale thin films in a wide range of renewable energy devices to improve their performance, solar cell/module reliability, as well as financial and performance modeling of gigascale solar farms. He published over 250 scientific papers and is best known for his groundbreaking work on aluminum oxide for crystalline silicon surface passivation, which is now the de facto standard for industrial PERC and TOPCon solar cells. His work has received various international recognitions, including the 2008 SolarWorld Junior Einstein and 2016 IEEE PVSC Young Professional awards. Renewable Energy World listed him in the "Solar 40 under 40 list" globally in 2018.

**Dr. Md Anower Hossain** is a Postdoctoral Fellow in the School of Photovoltaic and Renewable Energy Engineering at the University of New South Wales, specializing in Photovoltaics. He holds a PhD in Materials Science and Engineering from the National University of Singapore. Dr. Hossain's expertise lies in developing materials for solar cells, batteries, and fuel cell catalysts. He has extensive knowledge in quantum dots, thin-film and silicon solar cells, and lithium-ion batteries. His research integrates first-principles calculations, material synthesis, and the use of machine learning and artificial intelligence to accelerate materials discoveries for photovoltaic applications.

# References

Allen, T.G., Bullock, J., Jeangros, Q. et al. (2017). A low resistance calcium/reduced titania passivated contact for high-efficiency crystalline silicon solar cells. *Advanced Energy Materials* 7 (12): 1602606. https://doi.org/10.1002/aenm.201602606.

Allen, T.G., Bullock, J., Yang, X. et al. (2019). Passivating contacts for crystalline silicon solar cells. *Nature Energy* 4 (11): 914–928. https://doi.org/10.1038/s41560-019-0463-6.

Baker-Finch, S.C., McIntosh, K.R., Yan, D. et al. (2014). Near-infrared free carrier absorption in heavily doped silicon. *Journal of Applied Physics* 116 (6): 063106. https://doi.org/10.1063/1.4893176.

Bivour, M., Temmler, J., Steinkemper, H., and Hermle, M. (2015). Molybdenum and tungsten oxide: high work function wide band gap contact materials for hole selective contacts of silicon solar cells. *Solar Energy Materials and Solar Cells* 142: 34–41. https://doi.org/10.1016/j.solmat.2015.05.031.

Brendel, R. and Peibst, R. (2016). Contact selectivity and efficiency in crystalline silicon photovoltaics. *IEEE Journal of Photovoltaics* 6 (6): 1413–1420. https://doi.org/10.1109/JPHOTOV.2016.2598267.

Bullock, J., Cuevas, A., Allen, T., and Battaglia, C. (2014). Molybdenum oxide MoOx: a versatile hole contact for silicon solar cells. 105 (23): 232109. https://doi.org/10.1063/1.4903467.

Bullock, J., Wan, Y., Hettick, M. et al. (2019). Dopant-free partial rear contacts enabling 23% silicon solar cells. *Advanced Energy Materials* 9 (9): 1803367. https://doi.org/10.1002/aenm.201803367.

Cao, S., Li, J., Zhang, J. et al. (2020). Silicon solar cells: stable MoOX-based heterocontacts for p-type crystalline silicon solar cells achieving 20% efficiency. *Advanced Functional Materials* 30 (49): 2070325. https://doi.org/10.1002/adfm.202070325.

Cao, L., Procel, P., Alcañiz, A. et al. (2022). Achieving 23.83% conversion efficiency in silicon heterojunction solar cell with ultra-thin MoOx hole collector layer via tailoring (i)a-Si:H/MoOx interface. *Progress in Photovoltaics: Research and Applications* 31 (12): 1245–1254. https://doi.org/10.1002/pip.3638.

Chen, Z., Lin, W., Liu, Z. et al. (2021). Yttrium fluoride-based electron-selective contacts for crystalline silicon solar cells. *ACS Applied Energy Materials* 4 (3): 2158–2164. https://doi.org/10.1021/acsaem.0c02646.

Chen, N., Cai, L., Xie, F. et al. (2022). Gadolinium fluoride as a high-thickness-tolerant electron-selective contact material for solar cells. *ACS Applied Energy Materials* 5 (4): 4351–4357. https://doi.org/10.1021/acsaem.1c03919.

Chistiakova, G., Macco, B., and Korte, L. (2020). Low-temperature atomic layer deposited magnesium oxide as a passivating electron contact for c-Si-based solar cells. *IEEE Journal of Photovoltaics* 10 (2): 398–406. https://doi.org/10.1109/JPHOTOV.2019.2961603.

Cho, J., Sivaramakrishnan Radhakrishnan, H., Recaman Payo, M. et al. (2020). Low work function ytterbium silicide contact for doping-free silicon solar cells. *ACS Applied Energy Materials* 3 (4): 3826–3834. https://doi.org/10.1021/acsaem.0c00256.

Ding, J., Zhou, Y., Dong, G. et al. (2018). Solution-processed ZnO as the efficient passivation and electron selective layer of silicon solar cells. *Progress in Photovoltaics: Research and Applications* 26 (12): 974–980. https://doi.org/10.1002/pip.3044.

Dréon, J., Jeangros, Q., Cattin, J. et al. (2020). 23.5%-efficient silicon heterojunction silicon solar cell using molybdenum oxide as hole-selective contact. *Nano Energy* 70: 104495. https://doi .org/10.1016/j.nanoen.2020.104495.

Du, G., Li, L., Yang, X. et al. (2021). Improved V2OX passivating contact for p-type crystalline silicon solar cells by oxygen vacancy modulation with a SiOX tunnel layer. *Advanced Materials Interfaces* 8 (22): 2100989. https://doi.org/10.1002/admi.202100989.

Du, G., Li, L., Zhu, H. et al. (2022). High-performance hole-selective V2OX/SiOX/NiOX contact for crystalline silicon solar cells. *EcoMat* 4 (3): e12175. https://doi.org/10.1002/eom2.12175.

Gao, M., Chen, D., Han, B. et al. (2018). Bifunctional hybrid a-SiOx(Mo) layer for hole-selective and interface passivation of highly efficient MoOx/a-SiOx(Mo)/n-Si heterojunction photovoltaic device. *ACS Applied Materials & Interfaces* 10 (32): 27454–27464. https://doi .org/10.1021/acsami.8b07001.

Gao, K., Bi, Q., Wang, X. et al. (2022). Progress and future prospects of wide-bandgap metal-compound-based passivating contacts for silicon solar cells. *Advanced Materials* 34 (26): 2200344. https://doi.org/10.1002/adma.202200344.

Gerling, L.G., Voz, C., Alcubilla, R., and Puigdollers, J. (2017). Origin of passivation in hole-selective transition metal oxides for crystalline silicon heterojunction solar cells. *Journal of Materials Research* 32 (2): 260–268. https://doi.org/10.1557/jmr.2016.453.

Greiner, M.T. and Lu, Z.-H. (2013). Thin-film metal oxides in organic semiconductor devices: their electronic structures, work functions and interfaces. *NPG Asia Materials* 5 (7): e55. https://doi.org/10.1038/am.2013.29.

Gupta, B., Hossain, M.A., Riaz, A. et al. (2022). Recent advances in materials design using atomic layer deposition for energy applications. *Advanced Functional Materials* 32 (3): 2109105. https://doi.org/10.1002/adfm.202109105.

Hossain, M.A., Zhang, T., Zakaria, Y. et al. (2021). Doped nickel oxide carrier-selective contact for silicon solar cells. *IEEE Journal of Photovoltaics* 11 (5):: 1176–117. https://doi.org/10 .1109/JPHOTOV.2021.3095458.

Huang, Z., Natu, G., Ji, Z. et al. (2011). p-Type dye-sensitized NiO solar cells: a study by electrochemical impedance spectroscopy. *The Journal of Physical Chemistry C* 115 (50): 25109–25114. https://doi.org/10.1021/jp205306g.

Islam, M.B., Yanagida, M., Shirai, Y. et al. (2017). NiOx hole transport layer for perovskite solar cells with improved stability and reproducibility. *ACS Omega* 2 (5): 2291–2299. https://doi .org/10.1021/acsomega.7b00538.

Jhaveri, J., Avasthi, S., Nagamatsu, K., and Sturm, J.C. (2014). Stable low-recombination n-Si/TiO2 hole-blocking interface and its effect on silicon heterojunction photovoltaics. In: *Paper Presented at the 2014 IEEE 40th Photovoltaic Specialist Conference (PVSC) (8-13 June 2014)*.

Kim, S.-H., Jung, J.-Y., Wehrspohn, R.B., and Lee, J.-H. (2020). All-room-temperature processed 17.25%-crystalline silicon solar cell. *ACS Applied Energy Materials* 3 (4): 3180–3185. https:// doi.org/10.1021/acsaem.0c00133.

Li, L., Du, G., Lin, Y. et al. (2021a). NiOx/MoOx bilayer as an efficient hole-selective contact in crystalline silicon solar cells. *Cell Reports Physical Science* 2 (12): 100684. https://doi.org/10 .1016/j.xcrp.2021.100684.

Li, L., Du, G., Zhou, X. et al. (2021b). Interfacial engineering of Cu2O passivating contact for efficient crystalline silicon solar cells with an Al2O3 passivation layer. *ACS Applied Materials & Interfaces* 13 (24): 28415–28423. https://doi.org/10.1021/acsami.1c08258.

Li, J., Guo, C., Bai, Y. et al. (2022). One-step formation of low work-function, transparent and conductive MgFxOy electron extraction for silicon solar cells. *Advanced Science* 9 (23): 2202400. https://doi.org/10.1002/advs.202202400.

Liao, B., Hoex, B., Aberle, A.G. et al. (2014). Excellent c-Si surface passivation by low-temperature atomic layer deposited titanium oxide. *Applied Physics Letters* 104 (25): 253903. https://doi.org/10.1063/1.4885096.

Liao, B., Hoex, B., Shetty, K.D. et al. (2015). Passivation of boron-doped industrial silicon emitters by thermal atomic layer deposited titanium oxide. *IEEE Journal of Photovoltaics* 5 (4): 1062–1066. https://doi.org/10.1109/JPHOTOV.2015.2434596.

Lin, W., Wu, W., Liu, Z. et al. (2018). Chromium trioxide hole-selective heterocontacts for silicon solar cells. *ACS Applied Materials & Interfaces* 10 (16): 13645–13651. https://doi.org/10.1021/acsami.8b02878.

Lin, W., Boccard, M., Zhong, S. et al. (2020). Degradation mechanism and stability improvement of dopant-free ZnO/LiFx/Al electron nanocontacts in silicon heterojunction solar cells. *ACS Applied Nano Materials* 3 (11): 11391–11398. https://doi.org/10.1021/acsanm.0c02475.

Liu, M., Zhou, Y., Dong, G. et al. (2019). SnO2/Mg combination electron selective transport layer for Si heterojunction solar cells. *Solar Energy Materials and Solar Cells* 200: 109996. https://doi.org/10.1016/j.solmat.2019.109996.

Liu, Y., Li, Y., Wu, Y. et al. (2020). High-efficiency silicon heterojunction solar cells: materials, devices and applications. *Materials Science and Engineering: R: Reports* 142: 100579. https://doi.org/10.1016/j.mser.2020.100579.

Liu, Y., Sang, B., Hossain, M.A. et al. (2021). A novel passivating electron contact for high-performance silicon solar cells by ALD Al-doped TiO2. *Solar Energy* 228: 531–539. https://doi.org/10.1016/j.solener.2021.09.083.

LONGi (2022). LONGi once again sets new world record for HJT solar cell efficiency. In: [Press release].

Macco, B., Bivour, M., Deijkers, J.H. et al. (2018). Effective passivation of silicon surfaces by ultrathin atomic-layer deposited niobium oxide. *Applied Physics Letters* 112 (24): 242105. https://doi.org/10.1063/1.5029346.

Masudy-Panah, S., Dalapati, G.K., Radhakrishnan, K. et al. (2014). Reduction of Cu-rich interfacial layer and improvement of bulk CuO property through two-step sputtering for p-CuO/n-Si heterojunction solar cell. *Journal of Applied Physics* 116 (7): 074501. https://doi.org/10.1063/1.4893321.

Matsui, T., Bivour, M., Hermle, M., and Sai, H. (2020). Atomic-layer-deposited tiox nanolayers function as efficient hole-selective passivating contacts in silicon solar cells. *ACS Applied Materials & Interfaces* 12 (44): 49777–49785. https://doi.org/10.1021/acsami.0c14239.

Mazzarella, L., Alcañiz, A., Procel, P. et al. (2021). Strategy to mitigate the dipole interfacial states in (i)a-Si:H/MoOx passivating contacts solar cells. *Progress in Photovoltaics: Research and Applications* 29 (3): 391–400. https://doi.org/10.1002/pip.3381.

Messmer, C., Bivour, M., Schön, J. et al. (2018). Numerical simulation of silicon heterojunction solar cells featuring metal oxides as carrier-selective contacts. *IEEE Journal of Photovoltaics* 8 (2): 456–464. https://doi.org/10.1109/JPHOTOV.2018.2793762.

Meyer, B.K., Polity, A., Reppin, D. et al. (2012a). Binary copper oxide semiconductors: from materials towards devices. *Physica Status Solidi B* 249 (8): 1487–1509. https://doi.org/10.1002/pssb.201248128.

Meyer, J., Hamwi, S., Kröger, M. et al. (2012b). Transition metal oxides for organic electronics: energetics, device physics and applications. *Advanced Materials* 24 (40): 5408–5427. https://doi.org/10.1002/adma.201201630.

Nayak, M., Mandal, S., Pandey, A. et al. (2019). Nickel oxide hole-selective heterocontact for silicon solar cells: role of SiOx interlayer on device performance. *Solar RRL* 3 (11): 1900261. https://doi.org/10.1002/solr.201900261.

Panigrahi, J., Vandana, V., Singh, R. et al. (2017). Crystalline silicon surface passivation by thermal ALD deposited Al doped ZnO thin films. *AIP Advances* 7 (3): 035219. https://doi.org/10.1063/1.4979326.

Qin, P., Fang, G., Sun, N. et al. (2011). Organic solar cells with p-type amorphous chromium oxide thin film as hole-transporting layer. *Thin Solid Films* 519 (13): 4334–4341. https://doi.org/10.1016/j.tsf.2011.02.013.

Quan, C., Tong, H., Yang, Z. et al. (2018). Electron-selective scandium–tunnel oxide passivated contact for n-type silicon solar cells. *Solar RRL* 2 (8): 1800071. https://doi.org/10.1002/solr.201800071.

Ravindra, P., Mukherjee, R., and Avasthi, S. (2017). Hole-selective electron-blocking copper oxide contact for silicon solar cells. *IEEE Journal of Photovoltaics* 7 (5): 1278–1283. https://doi.org/10.1109/JPHOTOV.2017.2720619.

Reinders, A., Verlinden, P., van Sark, W., and Freundlich, A. (2017). *Photovoltaic Solar Energy : From Fundamentals to Applications*. New York, United Kingdom: John Wiley & Sons, Incorporated.

Safari-Alamuti, F., Jennings, J.R., Hossain, M.A. et al. (2013). Conformal growth of nanocrystalline CdX (X = S, Se) on mesoscopic NiO and their photoelectrochemical properties. *Physical Chemistry Chemical Physics* 15 (13): 4767–4774. https://doi.org/10.1039/C3CP43613F.

Tokura, Y. and Nagaosa, N. (2000). Orbital physics in transition-metal oxides. *Science* 288 (5465): 462–468. https://doi.org/10.1126/science.288.5465.462.

Tran, H.N., Dao, D.Q., Yoon, Y.J. et al. (2021). Inverted polymer solar cells with annealing-free solution-processable NiO. *Small* 17 (31): 2101729. https://doi.org/10.1002/smll.202101729.

Wan, Y., Samundsett, C., Bullock, J. et al. (2017). Conductive and stable magnesium oxide electron-selective contacts for efficient silicon solar cells. *Advanced Energy Materials* 7 (5): 1601863. https://doi.org/10.1002/aenm.201601863.

Wan, Y., Karuturi, S.K., Samundsett, C. et al. (2018). Tantalum oxide electron-selective heterocontacts for silicon photovoltaics and photoelectrochemical water reduction. *ACS Energy Letters* 3 (1): 125–131. https://doi.org/10.1021/acsenergylett.7b01153.

Wang, W., Cai, L., Meng, L. et al. (2021). Cerous fluoride dopant-free electron-selective contact for crystalline silicon solar cells. *Physica Status Solidi RRL: Rapid Research Letters* 15 (12): 2100135. https://doi.org/10.1002/pssr.202100135.

Wang, Y., Zhang, S.-T., Li, L. et al. (2023). Dopant-free passivating contacts for crystalline silicon solar cells: progress and prospects. *EcoMat* 5 (2): e12292. https://doi.org/10.1002/eom2.12292.

Wu, W., Lin, W., Zhong, S. et al. (2020). Dopant-free back-contacted silicon solar cells with an efficiency of 22.1%. *Physica Status Solidi RRL: Rapid Research Letters* 14 (4): 1900688. https://doi.org/10.1002/pssr.201900688.

Würfel, U., Cuevas, A., and Würfel, P. (2015). Charge carrier separation in solar cells. *IEEE Journal of Photovoltaics* 5 (1): 461–469. https://doi.org/10.1109/JPHOTOV.2014.2363550.

Xue, M., Islam, R., Chen, Y. et al. (2018). Carrier-selective interlayer materials for silicon solar cell contacts. *Journal of Applied Physics* 123 (14): 143101. https://doi.org/10.1063/1.5020056.

Yang, X., Bi, Q., Ali, H. et al. (2016). High-performance TiO2-based electron-selective contacts for crystalline silicon solar cells. *Advanced Materials* 28 (28): 5891–5897. https://doi.org/10.1002/adma.201600926.

Yang, X., Weber, K., Hameiri, Z., and De Wolf, S. (2017). Industrially feasible, dopant-free, carrier-selective contacts for high-efficiency silicon solar cells. *Progress in Photovoltaics: Research and Applications* 25 (11): 896–904. https://doi.org/10.1002/pip.2901.

Yang, X., Aydin, E., Xu, H. et al. (2018). Tantalum nitride electron-selective contact for crystalline silicon solar cells. *Advanced Energy Materials* 8 (20): 1800608. https://doi.org/10.1002/aenm.201800608.

Yang, X., Liu, W., De Bastiani, M. et al. (2019). Dual-function electron-conductive, hole-blocking titanium nitride contacts for efficient silicon solar cells. *Joule* 3 (5): 1314–1327. https://doi.org/10.1016/j.joule.2019.03.008.

Yang, X., Lin, Y., Liu, J. et al. (2020a). A highly conductive titanium oxynitride electron-selective contact for efficient photovoltaic devices. *Advanced Materials* 32 (32): 2002608. https://doi.org/10.1002/adma.202002608.

Yang, X., Xu, H., Liu, W. et al. (2020b). Atomic layer deposition of vanadium oxide as hole-selective contact for crystalline silicon solar cells. *Advanced Electronic Materials* 6 (8): 2000467. https://doi.org/10.1002/aelm.202000467.

Yu, J., Fu, Y., Zhu, L. et al. (2018). Heterojunction solar cells with asymmetrically carrier-selective contact structure of molybdenum-oxide/silicon/magnesium-oxide. *Solar Energy* 159: 704–709. https://doi.org/10.1016/j.solener.2017.11.047.

Yu, J., Liao, M., Yan, D. et al. (2019). Activating and optimizing evaporation-processed magnesium oxide passivating contact for silicon solar cells. *Nano Energy* 62: 181–188. https://doi.org/10.1016/j.nanoen.2019.05.015.

Yu, J., Phang, P., Samundsett, C. et al. (2020). Titanium nitride electron-conductive contact for silicon solar cells by radio frequency sputtering from a TiN target. *ACS Applied Materials & Interfaces* 12 (23): 26177–26183. https://doi.org/10.1021/acsami.0c04439.

Zhang, X., Wan, Y., Bullock, J. et al. (2016). Low resistance Ohmic contact to p-type crystalline silicon via nitrogen-doped copper oxide films. *Applied Physics Letters* 109 (5): 052102. https://doi.org/10.1063/1.4960529.

Zhang, T., Hossain, M.A., Lee, C.-Y. et al. (2018). Atomic layer deposited ZnxNi1−xO: a thermally stable hole selective contact for silicon solar cells. *Applied Physics Letters* 113 (26): 262102. https://doi.org/10.1063/1.5056223.

Zhang, T., Hossain, M.A., Khoo, K.T. et al. (2019). *Atomic layer deposited AlxNiyO as hole selective contact for silicon solar cells*. In: *Paper Presented at the 2019 IEEE 46th Photovoltaic Specialists Conference (PVSC) (16-21 June 2019)*.

Zhao, J., Wang, A., and Green, M.A. (1999). 24·5% efficiency silicon PERT cells on MCZ substrates and 24·7% efficiency PERL cells on FZ substrates. *Photovoltaics* 7 (6): 471–474. https://doi.org/10.1002/(SICI)1099-159X(199911/12)7:6<471::AID-PIP298>3.0.CO;2-7.

Zhong, S., Dreon, J., Jeangros, Q. et al. (2020). Mitigating plasmonic absorption losses at rear electrodes in high-efficiency silicon solar cells using dopant-free contact stacks. *Advanced Functional Materials* 30 (5): 1907840. https://doi.org/10.1002/adfm.201907840.

Zhou, J., Huang, Q., Ding, Y. et al. (2022). Passivating contacts for high-efficiency silicon-based solar cells: from single-junction to tandem architecture. *Nano Energy* 92: 106712. https://doi.org/10.1016/j.nanoen.2021.106712.

# 9

# Carrier-Induced Degradation

*Michelle Vaqueiro Contreras*

*School of Photovoltaic and Renewable Energy Engineering, University of New South Wales, Sydney, NSW, Australia*

## 9.1 Introduction

In this chapter, a comprehensive overview of the major carrier-induced degradation phenomena affecting silicon solar cells is presented. These include boron-oxygen related carrier-induced recombination (BO-CID), light and elevated temperature-induced degradation (LeTID), copper-related light-induced degradation (Cu-LID), and surface-related degradation (SRD), all of which cause a reduction in device performance upon carrier injection. The significance of these degradation mechanisms varies, and while standard field operating conditions can sometimes restore device performance to its initial levels, this process may exceed the warranty time frame (typically 25 to 30 years) for a module, posing a significant challenge for the solar photovoltaic (PV) industry. Likewise, the impact of these degradation mechanisms can also be exacerbated by factors such as temperature fluctuations that vary depending on local climates around the world. This, in turn, can lead to further module efficiency degradation, thereby increasing the levelized cost of electricity (LCOE). Nevertheless, there has been significant progress in developing processes to mitigate or minimize the effects of degradation. Optimization studies and procedures have been developed by researchers and industry professionals, which are now routinely used to ensure the long-term stability of solar cells under standard field operating conditions.

This chapter presents a detailed compilation of the key findings from decades of literature, which shapes our current understanding of these degradation phenomena and their management.

## 9.2 Boron–Oxygen Related Recombination

The so-called boron–oxygen light-induced degradation or "BO-LID," or more appropriately "BO-CID," where C stands for "carrier" as will be explained later, causing up to $5\%_{rel}$ losses in boron-doped silicon substrates has been studied for decades and has been discussed by numerous authors in the literature. A key driver for such intensive research has been the dominance of boron-doped material in the PV market since its large-scale adoption. Despite this, undisputed identification of its composition and complete characterization of its electrical properties and formation does not exist. An extensive review of the progress in our understanding of this CID up to 2017 for the interested reader has been given by Niewelt et al. (2017). This section is dedicated to consolidating the essential discoveries from literature spanning numerous decades that have contributed to our present understanding.

The first observations of boron-doped Czochralski silicon (Cz-Si) solar cell degradation under illumination were reported as early as 1973. Fischer and Pschunder revealed three characteristic behaviors of a particular photoinduced degradation; (i) significant reduction of all solar cell parameters (i.e. power output, short-circuit current, and open-circuit voltages) up to a saturation level upon illumination, (ii) complete recovery of initial performance after short dark annealing treatments at 200 °C, and (iii) complete reversibility upon corresponding treatments (Fischer and Pschunder 1973) (see Figure 9.1a). Weizer et al. (2008) later revealed that the defect responsible for degradation was not activated by photon interactions but rather by the generated carriers. This was supported by the nearly identical degradation observed in cells under forward bias injection in the absence of light and the acceleration of the degradation kinetics with increased light intensity. Therefore, the term "carrier-induced degradation (CID)" is considered more appropriate than "LID." A significant discovery occurred in 2006 when Herguth et al. (2006) found that by increasing the temperature during excess carrier injection (CI), the degradation process could be altered from the saturation stage to a stage where little or no degradation occurs under standard testing conditions. This newly discovered stage, known as the "regenerated" state, naturally occurs during field operation conditions; however, it can be accelerated by increasing temperature and CI (see Figures 9.1a). It is now known that the defect generation rate is directly proportional to the square of the hole concentration, resulting in the acceleration of the degradation (Schön et al. 2015). This knowledge now serves as the foundation for the widely adopted BO-mitigation treatment in the industry. A representation of the three-stage transformation model of BO-CID is given in Figure 9.1b, and Table 9.1 presents a summary of activation energies and attempt frequencies of each corresponding reaction.

Noteworthy of the degradation rate analysis from an initial (annealed) state is the appearance of two distinct decays occurring at different timescales, a fast-forming recombination

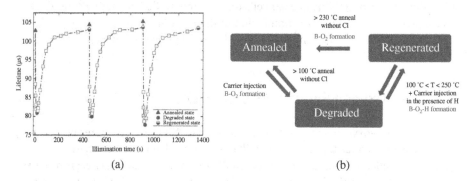

(a)                                        (b)

**Figure 9.1** (a) A BO-CID prone Cz: B sample upon accelerated light soaking and dark annealing cycles showcasing the BO-CID characteristic states (annealed, regenerated, and degraded). Source: Adapted from Wenham et al. (2018). (b) BO-CID state transition diagram.

**Table 9.1** Summary of activation energies ($E_a$) and attempt frequencies ($v$) for the BO-CID reactions.

| Reaction | State transition | $E_a$ (eV) | $v$ (s$^{-1}$) | Refs. |
| --- | --- | --- | --- | --- |
| Degradation | A → B | 0.47 | $4 \times 10^3$ | Bothe and Schmidt (2006) |
| Annihilation | B → A | 1.32 | $1 \times 10^{13}$ | Bothe and Schmidt (2006) |
| Regeneration | B → C | 0.98 | $1.25 \times 10^{10}$ | Wilking et al. (2014) |
| Destabilization | C → B | 1.25 | $1 \times 10^9$ | Wilking et al. (2014) |

**Figure 9.2** Modulation of BO-CID's fast and slow recombination states through pre-annealing without illumination. Source: Adapted from (Kim et al. 2017).

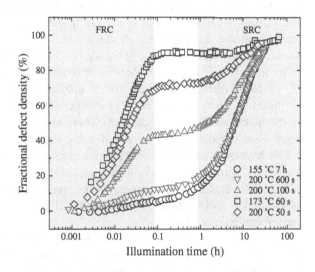

center (FRC) detected within the first few minutes of the degradation and a slow recombination centre (SRC) arising upon several hours of exposure (Bothe et al. 2003; Hashigami and Saitoh 2003). A representation of the relative changes in defect concentration increasing at two different rates is shown in Figure 9.2. Modulation of the extent of degradation of each is shown as the result of the dark anneal duration before the carrier injection. Such modulation has been used as evidence of a single defect being responsible for the degradation (Kim et al. 2017). In their study, Kim et al. (2018) found that the extent of the fast and slow degradation could be adjusted through brief dark annealing treatments on previously degraded wafers. Such treatments showed no impact on the capture cross-section ratios, and the reaction was fully reversible, indicating a strong correlation between degradation in both fast and slow timescales. These findings supported the notion that the degradation caused by B-O defects originates from a single defect rather than two separate defects. This theory has been further supported by a recent analysis conducted by Schmidt et al. (2020). Their study reconciles the previously contradictory theories regarding the origin of the fast and slow degradation components and proposes the involvement of a third species, X. According to their analysis, X serves as an activating impurity for the B-O complex and plays a role in adjusting each component.

From its initial discovery, the distinctive degradation observed in Cz substrates has been found exclusively in boron-doped n⁺p type samples, while gallium-doped and phosphorus-doped samples did not exhibit any degradation unless they were co-doped with boron (Schmidt et al. 1997; Schön et al. 2015). Subsequently, it was discovered that the degradation scaled with the boron concentration in the bulk and determined that oxygen was necessary for the degradation to occur in boron-doped wafers grown by the float-zone (FZ) method (Glunz et al. 1998). Such findings led to the identification of a quadratic correlation between [Oi] and the defect concentration. Established by Schmidt and Hezel (2002), this discovery resulted in the now widely accepted BO-defect model consisting of a substitutional boron atom and an oxygen dimer ($B_sO_{2i}$) – or simply the BO-defect.

Despite the many attempts at directly measuring the electronic properties of the responsible BO-defect, which is crucial for our understanding of the underlying recombination mechanism, a definite answer is still not found. To this date, most of the literature has focused on the extraction of Shockley–Read–Hall (SRH) parameters using minority carrier lifetime measurements. These studies have identified a defect of negative-U properties introducing an acceptor level at approximately $E_c - 0.41$ eV and a donor level at $E_v + 0.26$ eV, with a ratio of electron to hole capture cross sections for the donor level in the range 10 to 20. Yet, most of the BO-defect parameters appear highly scattered in the literature (Niewelt et al. 2017). There are less than a handful of relatively recent reports on the detection of a signal using deep-level transient spectroscopy (DLTS), which have been related to the BO-defect in its latent (annealed) (Markevich et al. 2019; Vaqueiro-Contreras et al. 2019; Zhou et al. 2022) and active states (Mchedlidze and Weber 2015). However, these studies are not in agreement, as the latter identifies two defect states placed at $E_c - 0.41$ eV and $E_v + 0.37$ eV of the silicon bandgap and relates them to an SRH recombination mechanism.

On the other hand, the former poses an alternative activation process driven by a trap-assisted Auger mechanism where, depending on the Fermi level, the BO defect acts as a non-recombination active deep donor level (annealed state) at $E_v + 0.56$ eV or a shallow acceptor level at about $E_v + 0.04$ eV producing a recombination-active state. Most of these studies, however, focus on the slow process (i.e. the SRC dominant state), partly because it has been primarily observed as the dominant degradation process but also due to the many conflicting theories on its true origin (Schmidt et al. 2020). There have been recent demonstrations of extent modulation by dark anneal duration prior to the carrier injection, as seen in Figure 9.2, which calls for a similar further examination of the fast degradation component. Regarding the deactivation mechanism of the BO-defect leading to the regenerated state, there have also been some disputes as to whether hydrogen plays a role, yet the current consensus is that its presence is required for a fast regeneration and higher final bulk minority carrier lifetimes. This is consistent with a theory very recently supported by DLTS studies suggesting that hydrogen (H⁺) bonds with the $B_sO_2^-$ complex, rendering it electrically inactive (Fattah et al. 2022).

Since it was recognized that there was a regenerated and much more stable state of the BO-defect achievable by CI at higher temperatures in 2006 (Herguth et al. 2006), rapid growth in optimization studies for industry-compatible stabilization processes followed (Niewelt et al. 2017). Accelerated regeneration using temperatures >175 °C and either several sun-equivalent illumination intensities or high applied currents led to outstanding stabilization of the defect in a matter of seconds (Hamer et al. 2015; Walter et al. 2014). Similar procedures are now routinely used in industry to mitigate, or at least reduce, the degradation impacts in finished cells. In 2019, many solar cell manufacturers rapidly transitioned from boron to gallium doping, primarily motivated by reducing CID. The market penetration of gallium doping increased from 10% to over 60% the following year (Fischer et al. 2020, 2021). One reason for the delayed adoption of Ga-doping was its earlier patent protection, which expired in May 2020 (Abe et al. 2004). However, some major manufacturers, such as LONGi and JA Solar, obtained licenses for the technology about a year before its expiration, which contributed to its wider adoption in the industry.

## 9.3 Light and Elevated Temperature Degradation (LeTID)

Often confused with BO-CID, another carrier-induced degradation phenomenon observed in solar cells made of silicon since 2012 is the so-called "light and elevated temperature induced degradation" (LeTID). This phenomenon was first reported by Ramspeck et al. (2012) when comparing light-induced failure modes at average module temperatures in the field ($\sim$75 °C) on the then industry-dominant aluminum back surface field (Al-BSF) cell concept with the emerging passivated emitter and rear cell (PERC) design. It was noted that this new degradation led to much worse performance losses than those produced by BO-CID, with reported values of up to $16\%_{rel}$ in preprocessed cells under laboratory testing conditions and, more recently, averaging $2\%_{rel}$ under field conditions even after stabilization processing during manufacturing.

Unlike BO-CID, the degradation onset proceeded at a rather slow timescale and often reached a maximum degradation over a month of continued 1000 W/m$^2$ illumination at 75 °C. Such a rate would translate to several years of continued degradation under average field conditions. This is longer than the three-year period shown in Figure 9.3 and in contrast to BO-CID, where maximum degradation is reached within the first year of operation. Nevertheless, this degradation rate can be significantly altered by the climate of the PV system installation. One example is also given in Figure 9.3, where the tracked relative loss in yield due to LeTID in mc-Si PERC modules was measured in the laboratory versus the losses in the field in two distinct locations – Germany and Cyprus (Kersten et al. 2017). In the same work, Kersten et al. developed a model to estimate the degradation extent of modules in the field in relation to laboratory testing conditions after a 4300 hours long laboratory examination. Their model suggested that such treatment is equivalent to 15 years of field degradation in Mediterranean climates. Given the designated warranty of over 25 years of any PV module in the market, this degradation can severely impact energy yield and, consequently, the levelized cost of electricity.

Relevant as it is, LeTID's root cause is still poorly understood. A wealth of literature describes phenomenological observations of the degradation, possible formation mechanisms, potentially responsible defects and numerous approaches for its suppression. A comprehensive review was recently published by Chen et al. (2021).

The exact reaction kinetics of LeTID have proven rather difficult to characterize as they are largely influenced by the thermal history of the silicon wafer. Nevertheless, more generally, the degradation has been described with a three-staged model analogous to

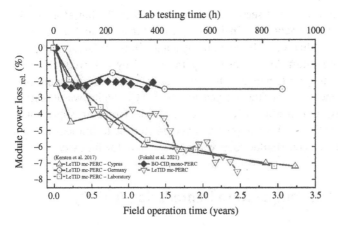

**Figure 9.3** (Bottom axis) Relative losses in mc-Si PERC module power due to LeTID in the field (Germany and Cyprus). (Top axis) Relative losses under laboratory testing conditions (75 °C, 1 sun eq.). Relative losses on a mono-Si PERC showing BO-CID are included for comparison. Source: Adapted from (Fokuhl et al. 2021; Kersten et al. 2017).

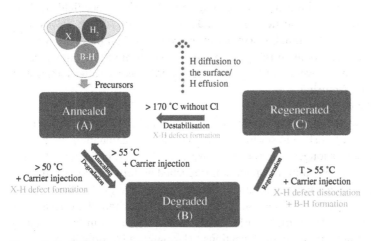

**Figure 9.4** LeTID state transition diagram.

some extent to the BO-CID model; a modified version of this model to fit new findings of the defect transformations is shown in Figure 9.4. Initially, just after the samples have gone through the high-temperature fast firing process (>750 °C for less than a minute), the LeTID defect is in State *A*, which is a non-recombination active state. Upon carrier injection, e.g. via above bandgap illumination or forward bias, at temperatures greater than 50 °C, the defect changes to a State *B* where it is recombination-active and decreases the minority charge carrier lifetime of the material. Upon further CI, the defect can go back to State *A* or move to a third State *C*, depending on the temperature at which the substrate is being held. For temperatures lower than about 55 °C the defect will go to State *A*, making cycling between State *A* and *B* possible indefinitely (Kwapil et al. 2020). Otherwise, the application of CI at *T* > 55 °C appears to move the system to State *C*, where it will remain

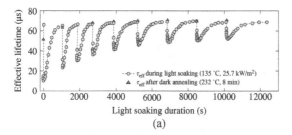
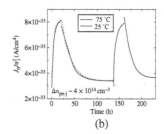

**Figure 9.5** (a) Effective minority carrier lifetime of an LeTID-prone mc-Si:B sample upon accelerated light soaking and dark annealing cycle kinetics. Source: Chen et al. (2021)/John Wiley & Sons/CC BY 4.0. (b) $n_i^2$ normalized dark saturation current ($J_0$) of an mc-PERC cell regenerated, showing the regeneration kinetics under low-temperature conditions. Source: Kwapil et al. (2020)/IEEE/CC by 4.0.

stable to subsequent CI treatments as long as no further annealing without CI is carried out at $T > 170\,°C$ (Fung et al. 2018) (Figure 9.5).

Similar behaving defects under CI and annealing treatments have been detected in several different silicon materials such as multi-crystalline (mc-Si) (Ramspeck et al. 2012), Czochralski (Cz-Si) (Chen et al. 2017) and float-zone (FZ-Si) (Niewelt et al. 2017; Sperber et al. 2017). Resemblances in response to (i) firing temperatures, $T$, and defect density, $N_T$, ($\uparrow T = \uparrow N_T$), (ii) firing profiles ($\uparrow$Cooling rate $= \uparrow N_T$), and (iii) dielectric material influence have been reported (see for example (Chen et al. 2021) and refs. therein). Further, similarities in capture cross-section ratios ($30 \pm 10$) among these silicon substrates have been reported in the literature (Chen et al. 2021). Figure 9.6 depicts the correlations between the maximum degradation extent and peak firing temperature of several dielectric structures on a variety of silicon substrates. Figure 9.6a suggests an exponential relationship between variables on samples coated with $SiN_x$ layers, almost irrespective of the base material. However, a more complex behavior is displayed on samples coated with aluminum oxide films, as shown in Figure 9.6b.

Up to now, there is little consensus on the true identity of the LeTID defect, however, various studies are ruling out the involvement of metallic impurities, e.g. the work of Vargas Castrillon (2019) and the fact that the degradation is also observed in FZ wafers. The only element that has been shown consistently to relate to LeTID is hydrogen. Whether its participation in the degradation is direct or indirect is still largely debated. Several reports suggest that a combination of boron-hydrogen (B-H) pairs and some form of hydrogen complexes (possibly dimers $H_2$) play a role in both the activation and the deactivation of the LeTID defect (de Guzman et al. 2022; Hammann et al. 2021; Kwapil et al. 2020). There is some consensus with regard to B-H acting as a "sink" for hydrogen for some parts of the LeTID process. Yet, there are contradicting arguments regarding the exact role of each complex during the degradation. Some authors propose an association model whereby $B^-$ bonds with $H_2$, resulting in a lifetime decrease (de Guzman et al. 2022), and some support a dissociation model where H detaches from another precursor, causing the degradation (Hammann et al. 2021).

It is possible that two precursor defects are coexisting in State $A$, one requiring excess carrier densities ($\Delta n$) over $\sim 1 \times 10^{13}$ cm$^{-3}$ at temperatures higher than $\sim 50\,°C$ (X-H$_{CI}$) and one activating at lower injection (in the order of $10^9$ cm$^{-3}$) but requiring temperatures greater

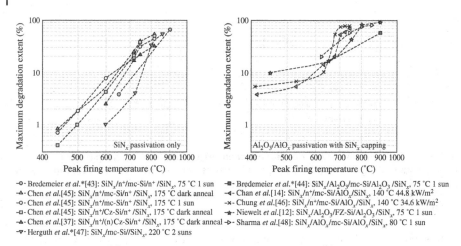

-○- Bredemeier *et al.*\*[43]: SiN$_x$/n$^+$/mc-Si/n$^+$ /SiN$_x$, 75 °C 1 sun  -■- Bredemeier *et al.*\*[44]: SiN$_x$/Al$_2$O$_3$/mc-Si/Al$_2$O$_3$ /SiN$_x$, 75 °C 1 sun
-▲- Chen *et al.*[45]: SiN$_x$/n$^+$/mc-Si/n$^+$ /SiN$_x$, 175 °C dark anneal  -◄- Chan *et al.*[14]: SiN$_x$/n$^+$/mc-Si/AlO$_x$/SiN$_x$, 140 °C 44.8 kW/m$^2$
-○- Chen *et al.*[45]: SiN$_x$/n$^+$/mc-Si/n$^+$ /SiN$_x$, 175 °C 1 sun  -×- Chung *et al.*[46]: SiN$_x$/n$^+$/mc-Si/AlO$_x$/SiN$_x$, 140 °C 34.6 kW/m$^2$
-□- Chen *et al.*[45]: SiN$_x$/n$^+$/Cz-Si/n$^+$ /SiN$_x$, 175 °C dark anneal  -★- Niewelt *et al.*[12]: SiN$_x$/Al$_2$O$_3$/FZ-Si/Al$_2$O$_3$ /SiN$_x$, 75 °C 1 sun
-▲- Chen *et al.*[37]: SiN$_x$/n$^+$/(n)Cz-Si/n$^+$ /SiN$_x$, 175 °C dark anneal  -▷- Sharma *et al.*[48]: SiN$_x$/AlO$_x$/mc-Si/AlO$_x$/SiN$_x$, 80 °C 1 sun
-▽- Herguth *et al.*\*[47]: SiN$_x$/mc-Si//SiN$_x$, 220 °C 2 suns

(a)                        (b)

**Figure 9.6** Graph showing the extent of maximum LeTID degradation as a function of peak firing temperature for two types of test structures: (a) SiN$_x$ passivated symmetrical test structures, and (b) structures with at least one side deposited AlO$_x$ or Al$_2$O$_3$ layers and both-side SiN$_x$ capping. Source: Chen et al. (2021)/John Wiley & Sons/CC BY 4.0.

than 125 °C (X-H$_{DA}$). This would explain some differences between the kinetics of LeTID studied under carrier injection (typically 75 °C, 1 kW/m$^2$) and dark annealing (typically 175 °C) see e.g. Ref. Luka et al. (2018).

According to the changes in calculated B-H concentration by Hammann et al. (2021), there is a linear correlation between [B-H] and [$N_T$] during the State $A{\to}B$ transition, but it seems unlikely that these pairs participate in the "temporary recovery" ($B{\to}A$), as they reach a saturation point coinciding with the point of maximum LeTID extent. Recently, de Guzman et al. (2022) reported the detection of an electron trap of donor character in FZ co-doped material with activation energy $E_c-0.175$ eV of apparent electron capture cross-section of $1 \times 10^{-15}$ cm$^2$, which they assigned to a B-H$_2$ complex and suggested to be responsible for the LeTID degradation. This trap is close to one of the traps reported by Mchedlidze and Weber (2019) on mc-Si degraded material (E166, $E_c-0.19$ eV), although no second trap related to the one at $E_c-0.34$ eV found in the same reference has been observed in FZ samples.

Even though LeTID continues to occur in a wide range of Si PV cells available in the market, remarkable progress in its reduction has been made in the industry to this date when mitigation procedures during manufacturing are utilized. Several approaches now widely applied in manufacturing include high-temperature treatments during carrier injection using high-throughput batch processing in "coin stackers," high-temperature and high illumination intensity (LASER or LED-based) treatments and/or gettering and dielectric optimization during the first stages of cell fabrication (Chen et al. 2021). However, as the Si PV technology evolves and dominant cell concepts are modified, restrictions and requirements for their manufacturing arise. Thus making the application of some proven mitigation strategies unhelpful and continued optimization and research essential.

## 9.4 Copper-Related Degradation

Copper is one of the most common impurities in silicon devices, and its properties have been studied intensively in the literature, demonstrating the electrical properties of its defects in singular or complex forms (Istratov and Weber 1998). Copper-related carrier-induced degradation (Cu-CID), also known as Cu-LID, is commonly observed in crystalline silicon in the presence of interstitial copper ($Cu_i$). This degradation can occur in silicon materials with $Cu_i$ concentrations as low as $10^{10}$ cm$^{-3}$ and cause significant lifetime degradation. It was first observed in 1998 by Henley et al. (1999) and Tarasov et al. (1998) upon illumination of Cu-contaminated B-doped Cz-Si samples. Since then, several studies have shown that Cu-CID can also occur on B- and P-doped Fz-Si, B-doped mc-Si and Ga-doped Cz-Si materials (Lindroos and Savin 2016). Still, the most susceptible material to Cu-CID has been found to be B-doped Cz-Si and related to its higher levels of boron and oxygen concentrations (Lindroos et al. 2013).

As a fast-diffusing impurity, Cu can be gettered during the conventional n$^+$ or p$^+$ diffusions in solar cell manufacturing, however, any thermal treatment over 800 °C (also typical in manufacturing) has been shown to release the gettered Cu from the diffused layer back to the bulk leading to Cu-CID (Nampalli et al. 2018). Cu-CID can coincide with other forms of CID, yet Cu-CID can be differentiated by the electronic properties of its responsible defect and its degradation kinetics. Figure 9.7 shows the difference in degradation kinetics of Cu-CID and BO-CID on two B-doped Cz-Si samples, one of which has been Cu-contaminated. Compared to the B-doped reference sample, the Cu-contaminated sample displays a three-staged degradation, one fast and two slower lifetime decays. In addition, and unlike BO-LID, Cu-CID does not show significant recovery upon 200 °C annealing treatments (Boulfrad et al. 2013; Inglese et al. 2015; Lindroos and Savin 2014). The degradation rate on the Cu sample is also significantly different, where the second decay occurs over a magnitude of order faster than the reference sample and leads to more severe degradation. These decays can be fitted to single or multiple exponentials depending on the silicon material under observation (Fz, Cz, or mc). However, the degradation rate in

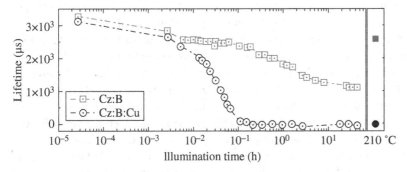

**Figure 9.7** Effective lifetime of two Cz-B samples, one of which has been contaminated with copper to display Cu-CID. The uncontaminated sample shows the characteristic BO-CID found on Cz:B samples. The testing conditions are 1-sun eq. illumination exposure at 27 °C for several hours and subsequent dark annealing at 210 °C for two minutes. Source: Adapted from (Lindroos and Savin 2014).

every case increases with temperature, illumination intensity, $Cu_i$ concentration and bulk micro-defect density (Lindroos and Savin 2016; Vahlman et al. 2017). Only in B-doped and Ga-doped Cz-Si an inverse correlation between doping concentration and degradation has been observed, which could be the result of fewer Cu-acceptor dissociation (Lindroos et al. 2013; Lindroos and Savin 2014).

While it is well known that the degradation is related to $Cu_i$ concentration (Väinölä et al. 2005), it is most likely that $Cu_i$ only acts as the Cu-CID defect precursor. This is because illumination has been shown to decrease the $Cu_i$ concentration (Belayachi et al. 2005), introducing a donor level with an activation energy of $E_c$–0.15 eV, which has been proven to be recombination inactive in concentrations as high as $10^{14}$ cm$^{-3}$ (Istratov and Weber 1998; Sachdeva et al. 2001). Although no conclusive studies exist determining the responsible defect of the degradation, most investigations suggest that Cu-CID stems from either copper precipitation (Vahlman et al. 2017; Väinölä et al. 2003) or the release of substitutional copper ($Cu_s$) through complex dissociation (Henley et al. 1999; Savin et al. 2009). The most accepted theory proposes that Cu-CID is caused by copper nano-precipitates, which are highly recombination active. $Cu_3Si$ precipitates have been shown to be severely detrimental to minority carrier lifetime in silicon and introduce defect bands between $E_c - 0.15$ and $E_c - 0.2$ eV, and $E_c - 0.4$ and $E_c - 0.58$ eV (Istratov and Weber 1998; Macdonald et al. 2003; Schröter et al. 2000). Unlike other impurities, copper precipitation cannot be explained by impurity supersaturation and can only precipitate when the Fermi level exceeds $E_c - 0.2$ eV (Flink et al. 2000), which leads $Cu_3Si$ to its neutrality state. Below this energy level, the $Cu_3Si$ complex is positively charged and repels other $Cu_i^+$ atoms, preventing precipitation. Given the dependence of Fermi level on doping concentration, $Cu_i^+$ concentration and temperature, these parameters determine the defects' precipitate size and distribution within the bulk (Istratov and Weber 1998). Changes in Fermi level due to carrier injection can also lead to $Cu_i^+$ trapping and could explain the observed increase in the Cu-CID degradation rate (Belayachi et al. 2005; Väinölä et al. 2003). Still, further research is needed to identify the responsible defect conclusively.

Several methods have been reported in the literature that attempt to mitigate the effects of Cu-CID on solar cells. Most applicable methods include copper gettering, copper out-diffusion or hydrogenation (Heikkinen et al. 2020). In defect-free silicon, copper diffuses to the surfaces to reach solubility equilibrium (Heiser et al. 2003), making it a straightforward process to remove copper precipitation from the surface via etching after out-diffusion occurs. Out-diffusion can be accelerated using continuous negative corona charging to attract the positive $Cu_i$ atoms (Heiser et al. 2003; Lindroos and Savin 2016) or through annealing in the 100–400 °C range (Kot et al. 2014). In the case of gettering, the application of heavily doped phosphorous, boron and aluminum layers has been largely studied for their known segregation gettering properties. However, even the most efficient methods of Cu removal through gettering using phosphorous diffusion gettering or Al gettering require intolerable temperatures and processing lengths for the industry, typically in the range of 800–900 °C for one to two hours (Bentzen et al. 2006; Buonassisi et al. 2005; Shabani et al. 2008). The most effective methods for preventing Cu-CID today are a negative surface charge followed by etching (Boulfrad et al. 2014) or combining a negative surface charge with illumination (Lindroos and Savin 2016). Although, it is still not fully understood how illumination prevents Cu back-diffusion and thus requires further analysis.

## 9.5   Surface-Related Degradation

With the substantial price drop in high-quality silicon materials used for solar cell fabrication and the ever-thinner wafers utilized, it has become increasingly important to focus on the decrease of surface recombination and contact resistance losses. Even though remarkable performances are achieved with mono-PERC/PERT/PERL solar cells at the global market level (>23% abs), surface and contact recombination continue to be efficiency-limiting factors for the conversion efficiency of these devices. Passivating contact structures that feature a full-area tunnel oxide passivated rear contact (known as the TOPCon cell) made of a heavily doped polysilicon (poly-Si) layer on a thin interfacial oxide layer and a high-quality top surface passivation have been showing great potential for enabling higher efficiency solar cells in an industrial setting since around 2019. However, in both cases, the increased performance arising from outstanding surface passivation comes at the price of a higher sensitivity to changes at the interfaces.

In contrast to the degradation mechanisms previously discussed, which primarily impact the bulk of the silicon material upon carrier injection, more recently, an additional form of degradation that affects the surface of the silicon material, known as SRD, has been identified. The initial comprehensive study on SRD was published in 2016 by Sperber et al. (2016). The study found that SRD affects all types of silicon materials, including FZ, Cz, and mc-Si, regardless of dopant species. However, it has most frequently been observed in P-doped silicon, which is predicted by the international technology roadmap for photovoltaic (ITRPV) to dominate the market within this decade (Fischer et al. 2022). Even though there had been some indications of SRD coming from dielectric passivation deterioration, it is now clear that the underlying defects are located directly at or very close to the interface between silicon and passivation layers since re-passivation is found to recover a sample from SRD after removing only a few nanometers of the sample surface (Sperber 2019).

It is important to note that SRD has been found to lead to much more significant efficiency losses on PERC cells than from LeTID in Cz material (>14%$_{rel.}$) (Herguth et al. 2018), albeit in a much longer time scale. Figure 9.8 shows the evolution of efficiency and FF losses tracked during 1 sun eq. illuminated annealing at 150 °C of a B:Cz PERC sample. It is observed that two distinct degradations occur on the sample at different time scales. The first degradation can be attributed to either BO-LID, LeTID or a combination of the two. In contrast, the second degradation is the result of increased recombination at the rear surface of the cell.

Observations show that SRD impacts various types of silicon wafers that are passivated with plasma-enhanced chemical vapor deposited (PECVD) silicon nitride (SiN$_x$:H), thin thermal oxides capped with nitride (SiO$_2$/SiN$_x$:H), and aluminum oxide (PECVD or atomic layer deposited) capped with silicon nitride (AlO$_x$:H/SiN$_x$:H, Al$_2$O$_3$/SiN$_x$:H) (Sperber et al. 2017; Sperber et al. 2016; Tan et al. 2021). All of these are relevant to the current mainstream technology, the PERC concept. Furthermore, a somewhat similar degradation mechanism arising from the near-surface region of structures using polysilicon/tunnel oxide passivating contacts has been recently reported (Chen et al. 2022; Winter et al. 2020).

Although there are not many studies on the topic in the literature, and several inconsistencies still exist, it is apparent that the thermal history of the wafers is one crucial

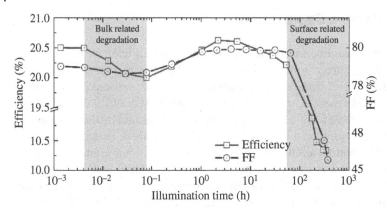

**Figure 9.8** Temporary evolution of the efficiency and fill factor of a Cz:B PERC showing a bulk-related degradation related to BO-CID or/and LeTID, followed by a severe surface degradation during standard testing conditions (150 °C, 1 sun eq. illumination). Source: Adapted from (Herguth et al. 2018).

factor in the evolution of SRD. Furthermore, degradation appears to be more pronounced in wafers that have undergone a high-temperature (e.g. firing) step in the presence of a hydrogen-containing dielectric. Therefore, it is plausible to assume that hydrogen, either directly or indirectly, is related to this degradation (Sperber et al. 2016; Sperber et al. 2017; Winter et al. 2020). One way that SRD may be explained is that defects at the interface, such as dangling bonds, are significantly affected by dielectric processing conditions but can be effectively passivated by hydrogen during high-temperature annealing. Nonetheless, the low-energy bonds produced may be easily broken, allowing H to move from the surface during high-temperature carrier-injection treatments, resulting in increased recombination activity. A contrasting but also the plausible explanation is for hydrogen defects to diffuse toward the surface and accumulate there during the CI treatments, forming hydrogen platelets (de Walle et al. 1989). Hammann et al. (2023) have recently provided evidence in support of this theory, showing that thinner samples experience accelerated and increased SRD, likely due to hydrogen's shorter path to the surface during degradation treatment. It is also suggested that faster and more severe SRD kinetics observed in samples cooled quickly during the firing step result from uneven distribution of hydrogen in the bulk of the material. Rapid cooling ramps prevent much of the hydrogen from diffusing into the bulk, leaving it concentrated at the surface and thus accelerating the degradation kinetics.

Similar to BO-LID and LeTID, SRD has been found to be carrier-injection and temperature dependent (Chen et al. 2022; Sperber 2019). However, despite high-intensity (up to 150 kW/m²) and high-temperature (up to 40 °C) treatments, the recovery of the degradation within a commercially viable time frame has not been achieved to date (Chen et al. 2022). It may be necessary to consider even higher temperatures at these high-intensity levels to achieve faster recovery. Chen et al. (2022) demonstrated that a full cycle of degradation and recovery is necessary for stabilizing SRD-susceptible samples, which is in accordance with Sperber (2019) and Tan et al. (2021), who showed stabilization of samples through 30-minute annealing treatments at temperatures above 400 °C.

**Table 9.2** Simplified comparison of CID's properties.

| | LeTID | BO-CID | Cu-CID | SRD |
|---|---|---|---|---|
| k-Value | 20–35 | 10–20 | 1.7–2.6 | n/a |
| Common testing conditions | 75°C, 1 sun/175°C DA | 25–30°C, 1 sun | 25–30°C, 1 sun | 75°C, 1 sun/150°C, 1 sun/150°C DA |
| Most susceptible materials | p-Type mc-Si | p-Type Cz | p-Type Cz | p-Type and n-type Cz |
| Observed in Ga-doped wafer | Yes | Not unless B co-doped | Yes | Unclear |
| Max. degradation time | >1000 h/>9 h | >48 h | >200 s | >1000 h/>50 h/>1000 h |
| Degradation rate ($R_{\text{deg}}$) | $R_{\text{deg}} \propto \Delta n^{0.86}$, increases with fast cooling and 125–200°C pre-DA | $R_{\text{deg}} \propto p_0^2$ | $R_{\text{deg}}$ decreases with $N_A$ | Increases with thinner wafers and faster cooling |
| Light intensity | Dependent | Dependent (up to ~2 suns) | Dependent | Dependent |
| Staged degradation | Two-stage | Two-stage | One or three-staged | Single exponential |
| Degradation temperature | >50°C | >25°C | >25°C | >75°C |
| Hydrogen's role | Precursor | Passivator | Passivator | Precursor |
| Processing at 200°C without carrier injection | Promotes defect formation and recombination | Full recovery | Partial or no recovery | Degradation |
| Cycling behavior | Observed | Observed | Not observed | Not observed |
| Recovery under testing conditions | Observed | Observed | Not observed | Observed |

*Note: DA* – dark annealing; *CI* – carrier injection; *K-value* – the extent of capture cross-section ($\sigma$) asymmetry of the defect state defined as $k = \sigma_n/\sigma_p$; $\Delta_n$ – excess carrier density; $p_0$ – hole density in thermal equilibrium; $N_A$ – acceptor dopant concentration.

## 9.6 Conclusions

Solar PV modules made of silicon have been observed to experience degradation throughout their operational lifetime. Mechanisms related to these degradations have been thus referred to as light-induced degradation (LID). However, these degradation mechanisms are also observed during carrier injection via applied bias or at high enough temperatures in darkness, therefore making them more accurately described as a form of CID. Despite extensive research, the defects causing the failure modes discussed in this chapter have not been definitively identified due to their complex nature.

In this chapter, a summary is presented regarding our present comprehension of various carrier-induced degradation phenomena that have a notable impact on silicon-based solar cells, including BO-CID, LeTID, Cu-CID, and SRD. In Table 9.2, a simplified comparison of each degradation is given. Various techniques have been developed to mitigate the impact of each type of degradation. For instance, to address BO-LID, high carrier injection through illumination or biasing, along with high-temperature annealing processes in the presence of hydrogen, has effectively stabilized the modules. Furthermore, the migration of the industry from boron-doped to gallium-doped base materials for solar cells has practically eliminated the effect of the BO-LID. In the same way, advancements in material selection, process modification, and post-cell production techniques have significantly mitigated the impact of LeTID, particularly in the most susceptible devices, such as mc-Si PERC cells. The originally reported efficiency loss of 10–12% has, on average, been reduced to less than 2% thanks to these innovations. Still, ongoing investigation is needed to eliminate this problem. It is uncertain if Cu-CID and SRD have a noticeable impact on current mainstream technologies. Copper contamination has been largely reduced in high-performing devices, and SRD onset takes longer than the already lengthy LeTID time frame under standard operating conditions. However, these mechanisms could become relevant if Cu-plated technology becomes more widely used or if changes in material and process conditions accelerate the onset of SRD, such as using thinner wafers or faster cooling conditions leading to higher H accumulation at the surface.

## Author Biography

**Dr Michelle Vaqueiro Contreras** is a Senior Research Associate at UNSW Sydney and an Australian Centre for Advanced Photovoltaics Fellow. She completed her master's and PhD degrees in 2018 at the University of Manchester (UK), where she was awarded the Distinguished Achievement Medal as Postgraduate of the Year in the Faculty of Engineering. Her work on silicon defects has led to significant progress in understanding and mitigation of impurities that reduce solar cell efficiency through carrier-induced degradation mechanisms such as light-induced degradation (LID) and light and elevated temperature degradation (LeTID). She has recently dedicated her efforts to supporting the Australian solar PV industry. Notably, she was one of the primary authors of the APVI-led Silicon to Solar (S2S) study. In this capacity, she conducted a thorough techno-economic analysis of the solar PV supply chain, aiming to the development of a credible, viable, and relevant roadmap for Australia's strategic engagement in the field.

## References

Abe, T., Hirasawa, T., Tokunaga, K. et al. (2004). Silicon single crystal and wafer doped with gallium and method for producing them. Google Patents. US Patent 6,815,605.

Belayachi, A., Heiser, T., Schunck, J.P., and Kempf, A. (2005). Influence of light on interstitial copper in p-type silicon. *Applied Physics A: Materials Science & Processing* 80: 201–204. https://doi.org/10.1007/s00339-004-3038-7.

Bentzen, A., Holt, A., Kopecek, R. et al. (2006). Gettering of transition metal impurities during phosphorus emitter diffusion in multicrystalline silicon solar cell processing. *Journal of Applied Physics* 99 (9): 093509. https://doi.org/10.1063/1.2194387.

Bothe, K. and Schmidt, J. (2006). Electronically activated boron-oxygen-related recombination centers in crystalline silicon. *Journal of Applied Physics* 99 (1): 013701. https://doi.org/10.1063/1.2140584.

Bothe, K., Schmidt, J., and Hezel, R. (2003). Comprehensive analysis of the impact of boron and oxygen on the metastable defect in Cz silicon. In: *Proceedings of the 3rd IEEE World Conference on Photovoltaic Energy Conversion*, vol. 2, 1077–1080.

Boulfrad, Y., Lindroos, J., Inglese, A. et al. (2013). Reduction of light-induced degradation of boron-doped solar-grade Czochralski silicon by corona charging. *Energy Procedia* 38: 531–535. https://doi.org/10.1016/J.EGYPRO.2013.07.313.

Boulfrad, Y., Lindroos, J., Wagner, M. et al. (2014). Experimental evidence on removing copper and light-induced degradation from silicon by negative charge. *Applied Physics Letters* 105 (18): 182108. https://doi.org/10.1063/1.4901533.

Buonassisi, T., Marcus, M.A., Istratov, A.A. et al. (2005). Analysis of copper-rich precipitates in silicon: chemical state, gettering, and impact on multicrystalline silicon solar cell material. *Journal of Applied Physics* 97 (6): 063503. https://doi.org/10.1063/1.1827913.

Chen, D., Kim, M., Stefani, B.V. et al. (2017). Evidence of an identical firing-activated carrier-induced defect in monocrystalline and multicrystalline silicon. *Solar Energy Materials and Solar Cells* 172: 293–300. https://doi.org/10.1016/J.SOLMAT.2017.08.003.

Chen, D., Vaqueiro Contreras, M., Ciesla, A. et al. (2021). Progress in the understanding of light- and elevated temperature-induced degradation in silicon solar cells: a review. *Progress in Photovoltaics: Research and Applications* 29 (11): 1180–1201. https://doi.org/10.1002/PIP.3362.

Chen, D., Madumelu, C., Kim, M. et al. (2022). Investigating the degradation behaviours of n+-doped poly-Si passivation layers: an outlook on long-term stability and accelerated recovery. *Solar Energy Materials and Solar Cells* 236: 111491. https://doi.org/10.1016/j.solmat.2021.111491.

Fattah, T.O.A., Markevich, V.P., de Guzman, J.A.T. et al. (2022). Interactions of hydrogen atoms with acceptor–dioxygen complexes in Czochralski-grown silicon. *Physica Status Solidi A* 219 (17): 2200176. https://doi.org/10.1002/PSSA.202200176.

Fischer, H. and Pschunder, W. (1973). Investigation of photon and thermal changes in silicon solar cells. In: *Proceedings of the 10th IEEE Photovoltaic Specialists Conference*. IEEE.

Fischer, M., Woodhouse, M., Herritsch, S., and Trube, J. (2020). *2020 International Technology Roadmap for Photovoltaics*. VDMA Photovoltaic Equipment.

Fischer, M., Woodhouse, M., Herritsch, S., and Trube, J. (2021). *2021 International Technology Roadmap for Photovoltaics*. VDMA Photovoltaic Equipment.

Fischer, M., Woodhouse, M., Herritsch, S., and Trube, J. (2022). *2022 International Technology Roadmap for Photovoltaics*. VDMA Photovoltaic Equipment.

Flink, C., Feick, H., McHugo, S.A. et al. (2000). Out-diffusion and precipitation of copper in silicon: an electrostatic model. *Physical Review Letters* 85 (23): 4900–4903. https://doi.org/10.1103/PHYSREVLETT.85.4900.

Fokuhl, E., Philipp, D., Mülhöfer, G., and Gebhardt, P. (2021). LID and LETID evolution of PV modules during outdoor operation and indoor tests. *EPJ Photovoltaics* 12: 9. https://doi.org/10.1051/EPJPV/2021009.

Fung, T.H., Kim, M., Chen, D. et al. (2018). A four-state kinetic model for the carrier-induced degradation in multicrystalline silicon: introducing the reservoir state. *Solar Energy Materials and Solar Cells* 184: 48–56. https://doi.org/10.1016/J.SOLMAT.2018.04.024.

Glunz, S.W., Rein, S., Warta, W. et al. (1998). On the degradation of Cz-silicon solar cells. Presented at the *2nd World Conference on Photovoltaic Solar Energy Conversion*.

de Guzman, J.A.T., Markevich, V.P., Coutinho, J. et al. (2022). Electronic properties and structure of boron–hydrogen complexes in crystalline silicon. *Solar RRL* 6 (5): 2100459. https://doi.org/10.1002/SOLR.202100459.

Hamer, P., Hallam, B., Abbott, M., and Wenham, S. (2015). Accelerated formation of the boron–oxygen complex in p-type Czochralski silicon. *Physica Status Solidi RRL: Rapid Research Letters* 9 (5): 297–300.

Hammann, B., Rachdi, L., Kwapil, W. et al. (2021). Insights into the hydrogen-related mechanism behind defect formation during light- and elevated-temperature-induced degradation. *Physica Status Solidi RRL: Rapid Research Letters* 15 (6): 2000584. https://doi.org/10.1002/PSSR.202000584.

Hammann, B., Assmann, N., Weiser, P.M. et al. (2023). The impact of different hydrogen configurations on light-and elevated-temperature-induced degradation. *IEEE Journal of Photovoltaics* 13 (2): 224–235.

Hashigami, H. and Saitoh, T. (2003). Carrier-induced degradation phenomena of carrier lifetime and cell performance in boron-doped Cz-Si. In: *Proceedings of the 3rd IEEE World Conference on Photovoltaic Energy Conversion*, vol. vol. 3, 2893–2898. IEEE.

Heikkinen, I.T.S., Wright, B., Soeriyadi, A.H. et al. (2020). Can hydrogenation mitigate Cu-induced bulk degradation in silicon? In: *2020 47th IEEE Photovoltaic Specialists Conference (PVSC)*, 2582–2585. IEEE.

Heiser, T., Belayachi, A., and Schunck, J.P. (2003). Copper behavior in bulk silicon and associated characterization techniques. *Journal of the Electrochemical Society* 150 (12): G831. https://doi.org/10.1149/1.1627351.

Henley, W.B., Ramappa, D.A., and Jastrezbski, L. (1999). Detection of copper contamination in silicon by surface photovoltage diffusion length measurements. *Applied Physics Letters* 74 (2): 278–280. https://doi.org/10.1063/1.123280.

Herguth, A., Schubert, G., Kaes, M., and Hahn, G. (2006). A new approach to prevent the negative impact of the metastable defect in boron doped Cz silicon solar cells. In: *Presented at the 4th IEEE World Conference on Photovoltaic Energy Conference*, vol. 1, 940–943.

Herguth, A., Derricks, C., and Sperber, D. (2018). A detailed study on light-induced degradation of Cz-Si PERC-type solar cells: evidence of rear surface-related degradation. *IEEE Journal of Photovoltaics* 8 (5): 1190–1201.

Inglese, A., Lindroos, J., and Savin, H. (2015). Accelerated light-induced degradation for detecting copper contamination in p-type silicon. *Applied Physics Letters* 107 (5): 052101. https://doi.org/10.1063/1.4927838.

Istratov, A.A. and Weber, E.R. (1998). Electrical properties and recombination activity of copper, nickel and cobalt in silicon. *Applied Physics A: Materials Science and Processing* 66 (2): 123–136. https://doi.org/10.1007/S003390050649.

Kersten, F., Fertig, F., Petter, K. et al. (2017). System performance loss due to LeTID. *Energy Procedia* 124: 540–546. https://doi.org/10.1016/J.EGYPRO.2017.09.260.

Kim, M., Abbott, M., Nampalli, N. et al. (2017). Modulating the extent of fast and slow boron-oxygen related degradation in Czochralski silicon by thermal annealing: evidence of a single defect. *Journal of Applied Physics* 121 (5): 053106. https://doi.org/10.1063/1.4975685.

Kim, M., Chen, D., Abbott, M. et al. (2018). Impact of interstitial iron on the study of meta-stable B-O defects in Czochralski silicon: further evidence of a single defect. *Journal of Applied Physics* 123 (16): 161586. https://doi.org/10.1063/1.5000323.

Kot, D., Kissinger, G., Sattler, A., and Müller, T. (2014). Development of a storage getter test for Cu contaminations in silicon wafers based on ToF-SIMS measurements. *Acta Physica Polonica A* 125 (4): 965–968. https://doi.org/10.12693/APHYSPOLA.125.965.

Kwapil, W., Schön, J., Niewelt, T., and Schubert, M.C. (2020). Temporary recovery of the defect responsible for light-and elevated temperature-induced degradation: insights into the physical mechanisms behind LeTID. *IEEE Journal of Photovoltaics* 10 (6): 1591–1603.

Lindroos, J. and Savin, H. (2014). Formation kinetics of copper-related light-induced degradation in crystalline silicon. *Journal of Applied Physics* 116 (23): 234901. https://doi.org/10.1063/1.4904197.

Lindroos, J. and Savin, H. (2016). Review of light-induced degradation in crystalline silicon solar cells. *Solar Energy Materials and Solar Cells* 147: 115–126. https://doi.org/10.1016/J.SOLMAT.2015.11.047.

Lindroos, J., Yli-Koski, M., Haarahiltunen, A. et al. (2013). Light-induced degradation in copper-contaminated gallium-doped silicon. *Physica Status Solidi RRL: Rapid Research Letters* 7 (4): 262–264. https://doi.org/10.1002/PSSR.201307011.

Luka, T., Turek, M., and Hagendorf, C. (2018). Defect formation under high temperature dark-annealing compared to elevated temperature light soaking. *Solar Energy Materials and Solar Cells* 187: 194–198. https://doi.org/10.1016/J.SOLMAT.2018.06.043.

Macdonald, D., Cuevas, A., Rein, S. et al. (2003). Temperature-and injection-dependent lifetime spectroscopy of copper-related defects in silicon. In: *Proceedings of the 3rd World Conference on Photovoltaic Energy Conversion, 2003*, vol. 1, 87–90. IEEE.

Markevich, V.P., Vaqueiro-Contreras, M., de Guzman, J.T. et al. (2019). Boron–oxygen complex responsible for light-induced degradation in silicon photovoltaic cells: a new insight into the problem. *Physica Status Solidi (A)* 216 (17): 1900315. https://doi.org/10.1002/PSSA .201900315.

Mchedlidze, T. and Weber, J. (2015). Direct detection of carrier traps in Si solar cells after light-induced degradation. *Physica Status Solidi RRL: Rapid Research Letters* 9 (2): 108–110.

Mchedlidze, T. and Weber, J. (2019). Location and properties of carrier traps in mc-Si solar cells subjected to degradation at elevated temperatures. *Physica Status Solidi (A)* 216 (17): 1900142. https://doi.org/10.1002/PSSA.201900142.

Nampalli, N., Laine, H.S., Colwell, J. et al. (2018). Rapid thermal anneal activates light induced degradation due to copper redistribution. *Applied Physics Letters* 113 (3): 032104. https://doi .org/10.1063/1.5029347.

Niewelt, T., Schon, J., Warta, W. et al. (2017). Degradation of crystalline silicon due to boron-oxygen defects. *IEEE Journal of Photovoltaics* 7 (1): 383–398. https://doi.org/10.1109/ JPHOTOV.2016.2614119.

Niewelt, T., Selinger, M., Grant, N.E. et al. (2017). Light-induced activation and deactivation of bulk defects in boron-doped float-zone silicon. *Journal of Applied Physics* 121 (18): 185702. https://doi.org/10.1063/1.4983024.

Ramspeck, K., Zimmermann, S., Nagel, H. et al. (2012). Light induced degradation of rear passivated mc-Si solar cells. *Proceedings of the 27th European Photovoltaic Solar Energy Conference and Exhibition.*

Sachdeva, R., Istratov, A.A., and Weber, E.R. (2001). Recombination activity of copper in silicon. *Applied Physics Letters* 79: 2937. https://doi.org/10.1063/1.1415350.

Savin, H., Yli-Koski, M., and Haarahiltunen, A. (2009). Role of copper in light induced minority-carrier lifetime degradation of silicon. *Applied Physics Letters* 95 (15): 152111. https://doi.org/10.1063/1.3250161.

Schmidt, J. and Hezel, R. (2002). Light-induced degradation in Cz silicon solar cells: fundamental understanding and strategies for its avoidance. Presented at the *12th Workshop on Crystalline Silicon Solar Cell Materials and Proceses*, 63.

Schmidt, J., Aberle, A.G., and Hezel, R. (1997). Investigation of carrier lifetime instabilities in Cz-grown silicon. In: *Presented at the 26th IEEE Photovoltaic Specialists Conference*, 13–18. IEEE.

Schmidt, J., Bothe, K., Voronkov, V.V., and Falster, R. (2020). Fast and slow stages of lifetime degradation by boron–oxygen centers in crystalline silicon. *Physica Status Solidi B* 257 (1): 1900167. https://doi.org/10.1002/PSSB.201900167.

Schön, J., Niewelt, T., Broisch, J. et al. (2015). Characterization and modelling of the boron-oxygen defect activation in compensated n-type silicon. *Journal of Applied Physics* 118 (24): 245702. https://doi.org/10.1063/1.4938569.

Schröter, W., Kveder, V., Seibt, M. et al. (2000). Atomic structure and electronic states of nickel and copper silicides in silicon. *Materials Science and Engineering B* 72 (2–3): 80–86. https:// doi.org/10.1016/S0921-5107(99)00499-7.

Shabani, M.B., Yamashita, T., and Morita, E. (2008). Metallic impurities in mono and multi-crystalline silicon and their gettering by phosphorus diffusion. *ECS Transactions* 16 (6): 179–193. https://doi.org/10.1149/1.2980302/XML.

Sperber, D. (2019). *Bulk and surface-related degradation phenomena in monocrystalline silicon at elevated temperature and illumination.* Ph.D. Thesis. University of Konstanz. Available at: https://kops.uni-konstanz.de/entities/publication/13070525-ff2a-40f9-86fe-c9bcc584d597 (Accessed 25 Apr 2024).

Sperber, D., Furtwängler, F., Herguth, A., and Hahn, G. (2016). On the stability of dielectric passivation layers under illumination and temperature treatment. In: *32nd European Photovoltaic Solar Energy Conference and Exhibition*, 523–526.

Sperber, D., Herguth, A., and Hahn, G. (2016). Instability of dielectric surface passivation quality at elevated temperature and illumination. *Energy Procedia* 92: 211–217. https://doi .org/10.1016/j.egypro.2016.07.061.

Sperber, D., Graf, A., Skorka, D. et al. (2017). Degradation of surface passivation on crystalline silicon and its impact on light-induced degradation experiments. *IEEE Journal of Photovoltaics* 7 (6): 1627–1634.

Sperber, D., Heilemann, A., Herguth, A., and Hahn, G. (2017). Temperature and light-induced changes in bulk and passivation quality of boron-doped float-zone silicon coated with SiNx:H. *IEEE Journal of Photovoltaics* 7 (2): 463–470. https://doi.org/10.1109/JPHOTOV .2017.2649601.

Tan, X., Chen, R., and Rougieux, F.E. (2021). The mechanism of surface passivation degradation in SiO 2/SiN x stack under light and elevated temperature. *IEEE Journal of Photovoltaics* 11 (6): 1380–1387.

Tarasov, I., Ostapenko, S., and Koveshnikov, S. (1998). Light induced defect reactions in boron-doped silicon: Cu versus Fe. In: *Proceedings of the 9th Workshop Role of Impurities and Defects in Silicon Device Processing*, 207–210.

Vahlman, H., Haarahiltunen, A., Kwapil, W. et al. (2017). Modeling of light-induced degradation due to Cu precipitation in p-type silicon. II. Comparison of simulations and experiments. *Journal of Applied Physics* 121 (19): 195704. https://doi.org/10.1063/1.4983455.

Väinölä, H., Yli-Koski, M., Haarahiltunen, A., and Sinkkonen, J. (2003). Sensitive copper detection in P-type CZ silicon using μPCD. *Journal of the Electrochemical Society* 150 (12): G790. https://doi.org/10.1149/1.1624845.

Väinölä, H., Saarnilehto, E., Yli-Koski, M. et al. (2005). Quantitative copper measurement in oxidized p-type silicon wafers using microwave photoconductivity decay. *Applied Physics Letters* 87 (3): 032109. https://doi.org/10.1063/1.1999008.

Vaqueiro-Contreras, M., Markevich, V.P., Coutinho, J. et al. (2019). Identification of the mechanism responsible for the boron oxygen light induced degradation in silicon photovoltaic cells. *Journal of Applied Physics* 125 (18): 185704. https://doi.org/10.1063/1 .5091759.

Vargas Castrillon, C. (2019). *Effects of Dark and Illuminated Anneals on Multi-crystalline Silicon Wafers and on Mono-crystalline Silicon Wafers Passivate with Carrier-Selective Contacts.* Sydney: UNSW.

de Walle, C.G., Denteneer, P.J.H., Bar-Yam, Y., and Pantelides, S.T. (1989). Theory of hydrogen diffusion and reactions in crystalline silicon. *Physical Review B* 39 (15): 10791.

Walter, D.C., Lim, B., Bothe, K. et al. (2014). Effect of rapid thermal annealing on recombination centres in boron-doped Czochralski-grown silicon. *Applied Physics Letters* 104 (4): 42111.

Weizer, V.G., Brandhorst, H.W., Broder, J.D. et al. (2008). Photon-degradation effects in terrestrial silicon solar cells. *Journal of Applied Physics* 50 (6): 4443. https://doi.org/10.1063/1.326437.

Wenham, A.C.N., Wenham, S., Chen, R. et al. (2018). Hydrogen-induced degradation. In: *2018 IEEE 7th World Conference on Photovoltaic Energy Conversion, WCPEC 2018 – A Joint Conference of 45th IEEE PVSC, 28th PVSEC and 34th EU PVSEC*, 1–8. https://doi.org/10.1109/PVSC.2018.8548100.

Wilking, S., Beckh, C., Ebert, S. et al. (2014). Influence of bound hydrogen states on BO-regeneration kinetics and consequences for high-speed regeneration processes. *Solar Energy Materials and Solar Cells* 131: 2–8. https://doi.org/10.1016/J.SOLMAT.2014.06.027.

Winter, M., Bordihn, S., Peibst, R. et al. (2020). Degradation and regeneration of n+-doped poly-Si surface passivation on p-type and n-type Cz-Si under illumination and dark annealing. *IEEE Journal of Photovoltaics* 10 (2): 423–430.

Zhou, Z., Vaqueiro-Contreras, M., Juhl, M.K., and Rougieux, F. (2022). Electronic properties of the boron-oxygen defect precursor of the light-induced degradation in silicon. *IEEE Journal of Photovoltaics* 12 (5): 1135–1141. https://doi.org/10.1109/JPHOTOV.2022.3190769.

# 10

# Hydrogenation

*Alison Ciesla*

*School of Photovoltaic and Renewable Energy Engineering, UNSW Sydney, Sydney, NSW, Australia*

## 10.1   Introduction

Hydrogen is the most abundant and lightest atom in the universe, with a standard atomic weight of 1.008. It has a single proton in its nucleus and an outer shell capacity of two electrons, allowing it to have 0, 1, or 2 electrons. This makes it negatively, neutrally or positively charged, respectively ($H^-$, $H^0$ and $H^+$), and highly reactive. With these properties, hydrogen can be very beneficial in silicon solar cells for bonding to and passivating electrically active defects that would otherwise capture electrical carriers, reducing wafer lifetimes and cell efficiencies. The benefit of hydrogen passivation or "hydrogenation" in silicon solar cells has been known for decades, stemming from earlier developments in the electronics industry. Some form of hydrogenation is used in almost all silicon solar cells ever made, even if not explicitly mentioned. Although the importance of hydrogenation is long recognized, the exact behavior of hydrogen and how to achieve effective passivation is still not fully understood for several reasons: hydrogen can interact with and form complexes with almost anything; its different forms and charge states have vastly different behavior; it is so light and abundant it is difficult to detect and prevent unintentional background incorporation; even minor changes in conditions impact it; and the defects to be passivated are also widely varying, often changing with time or exposure conditions. In the last decade, the realization of the importance of the hydrogen charge states for being able to passivate certain defects led to the development of "advanced hydrogenation" processes to manipulate the hydrogen and control its charge state. It is now becoming more important to passivate or reduce all forms of recombination in silicon solar cells as they are approaching the efficiency limits.

## 10.2   Typical Hydrogenation Methods

The most common source of recombination in solar cells, inherent even to the best quality wafers, is that at the surfaces or, rather, the silicon interfaces with the surface layers and contacts. Surface passivation layers are critical for reducing the surface recombination velocity (SRV) at these interfaces. The benefit of getting hydrogen to these interfaces has long been known and formed the basis for typical hydrogenation methods. Hydrogen readily ties up surface dangling bonds or interface states. Hydrogen can reduce the active surface state concentration ($D_{it}$) of an oxide ($SiO_2$) passivated surface by over an order of magnitude and thus reduce the SRV by the same amount. All the lowest SRVs measured include some hydrogen in their passivation (Bonilla et al. 2017).

To get effective hydrogen passivation at the interface, typically either a non-hydrogen-containing layer is grown or deposited and annealed while exposed to a hydrogen gas or plasma, or hydrogen-containing layer(s) is deposited followed by an anneal to free hydrogen within the layer and allow it to diffuse to the interface. Such thermal treatments can also allow hydrogen to diffuse into the silicon bulk for the passivation of some bulk defects.

### 10.2.1 Forming Gas Anneal and "Alneal"

Used in some of the earliest solar cells, these processes came from the electronics industry. A forming gas anneal (FGA) typically involves heating the samples to 350–450 °C, for >15 minutes in a hydrogen atmosphere of 4–5% $H_2$ with the remainder made up of argon or nitrogen. The samples must cool to less than ~100 °C in the presence of the FGA to avoid a reversing of the hydrogen passivation. Around 1990, FGA was used in both interdigitated back contact (IBC) solar cells (Swanson 1990; Verlinden et al. 1990) and passivated emitter and rear cells (PERCs) (Blakers et al. 1989).

The "anneal" process involves annealing at similar temperatures and times with an aluminum layer on a $SiO_2$ layer (Balk et al. 1965; Deal et al. 1969). Hydroxyl ions in or on the $SiO_2$ react with the aluminum to generate atomic hydrogen that improves passivation at the $Si/SiO_2$ interface. A hydrogen ambient is not required but can also be used.

A wafer with $SiO_2$ surface passivation and a starting lifetime of 14 μs has been shown to increase to 40 μs after FGA and increase further to 400 μs after an additional "anneal" process (Zhao et al. 1996).

### 10.2.2 Hydrogen Containing Dielectrics

Today, by far, the most dominant source of hydrogen in solar cells comes from the dielectric surface layers. Plasma-enhanced chemical vapor deposition (PECVD) with hydrogen-containing precursor gases, commonly silane, deposit heavily hydrogenated layers. Heterojunction cells (HJT), which are growing in industry significance and rely on very low SRV surfaces, use intrinsic and doped hydrogenated amorphous silicon (a-Si:H) layers. The industry dominant PERCs, the emerging tunneling oxide passivated contact (TOPCon) or polycrystalline silicon on oxide (POLO) cells, and the previously dominant aluminum back-surface field cells (Al-BSF) use a primary hydrogen source of hydrogenated silicon nitride ($SiN_x$:H). $SiN_x$ is not ideal for the passivation of p-type surfaces, so PERC and TOPCon cells which require passivation of both polarity surfaces, typically include a thin layer of aluminum oxide ($AlO_x$) at the p-type c-Si interface. $AlO_x$ can also contain large amounts of hydrogen if deposited by PECVD. Other deposition methods, such as low-pressure CVD (LPCVD), result in hydrogen-lean layers. Atomic layer deposition

(ALD) AlO$_x$ films contain some hydrogen, however, they have been shown to act as a barrier to hydrogen migrating from the SiN$_x$ to the silicon (Varshney et al. 2020).

### 10.2.2.1 Thermal Activation

A thermal process is generally required to free hydrogen from the dielectric layer and allow it to migrate to the interface (or silicon bulk) for passivation of defects. In Al-BSF, PERC and TOPCon cells that mostly have SiN$_x$:H layers and screen-printed contacts, "hydrogenation" occurs during the metal co-firing process. How much hydrogen is released and in what form depends largely on the Si–N ratio in the layer and peak firing temperature. Si—H bonds break at lower temps than N—H bonds. This means the more silicon-rich the film, the more hydrogen is released at lower temperatures. The extreme case of this is a-Si:H, which loses all hydrogen and degrades above 250 °C. This is a limiting factor in the thermal processing of HJT cells, but it also means that the anneal during the deposition itself can be enough for effective hydrogen passivation at the surface. Screen-printed cells that need to be fired at temperatures above 700 °C therefore use more nitrogen-rich layers. Such layers are denser, which has the benefit of encouraging the more reactive atomic hydrogen species to diffuse rather than form molecules. The ideal density to encourage atomic hydrogen release has been found with refractive indices 2.2–2.4, fortuitously aligned with ideal anti-reflection properties for the light-receiving cell surface (Bredemeier et al. 2019).

### 10.2.3 Hydrogen Plasma

Similar to FGA, the sample is annealed in a hydrogen ambient; however, in this case made up of hydrogen atoms broken down from a hydrogen-containing gas by a plasma source. Typically, a remote plasma is used to prevent additional surface damage. Although FGA is still used in some modern high-efficiency structures, hydrogen plasma treatments are becoming more common in record-efficiency cells. Both HJT and TOPCon cells, which rely on excellent surface passivation, have been shown to benefit from remote plasma hydrogen exposure to improve interfaces (Descoeudres et al. 2011; Glunz et al. 2015).

### 10.2.4 Bulk Silicon Hydrogen Passivation

During the thermal annealing or activation steps, some hydrogen can diffuse into the bulk of the silicon. Predicting the in-diffusion of hydrogen is not straightforward as there is, to date, no consensus on the diffusivity of hydrogen in silicon with widely varying reports (Hallam et al. 2020). Due to the difficulty of directly detecting hydrogen, reports of hydrogen behavior in silicon are based on the correlation of observed characteristics with theory and/or fundamental studies. Many factors affect the diffusivity, including the fractions of charge states or forms of hydrogen, dopants, defects, electric fields, any barrier or capping layers that affect in and out-diffusion. The diffusivity reported by (Van Wieringen and Warmoltz 1956) at high temperatures in defect-free intrinsic silicon is thought to be the upper limit, given by:

$$D = 9.4 \times 10^{-3} \cdot \exp\left(-\frac{11,000}{RT}\right) \text{cm}^2/\text{s}$$

At temperatures over 700 °C, as reached during co-firing process, hydrogen can diffuse throughout the entire thickness of a typical 150–200 µm thick wafer in ~one second. The benefit of this is particularly evident in multi-crystalline wafers. Unpassivated grain boundaries act like high recombination surfaces or interfaces and are readily passivated by hydrogen. After a firing process with a hydrogen-containing dielectric, the electrical activity of most grain boundaries can be entirely neutralized and passivated by the hydrogen (Hallam et al. 2020). For hydrogen to diffuse throughout the bulk of an HJT cell limited to <250 °C, hours-long processes may be required (Chen et al. 2019).

## 10.3 Advanced Hydrogenation

### 10.3.1 Background

It was long assumed that if enough hydrogen were available, then all defects that were capable of being passivated by hydrogen would be effectively passivated. However, not all defects are as effectively passivated as surfaces and grain boundaries by the typical methods discussed above. The hydrogen may not necessarily be in the right location or form needed for passivation. Depending on prior processes, the hydrogen may be trapped or stored in many different forms, including molecules, impurity complexes, bound to dopants or crystallographic defects, or trapped in surface layers. Reactive and mobile interstitial atomic hydrogen, as required for passivation reactions, is likely to exist only in small concentrations under equilibrium conditions. This is the reason that thermal processes are required in conjunction with a hydrogen source in the typical methods above: to free atomic hydrogen and allow it to migrate to and passivate defects. However, generating atomic hydrogen alone is not enough to guarantee the effective passivation of some defects.

The three different possible charge states of atomic hydrogen ($H^+$, $H^0$, and $H^-$) have very different behavior (Herring et al. 2001). The importance of the charge state of atomic hydrogen for defect passivation in solar cells was recognized in the last decade and led to the development of "advanced hydrogenation" processes to manipulate the hydrogen charge states (Wenham et al. 2013). Such processes are particularly important for improving the electrical quality of solar cells made using lower-quality wafers with a wide range of possible defects throughout the bulk of the silicon, notably boron–oxygen (BO) defects.

### 10.3.2 Atomic Hydrogen Behavior in Silicon

The mobility and reactivity of atomic hydrogen in silicon are heavily dependent on its charge state. Although there has been no consensus on the diffusivity of hydrogen in silicon, neutral hydrogen ($H^0$), which is unaffected by charge, is thought to have the highest diffusivity by up to several orders of magnitude (Hallam et al. 2020). Hydrogen's ability to bond with and passivate a given defect is also heavily dependent on their attraction and reactivity to each other. For example, if both the hydrogen and the defect are positively charged, they are unlikely to attract, and with missing electrons, they are unlikely to bond together. It is, therefore, likely to be beneficial to maximize concentrations of $H^0$, which has higher diffusivity and can bond by either giving or taking an electron.

However, interstitial atomic hydrogen is a negative-U impurity. This means that the $H^+$ donor energy level ($E_D = E_C - 0.16\,\text{eV}$) sits in the silicon band gap above the $H^-$ acceptor level ($E_A = E_V + 0.48\,\text{eV}$), making $H^0$ the highest energy state and never thermodynamically stable (Herring et al. 2001). The fractions of the respective charge states are determined by the position of the Fermi level ($E_F$). In p-type silicon with an abundance of holes, $H^+$ dominates. Conversely, $H^-$ dominates in n-type silicon with an abundance of electrons. $H^0$ only reaches significant concentrations around the transition energy level ($E_{H+/-}$), where $E_F$ is halfway between $E_D$ and $E_A$. At this point, there are also roughly equal concentrations of $H^+$ and $H^-$. This would be an ideal condition for being able to passivate defects of any charge state. This corresponds to an $E_F$ slightly above midgap and would occur in $\sim 2000\,\Omega\,\text{cm}$ n-type material. Since this is not a realistic doping level for most solar cell structures, a combination of temperature and carrier injection can be used to shift $E_F$ closer to the mid-gap.

### 10.3.3 The Impact of Temperature and Carrier Injection

Increasing the temperature and carrier injection not only alters the fractions of various charge states of atomic hydrogen, but it also increases diffusivity and can free up loosely bound hydrogen (in molecules or bound to dopants), thereby increasing the concentrations of atomic hydrogen.

Figure 10.1 shows how the fractions of atomic hydrogen charge states that exist in $1\,\Omega\,\text{cm}$ p- and n-type silicon at different (a) temperatures and (b) injection levels. Using temperature alone, $H^0$ concentrations can be increased significantly, especially around the temperatures used for screen-printed contact co-firing. However, there are always competing passivation and depassivation reactions. The fast cool from high-temperature firing processes or even lower temperature long processes in the dark are not ideal for passivation; bonds are more readily broken with the increased temperatures without the conditions needed for repassivation during cooling.

Using carrier injection at a moderate temp of $200\,°\text{C}$ (suitable even for HJT structures) the $H^0$ concentrations increase significantly above $10^{15}\,\text{cm}^{-3}$. Carriers can be generated using

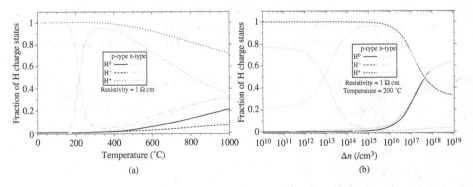

**Figure 10.1** Interstitial hydrogen charge state fractions (−/0/+) in $1\,\Omega\,\text{cm}$ p-type and n-type silicon under varied (a) temperature (b) carrier injection at $200\,°\text{C}$. Source: Figure created using model from (Hallam et al. 2020) based on (Sun et al. 2015a,b).

current injection or illumination. The illumination method can achieve higher carrier densities due to the ability to be performed in open circuit conditions. In contrast, the current injection method will be subject to significant series of resistance losses, which may result in non-uniform carrier injection. Another benefit of high-intensity illumination is that it can reduce the impact of spatial lifetime non-uniformities, which is particularly important for passivating wafers limited by non-uniformly distributed defects. In heavier doped or lower lifetime material, increased current injection or illumination is required to achieve the same carrier densities.

Since reactions are largely reversible, the highest temperature and last thermal treatment experienced will greatly impact the hydrogen availability and states within the silicon. To maintain effective passivation, a device should never be heated to temperatures in the vicinity of that used for a prior advanced hydrogenation process without using similar charge state control. In addition, too extensive thermal treatments can deplete the hydrogen sources.

### 10.3.4 The Benefit of Charge State Control

Figure 10.2 provides a good example to demonstrate the benefits of hydrogen charge state control for the passivation of some defects. The image is a photoluminescence (PL) image of a p-type upgraded metallurgical grade (UMG) Cz wafer prone to larger concentrations of impurities. The entire wafer had been coated with a hydrogenated ($SiO_xN_y$:H) layer and fired to reach 700 °C. After this typical hydrogenation process, the wafer was annealed at 250 °C, and the center was exposed to a 2 cm diameter laser spot of ~135 suns equivalent 938 nm photons for 60 seconds. The region that was illuminated to encourage the generation of minority charge states $H^0$ and $H^-$ shows more than four times as many PL counts as the surrounding regions, equivalent to more than 36 mV increase in expected device voltage. The regions annealed without charge state control still exhibit low PL counts and poor lifetime limited by unpassivated bulk defects despite having also been through a typical hydrogenation process.

Significant gains can be achieved from the application of an additional process to manipulate hydrogen (after a typical process) and can be applied to finished cells. (Hallam et al. 2017) showed 1.1%$_{abs}$ average efficiency gain on p-type Cz PERCs from 8 manufacturers. A similar process applied to p-type PERC cast-mono cells from Canadian Solar increased by 0.5%$_{abs}$. HJT cells on industrial p-type wafers showed 1.0–1.1%$_{abs}$ gain for boron doped

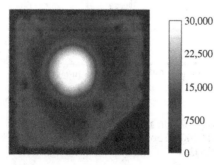

30,000

22,500

15,000

7500

0

**Figure 10.2** PL image of UMG silicon annealed after a typical hydrogenation process, the center spot was illuminated by a 938 nm laser ~135 suns equivalent for 60 seconds for local hydrogen charge state control.

wafers (Chen et al. 2019 and 0.3%$_{abs}$ for Ga doped with additional processes (Vicari Stefani et al. 2021).

N-type cells can also benefit, with a 0.64%$_{abs}$ gain observed on Jinko Solar's "HOT" TOPCon cells (Chen et al. 2021), 0.3%$_{abs}$ improvement seen in ECN's N-PERT cells (Coletti et al. 2014) and a 0.7%$_{abs}$ boost on HJT from CIE (Wright et al. 2019).

### 10.3.5 Industrial Advanced Hydrogenation Processes

There are several industrially suitable approaches to incorporate advanced hydrogenation processes into the production sequence. Many large-scale production tools exist based on various methods:

- LED (or other illumination source) built into the cooling region of a belt furnace, e.g. Schmid, Despatch Industries and Centrotherm (Derricks et al. 2019; Hallam et al. 2016; Pernau et al. 2015).
- A separate laser or LED-based anneal using a hotplate or heated ambient in an oven or furnace for the annealing, e.g. Asia Neo Tech and DR laser (Hallam et al. 2017).
- Current injection during annealing. Although slower due to the lower carrier densities possible, this approach is commonly used by industry as large batches of cells can be "coin-stacked" for processing (Herguth and Hahn 2013).
- Current injection (or more challenging illumination) can be incorporated into the laminator for charge state control while the cells are heated during module encapsulation (Lee et al. 2015).

## 10.4 Hydrogenation in Record Solar Cells

Table 10.1 shows how hydrogenation treatments have been incorporated into the current record solar cells.

## 10.5 Passivation of Specific Defects

Different types of wafers are prone to different defects. Boron-doped multi-crystalline wafers dominated the commercial market for decades. Such wafers have high densities of crystallographic defects and impurities. Cast-mono wafer technology was developed to reduce the crystallographic defects in cast ingots. The cast-mono (or quasi-mono) silicon ingots are grown from seeds to create an almost mono-crystalline silicon ingot in a low-cost casting process. However, cast-mono wafers are commonly affected by harmful stress dislocations and impurities, and as a result, this technology has never significantly taken off. In recent years, due to advances in Recharge Cz (RCz) or Continuous Cz (CCz)-pulling, boron-doped Cz wafers became cost-competitive and became preferred to multi. However, this growth method incorporates large concentrations of oxygen, which reacts with the boron under light to form harmful BO defects. In the last year, gallium-doped wafers, which are not subject to BO defects, have rapidly taken over and now totally dominate the

**Table 10.1**  Hydrogenation in record silicon cells (Green et al. 2017).

| Cell type | H source | Treatment | Efficiency | References |
|---|---|---|---|---|
| SHJ IBC (n-type) Kaneka | a-Si:H | Annealing <260 °C during deposition of a:Si and 2x dielectric layers | 26.7% | Yoshikawa et al. (2017) |
| POLO-IBC (p-type) ISFH | SiN$_x$:H | 30 min; 425 °C | 26.1% | Haase et al. (2018) |
| TOPCon (n-type) FhG-ISE | SiN$_x$:H | 1 h; 800 °C. Method referred to Feldmann et al. (2014), which added 30 min, 400 °C hydrogen plasma, not specified if used again | 25.8% | Richter et al. (2017) |
| SHJ (n-type) Hanergy | a-Si:H (10:1 H$_2$/SiH$_4$) | Not mentioned. Anneal, during deposition, most likely | 25.1% | Zhang et al. (2017) |
| PERC (p-type) LONGi | SiN$_x$:H | Fire at 810–830 °C. Then 4 s 220 °C laser process | 24.0% | Chen et al. (2020) |
| PERC (p-type multi) Canadian Solar | SiN$_x$:H | Firing. Then "advanced regeneration process" | 22.8% | Wu et al. (2019) |

Source: Adapted from Green et al. (2017).

commercial market since the patent by Shin Etsu Chemical expired in 2020 (Abe et al. 2000). We are also seeing small but significant increases in production volume of n-type wafers. The carrier lifetime in n-type wafers is less sensitive to common impurities (Fe, W, Cr, Ti, V, …), which are usually donor-like impurities. Therefore, n-type wafers are generally more suited for high-efficiency structures that require long minority carrier lifetimes. The n-type share of the market is expected to reach 50% by 2030 (ITRPV 2023).

### 10.5.1  Surfaces and Grain Boundaries

Surfaces (or interfaces with surface layers) are an unavoidable abrupt end to the silicon lattice. This leaves a plane of dangling bonds with a high density of trap states and an almost continuous distribution of energy levels across the band gap. Grain boundaries in multi-crystalline silicon have a similar lattice discontinuity. The properties of a grain boundary can vary depending on character, angle, and commination, which also affect its ability to be passivated (Chen et al. 2007). However, as discussed above, most are generally well-passivated by typical hydrogenation processes. Grain boundaries can also provide networks to accelerate hydrogen diffusion through a wafer (Dubé et al. 1984)

### 10.5.2  Dislocations

Dislocations of either line or screw types are also crystallographic defects but are not as readily passivated by typical hydrogenation methods. They form due to stresses from impurities or crystal growth. These are common in multicrystalline wafers from cast ingots; the larger the grain size, the greater the stress and the more prone to this type of defect. Cast-mono is an extreme example where the top of an ingot usually exhibits a high density of harmful dislocations. After a typical hydrogenation process, such defects have been shown to benefit from additional moderate temperature annealing processes both with and without charge state control (Wenham et al. 2017). Even after significant reductions in recombination around the dislocations, these defects still appear to be strongly recombination active. Dislocations can provide even faster paths for hydrogen diffusion through a silicon bulk than grain boundaries (Dube and Hanoka 1984).

### 10.5.3 Vacancies

Vacancies are missing silicon atoms in the crystal lattice. The ingot growth rate largely determines their prevalence. Like surface dangling bonds, they readily interact and bond with hydrogen. Not all vacancy hydrogen complexes are beneficial; a single vacancy can stably bond with 1, 2, 3, or 4 hydrogen atoms; however, any less than 4 and it remains recombination active (Myers et al. 1992). Vacancies can split $H_2$ molecules, and as a result, surface damage can increase the influx of atomic hydrogen (Sopori et al. 1996).

### 10.5.4 Process Induced Defects

Several defects can arise due to processing conditions. One example is structural defects created by laser damage during doping or ablation processes. These have been shown to be passivated by hydrogenation processes (Urueña et al. 2016). Another example is rings of oxygen precipitates that can form in Cz wafers during thermal processes and can cause a 4% abs efficiency loss. Advanced hydrogenation processes can completely remove their electrical recombination activity (Hallam et al. 2015)

### 10.5.5 Boron–Oxygen (BO) Defects

This defect forms from boron and oxygen impurities in the wafer when exposed to light (or excess carriers) (Glunz et al. 2001). Boron-doped Cz wafers, which are high in both boron and oxygen, are especially prone to such light-induced degradation (LID). After several years of solar cells plagued by LID and the introduction of hydrogen-rich $SiN_x$ to solar cells, a stable regenerated defect state was found to exist (Herguth et al. 2008; Münzer 2009). Like hydrogen, BO defects have negative-U behavior; with a donor level ($E_D = E_C - 0.41\,eV$) above an acceptor level ($E_A = E_V + 0.26\,eV$) (Niewelt et al. 2015), the neutral defect is unstable, and the dominant charge state is determined by the Fermi-level. In p-type silicon, where BO defects are prevalent, both BO defects and hydrogen are in the positive charge state, missing electrons. Charge state control can, therefore, be very effective at passivating these defects. Both current injection and illumination-based charge state control processes have been shown to be commercially suitable methods to effectively passivate BO defects (Hallam et al. 2017; Herguth and Hahn 2013).

### 10.5.6 Transition Metals

Transition metals (TMs) are common in solar-grade silicon and can be harmful to both p- and n-type wafers, particularly in interstitial form (Macdonald and Geerligs 2004). Iron, cobalt and chromium are most detrimental for p-type, while cobalt, chromium and nickel are worst for n-type (Schmidt et al. 2013). Interstitial TMs tend to be in the positive charge state in p-type silicon and are not likely, therefore, to attract and bond easily with the also predominantly positively charged hydrogen (Mullins et al. 2017). Therefore, advanced hydrogenation techniques could be very beneficial for the passivation of such impurities. The large benefit seen in the UMG sample of Figure 10.2 may be attributed to some passivation of TMs (Kobayashi et al. 2016).

The benefit of hydrogen passivation for TMs is not entirely clear since many of these impurities are also gettered by thermal processes. Iron (Fe) is a good example of this and a particularly well-known minority carrier trap in p-type silicon. Iron readily bonds to shallow p-type dopants. Fe-B pairs are dissociated quickly under light, freeing the high recombination active interstitial Fe and causing LID. Recently, it was shown that Ga-B pairs also exhibit recombination activity (Nærland et al. 2017). There are mixed reports on the ability of hydrogen to passivate iron. Thermal processes are shown to remove electrically active iron, with many studies attributing it to hydrogen passivation. However, more recent studies have questioned this, showing that Fe-H is not overly stable (Mullins et al. 2017) and suggesting that, instead, iron is gettered to the surface (Li et al. 2016).

A combination of gettering and hydrogenation processes has been shown to provide added benefit on top of either process alone, indicating a complementary effect (Sheoran et al. 2008). A $20\,\mu s$ lifetime Cz wafer with this combination enabled a device voltage of $>700\,mV$ (Chen et al. 2019).

## 10.6 Negative Effects of Hydrogen

Unfortunately, along with all the huge benefits that hydrogen can provide, it can also impact solar devices in negative ways. Large concentrations of hydrogen can break silicon–silicon bonds, passivate dangling bonds and form clusters of $H_2$ molecules between the lattice planes, causing extended defects known as platelets. The behavior of atomic hydrogen means that it can neutralize the background dopants required for the cell's electrical properties. It naturally takes the charge state that opposes the prevailing background doping, e.g. giving an electron and acting as a donor ($H^-$) in p-type, readily bonding to or neutralizing the dopants. Hydrogen can form recombination active complexes when it bonds with other species, including carbon, oxygen, carbon and oxygen combined, transition metals, and vacancies. Hydrogen has been implicated in causing degradation under field conditions known as light- and elevated temperature-induced degradation (LeTID) (Chen et al. 2020). However, it has recently been proposed that this defect is also passivated by hydrogen (Sharma et al. 2019). It appears that the forms and states of hydrogen are particularly important in LeTID, with very strong behavioral changes following thermal processes. Hydrogen that migrates to the surface of the solar cell during processing can also build up at surfaces or under contacts, causing degradation of surface passivation or high series resistance (Hamer et al. 2018).

## 10.7  Conclusion

Hydrogen is ubiquitous. It is essentially impossible to avoid and fortuitously can be extremely beneficial for passivating defects in silicon. Some form of hydrogenation has been incorporated into almost every solar cell ever made. Typically, processes such as FGAs, "alneals," hydrogen plasmas and hydrogenated dielectric layers are applied in conjunction with a thermal treatment with a focus on passivating defect sites at the silicon surface interface. All of the current record cells and lowest recombination surfaces incorporate some form of hydrogen passivation.

"Advanced hydrogen" processes developed in recent years involve additional thermal processes with carrier injection to control the charge state of the hydrogen. Such processes are particularly important for certain defects, such as BO complexes and lower-quality wafers with a range of defects with different charge states.

## Acknowledgments

I would like to thank Dr. Moonyong Kim for his help in creating the figures. I also acknowledge my father, Professor Stuart Wenham, who instigated and led a lot of the advanced hydrogenation developments in the final years of his life and who would have loved to author this chapter. The advanced hydrogenation work was supported by the Australian Government through the Australian Renewable Energy Agency (ARENA). Responsibility for the views, information, or advice expressed herein is not accepted by the Australian Government.

## Author Biography

**Alison Ciesla** is a Senior Scientia lecturer in the School of Photovoltaics and Renewable Energy Engineering at UNSW Sydney. She completed her PhD in Photovoltaics in 2017, focusing on laser doping and hydrogenation to reduce recombination and mitigate light-induced degradation. Since then, she has been developing new techniques to manipulate hydrogen in silicon and investigating the degradation and reliability of commercial silicon modules. She is an inventor on 10 related patents and has published over 70 research papers. The current focus of Alison's research is to maximize the long-term power output of commercial silicon solar cells and modules through enhanced efficiencies and improved reliability.

## References

Abe, Y., Hirasawa, T., Tokunaga, K. et al. (2000). Silicon single crystal and wafer doped with gallium and method for producing them. Patent No. US6815605B1.

Balk, P. et al. (1965). Effects of hydrogen annealing on silicon surfaces. *Electrochemical Society Spring Meeting* 14 (1): 237–240.

Blakers, A.W., Wang, A., Milne, A.M. et al. (1989). 22.8% efficient silicon solar cell. *Applied Physics Letters* 55 (13): 1363–1365.

Bonilla, R.S., Hoex, B., Hamer, P., and Wilshaw, P.R. (2017). Dielectric surface passivation for silicon solar cells: a review. *Physica Status Solidi A: Applications and Material Science* 214 (7): https://doi.org/10.1002/pssa.201700293.

Bredemeier, D., Walter, D.C., Heller, R., and Schmidt, J. (2019). Impact of hydrogen-rich silicon nitride material properties on light-induced lifetime degradation in multicrystalline silicon. *Physica Status Solidi RRL: Rapid Research Letters* 1900201. https://doi.org/10.1002/pssr .201900201.

Chen, J., Sekiguchi, T., and Yang, D. (2007). Electron-beam-induced current study of grain boundaries in multicrystalline Si. *Physica Status Solidi C: Current Topics in Solid State Physics* 4 (8): 2908–2917. https://doi.org/10.1002/pssc.200675435.

Chen, D., Kim, M., Shi, J. et al. (2019). Defect engineering of p-type silicon heterojunction solar cells fabricated using commercial-grade low-lifetime silicon wafers. *Progress in Photovoltaics: Research and Applications* 29 (11): 1165–1179. https://doi.org/10.1002/pip.3230.

Chen, D., Vaqueiro Contreras, M., Ciesla, A. et al. (2020). Progress in the understanding of light- and elevated temperature-induced degradation in silicon solar cells: a review. *Progress in Photovoltaics: Research and Applications* 29 (11): 1180–1201. https://doi.org/10 .1002/pip.3362.

Chen, R., Tong, H., Zhu, H. et al. (2020). 23.83% efficient mono-PERC incorporating advanced hydrogenation. *Progress in Photovoltaics: Research and Applications* 28 (12): 1239–1247. https://doi.org/10.1002/pip.3243.

Chen, R., Wright, M., Chen, D. et al. (2021). 24.58% efficient commercial n-type silicon solar cells with hydrogenation. *Prog Photovolt Res Appl.* 29: 1213–1218. https://doi.org/10.1002/ pip.3464.

Coletti, G., Manshanden, P., Bernardini, S. et al. (2014). Removing the effect of striations in n-type silicon solar cells. *Solar Energy Materials and Solar Cells* 130: 647–651. https://doi .org/10.1016/j.solmat.2014.06.016.

Deal, B.E., MacKenna, E.L., and Castro, P.L. (1969). Characteristics of fast surface states associated with SiO2-Si and Si3 N 4-SiO2-Si structures. *Journal of the Electrochemical Society* 116 (7): 997–1005.

Derricks, C., Herguth, A., Hahn, G. et al. (2019). Industrially applicable mitigation of BO-LID in Cz-Si PERC-type solar cells within a coupled fast firing and halogen lamp based belt-line regenerator – a parameter study. *Solar Energy Materials and Solar Cells* 195: 358–366. https://doi.org/10.1016/j.solmat.2019.03.020.

Descoeudres, A., Barraud, L., De Wolf, S. et al. (2011). Improved amorphous/crystalline silicon interface passivation by hydrogen plasma treatment. *Applied Physics Letters* 99 (12): 1–4. https://doi.org/10.1063/1.3641899.

Dube, C. and Hanoka, J.I. (1984). Hydrogen passivation of dislocations in silicon. *Applied Physics Letters* 45 (10): 1135–1137.

Dubé, C., Hanoka, J.I., and Sandstrom, D.B. (1984). Hydrogen diffusion along passivated grain boundaries in silicon ribbon. *Applied Physics Letters* 44 (4): 425–427. https://doi.org/10.1063/ 1.94797.

Feldmann, F., Simon, M., Bivour, M. et al. (2014). Efficient carrier-selective *p*-and *n*-contacts for Si solar cells. *Solar Energy Materials and Solar Cells* 131: 100–104. https://doi.org/10.1016/j.solmat.2014.05.039.

Glunz, S.W., Rein, S., Warta, W. et al. (2001). Degradation of carrier lifetime in Cz silicon solar cells. *Solar Energy Materials and Solar Cells* 65 (1): 219–229.

Glunz, S.W., Feldmann, F., Richter, A. et al. (2015). The irresistible charm of a simple current flow pattern-25% with a solar cell featuring a full-area back contact. In: *Proceedings of the 31st European Photovoltaic Solar Energy Conference*, 259–263.

Green, M.A., Hishikawa, Y., Warta, W. et al. (2017). Solar cell efficiency tables (version 50). *Progress in Photovoltaics: Research and Applications* 25 (7): 668–676. https://doi.org/10.1002/pip.2909.

Haase, F., Hollemann, C., Schäfer, S. et al. (2018). Laser contact openings for local poly-Si-metal contacts enabling 26.1%-efficient POLO-IBC solar cells. *Solar Energy Materials and Solar Cells* 186: 184–193. https://doi.org/10.1016/j.solmat.2018.06.020.

Hallam, B.J., Chan, C.E., Abbott, M.D., and Wenham, S.R. (2015). Hydrogen passivation of defect-rich n-type Czochralski silicon and oxygen precipitates. *Solar Energy Materials and Solar Cells* 141: 125–131. https://doi.org/10.1016/j.solmat.2015.05.009.

Hallam, B., Chan, C., Payne, D.N.R. et al. (2016). Techniques for mitigating light-induced degradation (LID) in commercial silicon solar cells. *Photovoltaics International* 33: 37–46.

Hallam, B.J., Chan, C.E., Chen, R. et al. (2017). Rapid mitigation of carrier-induced degradation in commercial silicon solar cells. *Japanese Journal of Applied Physics* 56: 08MB13.

Hallam, B.J., Hamer, P.G., Ciesla née Wenham, A.M. et al. (2020). Development of advanced hydrogenation processes for silicon solar cells via an improved understanding of the behaviour of hydrogen in silicon. *Progress in Photovoltaics: Research and Applications* 28 (12): 1217–1238. https://doi.org/10.1002/pip.3240.

Hamer, P., Chan, C., Bonilla, R.S. et al. (2018). Hydrogen induced contact resistance in PERC solar cells. *Solar Energy Materials and Solar Cells* 184: https://doi.org/10.1016/j.solmat.2018.04.036.

Herguth, A. and Hahn, G. (2013). Towards a high throughput solution for boron-oxygen related regeneration. In: *28th European Photovoltaic Solar Energy Conference*, 1507–1511. https://doi.org/10.4229/28thEUPVSEC2013-2BV.3.55.

Herguth, A., Schubert, G., Käs, M., and Hahn, G. (2008). Investigations on the long time behavior of the metastable boron-oxygen complex in crystalline silicon. *Progress in Photovoltaics: Research and Applications* 16 (2): 135–140.

Herring, C., Johnson, N.M., and de Walle, C.G. (2001). Energy levels of isolated interstitial hydrogen in silicon. *Physical Review B* 64 (12): 125209.

ITRPV (2023). International Technology Roadmap for Photovoltaic-2020 Results. https://itrpv.vdma.org/documents/27094228/29066965/20210ITRPV/08ccda3a-585e-6a58-6afa-6c20e436cf41 (accessed 20 February 2024).

Lee, K., Kim, M.-S., Lim, J.-K. et al. (2015). Natural recovery from LID: regeneration under field conditions? In: *31st European Photovoltaic Solar Energy Conference*, 1837.

Li, Z., Yang, Y., Zhang, X. et al. (2016). Pilot production of 6 IBC solar cells yielding a median efficiency of 23 % with a low-cost industrial process. In: *Proceedings of the 32nd European Photovoltaic Solar Energy Conference*, 571–574.

Macdonald, D. and Geerligs, L.J. (2004). Recombination activity of interstitial iron and other transition metal point defects in $p$-and $n$-type crystalline silicon. *Applied Physics Letters* 85 (18): 4061–4063.

Mullins, J., Leonard, S., Markevich, V.P. et al. (2017). Recombination via transition metals in solar silicon: the significance of hydrogen–metal reactions and lattice sites of metal atoms. *Physica Status Solidi A: Applications and Material Science* 214 (7): https://doi.org/10.1002/pssa.201700304.

Münzer, K. (2009). Hydrogenated silicon nitride for regeneration of light induced degradation. In: *Proceedings of the 24th European Photovoltaic Solar Energy Conference, Hamburg*, 1558–1561.

Myers, S.M., Baskes, M.I., Birnbaum, H.K. et al. (1992). Hydrogen interactions with defects in crystalline solids. *Reviews of Modern Physics* 64 (2): 559–617. https://doi.org/10.1103/RevModPhys.64.559.

Nærland, T.U., Bernardini, S., Haug, H. et al. (2017). On the recombination centers of iron-gallium pairs in Ga-doped silicon. *Journal of Applied Physics* 122 (8): 85703. https://doi.org/10.1063/1.5000358.

Niewelt, T., Schön, J., Broisch, J. et al. (2015). Electrical characterization of the slow boron oxygen defect component in Czochralski silicon. *Physica Status Solidi RRL: Rapid Research Letters* 9 (12): 692–696.

Pernau, T., Romer, O., Scheiffele, W. et al. (2015). Rather high speed regeneration of BO-defects: regeneration experiments with large cell batches. In: *31st European Photovoltaic Solar Energy Conference*, vol. 1, 918–920.

Richter, A., Benick, J., Feldmann, F. et al. (2017). n-Type Si solar cells with passivating electron contact: identifying sources for efficiency limitations by wafer thickness and resistivity variation. *Solar Energy Materials and Solar Cells* 173: 96–105. https://doi.org/10.1016/j.solmat.2017.05.042.

Schmidt, J., Lim, B., Walter, D. et al. (2013). Impurity-related limitations of next-generation industrial silicon solar cells. *IEEE Journal of Photovoltaics* 3 (1): 114–118.

Sharma, R., Chong, A.P., Li, J.B. et al. (2019). Role of post-metallization anneal sequence and forming gas anneal to mitigate light and elevated temperature induced degradation of multicrystalline silicon solar cells. *Solar Energy Materials and Solar Cells* 195: 160–167. https://doi.org/10.1016/j.solmat.2019.02.036.

Sheoran, M., Upadhyaya, A., and Rohatgi, A. (2008). Bulk lifetime and efficiency enhancement due to gettering and hydrogenation of defects during cast multicrystalline silicon solar cell fabrication. *Solid-State Electronics* 52 (5): 612–617.

Sopori, B.L., Deng, X., Benner, J.P. et al. (1996). Hydrogen in silicon: a discussion of diffusion and passivation mechanisms. *Solar Energy Materials and Solar Cells* 42: 159–169.

Sun, C., Liu, A.Y., Phang, S.P. et al. (2015a). Charge states of the reactants in the hydrogen passivation of interstitial iron in P-type crystalline silicon. *Journal of Applied Physics* 118: 085709. https://doi.org/10.1063/1.4929757.

Sun, C., Rougieux, F.E., and Macdonald, D. (2015b). A unified approach to modelling the charge state of monatomic hydrogen and other defects in crystalline silicon. *Journal of Applied Physics* 117 (4): 45702.

Swanson, R.M. (1990). Method of fabricating back surface point contact solar cells. US Patent 4,927,770.

Urueña, A., Aleman, M., Cornagliotti, E. et al. (2016). Progress on large area n-type silicon solar cells with front laser doping and a rear emitter. *Progress in Photovoltaics: Research and Applications* 24: 1149–1156. https://doi.org/10.1002/pip.

Van Wieringen, A. and Warmoltz, N. (1956). On the permeation of hydrogen and helium in single crystal silicon and germanium at elevated temperatures. *Physica* 22 (6): 849–865.

Varshney, U., Chan, C., Hoex, B. et al. (2020). Controlling light-and elevated-temperature-induced degradation with thin film barrier layers. *IEEE Journal of Photovoltaics* 10 (1): 19–27. https://doi.org/10.1109/JPHOTOV.2019.2945199.

Verlinden, P., Lafontaine, B., Jacquemin, P. et al. (1990). Super self-aligned technology for backside contact solar cells: a route to low cost and high efficiency. In: *Conference Record of the IEEE Photovoltaic Specialists Conference*, vol. 1, 257–262.

Vicari Stefani, B., Kim, M., Wright, M. et al. (2021). Stability study of silicon heterojunction solar cells fabricated with gallium-and boron-doped silicon wafers. *Solar RRL* 5 (9): 2100406. https://doi.org/10.1002/solr.202100406.

Wenham, S. R., Hamer, P. G., Hallam, B. J., Sugianto, A., Chan, C. E., Song, L., Lu, P. H., Wenham, A. M., Mai, L., Chong, C. M., Xu, G. X. & Edwards, M. B. (2013). *Advanced Hydrogenation of Silicon Solar Cells*. WO2013173867 A1.

Wenham, A., Song, L., Abbott, M. et al. (2017). Defect passivation on cast-mono crystalline screen-printed cells. *Frontiers in Energy* 11 (1): 60–66. https://doi.org/10.1007/s11708-016-0443-5.

Wright, M., Kim, M., Dexiang, P. et al. (2019). Multifunctional process to improve surface passivation and carrier transport in industrial n-type silicon heterojunction solar cells by 0.7% absolute. In: *15th International Conference on Concentrator Photovoltaic Systems (CPV-15)*, vol. 2149, 110006. https://doi.org/10.1063/1.5123882.

Wu, J., Wu, H., Chen, X. et al. (2019). 21.4% efficiency bifacial multi-Si PERC cells and 410W modules. In: *Conference Record of the IEEE Photovoltaic Specialists Conference*, vol. 199, 1466–1470. https://doi.org/10.1109/PVSC40753.2019.8980864.

Yoshikawa, K., Kawasaki, H., Yoshida, W. et al. (2017). Silicon heterojunction solar cell with interdigitated back contacts for a photoconversion efficiency over 26%. *Nature Energy* 2 (5): https://doi.org/10.1038/nenergy.2017.32.

Zhang, Y., Yu, C., Yang, M. et al. (2017). Significant improvement of passivation performance by two-step preparation of amorphous silicon passivation layers in silicon heterojunction solar cells. *Chinese Physics Letters* 34 (3): https://doi.org/10.1088/0256-307X/34/3/038101.

Zhao, J., Wang, A., Altermatt, P.P. et al. (1996). 24% efficient PERL silicon solar cell: recent improvements in high efficiency silicon cell research. *Solar Energy Materials and Solar Cells* 41: 87–99.

**Part Four**

**Perovskite Solar Cells**

# 11

# Perovskite Solar Cells

*Anita Ho-Baillie[1,2,3], Md Arafat Mahmud[1,2,3], and Jianghui Zheng[1,2,3]*

[1] School of Physics, The University of Sydney, Sydney, NSW, Australia
[2] Sydney Nano, The University of Sydney, Sydney, NSW, Australia
[3] Australian Centre for Advanced Photovoltaics (ACAP), School of Photovoltaic and Renewable Energy Engineering, University of New South Wales, Sydney, NSW, Australia

## 11.1 Introduction

Solar photovoltaics have become one of the cheapest means of electricity production and are projected to account for >10% of global electricity generation by the year 2030. The manufacturing cost of the incumbent silicon solar cell has reduced by 85% in the last decade alone. While further manufacturing cost reduction is possible, one effective way of reducing the levelized cost of energy (LCOE), which is the average net present cost of electricity generation for a generator over its lifetime, is by increasing its power conversion efficiency. This can be done by developing more efficient single junction solar cells which have a theoretical efficiency limit of ~34% (Ho-Baillie et al. 2021), or by "tandeming" cells with different bandgaps, thereby "sectioning" the solar spectrum for more efficient energy conversion. This provides larger room for cost reduction as theoretical efficiency limits for double junction tandems and triple junction tandems are ~45% and ~51%, respectively (Ho-Baillie et al. 2021), especially if the extra cost of manufacturing the additional junction(s) in a tandem is compensated by the increase in performance (Chang et al. 2021) and the extra cost of moduling/paneling is marginal, which will effectively reduce the cost in terms of $/W and LCOE. However, the rate of efficiency increase slows beyond three junctions (Bremner et al. 2016) with manufacturing complexity increasing.

One of the many material systems that can potentially serve as the next generation of low-cost single-junction, or high-performance multi-junction solar cell technology, is metal halide perovskites (Ho-Baillie et al. 2021; Green et al. 2014; Green and Ho-Baillie 2017; Reinders et al. 2017). Some key attributes of metal halide perovskite include ease of fabrication via low-temperature solution process, high absorption coefficient, high defect tolerance, high specific power (power to weight ratio, as high as 29.4 W/g) (Hu et al. 2021), tunable bandgap, and compatibility with flexible substrates, presenting great potentials for a wide range of fixed or portable applications in the form, standalone, or integrated designs for land, aerospace, or space deployment.

*Photovoltaic Solar Energy: From Fundamentals to Applications, Volume 2*, First Edition.
Edited by Wilfried van Sark, Bram Hoex, Angèle Reinders, Pierre Verlinden, and Nicholas J. Ekins-Daukes.
© 2024 John Wiley & Sons Ltd. Published 2024 by John Wiley & Sons Ltd.
Companion website: www.wiley.com/go/PVsolarenergy

## 11.2   Metal Halide Perovskites

Perovskite is a class of organic–inorganic semiconductors having the chemical formula of $ABX_3$ (Figure 11.1a). For typical perovskite solar cells, A is a cation, which can be organic [methyl ammonium: (MA; $CH_3NH_3$), formamidinium: (FA; $NH_2CH=NH_2$), ethylammonium (EA; $CH_3CH_2NH_3$)] or inorganic (Cs, Rb). B is a divalent metal (Pb, Sn) ion and X is a halogen (I, Br, Cl) anion (Mahmud et al. 2021). For three-dimensional (3D) perovskites (Figure 11.1a), "formability" depends on its tolerance factor, $t$ and octahedral factor, $\mu$, where $t$ is defined as the ratio of the distance A-X to the distance B-X in an idealized solid sphere model ($t = (R_A + R_X)/\{\sqrt{2}(R_B + R_X)\}$, where $R_A$, $R_B$, and $R_X$ are the ionic radii of the corresponding ions, and $\mu$ is defined as the ratio $R_B/R_X$ (Cattin et al. 2008). For 3D halide perovskites, typical ranges of $t$ and $\mu$ are: $0.81 < t < 1.11$ and $0.44 < \mu < 0.90$ for a cubic structure, or $0.89 < t < 1.0$ for the less symmetric tetragonal or orthorhombic structures (Green et al. 2014).

Discovery and research on metal halide perovskites started over a century ago (Topsöe 1884). In the 1990s, Mitzi and coworkers (Mitzi et al. 1995; Mitzi et al. 2001) investigated their use for thin-film transistors and light-emitting diodes (LEDs) (Mitzi et al. 2001), and it was not until a decade later that Miyasaka and coworkers (Kojima et al. 2009; Kojima et al. n.d.-b) reported their first applications for photovoltaics (Figure 11.2). The first reported perovskite cell was based on $MAPbBr_3$ having a power conversion efficiency of 2.2% (Kojima et al. n.d.-a). Subsequently, they replaced Br with I to increase the efficiency to 3.8% in 2009 (Kojima et al. 2009). However, the fabricated cells were quite unstable. In these structures, a lithium halide containing organic electrolyte and the respective halogen acted as the hole transporting medium (HTM) to facilitate hole transfer to the external electrode. Such design is due to the inherited legacy of dye-sensitized solar cells, but in 2012, Lee et al. (Lee et al. 2012; Kim et al. 2012) and Park et al. (Kim et al. 2012), separately reported solid-state cell demonstrations representing a breakthrough as they led to simultaneous enhancement in device stability and efficiency approaching 10%

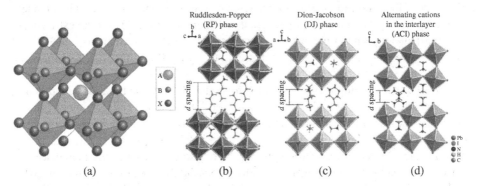

**Figure 11.1**   (a) Cubic perovskite crystal structure ($ABX_3$). For solar cell application, the large cation A is organic or inorganic, the small cation B is a divalent metallic cation, and the anion X is a halogen ion (Green et al. 2014). Lower dimensional layered 2D perovskite crystal structures: (b) Ruddlesden–Popper (RP) phase, (c) Dion–Jacobson (DJ) phase, and (d) alternating cations in the interlayer space (ACI) phase. Source: Adapted from Li et al. (2021).

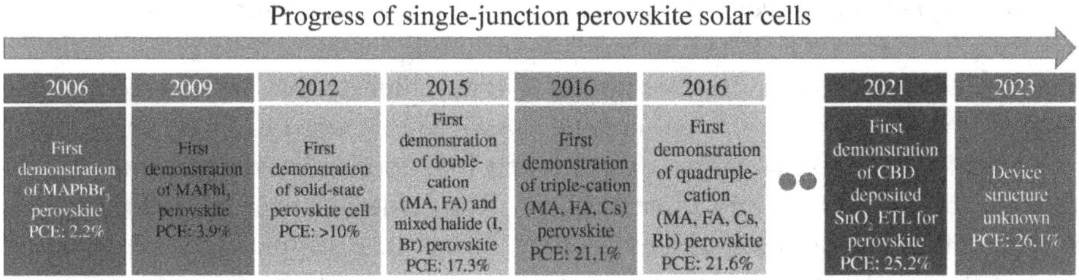

Progress of single-junction perovskite solar cells

| 2006 | 2009 | 2012 | 2015 | 2016 | 2016 | | 2021 | 2023 |
|------|------|------|------|------|------|------|------|------|
| First demonstration of MAPbBr₃ perovskite PCE: 2.2% | First demonstration of MAPbI₃ perovskite PCE: 3.9% | First demonstration of solid-state perovskite cell PCE: >10% | First demonstration of double-cation (MA, FA) and mixed halide (I, Br) perovskite PCE: 17.3% | First demonstration of triple-cation (MA, FA, Cs) perovskite PCE: 21.1% | First demonstration of quadruple-cation (MA, FA, Cs, Rb) perovskite PCE: 21.6% | ●● | First demonstration of CBD deposited SnO₂ ETL for perovskite PCE: 25.2% | Device structure unknown PCE: 26.1% |

**Figure 11.2** Progress of single-junction perovskite solar cells over the years (2006–2023).

(Kim et al. 2012). From then on, solid-state cells have become the mainstream, experiencing rapid efficiency improvement. The popular perovskite composition for the absorber has evolved from MAPbI₃ to FAPbI₃ with variations such as double (Pellet et al. 2014), triple (Saliba et al. 2016), or quadruple cation (Saliba et al. 2016) and mixed halide (Jeon et al. 2015) (e.g. $Cs_A MA_B FA_C Rb_{1-A-B-C} PbI_X Br_{3-X}$). FAPbI₃ has an optical bandgap (1.47 eV) closer to the optimal one for single junction cells, to achieve thermal dynamic electrical energy conversion efficiency limit is therefore more advantageous than MAPbI₃ in that regard (Pellet et al. 2014). However, a pure FAPbI₃ perovskite is unstable at room temperature as it can crystallize into photo-inactive, non-perovskite δ-phase (termed "yellow phase" due to its non-photoactivity as compared to the "black" photoactive α-phase). To overcome this challenge, double-cation or triple-cation perovskites, where a small amount of MA or a small amount of MA with Cs could induce a preferable crystallization of the black phase making it thermally and structurally stable (Pellet et al. 2014; Saliba et al. 2016). Quadruple cation perovskites having Cs, MA, FA, and Rb in the A-site followed suit shortly as Rb incorporation could also enhance voltage output and, therefore, device performance due to suppression of bulk and surface defects (Saliba et al. 2016). Compositional engineering was not confined to A-cation sites only. The first mixed halide perovskite (I/Br) cells were reported by Seok and coworkers for bandgap tuning (Noh et al. 2013) (useful for tandem application for example), for enhancing device performance (Jeon et al. 2015) and stability (Shi et al. 2020). However, a main challenge associated with mixed halide perovskites (e.g. with a bandgap >1.7 eV) is halide segregation during photoexcitation. Segregation-induced iodide-rich minority domains and bromide-rich majority domains act as trap centers for carrier recombination (Hoke et al. 2015). This also causes a reduction in the bandgap of the perovskite film or device and a reduction in quasi-Fermi-level splitting, increasing voltage losses. Although halide segregation is reversible in the dark, it has an adverse effect on the photostability of perovskite devices (Hoke et al. 2015).

To achieve a low bandgap, e.g., close to 1 eV as the bottom cell for tandem cell demonstration, Sn can be mixed into the B site for fabricating mixed Sn-Pb perovskite. Hayase and coworkers (Ogomi et al. 2014) reported the first demonstration of Sn–Pb perovskite cells, extending their absorption range to 1060 nm. However, a low bandgap Sn-based perovskite cell (1.2–1.5 eV) is unstable due to the tendency of $Sn^{2+}$ to oxidize into $Sn^{4+}$ (Ho-Baillie et al. 2021). This instability has been observed in complete devices, during cell fabrication (e.g. during the crystallization of Sn–Pb perovskite film), and even during the handling of perovskite precursor pre-fabrication. A number of antioxidant additives have been applied in perovskite precursor to solve the $Sn^{2+}$ oxidation problem, such as $SnF_2$ (Shirayama et al.

2016), SnF$_2$–pyrazine complex (Lee et al. 2016), hydroxybenzene sulfonic acid (Tai et al. 2019), metallic Sn (Lin et al. 2019), and surface-anchoring zwitterion (Xiao et al. 2020).

Recently, there has been an emergence of lower dimensional layered perovskites (Figure 11.1b), which contain large organic (alkyl- and/or aryl-) cation(s) that cannot be integrated into a 3D perovskite structure and thus form a layered structure. These phases can be categorized into three classes – Ruddlesden–Popper (RP) (Ruddlesden and Popper 1958) phase, Dion–Jacobson (DJ) (Mao et al. 2018) phase, and alternating cations in the interlayer space (ACI) (Soe et al. 2017) phase. Their general formulae are A$_2$A$'_{n-1}$B$_n$X$_{3n+1}$, AA$'_{n-1}$B$_n$X$_{3n+1}$, and AA$'_n$B$_n$X$_{3n+1}$, respectively (Liao et al. 2023), where A$'$ denotes the large organic cation, and $n$ refers to the number of inorganic slabs or metal-halide framework located between organic spacer layers (Figure 11.1c,d). Layered perovskites are more advantageous than their 3D counterparts for various optoelectronic applications due to their outstanding ambient and thermal stability, structural flexibility with tunable bandgap, and electronic confinement effect (Liao et al. 2023) and, therefore, have attracted increased research and development interests and effort.

## 11.3  Evolution of Perovskite Solar Cell Design

Figure 11.3 shows three typical device structures for perovskite solar cells. Earlier devices adopted the "regular" or n-i-p structure (Figures 11.3a,b), where the n-type carrier selective layer is deposited on transparent conductive electrode (TCE) first, followed by the intrinsic perovskite film, and the p-type carrier selective layer.

Figure 11.3a was the most popular choice in the first demonstrations where the mesoporous layer was of porous TiO$_2$ (Kojima et al. n.d.-a) or insulating Al$_2$O$_3$ scaffold (Lee et al. 2012) that served as a hole block or electron selective layer (ESL). Other candidates for ESL include compact TiO$_2$ (Liu et al. 2013), ZnO (Liu and Kelly 2014), SnO$_2$ (Ke et al. 2015), Zn$_2$SnO$_4$ (Lee et al. 2016), Nb$_2$O$_5$ (Shirayama et al. 2016), and WO$_3$ (Tai et al.

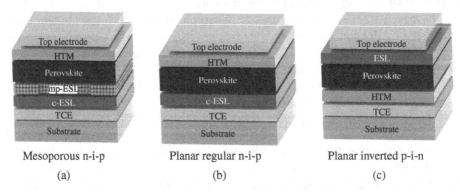

|                      |                      |                      |
| :------------------: | :------------------: | :------------------: |
| Mesoporous n-i-p     | Planar regular n-i-p | Planar inverted p-i-n |
| (a)                  | (b)                  | (c)                  |

**Figure 11.3**  Three typical device structures of perovskite solar cells: (a) mesoporous (n-i-p), (b) regular (n-i-p) planar structure, and (c) inverted (p-i-n) planar structure (Song et al. 2021). Substrates can be rigid (e.g. glass) or flexible (e.g. PET (polyethylene terephthalate), PEN (polyethylene naphthalate). TCE = transparent conductive electrode mostly in the form of transparent conductive oxides such as fluorine-doped tin oxide (FTO) or indium tin oxide (ITO). ESL = electron selective layer which can be compact ("c") or mesoporous ("mp"). HTM = hole transport material.

2019), enabling the formation of the planar structure for perovskite (Figure 11.3b). $TiO_2$ (whether mesoporous compact) generally requires high temperature (500 °C) processing and is therefore less suitable for temperature-sensitive demonstrations such as flexible substrates. The high-temperature process also reduces the conductivity of TCE's, necessitating different electrode designs (Zhang et al. 2018) to circumvent this problem. Over the years, alternative hole-blocking/electron-selective layers ($SnO_2$, $Zn_2SnO_4$, $Nb_2O_5$, $WO_3$) that can be processed at low-temperature (~100 °C) to realize planar structures have been demonstrated to be compatible with flexible cell demonstrations using roll-to-roll fabrication process for example.

In terms of p-type HTM (Kim et al. 2012), the first report of the use of 2,2′,7,7′-tetrakis (N,N-di-p-methoxyphenylamine)-9,9′-spirobifluorene (spiro-MeOTAD) was a significant breakthrough allowing researchers to reliably demonstrate perovskite cells in the laboratory to improve their efficiencies further. However, the stability of spiro-MeOTAD is limited (Lee et al. 2012). Other more stable HTM's including PTAA (Heo et al. 2013), CuSCN (Liu et al. 2013), copper phthalocyanine (Liu and Kelly 2014), CuI (Ke et al. 2015) have been reported.

In recent years, a planar structure using p-i-n or inverted structure (Jeng et al. 2013) (Figure 11.3c) producing high efficiencies are becoming popular, especially with the development of self-assembly monolayer or multi-layer (SAM) type HTM material (Al-Ashouri et al. 2019). SAM-based HTMs have greater flexibility for functionalization (Wang et al. 2023), producing enhanced hole extraction (Al-Ashouri et al. 2020). They are also ultra-thin (<5 nm), compared to conventional HTMs (~10–20 nm), requiring less material usage, making them cost-effective.

## 11.4 Optical Properties of Perovskites

The optical properties of perovskites are of considerable interest in photovoltaics not only for design optimization maximizing optical absorption but also for providing insights into fundamental material properties. Optically, a big advantage of perovskites is their high absorption coefficient, $\alpha$, in the order of $10^5$ cm$^{-1}$, making it possible to fabricate photovoltaic devices with the absorber in a thin film form, e.g., <1 μm (Green et al. 2015).

The optical properties of $CH_3NH_3PbI_3$ (MAPbI$_3$) and $CH_3NH_3PbBr_3$ (MAPbBr$_3$) have been previously reviewed (Green et al. 2015) highlighting relatively low refractive index and opposite bandgap variation compared to traditional inorganic tetrahedrally coordinated semiconductors. The low index is due to the low chemical valency of the halide atom while the opposite bandgap ($E_g$) variation is due to the antibonding nature of the highest energy valence band states in these perovskites (Green et al. 2015).

Kato et al. (2017) compared the optical properties between $HC(NH_2)_2PbI_3$ (FAPbI$_3$) and MAPbI$_3$. In terms of the influence of A-site cation on light absorption, there is a reduction in absorption coefficient ($\alpha$) (Figure 11.4b) and, therefore, a reduction in the imaginary part of the complex relative permittivity ($\varepsilon_2$) (Figure 11.4d) for $HC(NH_2)_2PbI_3$ (FAPbI$_3$) compared to MAPbI$_3$. Hence the effect of varying A-site cation is varied oscillator strength of the optical transition as the inter-band transition and the absorption strength in the visible region is modified by the strong A-X interaction (e.g. $\alpha$ reduction by strong anti-coupling effect induced by the shorter N-I distance due to the presence of two N atoms in FAPbI$_3$). In terms

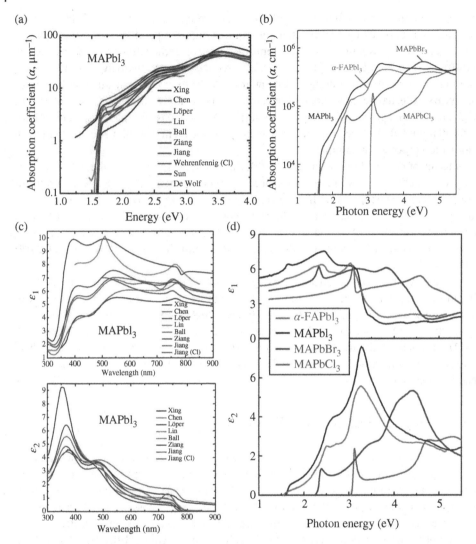

**Figure 11.4** Absorption coefficient of (a) MAPbI₃ at room temperature compiled by Green et al. (Green et al. 2015; Even et al. 2015) (b) α-FAPbI₃, and MAPbX₃ (X = I, Br, or Cl) compiled by Kato et al. Source: Kato et al. (2017)/American Institute of Physics. Real (ε1) and imaginary (ε2) parts of the dielectric functions of (c) MAPbI₃ at room temperature compiled by Green et al. (Green et al. 2015; Even et al. 2015) (d) a-FAPbI₃ and MAPbX₃ (X = I, Br, or Cl) compiled by Kato et al. Source: Kato et al. (2017)/American Institute of Physics.

of its implication on cell demonstrations, thicker FAPbI₃ absorbers (compared to MAPbI₃ solar cells) are required to compensate for the lower α. The X-site halogen, on the other hand, determines the transition energy. This can be seen by the shifting of the dielectric function towards higher energy as the halogen atom becomes lighter (I → Br → Cl) halogen atom (Figure 11.4b).

While the optical transitions for MAPbI₃ (Figure 11.4a,c) have been previously described (Green et al. 2015; Even et al. 2015), the band structure and various optical transitions of FAPbI₃ were calculated by Kato et al. as shown in Figure 11.5b,c (Kato et al. 2017).

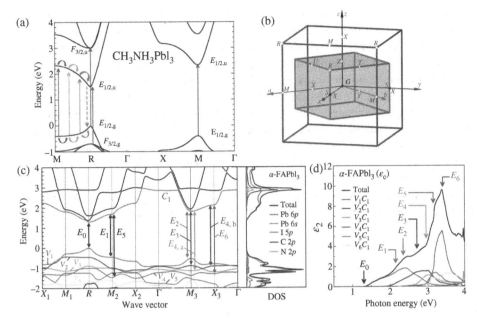

**Figure 11.5** Band structure of (a) MAPbI$_3$ cubic phase [43] and (b) α-FAPbI$_3$ pseudo-cubic crystal. (c) Relationship between the Brillouin zones associated with the cubic (outer cuboid) and orthorhombic (inner cuboid) crystalline phases. Source: Pavarini et al. (2005)/IOP Publishing. (d) Contributions of various inter-band transitions ($E_0$–$E_6$) to the $\varepsilon_2$ spectra. $V_j C_k$ denotes the inter-band transition from the *j*th valence band to the *k*th conduction band. Source: Kato et al. (2017)/American Institute of Physics.

The first valence band from the valence band maximum (VBM) $V_1$ mainly consists of the I 5p state, while the first conduction band from the conduction band minimum $C_1$ is dominated by the Pb 6p in α-FAPbI$_3$. It can be seen that α-FAPbI$_3$ is a direct semiconductor with the $E_0$ ($E_g$) transitions at the R point (cubic symmetry). Inter-band transitions are dominated by transitions to $C_1$ for visible light absorption and predominantly occur at the M points ($E_1$, $E_5$, $E_2$, $E_3$, $E_{4a}$). The difference between $E_1$ (at $M_2$) and $E_2$ (at $M_3$) can be due to the distortion of a-FAPbI$_3$ crystal resulting in a difference in the lattice parameters ($a = 6.416$ Å and $b = 6.236$ Å) and the resulting reciprocal lattices [a* → $M_3$ and b* → $M_2$ in the Brillouin zone].

In terms of the A-site influence on bandgap, the effect can be seen when a larger cation, e.g., FA compared to MA is introduced to the A-site, resulting in larger I—Pb—I bond angle, thereby reducing the bandgap. Cs in the A-site, on the other hand, increase the bandgap by the same principle. Revising the A-site influence on absorption coefficient in the case of Cs, Kato et al. carried out their analyses using the sum rule and predicted that (defect-free) CsPbI$_3$ will produce a high α due to reduced anti-coupling effect (Kato et al. 2017).

## 11.5 Defects and Defect Tolerance of Perovskite

Defects determine the physicochemical and photoelectrical properties of a semiconductor material. Low-temperature, solution-processed perovskite material is soft by nature, suffering from the rapid formation of defects during their fabrication (Ball and Petrozza 2016). Therefore, polycrystalline perovskite films demonstrate about five orders of magnitude higher defect density compared to silicon or single-crystal perovskites (Shi et al. 2015; Stranks et al. 2013). Metal perovskite exhibits high "defect tolerance", meaning the associated solar devices still produce outstanding energy conversion efficiencies. This is because of the antibonding nature of the electronic band edges and effective dielectric screening (Ran et al. 2018). Nevertheless, defects still act as bottlenecks for realizing the theoretical efficiency limit for perovskite solar cells. Defects influence kinetics and/or thermodynamics of photo-generated charge extraction or transfer, intrinsic ion migration, and degradation commencing at the interfaces, which ultimately influence the performance and durability of the photovoltaic devices (Chen and Zhou 2020).

The crystallographic defects of perovskite are similar to any conventional semiconductor. The common defects include point defects, such as (i) atomic vacancies when

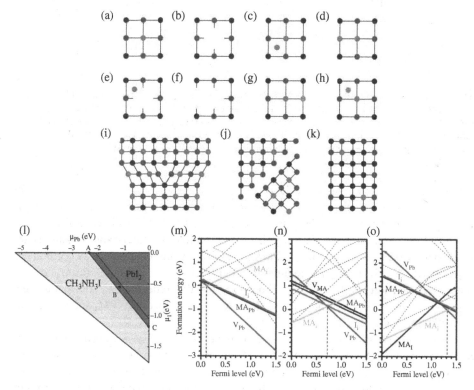

**Figure 11.6** Illustration of crystallographic defects in a crystal lattice. For perovskites, blue, black, and purple dots represent the A-, B-, and X-site ions, respectively. (a) Perfect lattice; (b) Vacancy; (c) Interstitial; (d) Anti-site substitution; (e) Frenkel; and (f) Schottky defects. (g) Substitutional and (h) interstitial impurities. (i) Edge dislocation. (j) Grain boundary. (k) Precipitate (Ball and Petrozza 2016). (l) Corresponding chemical potentials of constitute elements for forming $CH_3NH_3I$, $CH_3NH_3PbI_3$, and $PbI_2$. The range of potentials that allow thermodynamically stable growth for $CH_3NH_3PbI_3$ is marked by the red strip (Yin et al. 2014) (m, n, o). Calculated formation energies for intrinsic point defects in $CH_3NH_3PbI_3$ at points (m) A, (n) B, (o) C from (l) are plotted accordingly. Dashed lines represent defects with high formation energies. Source: Yin et al. (2014)/American Institute of Physics.

atom(s) are missing from the regular lattice position(s) (Figure 11.6b), (ii) interstitials when atom(s) occupy the space between atoms in the lattice (Figure 11.6c), (iii) anti-site substitutions when atom(s) occupy the wrong site in the lattice) (Figure 11.6d), (iv) Frenkel defects as interstitial and vacancy are created from the same ion (Figure 11.6e), (v) Schottky defects when cation and anion vacancies occur simultaneously (Figure 11.6f), (vi) substitutional impurity (Figure 11.6g), and (viii) interstitial impurity (Figure 11.6h), or higher-dimensional defects, such as (i) dislocations (Figure 11.6i), grain boundaries (Figure 11.6j), and precipitates (Figure 11.6k) (Ball and Petrozza 2016).

The innate point defects in basic MAPbI$_3$ perovskite crystal include the atomic vacancies $V_{MA}$, $V_{Pb}$, and $V_I$; the interstitials MA$_i$, Pb$_i$, and I$_i$; anti-site substitutionals Pb$_I$, I$_{Pb}$, I$_{MA}$, and MA$_I$; and cation substitutionals Pb$_{MA}$, and MA$_{Pb}$ (Ball and Petrozza 2016). FAPbI$_3$ perovskite lattice exhibits similar defects as MAPbI$_3$, except that anti-sites (FA$_I$ and I$_{FA}$) create deep levels within the bandgap of FAPbI$_3$ because of their much lower formation energies (Liu and Yam 2018). For α-phase CsPbI$_3$ (under Pb-rich conditions), $V_{Pb}$ and $V_I$ have the lowest formation energy among all acceptor and donor defects, respectively (Dunfield et al. 2020). For δ-phase CsPbI$_3$, the majority of the intrinsic defects induce deeper transition energy levels as compared to γ-phase CsPbI$_3$ (Huang et al. 2018).

It is generally believed that most defects in perovskites have low formation energies remaining at shallow energy levels near the valance band maximum or the conduction band minimum. However, some defects having high formation energies contribute to deep energy levels in the bandgap (Chen and Zhou 2020). The defects in perovskite materials not only deteriorate device performance but also adversely affect cell stability. The device lifetime is significantly reduced by internal (ion migration) and external stressors (oxygen, moisture, light irradiation, and thermal stress) in the presence of defects in perovskite materials (Lei et al. 2021). An understanding of the origin of defects and how they influence perovskite cell efficiency and stability is essential to develop appropriate passivation strategies to mitigate or even completely eliminate their harmful effects. There has been a plethora of literature reporting these studies, and notable ones can be found in Ball and Petrozza (2016), Ran et al. (2018), Chen and Zhou (2020), and Dunfield et al. (2020).

## 11.6 Outlook

Perovskite solar cells are currently the fastest-growing photovoltaic technology in terms of efficiency enhancement and number of publications. Single-junction perovskite solar cell has a theoretical efficiency limit of only ~31% (Pazos-Outón et al. 2018), and the full potential of perovskite solar cells capitalizing on their bandgap tunability can be achieved using tandem cell approach. Rapid progress has been made by perovskite tandem cells, whereby their energy conversion efficiencies have tripled in five years since their first reports. Such progress was in fact achieved by 2-terminal monolithic perovskite tandem devices, which may pose more challenges associated with tandem cell processing compared to 4-terminal tandems but offer more straightforward integration in conventional cell manufacturing and assembly lines. The performance of perovskite-Si, perovskite-CIGS (where CIGS = Cu(In,Ga)(S,Se)$_2$), perovskite-OPV (where OPV = organic photovoltaic), perovskite-perovskite tandems have already surpassed their single junction counterparts, e.g., Si, CIGS, OPV and perovskite single-junctions alone. One of the current challenges for pure perovskite tandems is synthesizing the suitable perovskite materials for low bandgap "bottom" perovskite sub-cells. The most common material used for the bottom cell is Sn-based perovskites, but they have a bandgap range of 1.2–1.5 eV, still above the ideal bottom cell bandgap for multi-junction tandems. (Bremner et al. 2016) In addition, Sn-based perovskites are unstable, as discussed above. Such instability is a major roadblock to the commercialization of perovskite–perovskite tandem cells with a reliance on Sn-based perovskites. On the other hand, high bandgap (>1.7 eV) perovskite cells suffer from higher bandgap-voltage offset (Almora 2022) compared to mid bandgap (1.5–1.7 eV) perovskite cells due to (i) non-radiative recombination within the perovskite bulk and at the surfaces, (ii) offset losses due to misalignment of charge transport layers, and (iii) halide segregation (Mahesh et al. 2020) warranting research and development effort into high bandgap perovskites. At present, steady progress is made on developing industry-relevant perovskite single junction and tandem cells with a focus on area scale-up and durability. The latter is a major hurdle for large-scale commercialization and industrialization, especially for terrestrial use. In recent years, applications of perovskite cells for building integrated photovoltaics (Bing et al. 2022; Batmunkh et al. 2020), for space (Ho-Baillie et al. 2021), and indoor applications (Polyzoidis et al. 2021), have also been explored.

## Author Biographies

**Anita Ho-Baillie** is the John Hooke Chair of Nanoscience at the University of Sydney, an Australian Research Council Future fellow, and an Adjunct Professor at University of New South Wales (UNSW). She completed her Bachelor of Engineering degree on a Co-op scholarship in 2001 and her PhD at UNSW in 2005. Her research interest is to engineer materials and devices at nanoscale for integrating solar cells onto all kinds of surfaces generating clean energy. She is a highly cited researcher from 2019 to 2023. In 2021, she was an Australian Museum Eureka Prize finalist and was named the top Australian sustainable-energy researcher by the Australian newspaper Annual-Research-Magazine. She won the Royal Society of NSW Warren Prize in 2022 for her pioneering work in the development of next generation solar cells. She is a fellow of the Australian Institute of Physics, the Royal Society of New South Wales and the Royal Society of Chemistry.

**Dr. Md Arafat Mahmud** is a research fellow at the School of Physics, at the University of Sydney. Prior to that, he worked as a research fellow at perovskite photovoltaics group at Australian National University (ANU). He obtained his PhD in Photovoltaic and Renewable Energy Engineering from University of New South Wales (UNSW), Australia, in 2018. His research focuses on developing high-efficiency and stable perovskite-based tandem solar cells. In 2023, he was awarded the School of Physics Research Award for Breakthrough Research in the Physical Sciences from the University of Sydney.

**Dr. Jianghui Zheng** is a research fellow at the School of Physics, at the University of Sydney. Prior to that, he worked as a postdoctoral research fellow at the School of Photovoltaic and Renewable Energy Engineering at UNSW since 2017. He obtained his B.E. degree in Materials Science and Engineering from Tongji University in 2011 and his PhD degree in Photovoltaic Engineering from Xiamen University in 2017. He was awarded the Australian Centre of Advanced Photovoltaics (ACAP) Research Fellowship in 2019 and the School of Physics Research Award for Breakthrough Research in the Physical Sciences from the University of Sydney in 2023. His current research interest focuses on developing multi-junction tandem photovoltaics based on perovskite and commercially available photovoltaics (e.g., Si or CIGS), and their applications.

# References

Al-Ashouri, A. et al. (2019). Conformal monolayer contacts with lossless interfaces for perovskite single junction and monolithic tandem solar cells. *Energy & Environmental Science* 12: 3356–3369, https://doi.org/10.1039/C9EE02268F.

Al-Ashouri, A. et al. (2020). Monolithic perovskite/silicon tandem solar cell with >29% efficiency by enhanced hole extraction. *Science* 370: 1300–1309, https://doi.org/10.1126/science.abd4016.

Almora, O. Device Performance of Emerging Photovoltaic Materials (Version 3) https://www.authorea.com/users/303476/articles/584106-device-performance-of-emerging-photovoltaic-materials-version-3?commit=0037e799700a0438e498d74c66b8828345289e34 (2022).

Ball, J.M. and Petrozza, A. (2016). Defects in perovskite-halides and their effects in solar cells. *Nature Energy* 1: 16149, https://doi.org/10.1038/nenergy.2016.149.

Batmunkh, M., Zhong, Y.L., and Zhao, H. (2020). Recent advances in perovskite-based building-integrated photovoltaics. *Advanced Materials* 32: 2000631.

Bing, J. et al. (2022). Perovskite solar cells for building integrated photovoltaics—glazing applications. *Joule* 6: 1446–1474.

Bremner, S.P., Yi, C., Almansouri, I. et al. (2016). Optimum band gap combinations to make best use of new photovoltaic materials. *Solar Energy* 135: 750–757, https://doi.org/10.1016/j.solener.2016.06.042.

Cattin, L. et al. (2008). Properties of NiO thin films deposited by chemical spray pyrolysis using different precursor solutions. *Applied Surface Science* 254: 5814–5821, https://doi.org/10.1016/j.apsusc.2008.03.071.

Chang, N.L. et al. (2021). A bottom-up cost analysis of silicon–perovskite tandem photovoltaics. *Progress in Photovoltaics: Research and Applications* 29: 401–413, https://doi.org/10.1002/pip.3354.

Chen, Y. and Zhou, H. (2020). Defects chemistry in high-efficiency and stable perovskite solar cells. 128: 060903, https://doi.org/10.1063/5.0012384.

Dunfield, S.P. et al. (2020). From defects to degradation: a mechanistic understanding of degradation in perovskite solar cell devices and modules. 10: 1904054, https://doi.org/10.1002/aenm.201904054.

Even, J. et al. (2015). Solid-state physics perspective on hybrid perovskite semiconductors. *The Journal of Physical Chemistry C* 119: 10161–10177, https://doi.org/10.1021/acs.jpcc.5b00695.

Green, M.A. and Ho-Baillie, A. (2017). Perovskite solar cells: the birth of a new era in photovoltaics. *ACS Energy Letters* 2: 822–830, https://doi.org/10.1021/acsenergylett.7b00137.

Green, M.A., Ho-Baillie, A., and Snaith, H.J. (2014). The emergence of perovskite solar cells. *Nature Photonics* 8: 506–514, https://doi.org/10.1038/nphoton.2014.134.

Green, M.A., Jiang, Y., Soufiani, A.M., and Ho-Baillie, A. (2015). Optical Properties of Photovoltaic Organic-Inorganic Lead Halide Perovskites. *J Phys Chem Lett* 4774–4785, https://doi.org/10.1021/acs.jpclett.5b01865.

Heo, J.H. et al. (2013). Efficient inorganic–organic hybrid heterojunction solar cells containing perovskite compound and polymeric hole conductors. *Nature Photonics* 7: 486–491, https://doi.org/10.1038/nphoton.2013.80.

Ho-Baillie, A.W.Y. et al. (2021). Recent progress and future prospects of perovskite tandem solar cells. *Applied Physics Review* 8: 041307. https://doi.org/10.1063/5.0061483.

Ho-Baillie, A.W.Y. et al. (2021). Deployment opportunities for space photovoltaics and the prospects for perovskite solar cells. *Advanced Materials Technologies* 7, https://doi.org/10.1002/admt.202101059.

Hoke, E.T. et al. (2015). Reversible photo-induced trap formation in mixed-halide hybrid perovskites for photovoltaics. *Chemical Science* 6: 613–617, https://doi.org/10.1039/C4SC03141E.

Hu, Y. et al. (2021). Flexible perovskite solar cells with high power-per-weight: progress, application, and perspectives. *ACS Energy Letters* 6: 2917–2943, https://doi.org/10.1021/acsenergylett.1c01193.

Huang, Y., Yin, W.-J., and He, Y. (2018). Intrinsic point defects in inorganic cesium lead iodide perovskite $CsPbI_3$. *The Journal of Physical Chemistry C* 122: 1345–1350, https://doi.org/10.1021/acs.jpcc.7b10045.

Jeng, J.-Y. et al. (2013). $CH_3NH_3PbI_3$ perovskite/fullerene planar-heterojunction hybrid solar cells. 25: 3727–3732, https://doi.org/10.1002/adma.201301327.

Jeon, N.J. et al. (2015). Compositional engineering of perovskite materials for high-performance solar cells. *Nature* 517: 476–480.

Kato, M. et al. (2017). Universal rules for visible-light absorption in hybrid perovskite materials. *Journal of Applied Physics* 121, https://doi.org/10.1063/1.4978071.

Ke, W. et al. (2015). Efficient hole-blocking layer-free planar halide perovskite thin-film solar cells. *Nature Communications* 6: 6700, https://doi.org/10.1038/ncomms7700.

Kim, H.-S. et al. (2012). Lead iodide perovskite sensitized all-solid-state submicron thin film mesoscopic solar cell with efficiency exceeding 9%. *Scientific Reports* 2: 591, https://doi.org/10.1038/srep00591.

Kojima, A., Teshima, K., Miyasaka, T. & Shirai, Y.ECS Meeting Abstracts. 397 (IOP Publishing), n.d.-a.

Kojima, A., Teshima, K., Shirai, Y., and Miyasaka, T. (2009). Organometal halide perovskites as visible-light sensitizers for photovoltaic cells. *Journal of the American Chemical Society* 131: 6050–6051.

Kojima, A., Teshima, K., Shirai, Y. & Miyasaka, T. (2007) ECS Meeting Abstracts. 27 (IOP Publishing).

Lee, M.M., Teuscher, J., Miyasaka, T. et al. (2012). Efficient hybrid solar cells based on meso-superstructured organometal halide perovskites. 338: 643–647, https://doi.org/10.1126/science.1228604.

Lee, S.J. et al. (2016). Fabrication of efficient formamidinium tin iodide perovskite solar cells through $SnF_2$–pyrazine complex. *Journal of the American Chemical Society* 138: 3974–3977, https://doi.org/10.1021/jacs.6b00142.

Lei, Y., Xu, Y., Wang, M. et al. (2021). Origin, influence, and countermeasures of defects in perovskite solar cells. 17: 2005495, https://doi.org/10.1002/smll.202005495.

Li, X., Hoffman, J.M., and Kanatzidis, M.G. (2021). The 2D halide perovskite rulebook: how the spacer influences everything from the structure to optoelectronic device efficiency. *Chem. Rev.* 121: 2230–2291.

Liao, C.-H., Mahmud, M.A., and Ho-Baillie, a.A.W.Y. (2023). Recent progress in layered metal halide perovskite for solar cells, photodetectors, and field-effect transistors. *Nanoscale* 15: 4219–4235, https://doi.org/10.1039/D2NR06496K.

Lin, R. et al. (2019). Monolithic all-perovskite tandem solar cells with 24.8% efficiency exploiting comproportionation to suppress Sn(ii) oxidation in precursor ink. *Nature Energy* 4: 864–873, https://doi.org/10.1038/s41560-019-0466-3.

Liu, D. and Kelly, T.L. (2014). Perovskite solar cells with a planar heterojunction structure prepared using room-temperature solution processing techniques. *Nature Photonics* 8: 133–138, https://doi.org/10.1038/nphoton.2013.342.

Liu, M., Johnston, M.B., and Snaith, H.J. (2013). Efficient planar heterojunction perovskite solar cells by vapour deposition. *Nature* 501: 395–398, https://doi.org/10.1038/nature12509.

Liu, N. and Yam, C. (2018). First-principles study of intrinsic defects in formamidinium lead triiodide perovskite solar cell absorbers. *Physical Chemistry Chemical Physics* 20: 6800–6804, https://doi.org/10.1039/C8CP00280K.

Mahesh, S. et al. (2020). Revealing the origin of voltage loss in mixed-halide perovskite solar cells. *Energy & Environmental Science* 13: 258–267, https://doi.org/10.1039/C9EE02162K.

Mahmud, M.A. et al. (2021). Origin of efficiency and stability enhancement in high-performing mixed dimensional 2D–3D perovskite solar cells: a review. *Advanced Functional Materials* 2009164. https://doi.org/10.1002/adfm.202009164.

Mao, L. et al. (2018). Hybrid Dion–Jacobson 2D lead iodide perovskites. *Journal of the American Chemical Society* 140: 3775–3783.

Mitzi, D.B., Chondroudis, K., and Kagan, C.R. (2001). Organic-inorganic electronics. *IBM Journal of Research and Development* 45: 29–45, https://doi.org/10.1147/rd.451.0029.

Mitzi, D.B., Wang, S., Feild, C.A. et al. (1995). Conducting layered organic-inorganic halides containing <110>-oriented perovskite sheets. 267: 1473–1476, https://doi.org/10.1126/science.267.5203.1473.

Noh, J.H., Im, S.H., Heo, J.H. et al. (2013). Chemical management for colorful, efficient, and stable inorganic–organic hybrid nanostructured solar cells. *Nano Letters* 13: 1764–1769, https://doi.org/10.1021/nl400349b.

Ogomi, Y. et al. (2014). $CH_3NH_3Sn_xPb_{(1-x)}I_3$ perovskite solar cells covering up to 1060 nm. *The Journal of Physical Chemistry Letters* 5: 1004–1011, https://doi.org/10.1021/jz5002117.

Pavarini, E., Yamasaki, A., Nuss, J., and Andersen, O.K. (2005). How chemistry controls electron localization in 3d1perovskites: a Wannier-function study. *New Journal of Physics* 7: 188–188, https://doi.org/10.1088/1367-2630/7/1/188.

Pazos-Outón, L.M., Xiao, T.P., and Yablonovitch, E. (2018). Fundamental efficiency limit of lead iodide perovskite solar cells. *The Journal of Physical Chemistry Letters* 9: 1703–1711, https://doi.org/10.1021/acs.jpclett.7b03054.

Pellet, N. et al. (2014). Mixed-organic-cation perovskite photovoltaics for enhanced solar-light harvesting. 53: 3151–3157, https://doi.org/10.1002/anie.201309361.

Polyzoidis, C., Rogdakis, K., and Kymakis, E. (2021). Indoor perovskite photovoltaics for the internet of things—challenges and opportunities toward market uptake. *Advanced Energy Materials* 11: 2101854.

Ran, C., Xu, J., Gao, W. et al. (2018). Defects in metal triiodide perovskite materials towards high-performance solar cells: origin, impact, characterization, and engineering. *Chemical Society Reviews* 47: 4581–4610, https://doi.org/10.1039/C7CS00868F.

Reinders, A., Verlinden, P., Van Sark, W., and Freundlich, A. (2017). *Photovoltaic Solar Energy: From Fundamentals to Applications*. John Wiley & Sons.

Ruddlesden, S. and Popper, P. (1958). The compound $Sr_3Ti_2O_7$ and its structure. *Acta Crystallographica* 11: 54–55.

Saliba, M. et al. (2016). Cesium-containing triple cation perovskite solar cells: improved stability, reproducibility and high efficiency. *Energy & Environmental Science* 9: 1989–1997, https://doi.org/10.1039/C5EE03874J.

Saliba, M. et al. (2016). Incorporation of rubidium cations into perovskite solar cells improves photovoltaic performance. 354: 206–209, https://doi.org/10.1126/science.aah5557.

Shi, D. et al. (2015). Low trap-state density and long carrier diffusion in organolead trihalide perovskite single crystals. 347: 519–522, https://doi.org/10.1126/science.aaa2725.

Shi, L. et al. (2020). Gas chromatography–mass spectrometry analyses of encapsulated stable perovskite solar cells. 368: eaba2412, https://doi.org/10.1126/science.aba2412.

Shirayama, M. et al. (2016). Optical transitions in hybrid perovskite solar cells: ellipsometry, density functional theory, and quantum efficiency analyses for $CH_3NH_3PbI_3$. *Physical Review Applied* 5, https://doi.org/10.1103/PhysRevApplied.5.014012.

Soe, C.M.M. et al. (2017). New type of 2D perovskites with alternating cations in the interlayer space, $(C(NH_2)_3)(CH_3NH_3)_nPb_nI_{3n+1}$: structure, properties, and photovoltaic performance. *Journal of the American Chemical Society* 139: 16297–16309.

Song, J.-X., Yin, X.-X., Li, Z.-F., and Li, Y.-W. (2021). Low-temperature-processed metal oxide electron transport layers for efficient planar perovskite solar cells. *Rare Metals* 40: 2730–2746, https://doi.org/10.1007/s12598-020-01676-y.

Stranks, S.D. et al. (2013). Electron-hole diffusion lengths exceeding 1 micrometer in an organometal trihalide perovskite absorber. 342: 341–344, https://doi.org/10.1126/science .1243982.

Tai, Q. et al. (2019). Antioxidant grain passivation for air-stable tin-based perovskite solar cells. 58: 806–810, https://doi.org/10.1002/anie.201811539.

Topsöe, H.J. (1884). Z. f. K. Krystallographisch-chemische untersuchungen homologer verbindungen. 8: 246–296.

Wang, G. et al. (2023). Molecular engineering of self-assembled monolayer hole selective layer for high bandgap perovskite and highly efficient and stable perovskite-silicon tandem solar cells. *Joule* 7: 2583–2594, https://doi.org/10.1016/j.joule.2023.09.007.

Xiao, K. et al. (2020). All-perovskite tandem solar cells with 24.2% certified efficiency and area over 1 cm² using surface-anchoring zwitterionic antioxidant. *Nature Energy* 5: 870–880, https://doi.org/10.1038/s41560-020-00705-5.

Yin, W.-J., Shi, T., and Yan, Y. (2014). Unusual defect physics in $CH_3NH_3PbI_3$ perovskite solar cell absorber. 104: 063903, https://doi.org/10.1063/1.4864778.

Zhang, M. et al. (2018). Electrode design to overcome substrate transparency limitations for highly efficient 1 cm² mesoscopic perovskite solar cells. *Joule*, https://doi.org/10.1016/j.joule .2018.08.012 .

# Part Five

# Tandem Structures

# 12

## Perovskite/Silicon Tandem Photovoltaics

*Thomas G. Allen[1], Erkan Aydin[1], Anand S. Subbiah[1], Michele De Bastiani[2], and Stefaan De Wolf[1]*

[1] *KAUST Solar Center (KSC), Physical Sciences and Engineering Division (PSE), King Abdullah University of Science and Technology (KAUST), Thuwal, Kingdom of Saudi Arabia*
[2] *Department of Chemistry, INSTM Università di Pavia, Pavia, Italy*

## 12.1 Introduction

Crystalline silicon (c-Si) solar cell technologies, notably the aluminum back surface field (Al-BSF) and, more recently, passivated emitter and rear cells (PERC), have dominated the photovoltaics (PV) industry since the 1970s. The transfer of knowledge from the integrated circuits industry, the relative stability of c-Si, native innovations within the c-Si PV research and development community, and industrially scaled processing have led to consistent reductions in manufacturing costs and increases in power conversion efficiencies (PCEs) (Battaglia et al. 2016; Ballif et al. 2022). The outcome of this has been an approximate cell PCE increase of ~0.5%/year and a 10-fold reduction in cost in $/W terms over the past decade, leading to the rapid adoption of PV technologies (predicted to reach an annual production rate of 1 TW/year by or before 2030) (Haegel et al. 2019).

The development of the PERC architecture, culminating in the 25% cell from UNSW in 1999 (Zhao et al. 1999; Green 2009), and more recent higher efficiency passivating contact technologies, such as silicon heterojunction (SHJ) and tunnel oxide passivating contact (TOPCon) solar cells, have provided clear technological direction for the c-Si PV manufacturing industry to develop and produce devices with increasing performance (Allen et al. 2019; Yan et al. 2021; Razzaq et al. 2022). However, even on industrial wafer formats measured over the entire cell area, c-Si cell technologies are approaching their practical upper limit efficiencies (26.8% for SHJ solar cells; 24% for PERC cells) (Green et al. 2023). It is widely acknowledged that tandem cell designs are the most likely high-efficiency solar cell concept to be adopted by industry to continue the rate of efficiency improvements beyond the practical upper limits of single-junction (1-J) c-Si technologies, which are now rapidly coming into view (Figure 12.1).

Tandem solar cells offer a pathway to PCEs higher than 1-J solar cells by more effectively utilizing the solar spectrum. Compared to wide bandgap 1-J solar cells, tandem cells

*Photovoltaic Solar Energy: From Fundamentals to Applications, Volume 2*, First Edition.
Edited by Wilfried van Sark, Bram Hoex, Angèle Reinders, Pierre Verlinden, and Nicholas J. Ekins-Daukes.

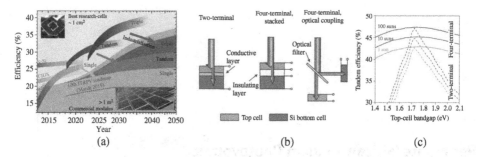

**Figure 12.1** (a) Predicated transition from single-junction to tandem PVs, illustrating the potential increase in PCE (b) and (c) Illustration of the 2T and 4T device designs, including radiative efficiency limits. Source: (a) Jošt et al. (2020)/John Wiley & Sons/CC BY 4.0. (b and c) From Yu et al. (2016).

utilize a broader range of the solar spectrum, reducing transparency losses, and compared to narrower bandgap 1-J solar cells, tandems minimize thermalization losses. Optimizing the top and bottom cell bandgaps leads to a thermodynamic efficiency limit of 45.7% for a monolithic, two terminal (2T), two junction tandem cell, while restricting the bottom cell to silicon leads to an efficiency limit of 45%, illustrating the suitability of c-Si as a bottom cell for tandem applications (Almansouri et al. 2015).

Metal halide perovskite solar cells (PSCs) have come to prominence in the past decade as a promising top cell technology to pair with a c-Si bottom cell in a perovskite/silicon tandem device design. This is owing to the material properties of metal halide perovskites, notably the tunable bandgap (largely by varying the I/Br ratio in the perovskite composition), high quantum yield, sharp band edge, and high absorption coefficient. Two-junction perovskite/silicon tandem solar cell PCEs have gone from 13.7% to 33.9% in less than 10 years (Mailoa et al. 2015; Green et al. 2023; NREL 2023), representing an extraordinary average cell PCE increase of > 2%/year over this period.

This success can largely be ascribed to the progress in PCE of 1-J PSCs, which has enabled their application in tandem cells in conjunction with c-Si bottom cells, combined with the fact that c-Si solar cells are near-optimal bottom cell candidates (Almansouri et al. 2015). Prior to the development of hybrid organic/inorganic halide perovskites, tandems with c-Si were limited to III-V compound semiconductors as solar cells based on thin films lacked sufficient efficiencies that would compensate for the loss of energy that would otherwise be absorbed in the silicon device (White et al. 2014; Almansouri et al. 2015). However, in addition to being prohibitively expensive, III-Vs on silicon suffer from incompatible crystal structures, which have hindered their monolithic integration. Mechanically stacked four terminal (4T) III-V on Si tandems have achieved high efficiencies (35.9% for a triple junction for a GaInP/GaInAsP/Si device), however, no reasonable pathway to scaled production has been outlined (Essig et al. 2017). In contrast, perovskite/silicon tandems have progressed to high PCEs in the monolithic 2T architecture, with device PCEs exceeding 30% now reported by several research groups, albeit on small device areas (~1 cm²). In this chapter, we will discuss the progress of this technology from its inception to the present-day state-of-the-art.

## 12.2 Monolithic Tandems: Evolution of Record Devices and Key Developments

Rapid progress in the PCE of perovskite/silicon tandem cells has been reported in the last decade following the first publication by Mailoa et al. in 2015, which demonstrated a modest PCE of 13.7% (Mailoa et al. 2015). Major reasons for this limited performance were the poor optical device design due to the use of single-side polished (SSP) c-Si bottom cells and high parasitic absorption in the contact layers, the use of non-ideal $CH_3NH_3PbI_3$ (the archetypical early generation perovskite absorber), and the lack of effective bulk and contact passivation strategies both for the top and bottom cell.

These early tandems were in the so-called *n-i-p* architecture, implying that for the top cell first the electron transport layer (ETL, *n*-type) is deposited, followed by the perovskite (*i*-layer) and hole transport layer (HTL, *p*-type). As a result, hole collection occurs on the light-incident side in *n-i-p* perovskite/silicon tandems. The main reason for the implementation of this device architecture was that initial 1-J PSCs were also predominantly in the *n-i-p* configuration, which was a direct legacy of the emergence of PSCs from the dye-sensitized solar cell community (Snaith 2013). In these first perovskite/silicon tandems, the reliance on sintered $TiO_2$ as the ETL prevented the use of a SHJ bottom cell, as the high sintering temperature would degrade the passivation of the hydrogenated amorphous silicon contact layers (De Wolf and Kondo 2009). However, the application of SHJ technology for the bottom cells became possible via the development of a low-temperature perovskite top cell process, for instance by applying PCBM or low-temperature processable $SnO_2$ as the bottom ETL, enabling tandem PCEs >20% (Werner et al. 2016a; Albrecht et al. 2016).

Next, again as a legacy of early generation 1-J PSCs, spiro-OMeTAD was used as the HTL (Kim et al. 2012). To avoid damage to these sensitive polymeric films, featured on the light-facing side in *n-i-p* tandems, a softly deposited transparent top electrode was required, for which in the first tandem a network of spin-coated Ag nanowires were used (Mailoa et al. 2015). However, the moderate optoelectronic properties of Ag nanowires, notably parasitic optical absorption, motivated the application of transparent conductive oxides (TCOs) as the transparent top electrodes. However, most TCOs are deposited by magnetron sputtering, which would invoke irreparable damage when directly applied to organic charge transport layers (Aydin et al. 2021a). This problem was resolved by the introduction of softly deposited metal-oxide buffer layers, usually by atomic layer deposition (ALD) or thermal evaporation. For example, the high work function metal oxide $MoO_x$ was shown to be an effective buffer layer onto spiro-OMeTAD (Löper et al. 2015a). Next, most TCOs require a moderate annealing step (at temperatures of about 200 °C) to crystallize and activate their optoelectronic properties (Morales-Masis et al. 2017). This would be catastrophic for the underlying layers in PSCs, however. Among available TCO options, amorphous indium zinc oxide (IZO) is a remarkable exception, which, even in its as-deposited state, features excellent optoelectronic properties without the need for post-deposition annealing (Morales-Masis et al. 2015). The combination of an ALD buffer layer with IZO was thus soon adapted in semi-transparent PSCs (Werner et al. 2015) and became the standard front contact for monolithic tandems thereafter, irrespective of the tandem polarity.

In parallel, optical device simulations, based on measured $n$ and $k$ data via spectroscopic ellipsometry of all relevant device layers (Löper et al. 2015b), indicated that with these early tandems, PCEs higher than 30% would be possible (Filipič et al. 2015). However, a critical performance limiting factor was found in the use of spiro-OMeTAD on the light-facing side due to its significant parasitic absorption. In the absence of a reasonable alternative HTL, to overcome this limitation, the research community turned its focus to so-called inverted PSCs, i.e. tandems in the *p-i-n* configuration. This was largely enabled by the development of single-junction *p-i-n* PSCs, which initially used PEDOT:PSS as the HTL (Jeng et al. 2013). For *p-i-n* perovskite/silicon tandems, the initial HTL of choice was $NiO_x$, deposited from solution (Bush et al. 2017). Soon after, it was found that room temperature sputtered $NiO_x$ is also a suitable HTL (Aydin et al. 2018), enabling conformal deposition on complex surfaces such as double-side textured (DST) wafers (Hou et al. 2020). For *p-i-n* tandems, the transparent front contact now needs to collect electrons, where a thermally evaporated thin (<20 nm) fullerene layer as ETL was shown to enable a high-performing tandem (Bush et al. 2017). However, again, as fullerenes are organic, this ETL is susceptible to damage from energetic particles during the sputtering of transparent electrodes. Here, ALD $SnO_2$ films were introduced as a buffer layer, which enabled damage-free *p-i-n* tandem solar cells (Bush et al. 2017).

Meanwhile, thanks to the improved optics of the devices, much wider bandgap perovskites (1.65–1.68 eV) with mixed halide/mixed cation compositions could be employed to optimize the current generation in the sub-cells, also leading to enhanced device voltages. Today, perovskite/silicon tandem cells can reach as high as ~2 V, which implies >1.25 V perovskite sub-cells for a top-cell bandgap in the range of 1.68–1.69 eV (Table 12.1).

Turning back to the first monolithic tandems, the choice for polished bottom cells was motivated by the ease of perovskite solution processing via spin coating on planar surfaces. However, using polished wafers is prohibitively expensive for practical applications and does not allow for optimal coupling of long wavelength photons in the bottom cell. A first step to enhance light trapping in the silicon bottom cell consisted of applying stamped PDMS antireflection foils, which gave insight into the practical optical limitations of experimental tandems (Werner et al. 2016a). However, such foils lose their light-trapping properties at the module level, as they feature similar refractive indices as the conventional lamination materials used in PV modules. The next step towards enhancing the light trapping in perovskite/silicon tandems was via the reapplication of front-side polished/rear-side textured bottom cells (Mailoa et al. 2015; Werner et al. 2016b) to accommodate easy solution processing of the top cell with gains in light trapping in the bottom cell.

One critical milestone to further increase the current in perovskite/silicon tandem cells was using DST bottom cells. Here, the critical challenge was developing processes to deposit a high-quality perovskite top cell on such a complex surface morphology. In a first approach, conformally deposited films through vacuum-based techniques were developed. Here, the first successful implementation was via a hybrid process for the formation of the perovskite layer, which combines an initial thermal evaporation process to grow a conformal template, followed by solution processing to then convert this template into the desired perovskite phase (Sahli et al. 2018; Aydin et al. 2020). With this approach, perovskites conformally covering micron-sized pyramids of the bottom cell can be achieved. More recently, this

**Table 12.1** Important milestones and certified cell reports for monolithic perovskite/silicon tandem solar cells development. Here, DSP, RST, and DST refer to double-side polished, rear-side textured front-side planar, and DST, respectively. The given PCEs are for the reverse scan direction.

| Progress definition | Year | PCE (%) | $V_{oc}$ (V) | $J_{sc}$ (mA cm$^{-2}$) | FF | Si wafer type | Si cell type | Subcell surface finish | Area (cm$^2$) | Certified by | Laboratory | Refs. |
|---|---|---|---|---|---|---|---|---|---|---|---|---|
| Record PCE as of July 2023; also highest reported current density | 2023 | 33.7 | 1.974 | 20.99 | 81.3 | FZ | SHJ | DST | 1.0035 | JRC-ESTI | KAUST | Green et al. (2023) |
| Highest reported $V_{oc}$ | 2022 | 32.5 | 1.980 | 20.24 | 81.2 | FZ | SHJ | RST | 1.014 | JRC-ESTI | HZB | HZB (2022) |
| Highest tandem PCE for hybrid processed PVK | 2022 | 31.25 | 1.913 | 20.473 | 79.8 | FZ | SHJ | DST | 1.167 | NREL | EPFL/CSEM | CSEM (2022) |
| Overcoming the limitation of the C$_{60}$/perovskite interface, first damp-heat stable tandems | 2022 | 29.3 | 1.91 | 19.8 | 77.6 | FZ | SHJ | DST | 1.03 | FhG-ISE | KAUST | Liu et al. (2022a, b) |
| Tandem on industrial TOPCon bottom cell | 2022 | 28.7 | 1.907 | 19.29 | 78.3 | Cz | PERx/TOPCon | RST | 1 | n/a | HZB/QCells | Sveinbjörnsson et al. (2022) |
| Mixed SAM HTLs | 2022 | 28.3 | 1.87 | 19.1 | 79.1 | Cz | SHJ | RST | 0.15 | FhG-ISE | Kaneka | Mishima et al. (2022) |
| $J_{sc}$ improvement via rear reflectors | 2021 | 29.8 | 1.92 | 19.48 | 79.4 | FZ | SHJ | RST | 1.016 | FhG-ISE | HZB | Tockhorn et al. (2022) |
| First 40 days of outdoor stable cell | 2021 | 28.2 | 1.87 | 19.6 | 78.6 | FZ | SHJ | DST | 1.03 | JET | KAUST | Liu et al. (2021) |
| Largest certified cell | 2021 | 26.8 | 1.891 | 17.84 | 79.4 | Cz | n/a | n/a | 274 | FhG-ISE | Oxford PV | Green et al. (2023) |

*(Continued)*

**Table 12.1** (Continued)

| Progress definition | Year | PCE (%) | $V_{oc}$ (V) | $J_{sc}$ (mA cm$^{-2}$) | FF | Si wafer type | Si cell type | Subcell surface finish | Area (cm$^2$) | Certified by | Laboratory | Refs. |
|---|---|---|---|---|---|---|---|---|---|---|---|---|
| First bifacial perovskite/silicon tandem cell | 2021 | 25.21 | 1.8 | 18.45 | 75.58 | FZ | SHJ | DST | 0.832 | FhG-ISE | KAUST/U.Toronto | De Bastiani et al. (2021a) |
| PCE update | 2020 | 29.5 | 1.884 | 20.26 | 77.3 | FZ | n/a | n/a | 1.121 | NREL | Oxford PV | OxfordPV 2020 |
| Self-assembled HTLs are introduced | 2020 | 29.15 | 1.897 | 19.75 | 79.9 | FZ | SHJ | RST | 1.06 | FhG-ISE | HZB | Al-Ashouri et al. (2020) |
| First nip tandem on DST, and giving $J_{sc}$ higher than 19 mA/cm$^2$ | 2020 | 27.1 | 1.828 | 19.5 | 75.9 | FZ | SHJ | DST | 1.03 | n/a | KAUST | Aydin et al. (2021b) |
| First scalable slot-die coated perovskite-based tandem cells | 2020 | 23.8 | 1.76 | 19.2 | 70 | FZ | SHJ | DST | 1.03 | n/a | KAUST | Subbiah et al. (2020) |
| ALD SnO$_2$ buffer layer free TCO front contact | 2020 | 26.2 | 1.818 | 18.86 | 76.4 | FZ | SHJ | RST | 1 | NREL | KAIST/SNU/NREL | Kim et al. (2020) |
| First time triple-halide perovskite on tandems | 2020 | 25.75 | 1.8672 | 18.306 | 75.3 | FZ | SHJ | RST | 1 | NREL | U.Colorado/NREL/ASU | Xu et al. (2020) |
| First solution-processed perovskite on textured | 2020 | 25.71 | 1.793 | 19.07 | 75.36 | FZ | SHJ | DST | 0.832 | FhG-ISE | U.Toronto/KAUST | Hou et al. (2020) |
| MA-free solution-based perovskite | 2020 | 25.1 | 1.770 | 17.7 | 80.83 | FZ | SHJ | RST | 0.25 | FhG-ISE | Fraunhofer ISE | Schulze et al. (2020) |

| | | | | | | | | | | | | |
|---|---|---|---|---|---|---|---|---|---|---|---|---|
| First outdoor test, revealing ideal bandgap <1.68 eV | 2020 | 25 | 1.73 | 19.8 | 73.1 | FZ | SHJ | DST | 0.832 | FhG-ISE | KAUST | Aydin et al. (2020) |
| Introducing nc-SiOx RJs | 2019 | 25.43 | 1.792 | 19.02 | 74.60 | FZ | SHJ | RST | 1.1 | Yes | HZB/Oxford | Mazzarella et al. (2019) |
| PCE update | 2018 | 28 | 1.802 | 19.75 | 78.7 | FZ | n/a | n/a | 1.03 | NREL | Oxford PV | OxfordPV 2018 |
| First conformal perovskite on textured interface | 2018 | 25.2 | 1.788 | 19.5 | 73.1 | FZ | SHJ | DST | 1.42 | FhG-ISE | EPFL | Sahli et al. (2018) |
| $J_{sc}$ enhancement via anti-reflection foil | 2018 | 25 | 1.76 | 18.5 | 78.5 | FZ | SHJ | RST | 0.765 | Yes | HZB | Jošt et al. (2018) |
| First certified cell, and first efficient cell in pin configuration | 2017 | 23.6 | 1.65 | 18.1 | 79 | FZ | SHJ | RST | 1 | NREL | Stanford/ASU | Bush et al. (2017) |
| First usage of PERC bottom cells | 2017 | 22.7 | 1.75 | 17.6 | 73.8 | FZ | PERC | RST | 1 | n/a | ANU | Wu et al. (2017) |
| First >1 cm$^2$ cell | 2016 | 19.2 | 1.701 | 16.1 | 70.1 | FZ | SHJ | DSP | 1.22 | n/a | EPFL | Werner et al. (2016a) |
| First report for perovskite/silicon tandems | 2015 | 13.7 | 1.58 | 11.5 | 75 | FZ | – | DSP | 1 | n/a | MIT/Stanford | Mailoa et al. (2015) |

technique was extended to full evaporation of the perovskite in a single step (Ross et al. 2021). A second approach relies on conventional solution processing (e.g. spin-coating, blade coating, and slot-die coating) of a perovskite precursor solution, followed by film crystallization, triggered by anti-solvent treatments or gas quenching. This technique is mainly used on smaller-sized pyramid textures, with a pyramid height limited to ~1 μm, to ensure the absence of local shunts. By minimizing the size of the pyramids while retaining the optical benefits for the bottom cell, the roadblock to utilizing solution processing on textured interfaces was removed, with successful examples using spin coating (Hou et al. 2020), blade-coating (Chen et al. 2020b), and slot-die-coating for the perovskite deposition (Subbiah et al. 2020). In conclusion, by using DST wafers, excellent optical light trapping can be obtained, where on the silicon side, besides employing passivating contacts, efforts are increasingly made to improve the rear reflector (Holman et al. 2013; Bush et al. 2017) (Figure 12.2).

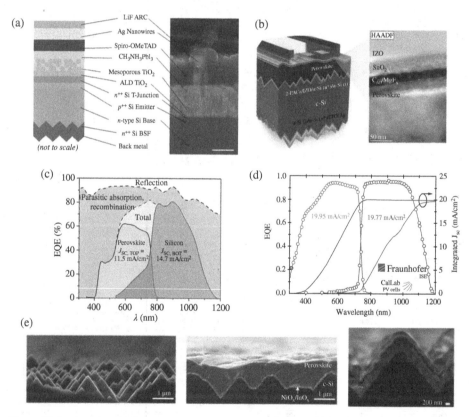

**Figure 12.2** (a and b) Device architectures and SEM images of the device stacks, and (c and d) corresponding EQEs, indicating the advances in perovskite/silicon tandem cell designs (e) SEM cross-sections of (from left to right) textured silicon, solution-processed perovskite/silicon tandem on textured silicon, and a conformal, hybrid processed perovskite/silicon tandem on textured silicon. Source: (a and c) From Mailoa et al. (2015)/AIP Publishing. (b and d) From Liu et al. (2022a)/American Association for the Advancement of Science – AAAS. (e) From Hou et al. (2020) and Aydin et al. (2020)/Springer Nature.

For the perovskite top cell, recent work focused on bulk and contact passivation and interfacial modifications to enhance charge extraction and minimize voltage losses. For instance, the defective nature of $NiO_x$ and its tendency to degrade the perovskite interface can be resolved through molecular engineering (Isikgor et al. 2023), such as through the application of dyes (Zhumagali et al. 2021). Via molecular engineering, it is even possible to eliminate the traditional p-type HTLs, for instance by applying self-assembled monolayers (SAMs) on top of the TCO transparent contact of the silicon bottom cell. This approach has been found to minimize voltage losses, providing low contact resistivity and a better energetic alignment with the perovskite (Al-Ashouri et al. 2020). The perovskite/$C_{60}$ interface is also passivated in state-of-the-art devices by contact displacers, notably alkali and alkaline earth metal fluorides, such as LiF and $MgF_2$ (Bush et al. 2017; Liu et al. 2022a, b). To further improve the current, the thickness of the front electrodes ($C_{60}$ and the front TCO) has progressively been reduced to minimize parasitic absorption losses, guided by in-depth optical analysis (Heydarian et al. 2023). Furthermore, via additive engineering, improved perovskite bulk passivation has been obtained, for instance via grain boundary passivation using small molecules (Isikgor et al. 2021; Liu et al. 2021).

As of today, the high PCE potential of 2T perovskite/silicon tandem solar cells has been adequately demonstrated, and further emphasis on scaling up the technology to larger device areas, device stability, module integration, and field testing is increasingly needed as a focus of the R&D community to progress perovskite/silicon tandems towards market entry. To this end, mainstream industrial c-Si cell technologies (e.g. PERC, TOPCon) have also been used as bottom cells, with acceptable performance penalties compared to SHJ-bottom cell-based tandems (Mariotti et al. 2022; Sveinbjörnsson et al. 2022; Wu et al. 2022).

## 12.3 Four Terminal Tandems

In contrast to 2T monolithic tandem solar cells discussed above, four-terminal (4T) tandem solar cells electrically isolate but optically couple both sub-cells. The 4T technology circumvents some critical challenges of monolithic 2T tandems, such as current matching/bandgap tuning, fabrication challenges on textured surfaces, recombination junction efficacy, and fill-factor losses. Critically, since the two sub-cells operate independently (i.e. they are not series connected via a recombination junction), the sub-cells in 4T tandems are relatively free of constraints on bandgap choices and are less sensitive to the influence of spectral variations on the energy yield. Also, 4T tandems can be integrated into any existing commercial PV technology such as PERC, SHJ, TOPCon, IBC cells, and thin-film technologies with almost no significant modifications for the bottom sub-cell (see Table 12.2). The top, wide bandgap PSC is also free to be in the *n-i-p* or *p-i-n* configuration or to employ the necessary fabrication techniques and materials while ensuring high transmissivity for the low-energy photons illuminating the bottom sub-cell. Notably, due to the relaxation of the current matching constraint in 2T tandems, the 4T architecture makes it easy to implement bifacial tandems, further boosting the energy yield (De Bastiani et al. 2021a, 2022; Babics et al. 2022).

**Table 12.2** Summary of 4T perovskite/silicon tandem solar cell results, ordered by date.

| Year | PSC top cell | Si bottom cell | $J_{sc}$ top (mA/cm²) | $J_{sc}$ bottom (mA/cm²) | $V_{oc}$ top (V) | $V_{oc}$ bottom (V) | FF top | FF bottom | PCE top (%) | PCE bottom (%) | PCE total (%) | Area top (cm²) | Area bottom (cm²) | Ref. |
|---|---|---|---|---|---|---|---|---|---|---|---|---|---|---|
| 2014 | MAPbI$_3$ | SHJ | 14.5 | 13.7 | 0.821 | 0.689 | 0.519 | 0.767 | 6.2 | 7.2 | 13.4 | 0.25 | 4 | Löper et al. (2015a, b) |
| 2014 | MAPbI$_3$ | mc-Si | 17.5 | 11.1 | 1.102 | 0.547 | 0.667 | 0.704 | 12.7 | 4.3 | 17 | 0.39 | 0.39 | Bailie et al. (2015) |
| 2015 | MAPbI$_3$ | SHJ | 17.5 | 17 | 1.034 | 0.687 | 0.77 | 0.77 | 14 | 9.02 | 25.02 | 0.25 | 4 | Werner et al. (2016a) |
| 2016 | FACsPbI$_{3-x}$Br$_x$ | SHJ | 19.9 | 13.9 | 1.1 | 0.69 | 0.707 | 0.764 | 12.5 | 7.3 | 19.8 | 0.09 | N/R | McMeekin et al. (2016) |
| 2016 | MAPbI$_3$ | PERL | 18.8 | 16.9 | 0.95 | 0.64 | 0.69 | 0.73 | 12.2 | 7.9 | 20.1 | 0.25 | 4 | Duong et al. (2016) |
| 2016 | MAPbI$_3$ | SHJ | 20.6 | 12.3 | 1.08 | 0.679 | 0.741 | 0.779 | 16.5 | 6.5 | 23 | 0.075 | 4 | Chen et al. (2016) |
| 2016 | MAPbI$_3$ | SHJ | 20.1 | 15.98 | 1.072 | 0.693 | 0.74 | 0.79 | 16.1 | 8.8 | 24.9 | 0.25 | 4 | Werner et al. (2016b) |
| 2016 | CsMAFAPbI$_{3-x}$Br$_x$ | IBC | 21.5 | 14.2 | 1.1 | 0.69 | 0.735 | 0.81 | 17.4 | 7.9 | 24.5 | 0.36 | 4 | Peng et al. (2016) |
| 2017 | MAPbI$_3$ | IBC | 18.2 | 15.3 | 0.6396 | 0.66 | 0.72 | 0.81 | 12 | 8.2 | 20.2 | 4 | 4 | Jaysankar et al. (2017) |
| 2018 | MAPbI$_3$ | PERL | 21 | 17.7 | 1.098 | 0.674 | 0.741 | 0.801 | 17.1 | 9.6 | 26.7 | 0.104 | 1 | Quiroz et al. (2018) |
| 2018 | CsMAFAPbI$_{3-x}$Br$_x$ | IBC | 19.4 | 17.7 | 1075 | 705 | 0.781 | 0.8 | 16.3 | 10 | 26.3 | 0.09 | 153 | Zhang et al. (2018) |
| 2019 | CsMAFAPbI$_{3-x}$Br$_x$ | TOPCon | 21.5 | 15.2 | 1.06 | 0.63 | 0.775 | 0.811 | 17.7 | 7.8 | 25.5 | 0.09 | 1 | Dewi et al. (2018) |
| 2019 | MAPbI$_3$ | SHJ | 19.8 | 18.8 | 1.008 | 0.708 | 0.78 | 0.793 | 15.6 | 10.6 | 26.2 | 0.16 | 4 | Aydin et al. (2019) |
| 2019 | MAPbI$_3$ | SHJ | 19.8 | 15.6 | 1.156 | 0.698 | 0.799 | 0.8 | 18.3 | 8.7 | 27 | 0.096 | 1 | Wang et al. (2019) |
| 2020 | CsFAMAPbI$_{3-x}$Br$_x$ | TOPCon | 20.5 | 16.6 | 1.11 | 0.675 | 0.786 | 0.796 | 17.9 | 8.9 | 26.8 | 0.059 | 4 | Rohatgi et al. (2020) |
| 2020 | RbCsMAFAPbI$_2$Br | IBC | 18 | 19.6 | 1.205 | 0.697 | 0.789 | 0.78 | 17.1 | 10.7 | 27.7 | 0.21 | 4 | Duong et al. (2020) |
| 2020 | CsFAMAPbI$_{3-x}$Br$_x$ | SHJ | 22.3 | 17.2 | 1.12 | 0.7 | 0.77 | 0.76 | 19.4 | 9.2 | 28.6 | 0.049 | 0.049 | Chen et al. (2020a) |
| 2021 | CsFAPbI$_3$ | SHJ | 21.9 | 14.5 | 1.13 | 0.69 | 0.797 | 0.83 | 19.8 | 8.5 | 28.3 | 6.25 | 1 | Yang et al. (2021) |
| 2022 | CsFAMAPbI$_{3-x}$Br$_x$ | SHJ | 20.3 | 18.3 | 1.189 | 0.686 | 0.82 | 0.79 | 19.9 | 9.9 | 29.8 | 1.04 | 1.04 | Liu et al. (2022a, b) |
| 2022 | 2D modified FAPbI$_3$ | PERL | 24 | 13.7 | 1.131 | 0.678 | 0.842 | 0.825 | 22.9 | 7.7 | 30.3 | 0.25 | 4 | Duong et al. (2023) |

While 4T tandem solar cells can be configured in various ways, the focus lies on conventional mechanically stacked 4T tandems, as recent energy yield calculations show that they can outperform 2T tandems in terms of energy yield in both monofacial and bifacial modes (Lehr et al. 2020; Patel et al. 2023). However, the additional balance of system (BOS) cost and added optical losses are some of the key disadvantages of 4T tandems compared to 2T tandems, specifically, the need for discrete inverters for individual sub-cells increases the BOS.

The necessary modification that combines standalone perovskite and c-Si PV technologies as one 4T tandem technology is the replacement of the metal contact in top-cell PSCs with a semitransparent contact. The challenge lies in extracting maximum performance from the top PSC cell while keeping the average transmission of NIR light high for the silicon bottom cell. Transparent electrodes need a wide bandgap and minimal free carrier absorption (FCA) to avoid parasitic absorption in the blue and red parts of the solar spectrum, respectively. As the semitransparent PSCs employ a double TCO stack for both front and rear contacts, a balance lies between the sheet resistance (controlled via doping of TCOs) and the parasitic FCA that limits NIR transmission to the bottom cell (Morales-Masis et al. 2017; Mujahid et al. 2020; Giuliano et al. 2021). As the carrier density of the TCO decreases, strategies to increase carrier mobility are essential to maintain a low sheet resistivity. In the earlier version of perovskite/silicon 4T tandems, the semitransparent contact employed a $MoO_x$ buffer layer with ITO (Löper et al. 2015a). TCOs based on doped $In_2O_3$ such as IZO, and H-doped $In_2O_3$ (IO:H) have shown promising improvement for 4T tandems by reducing FCA losses (Werner et al. 2016a; Chen et al. 2020a). Soft landing sputtering techniques are also being explored to limit or mitigate the effect of sputter damage, without the need for additional buffer layers between the charge transport layer and the TCO (Aydin et al. 2019, 2021a).

## 12.4 Packaging, Stability, and Field Testing

The module configuration of 2T perovskite/silicon tandems is like the configuration of c-Si technologies – each tandem cell is interconnected in series or parallel, forming strings of (perovskite/silicon tandem) cells. These strings are then laminated between two glass sheets using two polymer foils as an encapsulant to seal the module. In the 4T configuration, the perovskite submodule is realized in a superstrate configuration following the thin-film approach typical of CdTe or CIGS modules, with laser scribing (P1, P2, P3) for the division and interconnection of the thin film cells on the substrate. Local shunts in the perovskite cell, resulting from processing defects and pinholes in the perovskite film, can be further isolated with an additional P4 scribe. Once the perovskite submodule is complete, it is then laminated on top of the silicon submodule. Here, the insulating character of the encapsulant is important for electrical insulation between the two submodules, while the optical properties of the encapsulant should ensure effective optical coupling. 4T tandem modules can then be interconnected on separate circuits, or the perovskite and silicon circuits can be connected within the module's junction box, a configuration that requires voltage matching of the two modules but reduces the BOS costs associated with 4T tandem systems.

Either in the 2T or 4T configuration, a glass–glass module design with polymeric encapsulants and edge sealing will likely be required to provide a strong barrier against permeation of atmospheric moisture (Cheacharoen et al. 2018; De Bastiani et al. 2021b). The sealing of the glass–glass module via vacuum lamination prevents unwanted inter-actions and subsequent degradation induced by the moisture trapped in the module components. The choice of encapsulant polymers is critical to achieving optimal pack-aging, which is limited by the constraints of the perovskite in terms of temperature and chemical interaction. Indeed, the temperature of the vacuum lamination should not induce degradation of the perovskite bulk (particularly the organic volatile cations) or the contact layers. This often constrains the lamination process to modest temperatures around 120–130 °C. Alternatively, perovskite devices with higher temperature stability should be developed.

Moreover, the lamination process should not release chemical by-products that are harm-ful to the perovskite film. In the c-Si PV industry, the most common encapsulant is ethylene vinyl acetate (EVA). Unfortunately, the high sealing temperature (140–150 °C), and the release of by-products that interact with the perovskite during the cross-linking process, make EVA unsuitable for the packaging of perovskite/silicon tandem modules. Alternative encapsulants are polyolefins and polyurethane. These thermoplastic materials are charac-terized by lower temperature processing between 100–120 °C. Moreover, the absence of a cross-linking process avoids the release of by-products, making them the preferred candi-date for the packaging of perovskite/silicon tandems (De Bastiani et al. 2021b). Finally, to avoid moisture percolation from the edges of the module, it is useful to add between 0.5 and 1 cm of edge sealant that runs along the module perimeter. Butyl rubber (BR) and poly-isobutylene (PIB) are typically used for this purpose and are embedded directly into the lamination process.

The stability of hybrid perovskites is of great concern for the industrialization of per-ovskite PVs, including perovskite/silicon tandems. To be competitive with c-Si modules in conventional PV markets, it is imperative that perovskite/silicon tandem modules match the incumbent stability standards. Nowadays, commercial silicon modules are sold with 25-year warranties, however, it is not known how to reliably predict the performances of tandem modules after years of outdoor operation, given the dearth of stability and outdoor data for perovskite/silicon tandem modules. As a minimum, perovskite/silicon tandem modules will be subject to the same IEC (International Electrotechnical Commission) standards as c-Si modules, including the accelerated degradation tests of the IEC61215 protocol. However, additional tests may be required to ensure the stability of the perovskite top cell. Further accelerated degradation protocols, like those based on the International Summit on Organic Photovoltaic Stability (ISOS) consensus (Khenkin et al. 2020), and from the PACT (Perovskite PV Accelerator for Commercializing Technologies) consortium (PACT 2023), offer initial steps towards the formation of perovskite-specific stability and reliability testing.

Aside from the accelerated stability tests, outdoor testing of tandem modules is also of paramount importance. Collecting operational data allows for the acquisition of large datasets of information that can be used for different analyses. First, these data allow for a refining of energy yield predictions and more precise LCOE calculations. Second, the analysis of the outdoor performance provides insights into the role played by spectral variations, cloud cover, temperature variations, soiling accumulation, wind cooling, and albedo variation (for bifacial technologies) on the energy yield. Finally, and critically, extensive and widespread outdoor testing can give insights into module degradation rates (De Bastiani et al. 2021c; Babics et al. 2023), and degradation pathways in the perovskite top cell that involve multiple stressors that are currently unknown or poorly understood.

## 12.5 Outlook

The advent of perovskite PV has finally enabled the implementation of a thin-film PV technology with c-Si capable of high efficiencies. Great strides have been taken over the past decade to enable PCEs to over 30%, and thereby demonstrate the viability of perovskite/silicon tandem PV as a potential path forward for the PV industry to progress the PCE of c-Si PVs beyond the fundamental and practical upper limits on PCE. Significant roadblocks remain, however, especially regarding scaling up and stability, and demonstrable outdoor testing showing high energy yields and minimal module degradation rates.

## Abbreviations and Acronyms

| | |
|---|---|
| 1-J | single junction |
| 2T | two terminal |
| 4T | four terminal |
| Al-BSF | aluminum back surface field |
| ALD | atomic layer deposition |
| BOS | balance of system |
| BR | butyl rubber |
| c-Si | crystalline silicon |
| Cz | Czochralski |
| DSP | double-side polished |
| DST | double-side textured |
| ETL | electron transport layer |
| EVA | ethylene vinyl acetate |
| FCA | free-carrier absorption |
| FZ | float zone |
| HTL | hole transport layer |
| IBC | interdigitated back contact |
| IEC | International Electrotechnical Commission |
| ISOS | International Summit on Organic Photovoltaic Stability |
| IZO | indium zinc oxide |
| LCOE | levelized cost of energy |
| NIR | near infrared |
| PCE | power conversion efficiency |
| PERC | passivated emitter and rear cell |
| PIB | polyisobutylene |

| | |
|---|---|
| PSC | perovskite solar cell |
| PV | photovoltaics |
| RST | rear side textured |
| SAM | self-assembled monolayer |
| SHJ | silicon heterojunction |
| SSP | single-side polished |
| TCO | transparent conductive oxide |
| TOPCon | tunnel oxide passivating contact |

## Author Biographies

**Thomas G. Allen** was awarded his Ph.D. degree from the Australian National University (ANU) in 2017. Since then, he has been at the King Abdullah University of Science and Technology (KAUST), Saudi Arabia, first as a postdoctoral researcher and now as a research scientist in the KPV Lab. His research focuses on the development and characterization of silicon heterojunction and perovskite/silicon tandem solar cells.

**Erkan Aydın** earned his Ph.D. (2016) degree from the Micro and Nanotechnology Program at TOBB University of Economics and Technology, Ankara, Türkiye. Since 2016, he has been continuing his postdoctoral research at the KAUST (King Abdullah University of Science and Technology) campus in Jeddah, Kingdom of Saudi Arabia. Dr. Aydın's expertise and current research interest lie in developing high-efficiency and realistic single-junction perovskite and perovskite/silicon tandem solar cells for terrestrial and extraterrestrial applications.

**Anand S. Subbiah** obtained his MTech (2012) in Nanotechnology from PSG CT (TN, India) and his Ph.D. (2019) in Energy Engineering from IIT Bombay, India. After graduating, he joined the KPV Lab, at King Abdullah University of Science and Technology (KAUST), Saudi Arabia. His current research focuses on scalable solution processing techniques for perovskite fabrication towards tandem applications.

**Michele De Bastiani** obtained his Ph.D. degree in 2016 from the University of Padua and the Italian Institute of Technology. In 2016, he joined King Abdullah University of Science and Technology (KAUST) as a postdoctoral fellow in the group of Prof. Osman Bakr. From 2017 to 2022, he had been a member of the KPV Lab team, first as a postdoctoral fellow and later as a research scientist. Since 2022 he is an assistant professor (RTDa) at the University of Pavia. His research focuses on the reliability and scalability of perovskite photovoltaics.

**Stefaan De Wolf** received his Ph.D. degree in 2005 from the Katholieke Universiteit Leuven and imec in Belgium. From 2005 to 2008, he was with the National Institute of Advanced Industrial Science and Technology (AIST), Tsukuba, Japan. In 2008, he joined Ecole Polytechnique Federale de Lausanne (EPFL), Switzerland, as a team leader. At present, he is a Professor at the King Abdullah University of Science and Technology (KAUST) in Saudi Arabia, focusing on silicon, perovskite, and perovskite/silicon tandem solar cells, as well as photovoltaics for sunny and hot climates.

# References

Al-Ashouri, A., Kohnen, E., Li, B. et al. (2020). Monolithic perovskite/silicon tandem solar cell with >29% efficiency by enhanced hole extraction. *Science* 370 (6522): 1300–1309.

Albrecht, S., Saliba, M., Correa Baena, J.P. et al. (2016). Monolithic perovskite/silicon-heterojunction tandem solar cells processed at low temperature. *Energy & Environmental Science* 9: 81.

Allen, T.G., Bullock, J., Yang, X. et al. (2019). Passivating contacts for crystalline silicon solar cells. *Nature Energy* 4: 914–928.

Almansouri, I., Ho-Baillie, A., Bremner, S.P., and Green, M.A. (2015). Supercharging silicon solar cell performance by means of multijunction concept. *IEEE Journal of Photovoltaics* 5 (3): 968–976.

Aydin, E., Troughton, J., De Bastiani, M. et al. (2018). Room-temperature-sputtered nanocrystalline nickel oxide as hole transport layer for p-i-n perovskite solar cells. *ACS Applied Energy Materials* 1 (11): 6227.

Aydin, E., De Bastiani, M., Yang, X. et al. (2019). Zr-doped indium oxide (IZrO) transparent electrodes for perovskite-based tandem solar cells. *Advanced Functional Materials* 29 (25): 1901741.

Aydin, E., Allen, T.G., De Bastiani, M. et al. (2020). Interplay between temperature and bandgap energies on the outdoor performance of perovskite/silicon tandem solar cells. *Nature Energy* 5: 851–859.

Aydin, E., Altinkaya, C., Smirnov, Y. et al. (2021a). Sputtered transparent electrodes for optoelectronic devices: Induced damage and mitigation strategies. *Matter* 4 (11): 3549–3584.

Aydin, E., Liu, J., Ugur, E. et al. (2021b). Ligand-bridged charge extraction and enhanced quantum efficiency enable efficient n-i-p perovskite/silicon tandem solar cells. *Energy & Environmental Science* 14: 4377–4390.

Babics, M., De Bastiani, M., Balawi, A.H. et al. (2022). Unleashing the full power of perovskite/silicon tandem modules with solar trackers. *ACS Energy Letters* 7 (5): 1604–1610.

Babics, M., De Bastiani, M., Ugur, E. et al. (2023). One-year outdoor operation of monolithic perovskite/silicon tandem solar cells. *Cell Reports Physical Science* 4: 101280.

Bailie, C.D., Christoforo, M.G., Mailoa, J.P. et al. (2015). Semi-transparent perovskite solar cells for tandems with silicon and CIGS. *Energy & Environmental Science* 8: 956–963.

Ballif, C., Haug, F.-J., Boccard, M. et al. (2022). Status and perspectives of crystalline silicon photovoltaics in research and industry. *Nature Reviews Materials* 7: 597–616.

Battaglia, C., Cuevas, A., and De Wolf, S. (2016). High efficiency crystalline silicon solar cells: status and perspectives. *Energy & Environmental Science* 9: 1552–1576.

Bush, K.A., Palmstrom, A., Yu, Z. et al. (2017). 23.6%-efficient monolithic perovskite/silicon tandem solar cells with improved stability. *Nature Energy* 2: 17009.

Cheacharoen, R., Boyd, C.C., Burkhard, G.F. et al. (2018). Encapsulating perovskite solar cells to withstand damp heat and thermal cycling. *Sustainable Energy & Fuels* 2: 2398–2406.

Chen, B., Bai, Y., Yu, Z. et al. (2016). Efficient semitransparent perovskite solar cells for 23.0%-efficiency perovskite/silicon four-terminal tandem cells. *Advanced Energy Materials* 6 (19): 1601128.

Chen, B., Baek, S.-W., Hou, Y. et al. (2020a). Enhanced optical path and electron diffusion length enable high-efficiency perovskite tandems. *Nature Communications* 11: 1257.

Chen, B., Yu, Z.J., Manzoor, S. et al. (2020b). Blade-coated perovskites on textured silicon for 26%-efficient monolithic perovskite/silicon tandem solar cells. *Joule* 4 (4): 850–864.

CSEM (2022). New World Records: Perovskite-On-Silicon-Tandem Solar Cells. Retrieved March 21, 2023, from: https://www.csem.ch/press/new-world-records-perovskite-on-silicon-tandem-solar?pid=172296

De Bastiani, M., Mirabelli, A.J., Hou, Y. et al. (2021a). Efficient bifacial monolithic perovskite/silicon tandem solar cells via bandgap engineering. *Nature Energy* 6: 167–175.

De Bastiani, M., Babics, M., Aydin, E. et al. (2021b). All set for efficient and reliable perovskite/silicon tandem photovoltaic modules? *Solar RRL* 6 (3): 2100493.

De Bastiani, M., Van Kerschaver, E., Jeangros, Q. et al. (2021c). Toward stable monolithic perovskite/silicon tandem photovoltaics: a six-month outdoor performance study in a hot and humid climate. *ACS Energy Letters* 6 (8): 2944–2951.

De Bastiani, M., Subbiah, A.S., Babics, M. et al. (2022). Bifacial perovskite/silicon tandem solar cells. *Joule* 6: 1431–1445.

De Wolf, S. and Kondo, M. (2009). Nature of doped a-Si:H/c-Si interface recombination. *Journal of Applied Physics* 105 (10): 103707.

Dewi, H.A., Wang, H., Li, J. et al. (2018). Highly efficient semitransparent perovskite solar cells for four terminal perovskite-silicon tandems. *ACS Applied Materials & Interfaces* 11: 34178–34187.

Duong, T., Lal, N., Grant, D. et al. (2016). Semitransparent perovskite solar cell with sputtered front and rear electrodes for a four-terminal tandem. *IEEE Journal of Photovoltaics* 6 (3): 679–687.

Duong, T., Pham, H., Kho, T.C. et al. (2020). High efficiency perovskite-silicon tandem solar cells: effect of surface coating versus bulk incorporation of 2D perovskite. *Advanced Energy Materials* 10 (9): 1903553.

Duong, T., Nguyen, T., Huang, K. et al. (2023). Bulk incorporation with 4-methylphenethylammonium chloride for efficient and stable methylammonium-free perovskite and perovskite-silicon tandem solar cells. *Advanced Energy Materials* 13 (9): 2203607.

Essig, S., Allebé, C., Remo, T. et al. (2017). Raising the one-sun conversion efficiency of III–V/Si solar cells to 32.8% for two junctions and 35.9% for three junctions. *Nature Energy* 2: 17144.

Filipič, M., Löper, P., Niesen, B. et al. (2015). $CH_3NH_3PbI_3$ perovskite/silicon tandem solar cells: characterization based optical simulations. *Optics Express* 23 (7): A263.

Giuliano, G., Bonasera, A., Arrabito, G., and Pignataro, B. (2021). Semitransparent perovskite solar cells for building integration and tandem photovoltaics: design strategies and challenges. *Solar RRL* 5 (12): 2100702.

Green, M.A. (2009). The path to 25% silicon solar cell efficiency: history of silicon cell evolution. *Progress in Photovoltaics: Research and Applications* 17: 183–189.

Green, M.A., Dunlop, E.D., Yoshita, M. et al. (2023). Solar cell efficiency tables (Version 62). *Progress in Photovoltaics: Research and Applications* 31 (7): 651–663.

Haegel, N.M., Atwater, H., Barnes, T. et al. (2019). Terawatt-scale photovoltaics: transform global energy. *Science* 364 (6443): 836–838.

Heydarian, M., Messmer, C., Bett, A.J. et al. (2023). Maximizing current density in monolithic perovskite silicon tandem solar cells. *Solar RRL* 2200930.

Holman, Z.C., Filipic, M., Descoeudres, A. et al. (2013). Infrared light management in high-efficiency silicon heterojunction and rear-passivated solar cells. *Journal of Applied Physics* 113 (1): 013107.

Hou, Y., Aydin, E., De Bastiani, M. et al. (2020). Efficient tandem solar cells with solution-processed perovskite on textured crystalline silicon. *Science* 367: 1135–1140.

HZB (2022). World record back at HZB: Tandem solar cell achieves 32.5 percent efficiency. Retrieved March 21, 2023, from: https://www.helmholtz-berlin.de/pubbin/news_seite?nid=24348;sprache=en.

Isikgor, F.H., Furlan, F., Liu, J. et al. (2021). Concurrent cationic and anionic perovskite defect passivation enables 27.4% perovskite/silicon tandems with suppression of halide segregation. *Joule* 5 (6): 1566.

Isikgor, F.H., Zhumagali, S., Merino, L.V.T. et al. (2023). Molecular engineering of contact interfaces for high-performance perovskite solar cells. *Nature Reviews Materials* 8 (2): 89.

Jaysankar, M., Qui, W., van Eerden, M. et al. (2017). Four-terminal perovskite/silicon multijunction solar modules. *Advanced Energy Materials* 7 (15): 1602807.

Jeng, J.-Y., Chiang, Y.-F., Lee, M.-H. et al. (2013). $CH_3NH_3PbI_3$ perovskite/fullerene planar-heterojunction hybrid solar cells. *Advanced Materials* 25 (27): 3727.

Jošt, M., Köhnen, E., Morales-Vilches, A.B. et al. (2018). Textured interfaces in monolithic perovskite/silicon tandem solar cells: advanced light management for improved efficiency and energy yield. *Energy & Environmental Science* 11: 3511–3523.

Jošt, M., Kegelmann, L., Korte, L., and Albrecht, S. (2020). Monolithic perovskite tandem solar cells: a review of the present status and advanced characterization methods toward 30% efficiency. *Advanced Energy Materials* 10 (26): 1904102.

Khenkin, M.V., Katz, E.A., Abate, A. et al. (2020). Consensus statement for stability assessment and reporting for perovskite photovoltaics based on ISOS procedures. *Nature Energy* 5: 35–49.

Kim, H.-S., Lee, C.-R., Im, J.-H. et al. (2012). Lead iodide perovskite sensitized all-solid-state submicron thin film mesoscopic solar cell with efficiency exceeding 9%. *Scientific Reports* 2 (1): 591.

Kim, D., Jung, H.J., Park, I.K. et al. (2020). Efficient, stable silicon tandem cells enabled by anion-engineered wide-bandgap perovskites. *Science* 368: 155–160.

Lehr, J., Langehorst, M., Schmager, R. et al. (2020). Energy yield of bifacial textured perovskite/silicon tandem photovoltaic modules. *Solar Energy Materials and Solar Cells* 208: 110367.

Liu, J., Aydin, E., Yin, J. et al. (2021). 28.2%-efficient, outdoor-stable perovskite/silicon tandem solar cell. *Joule* 5 (12): 3169–3186.

Liu, J., De Bastiani, M., Aydin, E. et al. (2022a). Efficient and stable perovskite-silicon tandem solar cells through contact displacement by $MgF_x$. *Science* 377 (6603): 302–306.

Liu, Z., Zhu, C., Luo, H. et al. (2022b). Grain regrowth and bifacial passivation for high-efficiency wide-bandgap perovskite solar cells. *Advanced Energy Materials* 13 (2): 2203230.

Löper, P., Moon, S.-J., Martin de Nicolas, S. et al. (2015a). Organic-inorganic halide perovskite/crystalline silicon four-terminal tandem solar cells. *Physical Chemistry Chemical Physics* 17: 1619.

Löper, P., Stuckelberger, M., Niesen, B. et al. (2015b). Complex refractive index spectra of $CH_3NH_3PbI_3$ perovskite thin films determined by spectroscopic ellipsometry and spectrophotometry. *Journal of Physical Chemistry Letters* 6 (1): 66.

Mailoa, J.P., Bailie, C.D., Johlin, E.C. et al. (2015). A 2-terminal perovskite/silicon multijunction solar cell enabled by a silicon tunnel junction. *Applied Physics Letters* 106 (12): 121105.

Mariotti, S., Jäger, K., Diederich, M. et al. (2022). Monolithic perovskite/silicon tandem solar cells fabricated using industrial p-type polycrystalline silicon on oxide/passivated emitter and rear cell silicon bottom cell technology. *Solar RRL* 6 (4): 2101066.

Mazzarella, L., Lin, Y.-H., Kirner, S. et al. (2019). Infrared light management using a nanocrystalline silicon oxide interlayer in monolithic perovskite/silicon heterojunction tandem solar cells with efficiency above 25%. *Advanced Energy Materials* 9 (14): 1803241.

McMeekin, D.P., Sadoughi, G., Rehman, W. et al. (2016). A mixed-cation lead mixed-halide perovskite absorber for tandem solar cells. *Science* 351 (6269): 151–155.

Mishima, R., Hino, M., Kanematsu, M. et al. (2022). 28.3% efficient perovskite-silicon tandem solar cells with mixed self-assembled monolayers. *Applied Physics Express* 15: 076503.

Morales-Masis, M., De Nicolas, S.M., Holovsky, J. et al. (2015). Low-temperature high-mobility amorphous IZO for silicon heterojunction solar cells. *IEEE Journal of Photovoltaics* 5 (5): 1340.

Morales-Masis, M., De Wolf, S., Woods-Robinson, R. et al. (2017). Transparent electrodes for efficient optoelectronics. *Advanced Electronic Materials* 3 (5): 1600529.

Mujahid, M., Chen, C., Zhang, J. et al. (2020). Recent advances in semitransparent perovskite solar cells. *InfoMat* 3 (1): 101–124.

NREL (2023). Interactive Best Research-Cell Efficiency Chart. Retrieved Dec 20, 2023, from: https://www.nrel.gov/pv/interactive-cell-efficiency.html

OxfordPV (2018). Oxford PV perovskite solar cell achieves 28% efficiency. Retrieved March 21, 2023, from: https://www.oxfordpv.com/news/oxford-pv-perovskite-solar-cell-achieves-28-efficiency.

OxfordPV (2020). Oxford PV hits new world record for solar cell. Retrieved March 21, 2023, from: https://www.oxfordpv.com/news/oxford-pv-hits-new-world-record-solar-cell.

PACT (2023). PACT – publications and protocols. Retrieved March 21, 2023, from: https://pvpact.sandia.gov/publications-and-protocols/

Patel, M.T., Asadpour, R., Jahangir, J.B. et al. (2023). Current-matching erases the anticipated performance gain of next-generation two-terminal perovskite-Si tandem solar farms. *Applied Energy* 329: 120175.

Peng, J., Duong, T., Zhou, X. et al. (2016). Efficient indium-doped $TiO_x$ electron transport layers for high-performance perovskite solar cells and perovskite-silicon tandems. *Advanced Energy Material* 7 (4): 1601768.

Quiroz, C.O.R., Shen, Y., Salvador, M. et al. (2018). Balancing electrical and optical losses for efficient 4-terminal Si-perovskite solar cells with solution processed percolation electrodes. *Journal of Materials Chemistry A* 6: 3583–3592.

Razzaq, A., Allen, T.G., Liu, W. et al. (2022). Silicon heterojunction solar cells: techno-economic assessment and opportunities. *Joule* 6 (3): 514.

Rohatgi, A., Zhu, K., Tong, J. et al. (2020). 26.7% efficient 4-terminal perovskite–silicon tandem solar cell composed of a high-performance semitransparent perovskite cell and a doped poly-Si/SiO_x passivating contact silicon cell. *IEEE Journal of Photovoltaics* 10 (2): 417–422.

Ross, M., Severin, S., Stuz, M.B. et al. (2021). Co-evaporated formamidinium lead iodide based perovskites with 1000 h constant stability for fully textured monolithic perovskite/silicon tandem solar cells. *Advanced Energy Material* 11 (35): 2101460.

Sahli, F., Werner, J., Kamino, B.A. et al. (2018). Fully textured monolithic perovskite/silicon tandem solar cells with 25.2% power conversion efficiency. *Nature Materials* 17: 820–826.

Schulze, P.S.C., Bett, A.J., Bivour, M. et al. (2020). 25.1% high-efficiency monolithic perovskite silicon tandem solar cell with a high bandgap perovskite absorber. *Solar RRL* 4 (7): 2000152.

Snaith, H.J. (2013). Perovskites: the emergence of a new era for low-cost, high-efficiency solar cells. *Journal of Physical Chemistry Letters* 4 (21): 3623.

Subbiah, A.S., Isikgor, F.H., Howells, C.T. et al. (2020). High-performance perovskite single-junction and textured perovskite/silicon tandem solar cells via slot-die-coating. *ACS Energy Letters* 5 (9): 3034–3040.

Sveinbjörnsson, K., Li, B., Mariotti, S. et al. (2022). Monolithic perovskite/silicon tandem solar cell with 28.7% efficiency using industrial silicon bottom cells. *ACS Energy Letters* 7: 2654–2656.

Tockhorn, P., Sutter, J., Cruz, A. et al. (2022). Nano-optical designs for high-efficiency monolithic perovskite–silicon tandem solar cells. *Nature Nanotechnology* 17: 1214–1221.

Wang, Z., Zhu, X., Zuo, S. et al. (2019). 27%-efficiency four-terminal perovskite/silicon tandem solar cells by sandwiched gold nanomesh. *Advanced Functional Materials* 30 (4): 1908298.

Werner, J., Dubuis, G., Walter, A. et al. (2015). Sputtered rear electrode with broadband transparency for perovskite solar cells. *Solar Energy Materials and Solar Cells* 141: 407.

Werner, J., Weng, C.-H., Walter, A. et al. (2016a). Efficient monolithic perovskite/silicon tandem solar cell with cell area >1 cm². *Journal of Phyical Chemistry Letters* 7 (1): 161.

Werner, J., Barraud, L., Walter, A. et al. (2016b). Efficient near-infrared-transparent perovskite solar cells enabling direct comparison of 4-terminal and monolithic perovskite/silicon tandem cells. *ACS Energy Letters* 1: 474–480.

White, T.P., Lal, N.N., and Catchpole, K.R. (2014). Tandem solar cells based on high-efficiency c-Si bottom cells: top cell requirements for >30% efficiency. *IEEE Journal of Photovoltaics* 4 (1): 208–214.

Wu, Y., Yan, D., Peng, J. et al. (2017). Monolithic perovskite/silicon-homojunction tandem solar cell with over 22% efficiency. *Energy & Environmental Science* 10 (11): 2472–2479.

Wu, Y., Zheng, P., Peng, J. et al. (2022). 27.6% perovskite/c-Si tandem solar cells using industrial fabricated TOPCon device. *Advanced Energy Materials* 12: 2200821.

Xu, J., Boyd, C.C., Yu, Z.J. et al. (2020). Triple-halide wide-band gap perovskites with suppressed phase segregation for efficient tandems. *Science* 367 (6482): 1097–1104.

Yan, D., Cuevas, A., Ibarra Michel, J. et al. (2021). Polysilicon passivated junctions: the next technology for silicon solar cells? *Joule* 5 (4): 811.

Yang, D., Zhang, X., Hou, Y. et al. (2021). 28.3%-efficiency perovskite/silicon tandem solar cell by optimal transparent electrode for high efficient semitransparent top cell. *Nano Energy* 84: 105934.

Yu, Z., Leilaeioun, M., and Holman, Z. (2016). Selecting tandem partners for silicon solar cells. *Nature Energy* 1: 16137.

Zhang, D., Najafi, M., Zardetto, V. et al. (2018). High efficiency 4-terminal perovskite/c-Si tandem cells. *Solar Energy Materials and Solar Cells* 188: 1–5.

Zhao, J., Wang, A., and Green, M.A. (1999). 24.5% efficiency silicon PERT cells on MCZ substrates and 24.7% efficiency PERL cells on FZ substrates. *Progress in Photovoltaics: Research and Applications* 7: 471–474.

Zhumagali, S., Isikgor, F.H., Maity, P. et al. (2021). Linked nickel oxide/perovskite interface passivation for high-performance textured monolithic tandem solar cells. *Advanced Energy Materials* 11 (40): 2101662.

# 13

## An Overview of Chalcogenide Thin Film Materials for Tandem Applications

*Bart Vermang*

Hasselt University and Imec, Hasselt, Belgium

## 13.1 Chalcogenide Thin Film Materials With Tuneable Bandgap

For over 135 years, numerous chalcogenides have been investigated for their remarkable photovoltaic (PV) material properties. Chalcogenide PV materials have the longest history, with the first solar cells dating back to 1883 when Charles Fritts made a selenium (Se) solar cell using a metal foil coated with selenium and a thin layer of gold (Fritts 1883). In the meanwhile, many other materials have been explored, ranging from single elemental (i.e. Se) to binary [e.g. CdTe, $Sb_2(S,Se)_3$] and multinary compounds [e.g. chalcopyrite $Cu(In,Ga)(S,Se)_2$ (CIGSSe) and kesterite $Cu_2ZnSn(S, Se)_4$ (CZTSSe)] (Green et al. 2023; Zakutayev et al. 2021). These materials are direct bandgap semiconductors, which are suitable absorber materials for solar cells thanks to their high absorption coefficients, favorable defect properties and solar cells, which are intrinsically stable in operation (Wilson et al. 2020).

Despite their unique features, market penetration remains challenging for commercial single-junction chalcogenide PV technologies. Both CIGSSe and CdTe have shown high solar cell efficiencies [i.e. 23.35% and 22.1%, respectively (Green et al. 2023)], combined with high energy yield and reliability, and at low material consumption, short energy payback times (Celik et al. 2017) and low carbon footprint (Gibon et al. 2022). Based on their intrinsic advantages, these thin film technologies are suitable for various applications, such as utility-scale, rooftop and (building) integrated PV using colored, patterned and/or flexible products. Fully vertically integrated production facilities permit low-cost production on a GW scale, eventually even competitive with silicon (Si) technologies (Heske et al. 2019). However, the chalcogenide thin-film PV market share remains small, i.e. 8.4 GWp for CdTe and 1.6 GWp for CIGSSe in 2021, led by First Solar and Solar Frontier (SF), respectively (Philipps and Warmuth 2023).

Today, their tunable bandgap puts chalcogenide thin film materials at the forefront of tandem research. As a key example (see Sections 13.3 and 13.4), Figure 13.1 gives an

☆ Partners in EnergyVille and Solliance.

*Photovoltaic Solar Energy: From Fundamentals to Applications*, *Volume 2*, First Edition.
Edited by Wilfried van Sark, Bram Hoex, Angèle Reinders, Pierre Verlinden, and Nicholas J. Ekins-Daukes.
© 2024 John Wiley & Sons Ltd. Published 2024 by John Wiley & Sons Ltd.
Companion website: www.wiley.com/go/PVsolarenergy

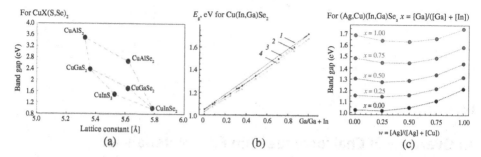

**Figure 13.1** Bandgap versus alloy composition, based on experimental and/or theoretical data. It is shown that the bandgap is determined by chalcopyrite absorber composition, i.e. the ratio of (a) Se to S, (b) Ga to In, or (c) Ag to Cu. Source: Taken from Keller et al. (2020a), Klenk et al. (1993), and Mudryi et al. (2010).

overview of possible bandgap tuning for various chalcopyrite semiconductor alloys, where the bandgap is determined by absorber composition, i.e. the ratio of Se to S as well as Ga to In or even Ag to Cu (Keller et al. 2020a; Klenk et al. 1993; Mudryi et al. 2010). This property is very appealing for optimization of tandem solar cells, where two absorber materials with different bandgaps are combined to reach higher efficiencies. As they do not induce additional area-related balance-of-system costs, such high-efficiency tandem solar cells can enable a reduction of the levelized cost of energy (Jošt et al. 2020). Chalcogenide-based high-efficiency tandem technologies would be attractive for low-weight and flexible applications, whereas Si wafer-based heavy and rigid modules have their limitations (Lincot et al. 2021).

## 13.2 Tandem Applications

There are two leading tandem architectures, i.e. two-terminal (2T) monolithically integrated tandems, which are commercially more interesting, and mechanically stacked 4T tandems, which are possibly more appropriate as study devices. Figure 13.2a provides sketches of both concepts. 2T advantages are reduced material usage, lower parasitic losses (due to a reduced number of transparent conductive oxides), and the use of a single electric circuit. Its disadvantage is the need for current matching between the top and bottom cells. Conventional 2T devices are realized by monolithic cell interconnection, which requires a stable and efficient tunnel/recombination layer and a bottom cell that is stable during the top cell deposition process. The 4T concept allows independent manufacturing of both cells and thus simplification of the growth process. It is also less dependent on the incoming spectrum or asymmetric degradation of both cells. However, material usage, optical losses, and complexity in wiring, voltage conversion, and panel-level electronics are augmented (Feurer 2019; Todorov et al. 2018).

The anticipated ideal top cell band gap range is 1.5–1.9 eV for 4T and even 1.6–1.8 eV in the case of 2T, while for the bottom cell, this range is 0.9–1.1 eV for 4T and very close to 1 eV for 2T. Figure 13.2b provides calculated efficiencies for 2T and 4T tandem devices as a

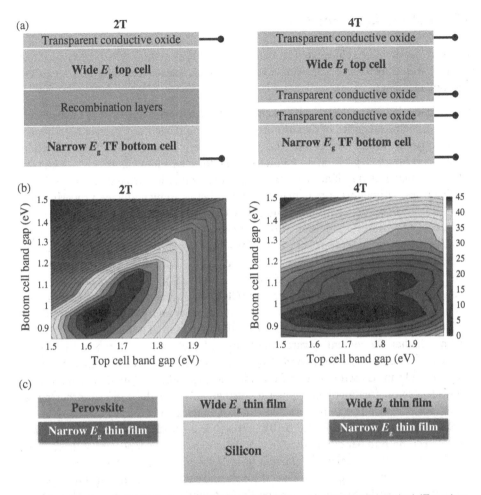

**Figure 13.2** (a) Monolithically integrated 2-terminal (2T) and mechanically stacked 4T tandem concepts. (b) Calculated efficiencies using the detailed, balanced limit (without optical losses) for both concepts. (c) Considered tandem architectures are incorporating chalcogenide thin film materials. Source: (b) Feurer (2019)/ETH Zurich.

function of top and bottom cell bandgaps. This calculation shows a narrow region of highest efficiency for 2T devices (due to current matching), while for 4T, the bandgap restrictions on suitable partners are more relaxed. However, in both configurations, a narrow band gap bottom cell is preferable to enable high tandem efficiencies (Feurer 2019). Figure 13.2c gives an overview of the considered tandem architectures with chalcogenide top and/or bottom cell(s), i.e. combining narrow bandgap ($E_g$) chalcogenide bottom and perovskite top cells, Si bottom and wide bandgap chalcogenide top cells, or narrow and wide bandgap chalcogenide cells. The first and last approaches enable full thin film tandems and, thus, low-weight and flexible applications.

## 13.3 Bottom Cell Candidates

A list of narrow bandgap chalcogenide materials suitable as bottom cells are provided and discussed, with a focus on record single junction efficiencies – see Table 13.1 – and key fabrication elements.

(1) Pure selenide Cu(In,Ga)Se$_2$ (CIGSe) chalcopyrite can have a bandgap of 1.1 eV. There are two main categories of deposition techniques employed to fabricate high-quality chalcopyrites, i.e. a 1-step reactive co-deposition process or a two-step process where metal precursor deposition is followed by sulfo-selenization. Solar cells based on Cu-poor material (i.e. material with a sub-stoichiometric Cu content) show better efficiencies, although Cu-rich material shows better semiconductor properties (i.e. with lower defect concentrations and better transport properties) but typically lower efficiency due to recombination at or near the interface with the n-type buffer layer (hence the importance of surface passivation and buffer layer optimization) (Siebentritt 2017a). Best co-deposition materials enabling an efficiency of 22.6% are fabricated at the Center for Solar Energy and Hydrogen Research (ZSW) (Jackson et al. 2016). They co-evaporate Cu, In, Ga, and Se to grow the CIGSe in a multistage process. This process is based on the interdiffusion of the different elements and yields better crystallinity of the absorber layer. It naturally results in the formation of a double grading profile with a higher Ga content towards the front and the back interfaces and lower Ga content in the central-front region. This Ga grading profile (which is directly related to the band gap) is used to reduce front and rear surface recombination, and an alkali post-deposition treatment is applied to further passivate the front absorber/buffer interface. The best two-step processed material enables an efficiency of 23.35% and is produced at SF (Nakamura et al. 2019). In this process, first CuGa/In metal precursors with a certain number of additives (e.g. Na) are sequentially deposited, followed by a sulfurization-after-selenization process to turn them into CIG(S)Se absorber layers. These layers have Ga accumulation at the back and sulfur incorporation at the front for rear and front surface passivation, respectively. Additionally, they use Zn(O,S)/ZnMgO double buffer layers as an alternative to the conventional CdS buffer layer. A best efficiency of 21.4% on a flexible polyimide substrate is achieved by the Swiss Federal

**Table 13.1** Reported champion efficiencies of important single-junction narrow bandgap chalcogenide solar cells, measured under standard test conditions. See text for details.

| Absorber | Bandgap (eV) | Substrate | $J_{SC}$ (mA/cm$^2$) | $V_{OC}$ (mV) | FF (%) | Eff. (%) | Institute |
|---|---|---|---|---|---|---|---|
| CIGSe | 1.1 | Glass | 37.8 | 741 | 80.6 | 22.6 | ZSW (Jackson et al. 2016) |
| CIG(S)Se | 1.1 | Glass | 39.6 | 734 | 80.4 | 23.4 | SF (Nakamura et al. 2019) |
| CIGSe | 1.1 | Polymer | 37.4 | 747 | 76.7 | 21.4 | EMPA (Carron et al. 2022) |
| CISe | 1.0 | Glass | 42.3 | 609 | 74.6 | 19.2 | EMPA (Feurer et al. 2019) |
| CZTSe | 1.0 | Glass | 37.4 | 491 | 68.2 | 12.5 | UNSW (Li et al. 2020) |
| Sb$_2$Se$_3$ | 1.2 | Glass | 32.6 | 400 | 70.3 | 9.2 | HKL (Li et al. 2019) |

Laboratories for Materials Science and Technology (EMPA) (Carron et al. 2019, 2022). They employed a very similar multistage co-evaporation process as ZSW at lower temperatures to protect the substrate, but applied a comparable heavy alkali treatment and an optimized heat-light soaking treatment. Recently, they increased this flexible record efficiency to 22.2% (Bellini 2022), most likely by adding a small amount of Ag to widen the absorber deposition temperature window of high-performance CIGS absorber layers (Yang et al. 2021).

(2) Gallium-free $CuInSe_2$ (CISe) chalcopyrite even has a more favorable 1.0 eV bandgap. Figure 13.2b confirms that such a 1.0 eV bottom cell is preferable to enable (i) the highest tandem efficiencies and (ii) using slightly lower bandgap top cells, which typically also have higher efficiencies. This opportunity to tune the bandgap of chalcogenide materials to 1.0 eV is a clear advantage compared to Si as the bottom cell candidate, which will always have a 1.1 eV bandgap. EMPA has been focusing on developing CISe solar cells and already reached 19.2% efficiency (Feurer et al. 2018, 2019). Optimizations they used are Cu-rich absorber material with high doping concentration and low defect density, single bandgap grading by adding Ga close to the back as rear surface passivation, a heavy alkali (RbF) treatment to passivate the front surface, and a high mobility front transparent conductive oxide. Alloying CISe with Ag is a feasible approach to further improve J–V parameters while maintaining the 1.0 eV bandgap (Valdes et al. 2019). Such Ag alloyed devices show slightly increased $J_{sc}$ attributed to an improved current collection at long wavelengths (because of increased space charge width) and are thus beneficial for potential use in tandem solar cells.

(3) $Cu_2(Zn,Sn)Se_4$ (CTZSe) kesterite (1.0 eV) and other emerging materials (Zakutayev et al. 2021), e.g. $Sb_2Se_3$ (1.2 eV) are investigated as alternative materials since they are free of critical raw materials (CRMs). Indium, gallium, and tellurium are classified as CRM by the European Commission (Internal Market, Industry, Entrepreneurship and SMEs 2023). The Commission recommends their partial or total substitution in PV technologies and, hence, supports ongoing research and development of kesterites and other emerging materials. Note that an international group of leading chalcogenide research institutes has compiled information showing that indium is an earth-abundant element and thus should not be on the CRM list (Lincot et al. 2021). Pure selenide kesterite has a favorable 1.0 eV bandgap, and the best two-step processed CZTSe solar cell has been reported by the University of New South Wales (UNSW) with an efficiency of 12.5% (Li et al. 2020). They combined sputtering of precursors and a soft-selenization process to obtain high-quality absorber layers with a low intrinsic defect density. Alternatively, one could also think of $Sb_2Se_3$ solar cells, where the best efficiency is 9.2% using high-quality $Sb_2Se_3$ nanorod arrays, as obtained by the Hebei Key Laboratory (HKL) (Chen et al. 2022; Li et al. 2019).

## 13.4 Top Cell Candidates

Interesting wide bandgap chalcogenide materials suitable as top cells are also reviewed, first focusing on opaque single junction efficiencies – see Table 13.2 – and key fabrication elements.

**Table 13.2** Reported champion efficiencies of important and opaque single-junction wide bandgap chalcogenide solar cells, measured under standard test conditions. See text for details.

| Absorber | Bandgap (eV) | Substrate | $J_{sc}$ (mA/cm$^2$) | $V_{oc}$ (mV) | FF (%) | Eff. (%) | Institute |
|---|---|---|---|---|---|---|---|
| CIGS | 1.6 | Glass | 22.5 | 925 | 77.0 | 16.0 | IMN (Barreau et al. 2022) |
| CIGS | 1.6 | Glass | 24.8 | 954 | 71.5 | 16.9 | SF (Sugimoto et al. 2017) |
| CGSe | 1.6 | Glass | 17.5 | 1017 | 67.0 | 11.9 | UU (Larsson et al. 2017) |
| ACIGSe | 1.5 | Glass | 23.3 | 875 | 74.1 | 15.1 | UU (Keller et al. 2020a) |
| ACIGSe | 1.6 | Glass | 16.0 | 910 | 70.2 | 10.2 | UU (Keller et al. 2020b) |
| CdZnTe | 1.8 | Si | 15.9 | 1340 | 77.0 | 16.4 | EPIR Tech. (Carmody et al. 2010) |
| CdMgTe | 1.7 | InSb | 15.0 | 1176 | 63.5 | 11.2 | ASU (Becker et al. 2018) |
| CZTS | 1.5 | Glass | 21.7 | 731 | 69.3 | 11.0 | UNSW (Yan et al. 2018) |
| Sb$_2$S$_3$ | 1.7 | Glass | 16.1 | 711 | 65.0 | 7.5 | KRICT (Choi et al. 2014) |
| Sb$_2$(S,Se)$_3$ | 1.5 | Glass | 23.7 | 630 | 67.7 | 10.1 | CAS/UNSW (Tang et al. 2020) |

(1) Pure sulfide Cu(In,Ga)S$_2$ (CIGS) chalcopyrite can have a bandgap of 1.6 eV. Such solar cells are known to be limited by large $V_{oc}$ deficits due to bulk and surface recombination. This can be reduced by using Cu-poor and Ga-containing absorbers and a buffer layer with suitably high conduction band edge (Shukla et al. 2021). Best co-deposition materials have been fabricated at the Jean Rouxel Institute of Materials (IMN), obtaining a record cell efficiency of 16.0% (Barreau et al. 2022). They applied a modified three-stage process using graded In and Ga fluxes during the first stage. The resulting absorbers are single-phase and made of large grains extended throughout the entire film thickness. Replacing the standard CdS buffer with Zn(O,S) leads to increased voltage and current but lower performance due to decreased fill factor. Best two-step processed CIGS materials have been manufactured at SF by sulfurization with H$_2$S gas in a furnace, leading to a champion efficiency of 16.9% (Hiroi et al. 2016; Sugimoto et al. 2017). Similar enhancements for the co-deposition were used, i.e. optimization of the sulfurization temperature, the Ga grading profile, and a wide bandgap ZnMgO buffer layer. As compared to more performant CIGSe solar cells, there is still a sizeable efficiency gap due to large $V_{oc}$ deficit and high series resistance. The current understanding is that the short minority carrier lifetime and the low carrier density are challenging further efficiency improvement.

(2) Indium-free CuGaSe$_2$ (CGSe) and Ag-alloyed (Ag,Cu)(In,Ga)Se$_2$ (ACIGSe) are other chalcopyrite materials with bandgaps up to 1.6 eV. At Uppsala University (UU), using a 3-stage co-deposition process, they have shown CGSe solar cells exhibiting open-circuit voltages of up to 1017 mV and efficiencies of up to 11.9% (Larsson et al. 2017). This high $V_{oc}$ was accomplished by fitting the conduction band alignment very well to the n-type ZnSnO buffer layer, grown by atomic layer deposition. Temperature-dependent current-voltage measurements show that this high $V_{oc}$ can be explained by reduced interface recombination. It is assumed that further improvements in p–n junction

quality and CGSe growth techniques are still feasible (Ishizuka 2019). Using a similar process also record ACIGSe solar cells have been fabricated at UU (Keller et al. 2021, 2020a, 2020b), reaching efficiencies of 15.1 and 10.2% for 1.5 and 1.6 eV bandgaps, respectively. To reach bandgap energies suitable for top cells in a tandem device, a [Ga]/[Ga]+[In] ratio > 0.8 is necessary. Capacitance profiling reveals that the absorber doping gradually decreases toward stoichiometric composition, eventually leading to complete depletion. As only fully depleted stoichiometric samples show perfect carrier collection, a very low diffusion length is assumed to be the current bottleneck for these solar cells. A feasible way forward seems to be the introduction of heavy alkali elements to increase the carrier lifetime and optimization of the Ga profile to increase the "effective" diffusion length. Solar cells with $V_{oc}$'s up to 916 mV have already been achieved for 1.5 eV bandgap ACIGSe material.

(3) The conventional 1.4 eV band gap for CdTe is too low for a top cell, but by alloying with zinc (CdZnTe) or magnesium (CdMgTe), the bandgap can be increased to the 1.7–1.8 eV range. EPIR Technologies has reported above 16.4% efficiency for single-crystal CdZnTe-based solar cells grown epitaxially on a Si substrate (Carmody et al. 2010). They used molecular-beam epitaxy (MBE) to grow high-quality CdZnTe homojunction's of approximately 3 μm thick. In a very similar way, the Arizona State University (ASU) used MBE to grow high-quality monocrystalline MgCdTe solar cells with 11.2% efficiency on an InSb substrate (Becker et al. 2018). Both cells show very high open-circuit voltages close to the thermodynamic limit. A key reason is the use of a high-quality and thus expensive absorber growth technique, but it also shows the potential of these materials. Recently, SunPower announced that it is working with First Solar to develop a new type of tandem solar module pairing CdTe thin film with traditional silicon (Keller et al. 2021).

(4) Also, here, CRM-free $Cu_2(Zn,Sn)S_4$ (CTZS) kesterite (1.5 eV) and other emerging materials, e.g. $Sb_2S_3$ (1.7 eV) or Se ($\approx$1.95 eV), are considered. UNSW reported an 11% efficiency pure sulfide CZTS solar cell (Yan et al. 2018). The precursors were deposited by co-sputtering Cu/ZnS/SnS material and sulfurized within a combined sulfur and SnS atmosphere. They applied a post-heat treatment of the heterojunction to reduce heterojunction recombination, mainly due to more favorable conduction band alignment and contributing to reduced non-radiative recombination. As an alternative emerging material, one could consider $Sb_2S_3$ inorganic semiconductor-sensitized solar cells with a 7.5% record efficiency, as obtained by the Korea Research Institute of Chemical Technology (KRICT) (Shah et al. 2021; Tang et al. 2020). Chinese Academy of Sciences (CAS) Key Laboratory of Materials for Energy Conversion and UNSW developed a hydrothermal approach to deposit lower bandgap $Sb_2(S,Se)_3$ films. Varying the Se/S ratio and the post-deposition annealing temperature, they improved the film morphology, increased the grain size, and reduced the number of defects, achieving 10.1% champion efficiency (Tang et al. 2020). A final material choice with potentially very high bandgap could be Se. The Technical University of Denmark fabricated Se solar cells with an interesting record $V_{oc}$ of 990 mV and assessed that improving the bulk optoelectronic quality of selenium appears to be a more urgent need than optimizing its device structure (Nielsen et al. 2022).

## 13.5 Tandem Results

Integration of wide bandgap chalcogenide top cell candidates in 4T and 2T tandem concepts is still at a starting point. In the case of a 4T tandem with Si as bottom cell, the minimum top cell efficiency required has been evaluated in White et al.'s work (White et al. 2013). It is shown that targeting 30% tandem efficiency, this minimum top cell efficiency ranges from 22% for a bandgap of 1.5 eV to 15% for a bandgap of 1.9 eV. Such a top cell also needs to be semi-transparent, which entails a transparent back contact, as is indicated in Figure 13.2a. Currently, most research has been focused on wide bandgap chalcogenide material optimization (hence the focus on opaque cell efficiencies in Table 13.2) and not yet on the development of high-efficiency semi-transparent cells. A few noteworthy exceptions: (i) A study of $In_2O_3$:Sn (ITO) and $In_2O_3$:H (IOH) as transparent back contacts for ACIGS (1.44 eV) and CIGS (1.41 eV) solar cells. Using IOH, the record efficiency obtained for ACIGS is 12% and using IOH, the best CIGS device reaches 11.2% efficiency (Keller et al. 2022). (ii) A combination of an opaque 1.6 eV ACIGSe top cell with a 1.15 eV CIGSe bottom cell to reach 13.9% tandem efficiency applying beam splitting (Kim et al. 2018). And (iii) a 4T all antimony $Sb_2S_3$-on-$Sb_2Se_3$ tandem device delivering an efficiency of 7.93%, which outperforms the individually optimized top and bottom cells (Zhang et al. 2020). In the case of a 2T tandem approach, processing compatibility between top and bottom cells is required. This is a challenge due to the typically high temperature and harsh environment (e.g. Se) needed for growth of the top absorber material, i.e. on the bottom cell. Remarkable outcomes are 9.7% CGSe-on-Si and 16.8% CdZnTe-on-Si 2T tandem solar cells (Carmody et al. 2010; Jeong et al. 2017).

Very interesting results have been obtained using narrow bandgap CI(G)Se as bottom cell in 4T and 2T tandem approaches, as summarized in Table 13.3. Chalcogenide bottom cells can be rather easily combined with 1.6–1.7 eV perovskite top cells. Perovskite materials are typically crystallized at temperatures around 100 °C or less, not harming the chalcogenide material when integrated into 2T tandems (Jošt et al. 2020). The National Renewable Energy Laboratory (NREL) coupled a semi-transparent perovskite top cell with a standard 1.1 eV CIGSe bottom cell and achieved 25.9% 4T tandem efficiency (Kim et al. 2019). The semi-transparent perovskite top cell reached 17.1%, thanks to enhanced morphology ,reduced defect density and energetic disorder. Recently, the Karlsruhe Institute of Technology (KIT) and ZSW increased this record perovskite-on-CIGSe 4T tandem cell efficiency to 27.3%, mainly due to the use of an 18.5% perovskite top cell with improved light management and perovskite material composition (Ruiz-Preciado et al. 2022). Using spectral splitting, even 28.0% efficiency has already been shown for this tandem combination (Nakamura et al. 2020). All previous perovskite-on-CIGSe results have been obtained on glass, while the Netherlands Organisation for Applied Scientific Research (TNO) and MiaSolé joined forces to achieve a 23.0% 4T tandem efficiency on flexible substrates (Solliance 2019), which in the meanwhile has been improved to 26.5% (Solliance 2021). TNO also collaborated with EMPA, who provided a 1.0 eV CISe bottom sample with 18.7% single-cell efficiency, reaching 25.0% 4T tandem efficiency (Feurer et al. 2019). Excellent 2T perovskite-on-CIGSe tandem solar cells have been reported by Helmholtz-Zentrum Berlin (HZB), with a perovskite top cell fabricated directly on an as-grown, rough CIGSe bottom cell (Al-Ashouri et al. 2019; Green et al. 2023). To allow conformal perovskite growth on such a surface, molecules based

**Table 13.3** Notable perovskite-on-chalcogenide 4T and 2T tandem efficiencies, measured under standard test conditions. See text for details.

| Tandem concept | Absorbers | Bandgap (eV) | Substrate | $J_{SC}$ (mA/cm$^2$) | $V_{OC}$ (mV) | FF (%) | Eff. (%) | Institute(s) |
|---|---|---|---|---|---|---|---|---|
| 4T | Perovskite | 1.7 | Glass | 19.6 | 1120 | 77.8 | 17.1 | NREL |
|    | CIGSe | 1.1 | Glass | 15.6 | 715 | 79.2 | 8.8 | (Kim et al. 2019) |
|    | Tandem | | | | | | **25.9** | |
| 4T | Perovskite | 1.6 | Glass | 20.8 | 1172 | 79.3 | 19.3 | KIT |
|    | | | | | | | *18.5** | ZSW |
|    | CIGSe | 1.1 | Glass | 16.8 | 666 | 78.4 | 8.8 | (Ruiz-Preciado et al. 2022) |
|    | Tandem | | | | | | **27.3** | |
| 4T | Perovskite | 1.6 | Flexible | 19.9 | 1052 | 71.0 | 14.9 | TNO |
|    | CIGSe | 1.1 | Flexible | 15.5 | 675 | 77.4 | 8.1 | MiaSolé |
|    | Tandem | | | | | | **23.0** | (Solliance 2019) |
| 4T | Perovskite | 1.6 | Glass | 20.8 | 1034 | 79.8 | 17.2 | TNO |
|    | | | | | | | *16.9** | EMPA |
|    | CISe | 1.0 | Glass | 19.4 | 565 | 74.2 | 8.1 | (Feurer et al. 2019) |
|    | Tandem | | | | | | **25.0** | |
| 4T | Perovskite | 1.7 | Glass | 18.8 | 1230 | 77.3 | 17.9 | CityU |
|    | CZTSSe | 1.1 | Glass | 15.4 | 430 | 66.5 | 4.4 | NKU |
|    | Tandem | | | | | | **22.3** | (Wang et al. 2022) |
| 2T | Perovskite | 1.6 | Glass | 19.2 | 1768 | 72.9 | **24.2** | HZB |
|    | CIGSe | 1.1 | | | | | | (Al-Ashouri et al. 2019) |
| 2T | Perovskite | 1.6 | Glass | 21.1 | 1570 | 75.2 | **24.9** | KIT/TNO |
|    | CISe | 1.0 | | | | | | (Ruiz-Preciado et al. 2022) |

* After maximum power point tracking.

on carbazole bodies with phosphonic acid anchoring groups are used. These molecules can form self-assembled monolayers (SAMs) on various oxides, which are designed to create an energetically aligned interface to the perovskite absorber without non-radiative losses. This way, they obtained a 24.2% perovskite-on-CIGSe 2T tandem efficiency. KIT and TNO even obtained 24.9% 2T tandem efficiency (Ruiz-Preciado et al. 2022). They used a relatively planar surface profile and narrow band gap CISe bottom cell, which allowed exploiting the optoelectronic properties and photostability of a low-Br-containing perovskite top cell. Current matching was attained by proper tuning of the thickness and bandgap of the perovskite, along with the optimization of an antireflective coating for improved light in-coupling. Also for CRM-free alternative material CZTSSe, interesting results have already been obtained. City University of Hong Kong (CityU) and Nankai University (NKU) demonstrated a champion efficiency of 22.27% for all-solution-processed 4T perovskite on CZTSSe tandem solar cells (Wang et al. 2022), while IBM reported a 2T device with high $V_{oc}$ of 1350 mV but limited efficiency of 4.6% (Todorov et al. 2014).

## 13.6 Conclusions and Outlook

This overview summarizes high-potential top and bottom chalcogenide thin film materials for tandem solar cell applications. To start, the focus has been on reviewing remarkable single junction efficiencies and major fabrication aspects. Most interesting solar cells have been obtained using chalcopyrite materials since, for CdTe-based materials, typically too expensive processing is applied, and kesterite and other emerging materials lack voltage (Siebentritt 2017b). Attractive top cell candidates have been identified (i.e. CIGS, CGSe, and ACIGSe), but now the emphasis must shift to developing semi-transparent high bandgap top cells using processing parameters not harmful for the bottom cell. 4T tandem solar cells are more appropriate as study devices, i.e. to learn and optimize cell architectures, before integrating them into a 2T approach. Best bottom cell candidates (i.e. CIGSe and CISe) already followed this 2T after 4T methodology. When combining CIGSe bottom cells with wide bandgap perovskite top cells, very promising record 4T and 2T tandem cell efficiencies have been obtained, i.e. 27.3% and 24.9%, respectively. In this case, the emphasis must now change to the more favorable 1.0 eV CISe and CRM-free CZTSe bottom cells and 2T integration. Thin film tandem technologies are mostly attractive for low-weight and flexible applications. Hence, additional results employing flexible substrates are also essential. Currently, the best 4T perovskite-on-CIGSe tandem efficiency at mini-module scale ($3.8\,\text{cm}^2$) is 21.3% (Jaysankar et al. 2019), showing a rather large – and unfortunately typical for thin film PV – cell-to-module efficiency gap. Consequently, more effort into mini-module fabrication is also becoming important at this stage.

## Author Biography

**Bart Vermang** received an M.Sc. in physics from the University of Ghent and a Ph.D. in electrical engineering from the University of Leuven (both in Belgium). During his Ph.D., he studied silicon photovoltaics (PV) at Imec (Belgium), which was followed by two Postdoctoral fellowships to delve into thin film PV. A Marie Skłodowska-Curie fellowship to move to the University of Uppsala (Sweden) and a fellowship from the Flemish Research Foundation to return to Imec. In 2016, Bart acquired a European Research Council grant and consequently became a professor at Hasselt University (Belgium). Currently, he is coordinating the EU-funded project PERCISTAND, which concentrates on the development of perovskite-on-chalcogenide tandem technologies. Key research activities are the development of wide bandgap perovskite and narrow bandgap $CuInSe_2$ materials, suitable transparent conductive oxides, and integration into tandem configurations. The focus is on obtaining high efficiency, stability, and large-area manufacturability at low production cost and environmental footprint.

## References

Al-Ashouri, A., Magomedov, A., Roß, M. et al. (2019). Conformal monolayer contacts with lossless interfaces for perovskite single junction and monolithic tandem solar cells. *Energy & Environmental Science* 12 (11): 3356–3369.

Barreau, N., Bertin, E., Crossay, A. et al. (2022). Investigation of co-evaporated polycrystalline Cu (In, Ga) S2 thin film yielding 16.0% efficiency solar cell. *EPJ Photovoltaics* 13: 17.

Becker, J.J., Campbell, C.M., Tsai, C.-Y. et al. (2018). Monocrystalline 1.7-eV-bandgap MgCdTe solar cell with 11.2% efficiency. *IEEE Journal of Photovoltaics* 8 (2): 581–586.

Bellini, E. (2022). Swiss scientists achieve 22.2% efficiency for flexible CIGS solar cell. *pv magazine* .

Carmody, M., Mallick, S., Margetis, J. et al. (2010). Single-crystal II–VI on Si single-junction and tandem solar cells. *Applied Physics Letters* 96 (15).

Carron, R., Nishiwaki, S., Feurer, T. et al. (2019). Advanced alkali treatments for high-efficiency Cu (In, Ga) Se$_2$ solar cells on flexible substrates. *Advanced Energy Materials* 9 (24): 1900408.

Carron, R., Nishiwaki, S., Yang, S.-C., Ochoa, M., Sun, X., Feurer, T., et al. (2022). Heat-light soaking treatments for high-performance CIGS solar cells on flexible substrates. https://doi .org/10.21203/rs.3.rs-2116168/v1.

Celik, I., Philips, A.B., Song, Z. et al. (2017). Energy payback time (EPBT) and energy return on energy invested (EROI) of perovskite tandem photovoltaic solar cells. *IEEE Journal of Photovoltaics* 8 (1): 305–309.

Chen, C., Li, K., and Tang, J. (2022). Ten years of Sb2Se3 thin film solar cells. *Solar RRL* 6 (7): 2200094.

Choi, Y.C., Lee, D.U., Noh, J.H. et al. (2014). Highly improved Sb2S3 sensitized-inorganic–organic heterojunction solar cells and quantification of traps by deep-level transient spectroscopy. *Advanced Functional Materials* 24 (23): 3587–3592.

Feurer, T. (2019). *Narrow Band Gap Cu (In, Ga) Se₂ for Tandem Solar Cell Application*. ETH Zurich.

Feurer, T., Bissig, B., Weiss, T.P. et al. (2018). Single-graded CIGS with narrow bandgap for tandem solar cells. *Science and Technology of advanced MaTerialS* 19 (1): 263–270.

Feurer, T., Carron, R., Torres Sevilla, G. et al. (2019). Efficiency improvement of near-stoichiometric CuInSe2 solar cells for application in tandem devices. *Advanced Energy Materials* 9 (35): 1901428.

Fritts, C.E. (1883). On a new form of selenium cell, and some electrical discoveries made by its use. *American Journal of Science* 3 (156): 465–472.

Gibon, T., Menacho, Á.H., and Guiton, M. (2022). *Carbon Neutrality in the UNECE Region: Integrated Life-cycle Assessment of Electricity Sources*. United Nations Economic Commission for Europe (UNECE).

Green, M.A., Dunlop, E.D., Siefer, G. et al. (2023). Solar cell efficiency tables (Version 61). *Progress in Photovoltaics: Research and Applications* 31 (1): 3–16.

Heske, C., Daniel, L., Powalla, M., Salomé, P., Schlatmann, R., Tiwari, A. N., et al. (2019). CIGS White Paper 2019: CIGS-PV.

Hiroi, H., Iwata, Y., Adachi, S. et al. (2016). New world-record efficiency for pure-sulfide Cu (In, Ga) S$_2$ thin-film solar cell with Cd-free buffer layer via KCN-free process. *IEEE Journal of Photovoltaics* 6 (3): 760–763.

Internal Market, Industry, Entrepreneurship and SMEs. (2023). Retrieved 23/07/2027, from Europen Commision: https://single-market-economy.ec.europa.eu/sectors/raw-materials/ areas-specific-interest/critical-raw-materials_en.

Ishizuka, S. (2019). CuGaSe$_2$ thin film solar cells: challenges for developing highly efficient wide-gap chalcopyrite photovoltaics. *Physica Status Solidi (A)* 216 (15): 1800873.

Jackson, P., Wuerz, R., Hariskos, D. et al. (2016). Effects of heavy alkali elements in Cu (In, Ga) Se$_2$ solar cells with efficiencies up to 22.6%. *Physica Status Solidi (RRL)–Rapid Research Letters* 10 (8): 583–586.

Jaysankar, M., Paetel, S., Ahlswede, E. et al. (2019). Toward scalable perovskite-based multijunction solar modules. *Progress in Photovoltaics: Research and Applications* 27 (8): 733–738.

Jeong, A.R., Choi, S.B., Kim, W.M. et al. (2017). Electrical analysis of c-Si/CGSe monolithic tandem solar cells by using a cell-selective light absorption scheme. *Scientific Reports* 7 (1): 15723.

Jošt, M., Kegelmann, L., Korte, L., and Albrecht, S. (2020). Monolithic perovskite tandem solar cells: a review of the present status and advanced characterization methods toward 30% efficiency. *Advanced Energy Materials* 10 (26): 1904102.

Keller, J., Sopiha, K.V., Stolt, O. et al. (2020a). Wide-gap (Ag, Cu)(In, Ga) Se$_2$ solar cells with different buffer materials—a path to a better heterojunction. *Progress in Photovoltaics: Research and Applications* 28 (4): 237–250.

Keller, J., Stolt, L., Sopiha, K.V. et al. (2020b). On the paramount role of absorber stoichiometry in (Ag, Cu)(In, Ga) Se$_2$ wide-gap solar cells. *Solar RRL* 4 (12): 2000508.

Keller, J., Pearson, P., Shariati Nilsson, N. et al. (2021). Performance limitations of wide-gap (Ag, Cu)(In, Ga) Se$_2$ thin-film solar cells. *Solar RRL* 5 (9): 2100403.

Keller, J., Stolt, L., Donzel-Gargand, O. et al. (2022). Wide-gap chalcopyrite solar cells with indium oxide-based transparent back contacts. *Solar RRL* 6 (8): 2200401.

Kim, D.H., Muzzillo, C.P., Tong, J. et al. (2019). Bimolecular additives improve wide-band-gap perovskites for efficient tandem solar cells with CIGS. *Joule* 3 (7): 1734–1745.

Kim, K., Ahn, S.K., Choi, J.H. et al. (2018). Highly efficient Ag-alloyed Cu (In, Ga) Se$_2$ solar cells with wide bandgaps and their application to chalcopyrite-based tandem solar cells. *Nano Energy* 48: 345–352.

Klenk, R., Walter, T., Schock, H.W., and Cahen, D. (1993). A model for the successful growth of polycrystalline films of CuInSe$_2$ by multisource physical vacuum evaporation. *Advanced Materials* 5 (2): 114–119.

Larsson, F., Nilsson, N.S., Keller, J. et al. (2017). Record 1.0 V open-circuit voltage in wide band gap chalcopyrite solar cells. *Progress in Photovoltaics: Research and Applications* 25 (9): 755–763.

Li, J., Huang, Y., Huang, J. et al. (2020). Defect control for 12.5% efficiency Cu$_2$ZnSnSe$_4$ kesterite thin-film solar cells by engineering of local chemical environment. *Advanced Materials* 32 (52): 2005268.

Li, Z., Liang, X., Li, G. et al. (2019). 9.2%-efficient core-shell structured antimony selenide nanorod array solar cells. *Nature Communications* 10 (1): 125.

Lincot, D., Guillemoles, F., Tiwari, A. N., Bär, M., Vermang, B., Le Gleuher, M., et al. (2021). *Indium Availability for CIGS: CIGS-PV*.

Mudryi, A., Gremenok, V., Karotki, A. et al. (2010). Structural and optical properties of thin films of Cu (In, Ga) Se$_2$ semiconductor compounds. *Journal of Applied spectroscopy* 77: 371–377.

Nakamura, M., Yamaguchi, K., Kimoto, Y. et al. (2019). Cd-free Cu (In, Ga)(Se, S)$_2$ thin-film solar cell with record efficiency of 23.35%. *IEEE Journal of Photovoltaics* 9 (6): 1863–1867.

Nakamura, M., Tada, K., Kinoshita, T. et al. (2020). Perovskite/CIGS spectral splitting double junction solar cell with 28% power conversion efficiency. *Iscience* 23 (12).

Nielsen, R., Youngman, T.H., Moustafa, H. et al. (2022). Origin of photovoltaic losses in selenium solar cells with open-circuit voltages approaching 1 V. *Journal of Materials Chemistry A* 10 (45): 24199–24207.

Philipps, S. and Warmuth, W. (2023). *Photovoltaics Report*. Freiburg: Fraunhofer Institute for Solar Energy Systems.

Ruiz-Preciado, M.A., Gota, F., Fassl, P. et al. (2022). Monolithic two-terminal perovskite/CIS tandem solar cells with efficiency approaching 25%. *ACS Energy Letters* 7 (7): 2273–2281.

Shah, U.A., Chen, S., Khalaf, G.M.G. et al. (2021). Wide bandgap $Sb_2S_3$ solar cells. *Advanced Functional Materials* 31 (27): 2100265.

Shukla, S., Sood, M., Adeleye, D. et al. (2021). Over 15% efficient wide-band-gap Cu (In, Ga) $S_2$ solar cell: suppressing bulk and interface recombination through composition engineering. *Joule* 5 (7): 1816–1831.

Siebentritt, S. (2017a). Chalcopyrite compound semiconductors for thin film solar cells. *Current Opinion in Green and Sustainable Chemistry* 4: 1–7.

Siebentritt, S. (2017b). High voltage, please! *Nature Energy* 2 (11): 840–841.

Solliance (2019). *Record Breaking 23% Efficiency Proved for Flexible Perovskite/CIGS-Tandem*. Solliance.

Solliance (2021). *World Record Efficiency of 26.5% on a Tandem Solar Cell Based on a Flexible Cigs Solar Cell*. Solliance.

Sugimoto, H., Hiroi, H., Iwata, Y., and Yamada, A. (2017). Recent progress in high efficiency pure sulfide CIGS solar cells. In: *Paper Presented at the 27th International Photovoltaic Science and Engineering Conference*.

Tang, R., Wang, X., Lian, W. et al. (2020). Hydrothermal deposition of antimony selenosulfide thin films enables solar cells with 10% efficiency. *Nature Energy* 5 (8): 587–595.

Todorov, T., Gershon, T., Gunawan, O. et al. (2014). Perovskite-kesterite monolithic tandem solar cells with high open-circuit voltage. *Applied Physics Letters* 105 (17).

Todorov, T.K., Bishop, D.M., and Lee, Y.S. (2018). Materials perspectives for next-generation low-cost tandem solar cells. *Solar Energy Materials and Solar Cells* 180: 350–357.

Valdes, N., Lee, J., and Shafarman, W. (2019). Comparison of Ag and Ga alloying in low bandgap $CuInSe_2$-based solar cells. *Solar Energy Materials and Solar Cells* 195: 155–159.

Wang, D., Guo, H., Wu, X. et al. (2022). Interfacial engineering of wide-bandgap perovskites for efficient perovskite/CZTSSe tandem solar cells. *Advanced Functional Materials* 32 (2): 2107359.

White, T.P., Lal, N.N., and Catchpole, K.R. (2013). Tandem solar cells based on high-efficiency c-Si bottom cells: top cell requirements for >30% efficiency. *IEEE Journal of Photovoltaics* 4 (1): 208–214.

Wilson, G.M., Al-Jassim, M., Metzger, W.K. et al. (2020). The 2020 photovoltaic technologies roadmap. *Journal of Physics D: Applied Physics* 53 (49): 493001.

Yan, C., Huang, J., Sun, K. et al. (2018). $Cu_2ZnSnS_4$ solar cells with over 10% power conversion efficiency enabled by heterojunction heat treatment. *Nature Energy* 3 (9): 764–772.

Yang, S.-C., Sastre, J., Krause, M. et al. (2021). Silver-promoted high-performance (Ag, Cu)(In, Ga) Se$_2$ thin-film solar cells grown at very low temperature. *Solar RRL* 5 (5): 2100108.

Zakutayev, A., Major, J.D., Hao, X. et al. (2021). Emerging inorganic solar cell efficiency tables (version 2). *Journal of Physics: Energy* 3 (3): 032003.

Zhang, J., Lian, W., Yin, Y. et al. (2020). All antimony chalcogenide tandem solar cell. *Solar RRL* 4 (4): 2000048.

# Part Six

# Nanophotonics

# 14

# Nanoscale Photovoltaics

*Sander A. Mann[1] and Esther Alarcón-Lladó[2]*

[1] *Photonics Initiative, Advanced Science Research Center, City University of New York, New York, USA*
[2] *Center for Nanophotonics, AMOLF, Amsterdam, The Netherlands*

## 14.1 Introduction

Virtually all solar cells comprise bulk semiconductor materials, and – particularly silicon solar cells – operate in the ray optics regime, where traversed distances by radiation through the solar cell are many wavelengths long. Ray optics is a geometrical approximation of Maxwell's equations describing the interaction between light and materials, in which light is considered to travel in straight lines, except at the interface between materials with different refractive indices or when the refractive index inside a material varies. This approximation breaks down when material objects become comparable to the size of the wavelength, as it does not capture wave phenomena like interference and diffraction. This is the regime where nanophotovoltaic solar cells operate – they possess size features that are on the scale of the wavelength, which allows them to utilize wave optics effects not accessible to standard, bulky solar cells. While no nanoscale-based solar cells are commercially available yet, their fundamentally different interaction with light provides an interesting platform to gain insight into photovoltaics generally. In this chapter, we will first discuss general nanophotovoltaic devices from a theoretical standpoint, and then move on to discuss experimental results on nanowire solar cells, which are by far the most prevalent nanophotovoltaic device.

## 14.2 Absorption and Scattering by Nanoscale Objects

The goal of light management in solar cells is generally to reduce reflections, increase absorption in the active region, and reduce parasitic absorption. Without increasing the semiconductor thickness, absorption can be enhanced by increasing the distance that light traverses in the solar cell through surface texturing (compare Figure 14.1a,b) – instead of straight down, light is deflected into oblique angles. In the case of an ideal back reflector and weak absorption, the average path length can be enhanced by a factor of $4n^2$, where

*Photovoltaic Solar Energy: From Fundamentals to Applications, Volume 2*, First Edition.
Edited by Wilfried van Sark, Bram Hoex, Angèle Reinders, Pierre Verlinden, and Nicholas J. Ekins-Daukes.
© 2024 John Wiley & Sons Ltd. Published 2024 by John Wiley & Sons Ltd.
Companion website: www.wiley.com/go/PVsolarenergy

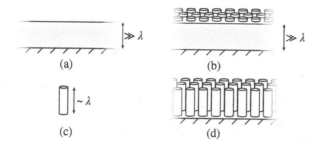

**Figure 14.1** Schematic depictions of (a) a bulk solar cell (without texturing), where the semiconductor layer is many wavelengths thick, (b) a bulk solar cell decorated with subwavelength size elements to achieve nanophotonic light trapping, (c) a single wavelength scale object, and (d) an array of nanoscale objects. This chapter will focus on (c) and (d).

$n$ is the refractive index of the semiconductor (Yablonovitch 1982). This so-called ergodic limit also assumes that the semiconductor layer is thick enough that the density of modes of an infinite medium with the same refractive index can be used. In the case of thin solar cells, however, this is no longer valid. A thin semiconductor layer supports a discrete set of guided modes, and rather than coupling to the continuum of modes in a thick layer, the surface texture now couples to these guided modes (Stuart and Hall 1997; Yu et al. 2010). In the nanophotovoltaic devices we consider here, this is taken one step further. There is no continuous semiconductor layer (except potentially a substrate), but the entire solar cell is on the order of the wavelength in all dimensions (Figure 14.1c), or consists of an arrangement of such nanoscale objects (Figure 14.1d).

Rather than speaking of the reflection and transmission that we are accustomed to in the case of "large" objects and planar interfaces, in the case of objects smaller than or on the order of the wavelength, it is more useful to use the scattering, absorption, and extinction cross-sections (Bohren and Huffman 1983) – $\sigma_{sca}$, $\sigma_{abs}$, and $\sigma_{ext} = \sigma_{sca} + \sigma_{abs}$. These quantities have units of area and enable calculation of, for example, the absorbed power $P_{abs}$ for a given incident intensity $I_{inc}$: $P_{abs} = \sigma_{abs} I_{inc}$. In a clear departure from ray optics, these cross-sections can be significantly larger than the geometrical cross-section (or projected area, $\sigma_{geo}$) of the object. An object can absorb more power than, naively speaking, is incident upon it (Bohren 1983). This can be understood by considering the interaction of light and nanostructures in more detail. As with textured thin films, the interaction of small objects with light is most easily understood in terms of a discrete set of modes or resonances. In the following subsections, we will make a distinction between two scenarios – the incident light is momentarily captured in a localized resonance, where it does not travel in any direction (Figure 14.2, left), or in a guided mode that travels along an extended direction of the nanostructure (Figure 14.2, right).

### 14.2.1 Localized Resonances

For most wavelength scale objects, the absorption and scattering must be calculated numerically, using time or frequency domain simulation techniques. Analytical solutions exist only for select objects, such as spheres and infinitely long nanowires (Bohren and Huffman 1983). For illustrative purposes, Figure 14.3a shows the absorption cross-section

**Figure 14.2** Comparison between a localized resonance (on the left), where incident light bounces back and forth without net momentum in any direction, and a guided mode (on the right), where light is guided by the nanostructure and travels along its extended direction. Both examples are a nanowire, with radiation incident transverse to (left) or along (right) its long axis.

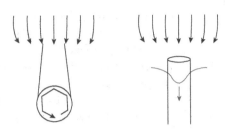

of an infinitely long amorphous silicon cylinder as a function of its diameter and the incident wavelength. Clear bright branches in the absorption are visible, which correspond to resonances, and as the cylinder increases in diameter, the resonance redshifts. The figure also shows dispersion of the resonances, labeled by their polarization ($TM_{ml}$ for electric field along the cylinder axis, $TE_{ml}$ for the electric field along the cylinder axis), and the number of angular nodes $m$ and radial nodes $l$ in the field profile or the resonance.

An intuitive framework to understand the absorbed and scattered power in these objects is based on a multipole expansion of the fields. Since the objects are small, they only interact with the first few terms – a nanowire with a diameter of 100 nm interacts only with the monopolar and quadrupolar harmonics. The absorption cross-section can conveniently be approximated using temporal coupled-mode theory, by considering each of the resonances supported by the nanostructure (Ruan and Fan 2010, 2012):

$$\sigma_{abs}(\omega) = \sum_{n=1}^{\infty} \sigma_{abs,n} \approx \sum_{n=1}^{\infty} \frac{2\lambda}{\pi} \frac{\gamma_{r,n}\gamma_{a,n}}{(\omega - \omega_{0,n})^2 + (\gamma_{r,n} + \gamma_{a,n})^2} \tag{14.1}$$

Here $\omega_{0,n}$ is the resonant frequency for each supported resonance and $\gamma_{r,n}$ and $\gamma_{a,n}$ are the radiative and absorptive loss rates, i.e. the rates at which power leaks out of the resonator or is absorbed inside of it. From this equation, it is clear that each resonance contributes most to the absorption cross-section when excited at the resonant frequency $\omega = \omega_0$ and when

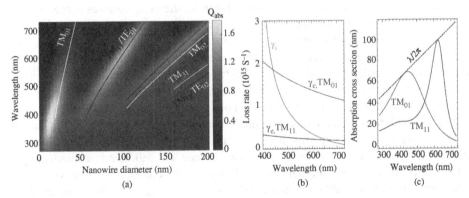

**Figure 14.3** (a) The absorption efficiency ($Q_{abs} = \sigma_{abs}/\sigma_{geo}$) of a cylinder made of amorphous silicon excited transverse to the cylinder axis (see Figure 14.2, left), with clear resonant absorption enhancement visible near the band gap. (b) The absorption loss rate in amorphous silicon and the radiative loss rates for the first two resonances in the TM polarization. (c) The absorption cross-section for each channel individually, reaching the maximum absorption approximately near the conditions expected for critical coupling. Source: Mann & Garnett, 2013/American Chemical Society.

$\gamma_{r,n} = \gamma_{a,n}$ – the so-called critical coupling condition (Haus 1984; Mann and Garnett 2013). This is shown in Figure 14.3b. The absorption loss rate (approximated as $\gamma_{a,n} = \gamma_a = \alpha c/n_g$, where $\alpha$ is the absorption coefficient, $c$ is the speed of light, and $n_g$ is the group index) is shown together with the radiative loss rates of the first two resonances in the TM polarization. Where they cross is where the resonance should reach the maximum cross-section, which is confirmed in Figure 14.3c. Similar analysis can be applied to TE resonances or higher order TM resonances, which appear in the unpolarized absorption in Figure 14.3a. To enhance the total cross-section, multiple critically coupled resonances should be aligned at the same frequency (Mann and Garnett 2013).

## 14.2.2 Guided Modes

In cases where the nanoscale object is extended along the direction of incidence, it is often more convenient to consider the guided modes of the structure (Figure 14.2, right). These modes are also discrete, as in the case of the localized resonances. The propagation constant of a guided mode is larger than the free space momentum, which means that light cannot leak out and travels along the nanostructure until absorbed or scattered out at the other end. While these guided modes typically do not form localized resonances (which would require the guided mode to bounce back and forth between a top and bottom interface as in a Fabry–Pérot resonance), they can still have absorption cross-sections significantly larger than their geometrical cross-section, as the guided modes extend outside of the waveguide.

An exemplary structure is the vertical nanowire, where light is incident along the long axis, strikes an end facet, and couples to modes that propagate along its length. Just as in the horizontal nanowire example in Section 14.2.1, one can classify the modes into TE and TM polarizations. Additionally, hybrid HE and EH modes exist, where neither the electric nor magnetic field is purely transverse. The TE and TM guided modes are, in fact, the same as in the horizontal nanowire, but with different momentum $\beta$ along the nanowire axis – in the horizontal case of the previous subsection, the excitation was perpendicular to the axis, in which case there is no momentum along the nanowire axis. The propagation constant along the nanowire axis can be increased until beyond *cutoff* $\beta > k_0$, where the guided mode momentum exceeds the free space momentum $k_0$, and the TE/TM modes become fully guided. Below cutoff, these modes have a complex propagation constant even for non-absorbing materials due to radiation losses, resulting in attenuation of the propagating mode amplitude with a $1/e$ decay length of $L_d = \text{Im}(\beta)$.

The lowest order guided mode, the $HE_{11}$ mode, does not have a cutoff and remains guided for all conditions. This means this mode can only be excited at an end facet. For small diameters, this guided mode is the only supported mode and will therefore dominate the response. Figure 14.4 shows the field distribution of the fundamental $HE_{11}$ mode for the same nanowire at three different wavelengths. At the shortest wavelength, the mode is almost fully confined inside the nanowire. As the wavelength increases, more and more of the fields are stored outside of the nanowire, and eventually, most of the field is stored outside of the nanowire surface, indicating very low confinement. This low confinement indicates that for longer wavelengths, excitation by a planewave will increase efficiency.

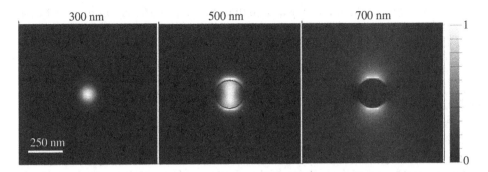

**Figure 14.4** Simulated distribution of the $HE_{11}$ eigenmode field intensity in the cross-section of an infinitely long 100 nm-thick GaAs nanowire surrounded by air at three different wavelengths, 300, 500, and 700 nm. The color scale indicates the intensity of the fields. The boundaries of the nanowire are indicated by the black circle.

We can quantify this using the overlap integral (Frederiksen et al. 2017; Mokkapati et al. 2015)

$$\sigma_{HE_{11}} \approx \frac{1}{I_{in}} \left| Re \left\{ \frac{\int (\mathbf{E}_{HE11} \times \mathbf{H}_{in}^*) \cdot \mathbf{e}_z dS \int (\mathbf{E}_{in}^* \times \mathbf{H}_{HE11}) \cdot \mathbf{e}_z dS}{\int (\mathbf{E}_{HE11}^* \times \mathbf{H}_{HE11}) \cdot \mathbf{e}_z dS} \right\} \right|, \tag{14.2}$$

where $\mathbf{e}_z$ is the surface normal, * indicates complex conjugation, $I_{in}$ is the incident intensity, and the subscripts "$HE_{11}$" and "in" refer to the guided mode and incoming excitation, respectively. As the field of a plane wave is uniform, the excitation efficiency depends on the lateral extent of the mode. The coupling efficiency $Q_{HE11} = \sigma_{HE11}/\sigma_{geo}$ calculated using Eq. (14.2) for an InP nanowire is shown in Figure 14.5a, which indeed shows that it increases exponentially towards longer wavelengths, as confinement decreases.

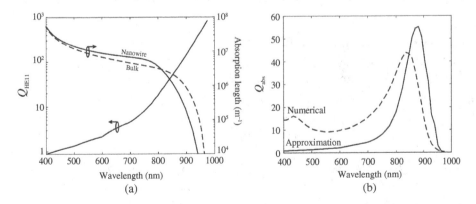

**Figure 14.5** (a) The coupling efficiency $Q_{HE11} = \sigma_{HE11}/\sigma_{geo}$ as a function of wavelength compared to the absorption length in an 80 nm radius InP nanowire, and the absorption length in both bulk InP and the 80 nm nanowire. (b) The absorption efficiency for a 2 µm long, 80 nm radius InP nanowire surrounded by air, calculated using the approximation in Eqs. (14.2) and (14.3) and using a numerical method.

In a finite wire, not all power coupled into the waveguide modes is absorbed, and what is not absorbed is scattered out at the other end facet. We can estimate the absorption cross-section as

$$\sigma_{abs} \approx e^{-2Im(\beta)L}\sigma_{HE_{11}}, \tag{14.3}$$

where $L$ is the nanowire length and $Im(\beta)$ is the $1/e$ attenuation length of the mode amplitude. While the coupling efficiency grows exponentially for longer wavelengths, $Im(\beta)$ decreases dramatically, as shown in Figure 14.5a (the real part of the propagation constant $Re(\beta)$ is not shown and describes phase propagation along the nanowire). For comparison, the absorption length of bulk InP is visualized as well. Note that the absorption length in a nanowire exceeds that of the bulk material at shorter wavelengths due to the lower group velocity of the guided wave, but it is significantly lower at longer wavelengths. This is a consequence of the low confinement – only a fraction of the electric field is localized inside the absorbing nanowire, while most is stored in its non-absorbing surroundings.

Figure 14.5b shows the approximate absorption cross-section of a 2 µm long, 80 nm radius InP nanowire compared with the absorption cross-section calculated numerically using a finite-difference time-domain (FDTD) solver, which discretizes space and time and solves Maxwell's equations numerically using finite difference differentiation. The competition between the increasing cross-section and increasing attenuation length results in a sharp peak at longer wavelengths. While there is qualitative agreement, differences between the approximation and the numerical result arise because the overlap integral in Eq. (14.2) is strictly speaking only valid for non-absorbing materials and only considers the contribution from one mode.

In the case of a localized resonance, discussed in the previous subsection, absorption peaks shift strongly with size. In the case of coupling to a guided mode, however, the length of the nanowire weakly affects the absorption, as shown in Figure 14.6a for a GaAs nanowire with a 75 nm radius. This is a direct consequence of the fact that it is not a resonant effect, but excitation of a guided mode. At very short lengths there is a small

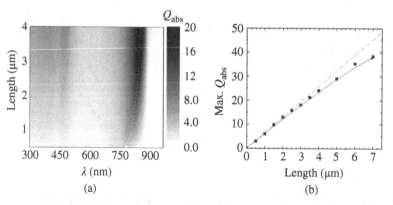

**Figure 14.6** (a) Simulated $Q_{abs}$ of a vertical single GaAs nanowire (diameter of 150 nm) as a function of wavelength and nanowire length. The absorption efficiency $Q_{abs}$ is obtained by normalizing the absorption cross-section by the geometrical cross-section. (b) Maximum $Q_{abs}$ at each length for the same nanowire. Dashed line is a linear extrapolation from short lengths to highlight the nonlinear nature of $Q_{abs}$ with nanowire length.

blueshift in the peak, due to the increasing importance of the absorption length. The maximum $Q_{abs}$ shows a monotonic increase as more and more of the power coupled into the HE11 mode is absorbed (Figure 14.6b). For very long wires, all power in the HE11 mode is absorbed, and additional absorption originates from light coupling into the nanowire through the sidewalls (Anttu 2019). Yet, the absorption peak in a nanowire due to coupling to a guided mode does have a similarly strong shift with nanowire diameter. This is due to the modulation of the mode confinement and light coupling with the nanowire diameter.

### 14.2.2.1 Arrays of Absorbing Nanoparticles

To create a large-scale nanophotovoltaic solar cell, the individual subwavelength objects are typically assembled in a periodic or random array with reasonable density (e.g. Figure 14.1d), since the goal is to achieve as much absorption as possible. Depending on the lattice structure and periodicity, different additional effects may occur, for example, when the periodicity is roughly equal to the incident wavelength, lattice resonances can occur (Brongersma et al. 2014). For denser lattices, when cross-sections start to overlap, individual particles start to interact and their modes will hybridize, similar to how atomic orbitals in crystals overlap to form band structures. While the absorption cross-section of a single particle may exceed its geometrical cross-section significantly, for an array the absorption is naturally limited to unity. As a rule of thumb, the required density of the array for full absorption depends on the absorption cross-section (if the spacing between particles is smaller than a wavelength), and the area of the unit cell should roughly be equal to the absorption cross-section. That means that, for a square array of nanowires of radius $r_{NW}$, the distance between nanowires should be $d \approx \sqrt{Q_{abs}\pi}\, r_{NW}$, where the absorption efficiency $Q_{abs} = \sigma_{abs}/\sigma_{geo}$. As an example, consider the case of a nanowire with $Q_{abs} = 10$ at a given wavelength. The spacing between nanowires should be below $\sqrt{10\pi}\, r_{NW} \approx 6 r_{NW}$ to absorb all light, which implies a filling fraction of roughly $1/Q_{abs} = 10\%$.

## 14.3 Nanophotovoltaics

The unique interaction of wavelength-scale objects with light has consequences for almost all aspects of a solar cell – ranging from the quantum efficiencies, short-circuit current, and open-circuit voltage to the conversion efficiency – which will be discussed in this section.

### 14.3.1 The Internal and External Quantum Efficiency

After light is absorbed and converted into charge carriers, these carriers must make it to the desired terminal contacts to contribute to the photocurrent. Once light is absorbed, the efficiency with which this occurs is called the *internal quantum efficiency* (IQE). We will distinguish between the spatially resolved $IQE_s(\omega, \mathbf{r})$ and the more commonly reported spatially weighted average $IQE(\omega)$. The *external quantum efficiency* (EQE) considers the likelihood of absorption of an incident photon as well, and therefore gives the efficiency of conversion of incident photon to electron collected at the contacts. In a macroscopic solar cell, the IQE and EQE are therefore typically considered simply related by the fraction of incident light that is absorbed, $EQE(\omega) = A(\omega)\, IQE(\omega)$. In a nanophotonic solar cell, however, this simple relationship does not hold, as, for example, the nanostructure can absorb more power than incident on its geometrical cross-section. Instead, the quantum efficiency cross-section should be considered (Mann et al. 2016):

$$\sigma_{EQE}(\omega, \theta, \varphi) = \frac{1}{I_{inc}} \int\!\!\int\!\!\int_V \frac{\omega}{2} Im(\varepsilon)|\mathbf{E}(\omega, \theta, \varphi)|^2 IQE_s(\mathbf{r})d^3\mathbf{r} \approx \sigma_{abs}(\omega, \theta, \varphi)IQE(\omega) \quad (14.4)$$

Here $I_{inc}$ is the incident intensity and $\varepsilon$ is the semiconductor permittivity. Note that this quantity has units of area, and therefore converts an incident flux into a current. Contrary to the macroscopic quantum efficiency, which at most is 100% (in absence of carrier multiplication), there is no strict upper limit to $\sigma_{EQE}$. Consider, for example, a single vertical nanowire as in (Krogstrup et al. 2013), where $\sigma_{EQE}$ exceeds the geometric'1al cross-section of the nanowire by a factor of nearly 15 (see Section 14.4.2).

### 14.3.2 The Short-Circuit Current

Just as with macroscopic solar cells, the short-circuit current is a function of the broadband absorption spectrum of the nanoscale solar cell. The short-circuit current is given by the integral over the incident flux and the quantum efficiency cross-section:

$$i_{sc} = q \int \sigma_{EQE}(\omega, \theta = 0)S(\omega)d\omega \quad (14.5)$$

Here, $S(\omega)$ is the AM1.5g solar spectrum, and the sun is assumed to be at $\theta = 0$ and subtend a neglible solid angle for simplicity. As with the EQE, a current density is not readily defined since it is unclear what the proper area to normalize the current to is. In certain cases, such as when a geometry is extended in one direction and current flow is along that direction (such as a nanowire) a cross-sectional current density can be defined, which will typically far exceed typical current densities in macroscopic solar cells.

### 14.3.3 The Open-Circuit Voltage

Just as in macroscopic solar cells, the idealized current–voltage relationship is given by:

$$(v) = i_0 e^{qv/k_B T} - i_{sc} \quad (14.6)$$

Here, $i_0$ is the dark current, $q$ is the elementary charge, $k_B$ is the Boltzmann constant, and $T$ is the temperature. The open-circuit voltage of the nanophotovoltaic device is therefore given by:

$$v_{oc} = \frac{k_B T}{q} \ln(i_{sc}/i_0) \quad (14.7)$$

As in macroscopic solar cells, the dark current is determined by nonradiative and radiative recombination. However, both are strongly affected by the nanoscale geometry, as we will discuss in the following subsections.

#### 14.3.3.1 Radiative Recombination

Radiative recombination must occur for the solar cell to maintain thermal equilibrium with its surroundings. The radiative dark current in equilibrium can be calculated by integrating the EQE cross-section over the blackbody spectrum at the ambient temperature ($S^{BB}(T, \omega)$) and over all angles (Kirchartz and Rau 2008; Sandhu et al. 2013):

$$i_0 = q \iiint S^{BB}(T, \omega)\sigma_{EQE}(\omega, \theta, \varphi)\theta(\omega)d\omega d\Omega \tag{14.8}$$

The dependence of the dark current very clearly depends on the spectral and angle-dependent properties of the absorption cross-section, which, in nanophotonic environments, can readily be manipulated, for example by modifying the object's size and shape (Sandhu et al. 2013). Radiative control of the $v_{oc}$ through the absorption cross-section can largely be divided into two effects (Mann et al. 2016a): (i) modification of the effective bandgap and (ii) directivity. Since the blackbody spectrum very strongly depends on frequency, the radiative dark current depends exponentially on the bandgap, resulting in a linear dependence of the $v_{oc}$ on the bandgap. Using nanophotonic engineering, the bandgap can effectively be raised above that of the constituent semiconductor, by suppressing absorption (and emission) near the bandgap. As a result, only carriers with enough energy can recombine. In principle, through nanophotonic engineering, the bandgap can be raised by an arbitrary amount.

### 14.3.3.2 Nonradiative Recombination

The impact of nonradiative recombination on the $v_{oc}$ can be quantified as follows (Kirchartz and Rau 2008):

$$v_{oc} = \frac{k_B T}{q}(\ln(i_{sc}/i_0) + \ln(\eta_{ext})) = v_{oc,rad} + \frac{k_B T}{q}\ln(\eta_{ext}) \tag{14.9}$$

Here, the first term is the open-circuit voltage in a purely radiatively limited solar cell, while the second term covers nonradiative recombination. The external radiative efficiency $\eta_{ext}$ is defined as the fraction of recombination that results in a photon outside of the solar cell (and is therefore also sometimes called the LED EQE or external luminescence efficiency) (Rau 2007):

$$\eta_{ext} = \frac{i_{0,r}}{i_0} = \frac{i_{0,r}}{i_{0,r} + i_{0,nr}} \tag{14.10}$$

Here, the subscripts $r$ and $nr$ refer to radiative and nonradiative. The external radiative efficiency is related to the internal radiative efficiency $\eta_{int}$ as (Steiner et al. 2013)

$$\eta_{ext} = \frac{\eta_{int} P_{esc}}{1 - \eta_{int} P_{abs}} \tag{14.11}$$

where $P_{esc}$ is the probability that a photon escapes after emission and $P_{abs}$ is the probability that it is reabsorbed in the semiconductor. Besides these two pathways, it is also possible that an emitted photon is parasitically absorbed in, for example, the back reflector, and $P_{esc} + P_{abs} + P_{par} = 1$. The internal radiative efficiency is given as $\eta_{int} = \Gamma_r/(\Gamma_r + \Gamma_{nr})$, where $\Gamma_r$ and $\Gamma_{nr}$ are the radiative and nonradiative recombination rates. We can thus write:

$$\eta_{ext} = \frac{\Gamma_r P_{esc}}{\Gamma_{nr} - (1 - P_{abs})\Gamma_r} \tag{14.12}$$

$\eta_{int}$ is often considered a material constant, indicative of its quality. However, in nanophotonic environments, the radiative recombination rate is not simply that of a bulk medium.

Instead, through manipulation of the photonic environment, the radiative decay rate may be enhanced or inhibited significantly (Purcell 1946; Yablonovitch 1987). On the other hand, the nonradiative recombination rate might also be affected, as the surface-to-volume ratio of nanostructured solar cells is significantly higher, while both the total volume and junction area can be considerably smaller. Finally, $p_{esc}$ in bulk solar cells is often limited due to total internal reflection – particularly in direct bandgap solar cells with antireflection coatings but no surface texturing. With a perfect back reflector, the escape probability in such cells is at best $1/(2n^2)$ (Steiner et al. 2013). In nanostructured solar cells, however, $p_{esc}$ can be made significantly larger, as, for example, momentum conversation at the top interface is no longer required (Mann et al. 2016b). This provides opportunity for higher efficiencies to be obtained with lower-quality materials.

### 14.3.4 Conversion Efficiency

The maximum power that the solar cell delivers is obtained at the maximum power point (MPP), which is at currents slightly below the short-circuit current and voltages slightly below the open-circuit voltage:

$$P_{max} = i_{sc} v_{oc} FF \tag{14.13}$$

Here, FF is the fill-factor, which captures how far below the product of $i_{sc}$ and $v_{oc}$ the maximum power output lies. For macroscopic solar cells, the conversion efficiency is straightforwardly calculated by dividing the maximum power output by the power input, $\eta = P_{out}/P_{in}$, where $P_{in}$ is given by the incident intensity of sunlight (1000 W/m² under standard test conditions), and the (exposed) area of the solar cell. For a single nanoscale solar cell, however, the conversion efficiency is undefined. While the incident intensity and output power can readily be determined, the incident power is undeterminable since it requires an area. As mentioned in Section 14.2.1, taking the geometrical area is unphysical as the absorption cross-section can exceed it significantly. Considering only the absorption cross-section results in a frequency-dependent area will also artificially enhance the efficiency, given that scattering is ignored.

#### 14.3.4.1 Efficiency Limits

The conversion efficiency for a macroscopic solar cell comprising nanoscale objects, as schematically shown in Figure 14.1d, naturally is well-defined. The limiting conversion efficiency can readily be calculated using the same approach as for regular solar cells, with the detailed balance method (Shockley and Queisser 1961). This approach only considers the solar cell as a black box, requires only the spectral and angular absorptivity, and assumes that there is no parasitic absorption or nonradiative recombination. As such, the limiting conversion efficiency of any solar cell made from nanoscale structures is the same as that of any bulk solar cell, which can be enhanced beyond the one-sun limit through concentration and angle restriction.

#### 14.3.4.2 Concentration

By placing a solar cell under a concentrating lens, the conversion efficiency can be increased beyond the one-sun detailed balance limit. This can be observed from Eq. (14.9).

The open-circuit voltage increases if the short-circuit current increases with respect to the dark current. The large absorption cross-section of nanoscale objects raises the question whether a similar effect occurs in nanoscale solar cells, and results in an intrinsic (i.e. without an external lens) efficiency boost. Upon closer inspection, this turns out to not be the case. Comparing Eqs. (14.5) and (14.8), we see that if the absorption cross-section of an object is increased equally for *all* angles of incidence, both the short-circuit current and the radiative dark current increase. As a result, the effect of nanoscale concentration cancels out. What is required, instead, is *directivity*. The absorption cross-section under normal incidence is enhanced while the cross-section for angles off-normal is reduced. The analogy can also be considered in the other direction – if the lens is considered part of the detailed balance "black box," no concentration occurs and the effective short-circuit current density is the regular one-sun current density. The effective radiative dark current density on the other hand is reduced. Thermal light incident normally is absorbed normally, while other angles of incidence are either focused next to the solar cell (resulting in no absorption) or are not focused (resulting in reduced absorption) (Mann et al. 2016b).

### 14.3.5 Angle Restriction

The absorption cross-section of nanoparticles can be engineered to be directive – strongly absorbing under normal incidence, but less so for other angles. This reduces the dark current with respect to the short-circuit current, and therefore also results in an enhanced open-circuit voltage. In macroscopic solar cells, this can be achieved using angular filters or parabolic mirrors (Kosten et al. 2014; Kosten et al. 2013), which results in enhanced recycling of emitted photons in bulk, thereby reducing the net recombination rate. Due to enhanced photon recycling, these approaches only work if $\eta_{int}$ is very high. This is because in macroscopic cells, the escape probability and directivity are exactly inversely proportional – a reduction of the emission cone also results in a reduction of the escape probability. Hence, if $\eta_{int}$ is low, $\eta_{ext} \approx p_{esc}\eta_{int}$, the reduction in the radiative dark current is offset by an increase in the nonradiative recombination rate due to photon recycling. If intrinsic directivity is achieved through nanophotonic engineering, however, the escape probability and directivity are uncoupled, resulting in a potential benefit. As long as $\eta_{int} > (1 + (D-1)p_{esc})^{-1}$, where $D$ is the reduction factor in the radiative recombination rate due to directivity, efficiencies beyond the one-sun detailed balance limit can still be achieved (Mann et al. 2016b).

## 14.4 Nanowire Solar Cells

Among nanoscale solar cells, nanowires are without doubt the most used platform. Nanowires are nanostructures with high aspect ratios, lengths ranging from a few to tens of microns, and diameters from tens to few hundreds of nanometers. Since the early 2000s, semiconductor nanowires were envisioned as promising building blocks for next generation photonics and photovoltaics due to strong light-matter interactions, the ability to integrate mismatched materials (Bao et al. 2008; Borg et al. 2014; Glas 2006; Gudiksen et al. 2002; Kästner and Gösele 2004; Russo-Averchi et al. 2012; Tomioka et al. 2008), and increased radiation tolerance for applications in outer space (Espinet-Gonzalez et al. 2019, 2020; Tan et al. 2017a, 2017b). The nanowire architecture and its unique optical properties also enable new concepts in tandem photovoltaics (Dorodnyy et al. 2015; Tavakoli and Alarcon-Llado 2019) and semi-transparent photovoltaics for building integration (Chen et al. 2021), where nanowire arrays can be made free-standing by embedding them in a polymer support matrix. The large absorption cross-section of vertical nanowires (as discussed in Section 14.2.2) results in nearly full absorption with a fraction of the volume required for thin-film solar cells, which is particularly relevant for solar cells made of scarce materials such as III–V or I–III–VI semiconductors. The low packing density (<15%) also enables nanowire solar cells to be fabricated as flexible, lightweight sheets without a substrate.

### 14.4.1  Nanowire Synthesis

The nanowire growth process and geometry allow for a larger lattice mismatch than planar epitaxy, suggesting the potential to grow defect free multijunction cells using a variety of materials and improving efficiency. Semiconductor nanowires can be synthesized using bottom-up and top-down methods (Dasgupta et al. 2014). Bottom-up methods are often gas-phase, such as vapor–liquid–solid (VLS) growth and metalorganic chemical vapor deposition (MOCVD), where a metal nanoparticle alloys with the semiconductor from a gas-phase precursor, until nucleation initiates semiconductor growth. Concerns about metal contamination in the grown nanowires has led to significant interest in direct deposition methods, including selective-area epitaxy (SAE), where a substrate is first patterned with nanoscale holes in a dielectric mask, from which semiconductor nanowires will grow epitaxially using for example molecular beam epitaxy (MBE) or MOCVD. In contrast, nanowires can also be created with a top-down method by etching a surface patterned with a mask of dots (rather than holes). Junctions are formed during the growth process, by changing the precursor composition and growth conditions. Formation of both homo and heterojunctions is possible, and these junctions can be arranged either along the nanowire axis (an axial junction) or perpendicular to it (a radial junction).

### 14.4.2  Single Nanowire Solar Cells

Single nanowire solar cells, complete devices constructed from just one nanowire, enable careful characterization of the entire photovoltaic process on a microscopic scale. Due to this scale, different characterization techniques are often required (Garnett et al. 2011). The first single nanowire photovoltaic devices were horizontal, as the horizontal architecture readily enables contacting (Kempa et al. 2008; Tian et al. 2007) and preceded the first array devices. After growth, wires are transferred to a substrate, after which selective contacts are applied to either end. The large and resonance-dependent absorption is directly observable in the photocurrent (Cao et al. 2009), as expected from the expression for the photocurrent, Eq. (14.5). To determine the IQE of such a device, knowledge of the absorption cross-section is required. The absorption of scattering samples is normally accurately determined using an integrating sphere, which for the purpose of single nanoscale devices must be integrated with a microscope (Mann et al. 2016). This enables simultaneous mapping of the

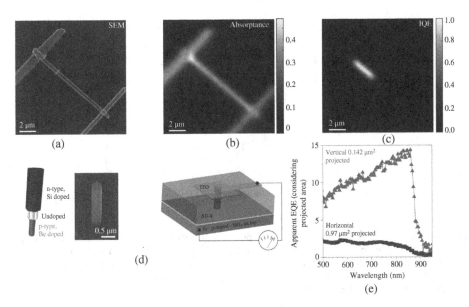

**Figure 14.7** Single nanowire solar cells. (a–c) Characterization of a single horizontal nanowire, using integrating sphere microscopy, which enables measurement of the IQE of a single nanophotovoltaic device. The dashed line in (a) indicates the approximate location of the junction (Mann et al. 2016/ American Chemical Society). (d–e) Characterization of a single vertical nanowire solar cell, which demonstrates a $\sigma_{EQE}$ significantly exceeding its geometrical cross-section by up to a factor of 15. Source: Krogstrup et al. (2013)/Springer Nature.

absorptance and IQE, as shown in Figure 14.7b,c. This measurement clearly shows that while the nanowire absorbs homogeneously along its length, carrier collection only happens efficiently near the junction. Using the same method, $\sigma_{EQE}$ can also be determined experimentally. As this InP nanowire has a relatively large diameter of 310 nm, with a 50 nm $SiO_2$ coating, the absorption cross-section does not extend significantly beyond its geometrical cross-section. This is very different in the case of a single GaAs vertical nanowire device (Krogstrup et al. 2013), also shown in Figure 14.7d,e. Here the apparent EQE reaches up to 15 times the projected area for longer wavelengths (i.e. $\sigma_{EQE}/\sigma_{geo} \approx 15$).

### 14.4.3 Nanowire Array Solar Cells

By proper tuning of the nanowire diameter and the density, both broadband and color-selective absorption can be achieved. Under the conditions of broadband absorption, enhancement of the $v_{oc}$ due to directionality is suppressed, and the array behaves as a thin-film in terms of emission patterns (Anttu 2015). Yet, there is great advantage in using nanowire arrays compared to thin-films in terms of material usage as shown in Figure 14.8a. The short-circuit current peaks at a surface coverage of roughly 25%, and for other materials full absorption can be achieved at even lower densities. As the nanowire density decreases from that of a thin-film, the competition for photons is lifted and, therefore, each individual nanowire rapidly reaches photocurrent density level orders of magnitude larger than in any thin-film device under 1 sun illumination. Figure 14.8b

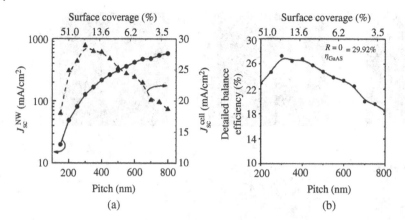

(a) (b)

**Figure 14.8** (a) Simulated photocurrent and (b) detailed balance efficiency in the radiative limit in a NW array based solar cell as a function of pitch distance. GaAs NWs of 150 nm in diameter is arranged in a square lattice in a free-standing fashion. Both the current of the total cell and that of each individual NW are plotted. The total absorption has been calculated by integrating the absorption spectra weighted by the AM1.5g solar spectrum.

shows the conversion efficiency of the same arrays, calculated using the detailed balance framework in the radiative limit. The maximum efficiency is ~26.8% with less than 13% of the surface covered, an efficiency that is not far from the limiting efficiency of bulk GaAs (29.92%).

Figure 14.9a shows a typical nanowire-based solar cell, which has a uniform array of vertical nanowires with an axial junction, covered by a transparent conductive oxide. In a common alternative geometry, the nanowires are embedded in a transparent matrix covered by a thin layer of transparent conducting oxide. This vertical nanowire geometry configuration was first proposed in 1992 as a light-emitting diode (Haraguchi et al. 1992)

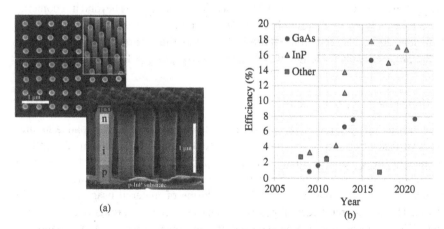

(a) (b)

**Figure 14.9** (a) Typical nanowire array solar cell (Wallentin et al. 2013/American Association for the Advancement of Science – AAAS). (b) Record efficiencies for III–V nanowire solar cells. The markers with a solid outline indicate that the nanowires were fabricated bottom-up. Markers with a dashed outline indicate that the nanowires were fabricated top-down.

and was embraced as a solar cell architecture a decade later (Wu et al. 2002). Early photovoltaic devices containing nanowires were nanorod-enabled organic and dye-sensitized solar cells, with efficiencies of less than 2% (Huynh et al. 2002; Law et al. 2005). The first solar cell, completely made of nanowires, was demonstrated in 2008 using a vertical array of silicon nanowires, with a power conversion efficiency of 0.1% (Stelzner et al. 2008). Since then, research in nanowire photovoltaics has grown rapidly, particularly for III–V semiconductors, motivated by high efficiencies of III–V thin film solars (Green et al. 2022) and the scarcity of raw materials. Reaching similar efficiencies using less raw material on cost-effective substrates (e.g. Si) thus has the potential to address the cost limitations of III–V solar cell technology.

Figure 14.9b shows the evolution of nanowire solar cell record efficiencies for GaAs (circles), InP (triangles), and other III–Vs (squares). Considering that the record efficiencies for bulk GaAs and InP are 29.1% and 24.2%, respectively (Green et al. 2022), one can see that III–V nanowire solar cells perform relatively well, although there is a lot to gain. Their performance has increased 17-fold since the first GaAs nanowire solar cell in 2009 (Czaban et al. 2009), with InP nanowire solar cells now achieving nearly 18% efficiency (Cui et al. 2016). It is worth noting that most record efficiencies are achieved with bottom-up fabrication methods of nanowires (shown by the markers with solid line shapes). This is particularly true for GaAs. GaAs have a much higher surface recombination velocity than InP, and thus GaAs suffer more from etching-induced surface roughening. Post-etching thermal treatments might potentially recover facet reconstruction. Yet, the large density of surface states existing in pristine GaAs surfaces makes this material more challenging for nanowire photovoltaics. Further studies on the optoelectronic properties in nanowires hold the potential for improved performance in nanowire-based optoelectronic devices.

Most high-efficiency cells are also based on axial junctions, but radial junctions were believed to be more beneficial due to closer proximity of photocarrier generation to the p-n junction and decoupling of light absorption and charge separation. However, controlling doping in both the core and shell while maintaining low defect density is challenging for bottom-up cells, and the large junction area in radial devices can reduce the efficiency due to an increase in the dark current (Haverkort et al. 2018; Raj et al. 2019). A notable exception is a demonstration of a near-record InP device efficiency for materials with a very short minority carrier diffusion length using a radial junction (Raj et al. 2019).

By analyzing the individual nanowire conductance with a conductive atomic force microscopy (c-AFM) mapping technique, they found that almost half of the nanowires in an array were electrically dead (Mikulik et al. 2017). For axial junction nanowire arrays grown by SAE, some of such unwanted inhomogeneities arise due to an undesired radial shell growth, which can give rise to electrical shortening in the solar cell (Mukherjee et al. 2021). In situ etching of the surface with HCl atmosphere in MOCVD growth, therefore, greatly benefits the performance (Wallentin et al. 2013). In SAE, in situ etching is not possible due to the high-vacuum conditions, and thus the successful realization of axial nanowire solar cells would require ex situ removal of the self-formed radial shell. Implementation of ex situ etching to axial p-i-n junction GaAs nanowire arrays epitaxially grown on a Si(111) substrate led to a highly rectifying device with largely improved photovoltaic performance (Mukherjee et al. 2021). Additionally, an additional ex situ piranha etch on InP nanowire arrays grown with in-situ etching removes the carbon contamination in the unintentionally grown sidewalls, which serve as a nonradiative recombination center and deteriorates $v_{oc}$ (Cui et al. 2013).

## 14.5 Conclusion

In conclusion, we have discussed the basics of nanoscale photovoltaics, where the photovoltaic device size is on the order of the wavelength. We started describing the interaction between light and nanoscale objects, making a distinction between localized resonances and guided modes. The way light and nanoscale objects interact has large implications for the operation of a photovoltaic device, which was discussed in Section 14.3. We then provided an overview of single and array nanowires, which are the most prevalent nanoscale photovoltaic devices. Nanoscale devices hold promise to achieve high efficiencies with lower quality materials, due to (i) decoupling of absorption and charge collection length scales, and (ii) manipulation of the external radiative efficiency beyond what is achievable in planar devices. In the case of nanowire solar cells, a rapid increase in the conversion efficiency has been observed over the past two decades, approaching those of bulk devices but with potential to improve further.

## Abbreviations and Acronyms

| | |
|---|---|
| TE | transverse electric |
| TM | transverse magnetic |
| HE | hybrid (or helical) magnetic-electric |
| EH | hybrid (or helical) electric-magnetic |
| NW | nanowire |
| IQE | internal quantum efficiency |
| AM1.5g | air mass 1.5 global standard spectrum |
| MPP | maximum power point |
| FF | fill factor |
| MOCVD | metal organic chemical vapor deposition |
| VLS | vapor–liquid–solid |
| MBE | molecular beam epitaxy |
| SAE | selective area epitaxy |
| c-AFM | conductive atomic force microscopy |

## Nomenclature

| | |
|---|---|
| $n$ | refractive index |
| $\sigma$ | cross-section |
| $P_{abs}$ | absorbed power |
| $Q_{abs}$ | absorption efficiency |
| $I_{inc}$ | incident power |

| | |
|---|---|
| $\gamma$ | loss rate |
| $\lambda$ | wavelength |
| $\omega$ | frequency |
| $\omega_{0,m}$ | natural frequency of mode $m$ |
| $\alpha$ | absorption coefficient |
| $c$ | speed of light |
| $n_g$ | group velocity |
| $\beta$ | momentum in the dielectric medium |
| $k_0$ | free space momentum |
| $L_d$ | decay length |
| $d$ | pitch |
| $r_{NW}$ | nanowire radius |
| $i$ | current |
| $q$ | elementary charge |
| $S(\omega)$ | photon flux |
| $\theta$ | angle of incidence |
| $\varphi$ | zenith angle |
| $v$ | voltage |
| $k_B$ | Boltzmann constant |
| $T$ | temperature |
| $\eta$ | efficiency |
| $p$ | probability |
| $\Gamma$ | recombination rate |
| $D$ | directivity |

## Author Biographies

**Sander A. Mann** received a B.Sc. degree in Physics (2010) and M.Sc. degree (2012) in Energy Science from Utrecht University, Utrecht, the Netherlands. He then pursued a Ph.D. at AMOLF and the University of Amsterdam (both in Amsterdam, the Netherlands) in the nanoscale solar cell group of Prof. Dr. Erik Garnett. Sander specialized in quantifying the limits and losses in single and arrays of nanoscale photovoltaic devices, using both experimental and theoretical techniques. Sander is currently a Research Scientist at the Advanced Science Research Center at CUNY in New York, USA.

**Esther Allarcón-Lladó** obtained her PhD in 2009 at the University of Barcelona, after which she joined Professor Tripathy's group at the Institute of Materials Research and Engineering (IMRE) in Singapore to work on efficient UV-light semiconductor devices. Thanks to an International Outgoing Marie Curie Fellowship, she moved to Berkeley for two years, working under the supervision of Prof. Joel W. Ager (LBNL) and subsequently joined Prof. Anna Fontcuberta i Morral's group (EPFL) afterward to work on different areas related to solar energy conversion, from 2014 to 2016 as an Ambizione Energy group leader. Esther currently leads the 3D Photovoltaics group at AMOLF in Amsterdam, the Netherlands.

# References

Anttu, N. (2015). Shockley–Queisser detailed balance efficiency limit for nanowire solar cells. *ACS Photonics* 2 (3): 446–453. https://doi.org/10.1021/ph5004835.

Anttu, N. (2019). Absorption of light in a single vertical nanowire and a nanowire array. *Nanotechnology* 30 (10): 104004. https://doi.org/10.1088/1361-6528/aafa5c.

Bao, X., Soci, C., Susac, D. et al. (2008). Heteroepitaxial growth of vertical GaAs nanowires on Si (111) substrates by metal–organic chemical vapor deposition. *Nano Letters* 8 (11): 3755.

Bohren, C.F. (1983). How can a particle absorb more than the light incident on it? *American Journal of Physics* 51 (4): 323–327. https://doi.org/10.1119/1.13262.

Bohren, C.E. and Huffman, D.R. (1983). *Absorption and Scattering of Light by Small Particles*. Wiley-VCH. https://doi.org/10.1002/9783527618156.

Borg, M., Schmid, H., Moselund, K.E. et al. (2014). Vertical III–V nanowire device integration on Si(100). *Nano Letters* 14 (4): 1914–1920. https://doi.org/10.1021/nl404743j.

Brongersma, M.L., Cui, Y., and Fan, S. (2014). Light management for photovoltaics using high-index nanostructures. *Nature Materials* 13 (5): 451–460. https://doi.org/10.1038/nmat3921.

Cao, L., White, J.S., Park, J.-S. et al. (2009). Engineering light absorption in semiconductor nanowire devices. *Nature Materials* 8 (8): 643–647. https://doi.org/10.1038/nmat2477.

Chen, Y., Hrachowina, L., Barrigon, E. et al. (2021). Semiconductor nanowire array for transparent photovoltaic applications. *Applied Physics Letters* 118 (19): 191107. https://doi.org/10.1063/5.0046909.

Cui, Y., Wang, J., Plissard, S., and Cavalli, A. (2013). Efficiency enhancement of InP nanowire solar cells by surface cleaning. *Nano Letters* 13 (9): 4113–4117.

Cui, Y., Van Dam, D., Mann, S.A. et al. (2016). Boosting solar cell photovoltage via nanophotonic engineering. *Nano Letters* 16 (10): 6467–6471. https://doi.org/10.1021/acs.nanolett.6b02971.

Czaban, J.A., Thompson, D.A., and LaPierre, R.R. (2009). GaAs core–shell nanowires for photovoltaic applications. *Nano Letters* 9 (1): 148–154. https://doi.org/10.1021/nl802700u.

Dasgupta, N.P., Sun, J., Liu, C. et al. (2014). 25th anniversary article: semiconductor nanowires – synthesis, characterization, and applications. *Advanced Materials* 26 (14): 2137–2184. https://doi.org/10.1002/adma.201305929.

Dorodnyy, A., Alarcon-Lladó, E., Shklover, V. et al. (2015). Efficient multiterminal spectrum splitting via a nanowire array solar cell. *ACS Photonics* 2 (9): 1284–1288. https://doi.org/10.1021/acsphotonics.5b00222.

Espinet-Gonzalez, P., Barrigón, E., Otnes, G. et al. (2019). Radiation tolerant nanowire array solar cells. *ACS Nano* 13 (11): 12860–12869. https://doi.org/10.1021/acsnano.9b05213.

Espinet-Gonzalez, P., Barrigón, E., Chen, Y. et al. (2020). Nanowire solar cells: a new radiation hard PV technology for space applications. *IEEE Journal of Photovoltaics* 10 (2): 502–507. https://doi.org/10.1109/JPHOTOV.2020.2966979.

Frederiksen, R., Tutuncuoglu, G., Matteini, F. et al. (2017). Visual understanding of light absorption and waveguiding in standing nanowires with 3D fluorescence confocal microscopy. *ACS Photonics* 4: 2235. https://doi.org/10.1021/acsphotonics.7b00434.

Garnett, E.C., Brongersma, M.L., Cui, Y., and Mcgehee, M.D. (2011). Nanowire solar cells. *Annual Review of Material Research* 269–295. https://doi.org/10.1146/annurev-matsci-062910-100434.

Glas, F. (2006). Critical dimensions for the plastic relaxation of strained axial heterostructures in free-standing nanowires. *Physical Review B* 74 (12): 121302. https://doi.org/10.1103/PhysRevB.74.121302.

Green, M.A., Dunlop, E.D., Hohl-Ebinger, J. et al. (2022). Solar cell efficiency tables (Version 60). *Progress in Photovoltaics: Research and Applications* 30 (7): 687–701. https://doi.org/10.1002/pip.3595.

Gudiksen, M.S., Lauhon, L.J., Wang, J. et al. (2002). Growth of nanowire superlattice structures for nanoscale photonics and electronics. *Nature* 415 (6872): 617–620. https://doi.org/10.1038/415617a.

Haraguchi, K., Katsuyama, T., Hiruma, K., and Ogawa, K. (1992). GaAs p-n junction formed in quantum wire crystals. *Applied Physics Letters* 60 (6): 745–747. https://doi.org/10.1063/1.106556.

Haus, H.A. (1984). *Waves and Fields in Optoelectronics*. Prentice-Hall https://doi.org/10.1080/716099690.

Haverkort, J.E.M., Garnett, E.C., and Bakkers, E.P.A.M. (2018). Fundamentals of the nanowire solar cell: optimization of the open circuit voltage. *Applied Physics Reviews* 5 (3): 031106. https://doi.org/10.1063/1.5028049.

Huynh, W.U., Dittmer, J.J., and Alivisatos, A.P. (2002). Hybrid nanorod-polymer solar cells. *Science* 295 (5564): 2425–2427. https://doi.org/10.1126/science.1069156.

Kästner, G. and Gösele, U. (2004). Stress and dislocations at cross-sectional heterojunctions in a cylindrical nanowire. *Philosophical Magazine* 84 (35): 3803–3824. https://doi.org/10.1080/1478643042000281389.

Kempa, T.J., Tian, B., Kim, D.R. et al. (2008). Single and tandem axial *p-i-n* nanowire photovoltaic devices. *Nano Letters* 8 (10): 3456–3460. https://doi.org/10.1021/nl8023438.

Kirchartz, T. and Rau, U. (2008). Detailed balance and reciprocity in solar cells. *Physica Status Solidi (A)* 205 (12): 2737–2751. https://doi.org/10.1002/pssa.200880458.

Kosten, E.D.E., Atwater, J.H.J., Parsons, J. et al. (2013). Highly efficient GaAs solar cells by limiting light emission angle. *Light: Science & Applications* 2 (1): 1–6. https://doi.org/10.1038/lsa.2013.1.

Kosten, E.D., Kayes, B.M., and Atwater, H.A. (2014). Experimental demonstration of enhanced photon recycling in angle-restricted GaAs solar cells. *Energy & Environmental Science* 7 (6): 1907. https://doi.org/10.1039/c3ee43584a.

Krogstrup, P., Jørgensen, H.I., Heiss, M. et al. (2013). Single-nanowire solar cells beyond the Shockley–Queisser limit. *Nature Photonics* 7: 1–5. https://doi.org/10.1038/nphoton.2013.32.

Law, M., Greene, L.E., Johnson, J.C. et al. (2005). Nanowire dye-sensitized solar cells. *Nature Materials* 4 (6): 455–459. https://doi.org/10.1038/nmat1387.

Mann, S.A. and Garnett, E.C. (2013). Extreme light absorption in thin semiconductor films wrapped around metal nanowires. *Nano Letters* 13 (7): 3173–3178. https://doi.org/10.1021/nl401179h.

Mann, S.A., Oener, S.Z., Cavalli, A. et al. (2016). Quantifying losses and thermodynamic limits in nanophotonic solar cells. *Nature Nanotechnology* 11 (12): 1071–1075. https://doi.org/10.1038/nnano.2016.162.

Mann, S.A., Grote, R.R., Osgood, R.M. et al. (2016a). Opportunities and limitations for nanophotonic structures to exceed the Shockley–Queisser limit. *ACS Nano* 10 (9): 8620–8631. https://doi.org/10.1021/acsnano.6b03950.

Mikulik, D., Ricci, M., Tutuncuoglu, G. et al. (2017). Conductive-probe atomic force microscopy as a characterization tool for nanowire-based solar cells. *Nano Energy* 41: 566–572. https://doi.org/10.1016/j.nanoen.2017.10.016.

Mokkapati, S., Saxena, D., Tan, H.H., and Jagadish, C. (2015). Optical design of nanowire absorbers for wavelength selective photodetectors. *Scientific Reports* 5: 15339. https://doi.org/10.1038/srep15339.

Mukherjee, A., Ren, D., Vullum, P.-E. et al. (2021). GaAs/AlGaAs nanowire array solar cell grown on Si with ultrahigh power-per-weight ratio. *ACS Photonics* 8 (8): 2355–2366. https://doi.org/10.1021/acsphotonics.1c00527.

Purcell, E.M. (1946). Spontaneous emission probabilities at radio frequencies. *Physics Review* 69: 681.

Raj, V., Fu, L., Tan, H.H., and Jagadish, C. (2019). Design principles for fabrication of InP-based radial junction nanowire solar cells using an electron selective contact. *IEEE Journal of Photovoltaics* 9 (4): 980–991. https://doi.org/10.1109/JPHOTOV.2019.2911157.

Raj, V., Vora, K., Fu, L. et al. (2019). High-efficiency solar cells from extremely low minority carrier lifetime substrates using radial junction nanowire architecture. *ACS Nano* 13 (10): 12015–12023. https://doi.org/10.1021/acsnano.9b06226.

Rau, U. (2007). Reciprocity relation between photovoltaic quantum efficiency and electroluminescent emission of solar cells. *Physical Review B – Condensed Matter and Materials Physics* 76 (8): 1–8. https://doi.org/10.1103/PhysRevB.76.085303.

Ruan, Z. and Fan, S. (2010). Superscattering of light from subwavelength nanostructures. *Physical Review Letters* 105 (1): 1–4. https://doi.org/10.1103/PhysRevLett.105.013901.

Ruan, Z. and Fan, S. (2012). Temporal coupled-mode theory for light scattering by an arbitrarily shaped object supporting a single resonance. *Physical Review A* 85 (4): 043828. https://doi.org/10.1103/PhysRevA.85.043828.

Russo-Averchi, E., Heiss, M., Michelet, L. et al. (2012). Suppression of three dimensional twinning for a 100% yield of vertical GaAs nanowires on silicon. *Nanoscale* 4 (5): 1486–1490. https://doi.org/10.1039/c2nr11799a.

Sandhu, S., Yu, Z., and Fan, S. (2013). Detailed balance analysis of nanophotonic solar cells. *Optics Express* 21 (1): 1209–1217. https://doi.org/10.1364/OE.21.001209.

Shockley, W. and Queisser, H.J. (1961). Detailed balance limit of efficiency of p-n junction solar cells. *Journal of Applied Physics* 32 (3): 510–519. https://doi.org/10.1063/1.1736034.

Steiner, M.A., Geisz, J.F., García, I. et al. (2013). Optical enhancement of the open-circuit voltage in high quality GaAs solar cells. *Journal of Applied Physics* 113 (12): 0–11. https://doi.org/10.1063/1.4798267.

Stelzner, T., Pietsch, M., Andrä, G. et al. (2008). Silicon nanowire-based solar cells. *Nanotechnology* 19 (29): 295203. https://doi.org/10.1088/0957-4484/19/29/295203.

Stuart, H.R. and Hall, D.G. (1997). Thermodynamic limit to light trapping in thin planar structures. *Journal of the Optical Society of America A* 14 (11): 3001. https://doi.org/10.1364/JOSAA.14.003001.

Tan, L.-Y., Li, F.-J., Xie, X.-L. et al. (2017a). Proton radiation effect on GaAs/AlGaAs core–shell ensemble nanowires photo-detector. *Chinese Physics B* 26 (8): 086202. https://doi.org/10.1088/1674-1056/26/8/086202.

Tan, L.-Y., Li, F.-J., Xie, X.-L. et al. (2017b). Study on irradiation-induced defects in GaAs/AlGaAs core–shell nanowires via photoluminescence technique. *Chinese Physics B* 26 (8): 086201. https://doi.org/10.1088/1674-1056/26/8/086201.

Tavakoli, N. and Alarcon-Llado, E. (2019). Combining 1D and 2D waveguiding in an ultrathin GaAs NW/Si tandem solar cell. *Optics Express* 27 (12): A909. https://doi.org/10.1364/OE.27.00A909.

Tian, B., Zheng, X., Kempa, T.J. et al. (2007). Coaxial silicon nanowires as solar cells and nanoelectronic power sources. *Nature* 449: 885–889. https://doi.org/10.1038/nature06181.

Tomioka, K., Motohisa, J., Hara, S., and Fukui, T. (2008). Control of InAs nanowire growth directions on Si. *Nano Letters* 8 (10): 3475–3480. https://doi.org/10.1021/nl802398j.

Wallentin, J., Anttu, N., Asoli, D. et al. (2013). InP nanowire array solar cells achieving 13.8% efficiency by exceeding the ray optics limit. *Science* 339 (6123): 1057–1060. https://doi.org/10.1126/science.1230969.

Wu, Y., Yan, H., and Yang, P. (2002). Semiconductor nanowire array: potential substrates for photocatalysis and photovoltaics. *Topics in Catalysis* 19 (2): 197–202.

Yablonovitch, E. (1982). Statistical ray optics. *Journal of the Optical Society of America* 72 (7): 899. https://doi.org/10.1364/JOSA.72.000899.

Yablonovitch, E. (1987). Inhibited spontaneous emission in solid-state physics and electronics. *Physical Review Letters* 58 (20): 2059–2062. https://doi.org/10.1103/PhysRevLett.58.2059.

Yu, Z., Raman, A., and Fan, S. (2010). Fundamental limit of nanophotonic light trapping in solar cells. *Proceedings of the National Academy of Sciences of the United States of America* 107 (41): 17491–17496. https://doi.org/10.1073/pnas.1008296107.

# 15

# Quantum Dots Solar Cells

*Han Wang and Maria Antonietta Loi*

*Photophysics and OptoElectronics Group, Zernike Institute for Advanced Materials, University of Groningen, Groningen, The Netherlands*

## 15.1 Introduction

Over the last two decades, the rapid development of solution-processable photovoltaics has been driven by the urgent need for easy-to-fabricate and cheap-to-produce renewable energy sources. Furthermore, the necessity of a swift transition from traditional energy sources to renewable ones requires their production with high-throughput and scalable techniques such as roll-to-roll (Kirmani et al. 2018; Sukharevska et al. 2021), inkjet printing (Sliz et al. 2019), spray coating (Choi et al. 2017; Song et al. 2019), etc.

When looking for semiconducting materials that can be deposited from solution, we have mostly three relevant classes: (i) organic semiconductors, where their solubility is determined by the molecular structure and the proper use of side chains; (ii) colloidal semiconductors, where the colloidal behavior depends on their nanometric size and the ligands decorating the surface; and (iii) the metal halide perovskite, which are soluble in very polar solvent thanks to the ionic nature of their bonds.

Each of these materials has shown its potential as solar cell active material, but are at a different stage of the development phase, while in this chapter we will concentrate our attention on colloidal semiconductors, we would like here to make a comparison among the three to better introduce the reader to the properties of colloidal semiconductors and of solar cells made with them.

Organic molecules with semiconductive behavior, were the first in time as a candidate in the race for solution-processable absorbers for solar cell applications. Organic semi-conductors have been investigated earlier on for applications in light emitting devices and display technology, and are now commercial. However, the solar cell field needed time after the seminal article of Tang (Tang 1986), who reported on the evaporation of organic molecules, to reach a certain critical mass of scientists that got convinced that solution processing was the missing ingredient in solar cell development. It took about 21 years for this community to go from the first technologically relevant report of a cell efficiency of 2.5% in 2001 (Shaheen et al. 2001) to the impressive 19.3% efficiency of

*Photovoltaic Solar Energy: From Fundamentals to Applications, Volume 2*, First Edition.
Edited by Wilfried van Sark, Bram Hoex, Angèle Reinders, Pierre Verlinden, and Nicholas J. Ekins-Daukes.
© 2024 John Wiley & Sons Ltd. Published 2024 by John Wiley & Sons Ltd.
Companion website: www.wiley.com/go/PVsolarenergy

today (Zhu et al. 2022). It is interesting to note that while several companies worldwide are busy producing organic solar panels, this market has been up until now mostly a niche up to now. While the expected physical limits of this class of materials have been extensively discussed in the literature in the last 20 years (Jiao et al. 2017; Kedikova et al. 2011), it is also worth mentioning that although the state-of-art organic absorbers show high optoelectronic performance, their chemical structures are becoming more and more complicated. This leads to an increase in synthesis cost, which casts doubts on the possibility of mass production of this technology.

The newest member of the solution-processable semiconductor family, but also the one performing better in terms of absolute power conversion efficiency, is the metal halide perovskite. The current record efficiency of this class of solution-processed solar cells in small-size devices is 25.7%, which is approaching single crystalline Si-based device efficiency, which is 26.3% (Green et al. 2021). This outstanding result obtained after 12 years of work of a large scientific community results from the outstanding physical properties of this class of materials. The large absorption coefficient, the number of photogenerated free charge carriers, and the long carrier diffusion length together contribute to making metal halide perovskite the most successful solution-processable semiconductor to date (Huang et al. 2017). Furthermore, recent experimental investigations have shown that crystalline metal halide perovskite is defect tolerant with the presence of only relatively shallow trap states, while in thin films, some deep traps appear (Ono et al. 2020). One of the main issues left for the community to be solved, is the severe degradation of perovskite active layers under ambient conditions, illumination, and temperature stress (Domanski et al. 2018; Rong et al. 2018). However, the flexibility of composition has provided a huge playfield for the optimization of their crystallization, with large consequences for the device performance and stability, giving an optimistic outlook on the fact that the stability issue will also be solved soon (Liu et al. 2022; Pitaro et al. 2022).

Besides the great success of both organics and metal halide perovskite solar cells, the band gaps of these materials are limited to about 900 nm (1.38 eV), leaving a large portion of the solar spectrum not used (Xi and Loi 2021). When thinking to develop solar cell technology towards multiple junctions, tandems, and beyond, active layers absorbing light at wavelengths longer than 900 nm will be necessary.

Colloidal semiconductors or more precisely colloidal quantum dots (CQD) were first demonstrated about 30 years ago with the idea of obtaining fully tunable semiconductors from the solar cell perspective; thus, a larger portion of the solar spectrum can be absorbed. It is also interesting to underline that the synthesis of CQDs through solution-based hot-injection methods ensures easy up-scaling of the active material.

In this chapter, we will give a short historical perspective of CQDs solar cells to concentrate then more closely on recent progress and in particular discuss the ink formulation and devices fabricated with one-step deposition. The role of surface chemistry, defect passivation, but also the different solar cell architectures will be discussed from the perspective of obtaining high-efficiency devices. Inks formulation, solvents, and stability of the inks will be critically reviewed from the perspective of possible future industrialization of solar cells based on CQDs.

## 15.2 Colloidal Quantum Dots Generalities

CQDs are very interesting for optoelectronic applications as they offer the opportunity to tune their bandgap through size adjustment (see Figure 15.1a,b). As schematically shown in Figure 15.1b, QDs of several nanometers in size are decorated by organic ligands, which passivate surface traps and determine their colloidal stability. Once the size of the crystalline nanocrystals is smaller than the corresponding exciton Bohr radius $a_B^*$ of the bulk semiconductor, these materials are named quantum dots due to the appearance of strong quantum confinement effects (Albaladejo-Siguan et al. 2021). The confinement of the electrons and holes in the zero-dimension potential barrier of these "artificial atoms" results in a discrete energy level spectrum. The band gap of CQDs $E_g$ depends heavily on the CQDs radius $R$ according to:

$$E_g = E_{g(bulk)} + \frac{\hbar^2 \pi^2}{2\mu R^2} - \frac{1.786e^2}{\varepsilon_r R} - R_y^* \tag{15.1}$$

where $E_{g(bulk)}$ is the intrinsic bandgap of the bulk semiconductor, $\varepsilon_r$ and $\mu$ are the relative dielectric constant and the reduced mass of the electron and hole, and $R_y^*$ is the Coulomb attraction between the electron and holes, which is proportional to Rydberg's energy (Yang 2021).

Among the many different types of CQDs that have been demonstrated in the last 30 years, the one based on lead sulfide (PbS) has been the most investigated for solar cell applications. Reasons for this are, on the one hand, the scalable and reproducible synthesis giving rise to highly mono-dispersed QDs (Li et al. 2010; Shrestha et al. 2016; Shrestha et al. 2016) showing a relatively limited number of defect states (Kahmann and Loi 2020). On the other hand, the large exciton Bohr radius in lead chalcogenide CQDs (PbS ~23 nm, PbSe ~46 nm) ensures the wide tunability of the size-dependent bandgap in the optimal range required for solar cells both as single, double and triple junctions. Therefore, as shown in Figure 15.1a, the absorption spectrum of these CQD materials can be broadly shifted to cover the whole near-infrared and short-wavelength infrared region without losing the quantum confinement. In addition, the air stability of the PbS CQDSC (CQD solar cell) is relatively good compared to PbSe ones (Ahmad et al. 2019; Zhang et al. 2016). This is not only due to the quality of the ligand passivation of the CQDs (see next section), but also to the intrinsic lower chemical reactivity of PbS toward oxygen. However, while the size variation of PbS CQDs can lead to tunable infrared absorption, it is also worth mentioning that the enlargement of dot diameter can lead to shape transition from [111]-only octahedron to [111]/[100] cuboctahedron, due to surface potential minimization (Balazs et al. 2015; Choi et al. 2013). Experimental reports show lower air stability of cuboctahedron CQDs derived from the unfavorable affinity of the organic ligands on the [100] facet (Kim et al. 2019; Roberge et al. 2020).

The large tuneability of the band gap of Pb chalcogenides CQDs is the basis of their interest in solar cell application, not only as simple absorbers but also for their potential to break the Shockley–Queisser efficiency limit for single-junction solar cells (Peng et al. 2012). When estimating the thermodynamic limit for a solar cell, the calculation shows that a solar cell absorber with the optimal bandgap exhibits 47% of spectral energy loss due

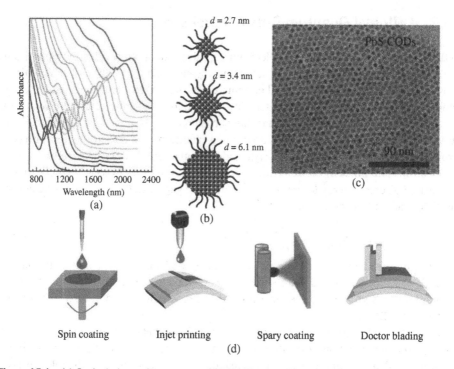

**Figure 15.1** (a) Optical absorption spectra of PbS CQDs of different sizes. Source: Moreels et al. (2012)/American Chemical Society. (b) Illustration of size-dependent PbS CQDs shape. Source: Roberge et al. (2020)/American Chemical Society. (c) TEM image of PbS CQDs synthesized by hot-injection method. (d) Schematic diagrams of four kinds of CQDs film fabrication approaches. Source: Xu et al. (2020)/John Wiley & Sons.

to incoming blackbody radiation from the sun, 2% of energy loss due to the recombination when trapped electrons from the conduction band are returned to the valence band, 18% of energy loss for photons of lower and higher energy than the band-gap of the absorber (Halim 2013). Therefore, the maximum conversion efficiency is around 33% for a solar cell made with an absorber of an optimal band gap of 1.34 eV.

The bandgap of PbS CQDs ranges from 0.38 to 2.0 eV, encompassing the optimal bandgap of 1.3 eV for the single junction device and 0.94 and 1.61 eV for the two-terminal tandem cell (Brown and Green 2002; Peng et al. 2012). Moreover, PbS CQDs have been proposed to exhibit effective carrier multiplication (also called multiple exciton generation, MEG), yielding external quantum efficiency for photons of energy >2Eg over 100% and theoretically exceeding the Shockley–Queisser limit with a conversion efficiency of 45% (Beard et al. 2010; Service 2008). A detailed schematic explaining the MEG physical process is shown in Figure 15.2b. Excitons generated with excessively large energy ($hv > 2E_g$) while decaying to the conduction band edge, can excite an extra electron from the valence band to the conduction band, thus producing an extra electron–hole pair. This process is named impact ionization. The effectiveness of this process in CQDs material far outweighs that in bulk materials as the discrete energy state retards the phonon-mediated cooling of excitons and extends the time window for the impact ionization (Shabaev et al. 2013).

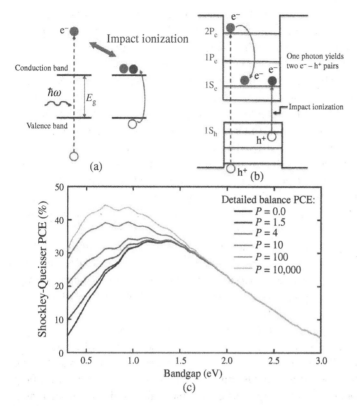

**Figure 15.2** Schematics of the charge-carrier pathway in case of impact ionization for (a) bulk materials and (b) CQD. Note that the impact ionization rate in the bulk material is very small. Source: Adapted from Goodwin et al. (2018). (c) Theoretical enhancement of the Shockley-Queisser limit for single junction solar cells through MEG. The parameter P relates the rates for MEG $k_{MEG}$ and hot carrier cooling $k_{cool}$ through the expression $P = k_{MEG}/k_{cool}$. Source: Beard et al. (2013)/American Chemical Society.

The first experimental proof of low-threshold MEG in the CQDs system was published in 2004 (Schaller and Klimov 2004). Departing from this finding, various investigations were done concerning more efficient MEG rate via the modification of CQDs material (Jaeger et al. 2012; Trinh et al. 2011), size and composition (Kumar et al. 2016; Midgett et al. 2013), shape (Cunningham et al. 2011; Sandberg et al. 2012), and surface chemistry (Jaeger et al. 2012; Jaeger et al. 2013).

While for years a large debate was stimulated by these reports (McGuire et al. 2010), the evidence of an enhancement of the photocurrent in devices for photon energy larger than 2Eg has been finally clarifying that the physical effect is occurring, but has little significance for solar cells due to the large photon energies required (Sambur et al. 2010).

One of the most prominent features of CQDs is the enormous surface-to-volume ratio of the dots. Around one-third of the atoms of a few nanometer CQDs are located at the surface, experiencing lower coordination than in bulk. While the lower coordination is definitely a problem for the occurrence of very detrimental surface traps, it also represents an opportunity, as the variation of the dielectric or surface chemical environment has a significant effect on the overall physical properties of CQDs (Balazs and Loi 2018). This tunability of the surface chemistry opens a tremendous potential for solar cell optimization by substituting ligands on the CQDs surface, referred to as the ligand exchange process.

Very early on, it was found that the charge carrier transport properties of CQDs can be enhanced after ligand exchange due to the availability of tunneling processes. In fact, the as-synthesized CQDs are usually decorated by long-chain organic ligands such as oleic acid (OA), which are insulating but allow long-term storage of the colloidal solution (Shrestha et al. 2019). As a consequence, the use of small-size ligands, mostly exchanged after film deposition (Ning et al. 2012, 2013), has been fundamental for the development of optoelectronic devices based on CQDs. By using this method, the first CQDs solar cells (Tang et al. 2011) but also photodetectors (Liu et al. 2010), and field effect transistors have been fabricated, depositing the active layer by spin-coating using a layer-by-layer technique (Klem et al. 2008; Liu et al. 2010; Liu et al. 2011). Samples deposited in this way display limited charge carrier mobilities even when surface traps are filled (vide infra) (Bisri et al. 2013), the charge transport is mostly governed by hopping (Shulga et al. 2018) due, among other things, to the lack of translational symmetry in the CQDs thin film (Murray et al. 2006). Recently, it was demonstrated that CQD superlattices, namely arrays with translational symmetry, display very high mobilities in field effect transistors using ionic gels as dielectrics (Pinna et al. 2022).

It is also interesting to note that, unlike bulk material, it is almost impossible to achieve stable doping in CQDs since dopants might break the chemical balance of the QDs and diffuse toward the surface. However, the ligand exchange process is found to result in the alteration of the energy band and shift of the Fermi level, which is in a way similar to doping and is beneficial for band alignment design. The energy band alteration can be accounted for by the difference in dipole moment between the surface CQDs atom and the different binding groups of ligands (Brown et al. 2014). Moreover, by altering the ligands on the surface, the stoichiometry of the QDs can be altered (Kim et al. 2013). Here, it is important to underline that PbS CQDs are synthetized off stoichiometry (Moreels et al. 2009) and, for this reason, exhibit n-type transport properties. It has been reported that by adding electron-abundant sulfur atoms, PbS CQDs can be tuned from n- to p-type (Balazs et al. 2017). In summary, the ligand exchange process provides wide tunability for manipulating the electronic properties of CQD solids according to the need of the application.

## 15.3 Ligand Exchange Methods

As mentioned above, the optoelectronic properties of PbS CQDs can be tuned by adjusting the surface chemistry and more in general their stoichiometry. The ligand exchange process can be achieved via two different methods, namely, the solid-state ligand exchange (SSLE) and the phase transfer ligand exchange (PTLE). The schematics of both processes are shown in Figure 15.3.

The SSLE process has been widely used in the fabrication of PbS CQDSC devices as there are no concerns about solvent erosion and solubility in this process (Cao et al. 2016; Lu et al. 2018; Szendrei et al. 2010; Tang et al. 2011). The procedure starts with the deposition of OA-capped PbS CQDs dissolved in a non-polar solvent; after this step, a polar solvent with

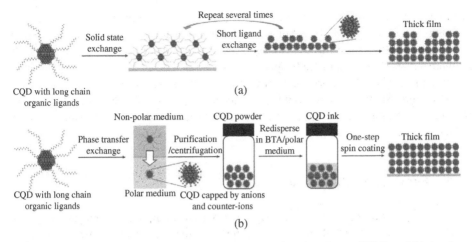

**Figure 15.3** Schematic illustration of (a) the solid-state ligand exchange (SSLE) and (b) the phase transfer ligand exchange (PTLE) process. Source: Mohan Yuan et al. (2022)/John Wiley & Sons.

shorter ligands such as 1,2-ethanedithiol (EDT), mercaptopropionic acid (MPA) or tetra-butylammonium iodide (TBAI) is deposited on the CQDs films leading to ligand exchange. The film can then be rinsed with the pure polar solvent for the removal of left-over OA or excessive new ligands. Due to the orthogonality between the solvent used for each treatment, the thickness of the SSLE process can be controlled by multiple iterations of the described procedure, a process that is generally called layer-by-layer deposition. This technique has allowed a relatively small community of scientists to investigate the potential of PbS QDs for solar cells, which started more than 10 years ago with an efficiency of 4% (Szendrei et al. 2010), reaching nowadays 12% (Ding et al. 2020).

Several limits of SSLE have been identified in the years: first, the insufficient passivation and thus the occurrence of trap states; second, the large shrinking of the interdot distance that generates nanoscopic cracks within the film, harming the overall active layer morphology and resulting in radiative and not radiative recombination; third, the energetic inhomogeneity determined by QD polydispersity, random packing, and organic leftovers can harm the carrier mobility (Gong et al. 2016; Guyot-Sionnest 2012; Jung et al. 2022); last, the repetitive deposition via spin coating is time- and material-consuming, and is not compatible with industrial processing for large scale production.

The PTLE process, on the other hand, is a relatively new method for CQD devices that enables single-step deposition of the active layer. Generally, the CQDs dispersed in nonpolar solvents are mixed with the corresponding polar solution containing the target ligands. After a stirring process aimed at facilitating a sufficient phase transfer reaction, post-transfer washing is required to remove residual organic ligands. It should be noted that colloidal stability is usually maintained during the phase transfer through different mechanisms, from steric hindrance in nonpolar solvents to electrostatic stabilization in polar solvents (Bederak et al. 2020; Yuan et al. 2022). CQDs can then be precipitated and separated from the solvent using an antisolvent and centrifugation. The resulting pellet would be re-dispersed in a polar solvent that can allow thin film deposition. With the PTLE method, a higher film quality can be achieved via single-step deposition, while the thickness is limited by the concentration of CQDs and the boiling point of the solvent. Generally speaking, the PTLE technique provides less defective film and better charge carrier transport than the SSLE and can be regarded as a more advanced technique, which has more perspective for industrial application.

Furthermore, given the more compact film with fully passivated CQDs, the PTLE film can be more resistant to oxidation-induced defects compared to the SSLE one. Therefore, we shall focus on the PTLE technique as the one that can provide further advancements in CQDs solar cells.

## 15.4 Evolution of CQDs Solar Cells

Besides the evolution of the ligands, it is also important to understand which structures have been used to make CQD solar cells and how ligands and structures are intertwined. The first CQDs solar cell was fabricated in 2008 by sandwiching a PbS CQDs thin film between two electrodes, as shown in Figure 15.4 (Johnston et al. 2008). The charge separation of this device was achieved by rectifying the Schottky junction formed between the PbS CQDs and the low-work-function metal electrode. The PCE of this simple device was reported to be 1.8% under AM1.5 illumination with $3\,mA/cm^2$ short-circuit current. Further optimization of the Schottky junction resulted in the highest PCE of 5.2% in 2013, still with a low short-circuit current due to the high recombination in trap states and poor charge transport (Piliego et al. 2013). However, the most critical effect of the use of the Schottky junction was the limited open circuit voltage ($V_{oc}$), which at max was reported to be even less than half of the active layer band gap (Speirs et al. 2016; Speirs et al. 2015; Szendrei et al. 2010).

The maximum achievable voltage in a Schottky-junction solar cell is equal to the built-in potential of the junction, which ideally depends on the metal work function (Sze and Ng 2006). However, in the case of high trap concentration, the Fermi-level pinning may determine the potential (Luther et al. 2008), setting it to less than half of the bandgap.

As mentioned previously, to allow charge transport, the ligand exchange of CQDs was deemed and demonstrated to be essential. The first attempt to minimize the interdot distance came with the substitution by SSLE of original OA ligands with shorter n-butylamine (BA) (Johnston et al. 2008). Simultaneously, more ambitious schemes were proposed, including the cross-linking of the CQDs with bifunctional molecules, again to further enforce the proximity of CQDs obtaining, as a result, also to make the layers insoluble after the treatment. Bidentate molecules with functional groups such as thiols (Luther et al. 2008), amines (Law et al. 2008), carboxylates (Zarghami et al. 2010), and hydrazine (Talapin and Murray 2005), etc., were chosen as these nucleophilic groups show a strong affinity to the lead-terminated surface (Moreels et al. 2012). Especially CQDSC using thiol capping ligands such as EDT (Barkhouse et al. 2008), MPA (Yuan et al. 2014), BDT (Piliego et al. 2013), and 1,3-propanedithiol (PDT) (Azmi et al. 2016) showed improved photovoltaic properties and were claimed in some cases to show strong p-type behavior. PCEs of 6.5% and 7.2% were reported for PbS-EDT and PbS-MPA as absorber materials, respectively (Ip et al. 2015; Zhang et al. 2014). The chemical structures of the most common ligands used for SSLE are reported in Figure 15.5.

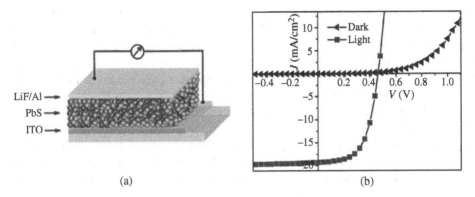

**Figure 15.4** The (a) schematic and (b) JV curves of the Schottky photovoltaic structure with the CQDs layer inserted between the ITO and the LiF/Al electrode. Source: Piliego et al. (2013)/The Royal Society of Chemistry.

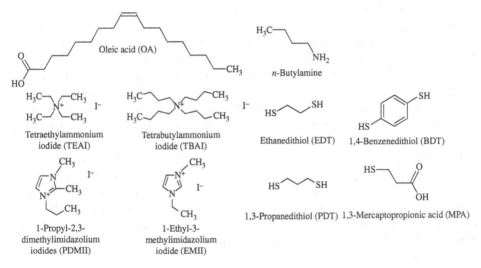

**Figure 15.5** Chemical structures of the most important capping and crosslinking ligands discussed in this chapter.

However, it was soon realized that these ligands did not provide full passivation of the surface (Speirs et al. 2015) and that also the p-type doping was mostly extrinsic and due to the oxygen exposure (Balazs et al. 2014). Furthermore, more sophisticated device structures were in the meantime introduced to substitute the Schottky junction, and the CQDs active layer was deposited on an n-type metal oxide layer to form the so-called depleted junction (Pattantyus-Abraham et al. 2010).

In 2011, salts featuring electron-abundant halide anions emerged as new atomic ligands to further reduce the CQDs spacing (Tang et al. 2011). For example, cetyltrimethylammonium bromide (CTAB) was utilized to decorate the CQDs surface via SSLE with Br anions. Devices fabricated with these CQDs showed an efficiency of up to 6%. Later research introduced iodide-based ligands such as tetra-*n*-butylammonium iodide (TBAI) (Liu et al. 2011), tetraethylammonium iodide (TEAI) (Park et al. 2019), 1-propyl-2,3-dimethylimidazolium iodides (PDMII) (Azmi et al. 2017), and 1-ethyl-3-methylimidazolium iodide (EMII) (Bashir et al. 2021) which have revolutionized the field of CQDs solar cells. It is very interesting to point out that while this passivation was considered to leave only iodide anions on the surface, a relatively large number of counter ions were proposed in the literature. In 2015, Balazs clarified the catalytic role of the counterion with the resulting control of the reaction kinetics (Balazs et al. 2015). The introduction of halide ligands on the CQDs surface has been reported to have various benefits, including erasing the mid-band trap with better passivation of the surface (Ning et al. 2012, 2013; Proppe et al. 2018), boosting the electron transfer to the electron transport layer (ETL) with a higher dielectric constant and carrier density (Zhitomirsky et al. 2012, enhancing the device stability in the air (Ning et al. 2014).

The next important boost to colloidal QD solar cells was achieved by improving the device structure using a p-n junction formed by a n-type CQDs film and p-type one (Ning et al. 2014). Since 2014, all the most efficient CQDs solar cells have been using a structure where an ethane-1,2-dithiol (EDT)-treated (and air exposed to improve the p-type doping) PbS CQDs layer is acting as p-type material, and an I-capped PbS CQDs layer is the n-type material, that together forms a p-n junction with a smooth interface and efficient carrier extraction. A PCE of ~8.55% using SSLE for fabricating the p-n structure was achieved already in 2014, and more recently, a certified PCE of almost 11% was reported using the same device structure and active layers exploiting SSLE (Wang et al. 2018). More details on the performances and limitations of this type of device can be found in the work of Speirs et al. (Speirs et al. 2016), who underline the need for a better p-type layer.

While the SSLE has dominated the early successes in CQDs solar cells, some activities on obtaining inks with PbS decorated with short ligands already started in 2006. Organic ligands such as BA (Clifford et al. 2009; Konstantatos et al. 2006; Konstantatos et al. 2007), TG (Fischer et al. 2013), and MPA (Aqoma and Jang 2018) were utilized. However, the defect concentrations of these inks were very high, most probably because of the steric hindrance of organic ligands, which prevented a good surface coverage (Ip et al. 2012; Kroupa et al. 2017; Yuan et al. 2022).

Inspired by the success of halide ligand SSLE and considering the smaller steric hindrance of the halide ions, the first solution-processed ink using TBAI was demonstrated by Ning et al. in 2014 using a nonpolar solvent (Ning et al. 2014). The CQDs concentration in the ink based on the nonpolar solvent was reported to be limited, insufficient for fabricating good photovoltaic thin-film device. Still, in 2014, Ning et al. first introduced the PTLE using MAI as ligand for PbS CQDs and fabricated devices displaying an efficiency of about 6% (Ning et al. 2014).

In 2019, Zhang et al. compared the two-ligand exchange methods in terms of ligand coverage, carrier mobility, stability, and photovoltaic performance (see Figure 15.6). This work showed the superiority of PTLE in all aspects (Zhang et al. 2019). After a decade of extensive investigation, PTLE solar cell currently exhibits the highest performance with PCE ~15% (Ding et al. 2022). In the following, we will discuss in detail the success stories and the limits and identify the most critical problems that need to be solved to further advance CQD solar cells.

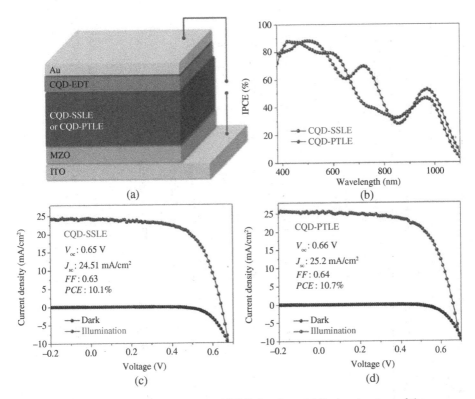

**Figure 15.6** Comparison between SSLE and PTLE absorbers. (a) Device structure of the SSLE-based PbS-TBAI solar cell or PTLE-based PbS-PbX$_2$ solar cells (X = I or Br). (b) EQE spectra of the two solar cells. (c,d) Photovoltaic performance of the (c) SSLE-based CQDs solar cell and (d) PTLE-based CQDs solar cell. Source: Zhang et al. (2019)/American Chemical Society.

## 15.5  Recent Progress in Solar Cells

Since it was demonstrated that CQDs treated with the optimized PTLE gave rise to comparable device performance to the one treated with the conventional SSLE (Liu et al. 2017), the PTLE process gradually replaced in the literature the SSLE one, mostly because of the facility of the deposition (one step instead of layer-by-layer deposition) and the possibility for industrialization this technique offers. In this section, we shall discuss the most common surface treatments performed with the PTLE process and the photovoltaic performances obtained with the CQDs inks.

Table 15.1 provides an incomplete overview of the solar cells reported (in chronological order) utilizing active layer CQDs inks obtained by PTLE. For each article, the device structure indicating the ligand type, the device parameters, and the reference is reported.

Following the first report of the PTLE process using MAI as ligand, lead halide PbX$_2$ (X = Br, I) ligands became of common use as it was reported that they could also be used in a PTLE process (Kiani et al. 2016). However, PbX$_2$ alone is found to have limited solubility and solvation for the CQDs, which leads to slow and insufficient PTLE and colloidal instability.

**Table 15.1** Non-exhaustive selection of the literature reporting CQDs solar cells where the active layer was deposited using an ink obtained by PTLE. Device structure, performances, and important features are reported for each work.

| CQDSC structure | PCE (%)[a] | $V_{oc}$ (V) | $J_{sc}$ (mA/cm²) | FF (%) | Comment | References |
|---|---|---|---|---|---|---|
| ITO/ZnO/PbS-PbX$_2$ + HI (1150 nm)/PbS-EDT/Au[b] | 0.86 | 0.42 | 3.16 | 64.8 | HI as additive for successful PTLE | Jo et al. (2018) |
| ITO/ZnO/PbS-PbX$_2$ + NaAc (1200 nm)/PbS-EDT/Au[b] | 0.9 | 0.43 | 3.2 | 65.2 | NaAc as additive for successful PTLE | Kim et al. (2019) |
| ITO/ZnO/PbS-PbX$_2$ + NaAc (1200 nm)/PbS-EDT/Au[b] | 1.17 | 0.46 | 4.4 | 58 | Using PbI$_2$ + PbBr$_2$ + PbCl$_2$ | Fan et al. (2019) |
| ITO/ZnO/PbS-PbX$_2$ + PbS-PbX2 + ME(1200 nm)/PbS-EDT/Au[b] | 1.07 | 0.43 | 5.5 | 57.9 | BHJ with good quality p-type CQD | Choi et al. (2020) |
| ITO/AZO/PbS-AI/PbS-EDT/Au | 11.4 | 0.65 | 26.6 | 66 | Using AI as ligand and PC as solvent | Jia et al. (2019) |
| ITO/ZnO/PbS-PbI$_2$(hybrid-amine)/PbS-EDT/Au | 12.01 | 0.647 | 29.04 | 63.8 | 2D matrix with hybrid amine solvent | Xu et al. (2018) |
| ITO/ZnO/PbS-PbX$_2$+Pb(SCN)$_2$/PbS-EDT/Au | 9.8 | 0.6 | 30.6 | 59 | Pb(SCN)$_2$ as additive | Sun et al. (2017) |
| ITO/ZnO/PbS-PbX$_2$+AA/PbS-EDT/Au | 11.28 | 0.611 | 27.23 | 67.8 | First AA additive report | Liu et al. (2017) |
| ITO/ZnO/PbS-PbX$_2$+TBAA:AA/PbS-EDT/Au | 10.9 | 0.7 | 25.3 | 62 | TBAA as additive for speed control | Jo et al. (2017) |
| ITO/MZO/PbS-PbX$_2$+KI$_3$/PbS-EDT/Au | 11.9 | 0.64 | 27.3 | 68 | KI$_3$ as additive | Hu et al. (2021) |
| ITO/ZnO/PbS-PbX$_2$+KI/PbS-EDT/Au | 12.6 | 0.64 | 28.8 | 68.5 | KI as additive | Choi et al. (2020) |
| ITO/ZnO/PbS-PbX$_2$+CPT/PbS-EDT/Au | 10.6 | 0.63 | 31.4 | 53.65 | Hybrid ligand CQDs using CPT | Mandal et al. (2021) |

| Device structure | PCE (%) | $V_{oc}$ (V) | $J_{sc}$ (mA cm$^{-2}$) | FF (%) | Notes | Reference |
|---|---|---|---|---|---|---|
| ITO/ZnO/PbS-PbX$_2$+MPA/PbS-EDT/Au | 11.18 | 0.61 | 27.42 | 66.82 | Hybrid ligand CQDs using MPA | Gu et al. (2019) |
| ITO/ZnO/PbS-PbX$_2$+MPE/PbS-EDT/MoO$_3$/Au/Ag | 9.6 | 0.61 | 30.3 | 52 | Hybrid ligand CQDs using MPE | Sharma et al. (2020) |
| ITO/ZnO/PbS-PbX$_2$+PbS-(PbX$_2$ + CTA)/PbS-EDT/Au | 13.3 | 0.65 | 30.2 | 68 | Record CQDs BHJ | Choi et al. (2020) |
| ITO/ZnO/PbS-MAPbI$_3$+PbS-TG/PbS-EDT/Au | 10.45 | 0.62 | 26.8 | 63.9 | First CQDs BHJ report | Yang et al. (2017) |
| ITO/ZnO/PbS-MAPbI$_3$/PbS-EDT/Au | 8.7 | 0.57 | 25 | 61 | Blade coating absorber with DFP solvent | Sukharevska et al. (2021) |
| ITO/ZnO/PbS-MAPbI$_3$/PbS-EDT/Au | 8.95 | 0.61 | 21.8 | 67.9 | First MAPbI$_3$ ligand report | Yang et al. (2015) |
| ITO/ZnO/PbS-FAPbX$_3$/PbS-EDT/Au | 13.8 | 0.65 | 30 | 71 | FAPbI$_3$ monolayer bridging | Sun et al. (2020) |
| FTO/ZnO/PCBM/PbS-FAPbI$_3$/PMMA-GO/PbS-EDT/Au | 15.45 | 0.66 | 31.5 | 74.3 | Novel FAPbI$_3$ exchange method | Ding et al. (2022) |
| ITO/ZnO/PbS-CsPbI$_3$/PbS-EDT/Au | 10.5 | 0.64 | 24.5 | 67 | First CsPbI$_3$ ligand report | Zhang et al. (2018) |

a) Device parameters listed here pertain to the highest performing device reported in the single article when tested with solar irradiation at AM 1.5.

b) The PCE of this work refers to the IRPCE, where a long-pass c-Si filter on 1100 nm is applied in front of the device.

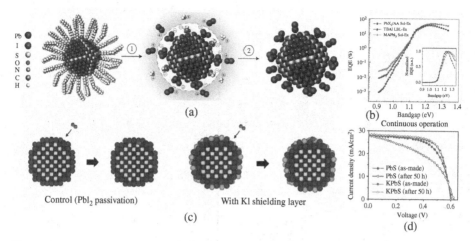

**Figure 15.7** Lead-halide PTLE. (a) The bulky oleic acid ligands are replaced by the $[PbX_3]^-$ anions with the aid of ammonium protons, both $[PbX]^+$ and $[NH_4]^+$ are expected to act as counter-ions. After ligand exchange, CQDs are precipitated via the addition of an anti-solvent and are separated by centrifugation. $NH_4Ac$ is taken away by the mixed solvent, and thus, clean $[PbX_3]^-/[PbX]^+$-capped CQDs solids are left without any organic residues. (b) High-dynamic-range (HDR) EQE measurement indicates the band gap tail states. Source: Liu et al. (2017)/Springer Nature. (c) PbS-$PbX_2$+KI CQDs schematics. The black dots represent Pb atoms, the yellow dots represent S atoms, the purple dots represent I atoms, the red dots represent O atoms, and the light-blue dots represent K atoms. (d) *J–V* characteristics of devices with and without KI shielding with MPP tracking in the air. Source: Choi et al. (2020)/John Wiley & Sons.

To address these issues, Liu et al. (2017) introduced ammonium acetate (AA) as an additive in the PTLE process, achieved by using this CQDs 11.28% PCE in solar cells. As illustrated in Figure 15.7a, the authors rationalize that the dissociation of AA making available $NH_4^+$ cations provides appropriate counterions to ensure colloidal stability during the phase transfer (Liu et al. 2017). The AA-assisted $PbX_2$ ligand exchange provides a very good passivation for the CQDs surface, as is evidenced by a narrower band tail around the bandgap (see Figure 15.7b) (Zhitomirsky et al. 2012). Besides the improved PCE, the performance of the unencapsulated devices also shows outstanding air stability, with only 20% reduction over 1000 h of exposure to ambient conditions owing to the complete passivation protecting the surface from environmental gasses.

As vigorous phase transfer reaction can evoke inhomogeneous CQDs fusion, tetrabutylammonium acetate (TBAA) was introduced to the process as an additive with larger cation size and weaker Lewis acidity to regulate the reactivity of the ligand exchange (Balazs et al. 2015; Jo et al. 2017). The solar cells based on this process exhibited a record $V_{oc}$ of 0.69 V thanks to the excellent monodispersity of the starting material and even flatter bandtail of the resultant CQDs ink.

Meanwhile, Choi et al. discovered that KI can be an alternative additive that increases the device stability in ambient conditions (Choi et al. 2020). The potassium cations have low electron affinity and large atom size and react more easily with iodide than oxygen. This means that the KI can form an additional shield protecting the CQD's surface from oxygenation. The resulting $O_2$ protection mechanism is shown in Figure 15.7d.

This approach was further utilized and optimized by Hu et al., who used $KI_3$ giving rise to $I_2$ and dissolved KI. The dissociative $I_2$ from the salt can remove the unfavorable surface vacancy on the [100] facet of the CQDs, improving the CQDs quality (Hu et al. 2021).

Both reports show devices with efficiency above 12% and prolonged air stability under MPP conditions, indicating the complete surface coverage and strong protective shielding of CQDs against oxygen and moisture.

One of the major drawbacks of the lead halide PTLE is the poor affinity between the iodide anions and the [100] surface facet due to lattice mismatching (Bi et al. 2018; Cao et al. 2016; Yin et al. 2022). The capped ligands on this facet can be easily stripped off by polar solvents during ligand exchange (Liu et al. 2021). This can severely undermine the surface coverage of larger CQDs (>5 nm), which are characterized by a higher [100]/[111] facet ratio, as shown in Figure 15.8a.

Furthermore, the remnant original ligands and the increased steric hindrance determined by those can obstruct the formation of a compact thin film and lower its solar cell performance and stability (Fan et al. 2017). It was proposed that using hydroiodic acid (HI) as an additive could promote the release of OA ligands (Jo et al. 2018). It was found that the remaining OA ligands were removed by binding with short acetate molecules ($CH_3COO^-$, binding group of OA), while its dissociative iodide anions achieved sufficient halide passivation on these extra vacancy sites. The resulting device shows 50% EQE on its

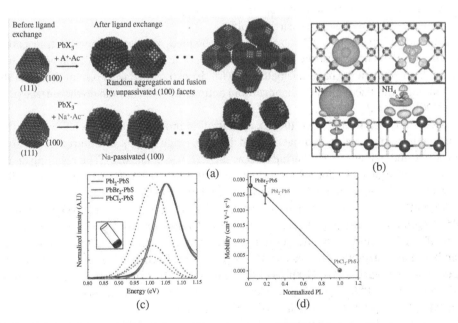

**Figure 15.8** (a) Schematic illustrations of ligand exchanged CQDs with and without Na-passivation on the [100] facets. (b) Differential charge densities of $Na^+$ and $NH_4^+$ on a PbS [100] surface. The Na, H, N, S, and Pb atoms are represented by the orange, pink, blue, yellow, and gray spheres, respectively. Source: Kim et al. (2019)/John Wiley & Sons. (c) Normalized photoluminescence (dotted lines) and absorbance (solid lines) spectra for different halide-passivated inks. (d) Film mobility versus normalized photoluminescence intensity for different halide-passivated CQDs. Source: Fan et al. (2019)/John Wiley & Sons.

1550 nm excitonic peak and superior air stability with 80% of the original PCE after 300 h of air exposure. As the major PV application for this large diameter CQDs is for tandem solar cells on top of a Si or a metal halide perovskite bottom devices to extend the solar wavelength usage, the group also reported the so-called IR-PCE (0.85%) measured under 1.5AM illumination and using an 1150 nm longpass Si-filter to estimate the additional PCE this device conveys to a silicon-based tandem.

Coming back to the difficulty of passivation of the different facets and in particular of the [100], NaAc was proposed as additive together with $PbI_2$ and $PbBr_2$ for the strong bonds that the small-size Na cation can make with this sulfur-dominated facet. The adsorption strength of $Na^+$ and $NH_4^+$ on a PbS (100) was investigated by density functional theory calculations, which confirmed that the electron exchange between $Na^+$ and PbS [100] is more favorable compared to that between $NH4^+$ and PbS [100] (see Figure 15.8b) (Kim et al. 2019). This additive reduces the degree of aggregation during the ligand exchange process and can even boost the number of halide ligands on the [111] facet, as confirmed by the XPS measurements.

This procedure was further improved in 2019 by combining lead iodide with other ligands in the lead halide family ($PbBr_2$ and $PbCl_2$) that use smaller sizes halogens and have a better affinity with the [100] facet that is already partially passivated with $Na^+$ (Fan et al. 2019). The best device fabricated with the 1200 nm CQDs inks prepared with this procedure shows an averaged IR-PCE of 1.17% and 80% EQE at the excitonic peak energy. As shown in Figure 15.8c,d, it is worth pointing out that better passivation using smaller halide anions also has an effect on the charge carrier mobility.

Further optimization of IR solar cells was obtained by altering the ITO transparent electrode and using a substrate with higher than 50% absorption in the IR range. Furthermore, fixing the energy band mismatching with the extracting layers originating from the shrinking of the size-dependent bandgap and optimizing the I/Br ratio on the surface (Kim et al. 2018; Li et al. 2022).

Recently, Ge et al. combined the ink optimization mentioned above with device structure optimization, using hydrogen-doped indium oxide (IHO) as infrared transparent electrode and magnetron-sputtering ZnO as the ETL (Ge et al. 2022). The resulting device shows IR-PCE of 1.25% and also a superior full-sun PCE of 11.15%, which is outstanding considering the non-optimal $V_{oc}$ achievable with large-size CQDs.

Recently, it was reported that the heteroepitaxial growth of a perovskite shell on PbS CQDs allows the complementary use of different properties of the two materials. The PbS CQDs and Pb-halide perovskites have an excellent lattice match due to the similar crystal structure and axis length ($a_{MAPbI_3} = 6.26$ Å; $a_{PbS} = 5.94$ Å), which enables excellent coherence in the [100] facet (Ning et al. 2015). The first publication showing $MAPbI_3$-capped PbS CQDs PTLE inks dates from 2015. Devices made with this ink showed a PCE of 8.95%, which was attributed by the authors to originate from the formation of a deep depletion region (Yang et al. 2015). The XPS analysis and TEM data for the annealed thin film are shown in Figure 15.9b and indicate the achievement of heteroepitaxial growth. Compared to the ionic ligands utilized to passivate CQDs facets, these so-called perovskite ligands have lower formation energy on the QD surface and are expected to provide better surface coverage, which should give rise to solar cells with a higher open circuit voltage. However, the type-I band alignment of the broad band-gap

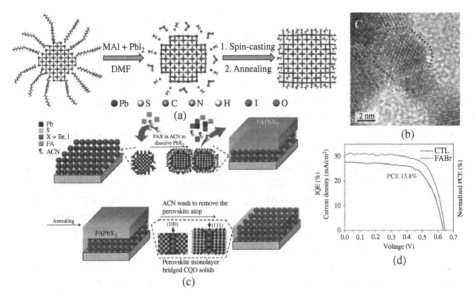

**Figure 15.9** (a) Schematic of solution ligand exchange with perovskite precursors and corresponding heteroepitaxial growth of thick perovskite shell on CQDs after deposition. (b) TEM image of the PbS-MAPbI$_3$ CQDs with dotted lines to guide the evidence of perovskite shell. (c) Schematic of perovskite monolayer that bridges neighboring CQDs. (d) Comparison of J-V characteristics for pristine CQDs (CTL) and CQDs after FABr treatment to form a thin FAPbX$_3$ shell. Source: Sun et al. (2020)/American Chemical Society; Yang et al. (2015)/American Chemical Society.

perovskite and the narrow band-gap CQDs severely deters the carrier transport, limiting the $J_{sc}$.

Besides the hybrid perovskites with organic molecules at the A-site, CsPbI$_3$ was also reported to have coherent lattice with PbS CQD. Samples based on this allowed to achieve a PCE of 10.5%, which is higher than the one obtained with MAPbI$_3$ but showed lower stability due to the intrinsic phase instability of this perovskite (Zhang et al. 2018). To obtain a stable cesium-based perovskite shell, Liu et al. introduced Br anions forming CsPbI$_x$Br$_{3-x}$. The addition of Br is known to relax the lattice tension and block the phase transition of CsPbI$_3$ (Liu et al. 2019). Moreover, the introduction of Br results in a reduction of the potential barrier for carrier transport, which partially solves the mobility issue of the perovskite-capped CQD. The resulting solar cell, in this case, shows a maximum PCE of 12.6% and enhanced photostability.

For further improving the transport, an ultrathin layer of perovskite shell is deemed necessary. Sun et al. achieved the self-programmed growth of a perovskite monolayer through a two-step deposition (Sun et al. 2020). The lead halide-capped CQDs are first deposited, followed by the soaking of the film in the A-cation precursor, such as FAX (X = I, Br). In this way, the heteroepitaxial growth is constrained on the superficial iodide sites. In addition, the monolayer growth also promotes bridging between adjacent dots, as shown in Figure 15.9d, which reduces the interdot distance by forming a sort of perovskite matrix covering the PbS dots and dramatically increasing the charge carrier mobility. This method enabled the fabrication of solar cells showing $J_{sc}$ up to 30 mA/cm$^2$ and PCE of

13.8%. Recently, Ding et al. proposed a variation to the method for achieving a single-step monolayer-perovskite capped PbS CQDs (Ding et al. 2022). The PbS-PbX$_2$ CQDs powder obtained after phase transfer, centrifugation, and drying was directly dispersed into the FAI solution to form the PbS-FAPbI$_3$ ink. Despite the lack of a clear investigation and conclusions proving for CQDs bridging in this work, the resulting films show enhanced mobility compared to its lead-halide capped reference. After some additional interface modification at the level of the device structure, the PCE of the photovoltaic device reached the record value of 15.45%, which is the highest value reached up until now for CQDs solar cells (Ding et al. 2022).

As we have discussed earlier, organic ligands with thiol functional groups have been the first utilized to achieve the PTLE of CQDs. Despite the nonoptimal performances (Sharma et al. 2020), all the previously discussed passivation schemes give rise to PbS displaying n-type characteristics, while organic PTLE still has unique position for substituting the p-type PbS commonly used, namely PbS-EDT, which is proven to be defective and problematic in many respects (Becker-Koch et al. 2020).

It is interesting to note that these organic molecules are also appropriate candidates as additives in the halide-based PTLE, ensuring the full passivation of the [100] facet and preventing the loss of Pb and I atoms of CQDs during the vigorous ligand exchange process (Sayevich et al. 2016). This hybrid ligand exchange has been reported various times using different organic ligands including 1-thioglycerol (TG) (Choi et al. 2020), 3-chloro-1-propanethiol (CPT) (Mandal et al. 2019), 3-methylmercapto propionate (MPE) (Sharma et al. 2020) and 2-mercaptoethanol (ME) (Choi et al. 2020). All reports show sharpened excitonic peaks in the absorption spectrum and FWHM comparable to the original peak of OA-capped PbS, indicating improved passivation during the PTLE process. However, it is often reported that a trade-off exists between carrier mobility and full passivation via organic ligands.

Another approach reported is the mixing of CQDs passivated with ligands of different natures, taking inspiration from the bulk-heterojunction (BHJ) approach used in organic solar cells. These CQDs' BHJ are obtained by mixing PbS-TG and PbS-MAPbI$_3$ inks. The resultant mixed-QD solid incorporated a nanoscale-distributed band offset that separates electrons and holes into different domains, preventing exciton recombination (Yang et al. 2017). After adjusting the D/A ratio for a balanced electron and hole extraction, the best device displayed a PCE of 10.4%. A follow-up work on BHJ CQDs used lead halide and cysteamine (CTA) hybrid ligands as the capping ligand of the p-type CQDs and lead halide capped CQDs as the n-type CQD (Choi et al. 2020). By achieving the complete passivation of both the blend components, a PCE of 13.3% was obtained with the corresponding solar cell.

## 15.6  Solvents for CQDs Inks

A very important prerequisite for the CQDs inks is the colloidal stability and the quality of the solvent, which should allow the use of large-area deposition methods, giving rise to good film quality. Obviously, the colloidal stability is determined by the interplay between the ligands and the solvent. For inorganic ligand-capped PbS CQD, the ionic shell surrounding the CQD's surface is the main contributor to the colloidal stability. Besides providing solubility in the corresponding solvent, the anionic ligands bonding on the surface and the cationic layer from the surrounding counterions form an electrical double layer (EDL) (Chatterjee and Maitra 2016; Vo et al. 2016). Obviously, the EDL will be stabilized in polar media with a large dielectric constant. As an example, the metal halide-capped CQDs can be colloidally stable for months in propylene carbonate (PC, $\varepsilon_r = 65$), $N$-methylformamide (NMF, $\varepsilon_r = 182$), and $N,N$-dimethylformamide (DMF, $\varepsilon_r = 37$) (Dirin et al. 2014; Nag et al. 2011). However, the boiling points of these solvents are too high for film fabrication utilizing techniques such as spin coating. It was argued in earlier publications that these solvents give rise to inks with concentrations that are too low for film fabrication. However, more recent reports demonstrated that PC is a feasible solvent for CQD deposition via blade coating, giving rise to high-quality films that can be used for field-effect transistors(Balazs et al. 2018).

It was also proposed that BA with a low boiling point of 78 °C could give rise to MAPbI$_3$-capped PbS CQDs inks of high concentration. Although the polarity of BA is low ($\varepsilon_r = 4.9$), its basic characteristic enables it to desorb MA+ ions partially from the CQDs surface, stabilizing them in solution via hydrogen bonds (Ning et al. 2014). However, it is important to underline that these inks are stable for a very limited time (Sukharevska et al. 2021).

Xu et al. found that by mixing the BA with longer-chain hexylamine as solvent for the PbS-PbI$_2$ inks, extended colloidal stability is gained, stemming from the enhanced steric repulsion of the longer amine surfactants (Xu et al. 2018). Another report shows the addition of $N$-methylformamide as co-solvent on the system, the butylammonium is formed through the proton exchange between NMF and BA, which is shown to strengthen the EDL and extend the lifetime of the ink (Jiang et al. 2019; Sayevich et al. 2017).

Though the formation of butylammonium seems to improve the colloidal stability, the stability for all the BA-based CQDs ink is still not satisfactory for long-term storage (Gu et al. 2019). PbS CQDs with short ligands (metal halides and perovskite ligands) in a solvent of low polarity, such as BA, typically start to agglomerate and precipitate within the first few hours after preparation (Bederak et al. 2020). Recently, Bederak et al. investigated PC and especially difluoropyridine (DFP) as a new solvent for stabilizing both MAPbI$_3$-capped and PbX$_2$-capped PbS CQD. The second solvent not only has a low boiling point, 120 °C, which allows using large area deposition techniques, but also has a high polarity ($\varepsilon_r = 108$) to maintain the colloidal stability of CQD inks for more than 120 days. MAPbI$_3$-capped PbS inks in DFP were deposited as homogeneous films with blade coating at low temperatures, obtaining a competitive PCE of 8.7% in solar cells (Figure 15.10) (Sukharevska et al. 2021).

## 15.7 Device Structure

### 15.7.1 Electron Transport Layer

Besides the work described above for the improvement of the photoactive layers, device structure modifications are crucial for achieving favorable band alignment and trap-free

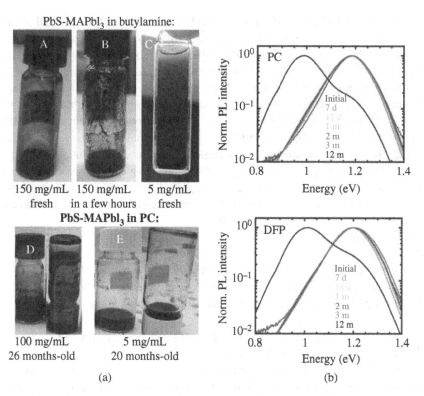

**Figure 15.10** (a) Photographs of the PbS-MAPbI3 inks in BA (top) and PC (bottom) in different concentrations under different storage times. The link in BA under low concentration only remains stable for half an hour. (b) Photoluminescence of PbS-MAPbI3 inks in DFP (top) and PC (bottom), indicating the remarkable (>3 months) colloidal stability. Source: Sukharevska et al. 2021/American Chemical Society.

interfaces in CQD solar cells. Table 15.2 gives an overview of the electron and hole transport layers, which have been lately utilized in QD solar cells. Metal oxides, such as ZnO (Hoye et al. 2015; Liu et al. 2016), $TiO_2$ (Liu et al. 2011), and $SnO_2$ (Khan et al. 2017; Li et al. 2020), have been largely used as ETL. Among them, ZnO is the most utilized ETL as $TiO_2$ suffers from low electron mobility and $SnO_2$ suffers from not optimal band alignment. ZnO nanocrystals (ZnO NCs) have shown superiority over other competitors due to their high conductivity (Jagadamma et al. 2014; Lan et al. 2016; Turner et al. 2002) and low required annealing temperature (Kim et al. 2015; Liu et al. 2016). Jang et al. have explicitly compared ZnO NCs to sol-gel synthesis, obtaining higher performance for NCs-based devices (Azmi et al. 2016). However, for achieving further breakthroughs, there are two major concerns regarding the use of ZnO NCs in CQDSCs: first, the band alignment can be troublesome with the narrow intrinsic bandgap of larger CQDs. Secondly, both the CQDs and ZnO NCs have defective surfaces. Therefore, the treatment of the interface between these layers is of great importance for preventing charge carrier recombination and enhancing the $V_{oc}$.

Choi et al. found that Cl-doping of ZnO NCs helps both to deepen the conduction band but also to passivate the surface (Choi et al. 2017). Cl-doping can also affect the quasi-Fermi

**Table 15.2** Non-exhaustive selection of CQDSC structures and performances with structural modification.

| CQDSC structure | PCE (%)[a] | $V_{oc}$ (V) | $J_{sc}$ (mA/cm²) | FF (%) | Comment | References |
|---|---|---|---|---|---|---|
| ITO/CAZO/PbS-PbX$_2$(1200 nm)/PbS-EDT/Au[b] | 0.99 | 0.391 | 3.84 | 65.71 | Doping ZnO with Al/Cl | Choi et al. (2018) |
| ITO/MZO/PbS-PbX$_2$+KI$_3$/PbS-EDT/Au | 11.9 | 0.64 | 27.3 | 68 | Doping ZnO with Mg | Hu et al. (2021) |
| ITO/ZnO(Cl)/PbS-PbX$_2$/PbS-EDT/Au | 11.63 | 0.63 | 28.45 | 65.36 | Doping ZnO with Cl | Choi et al. (2017) |
| ITO/AZO/Pbs-AI/PbS-EDT/Au | 11.4 | 0.65 | 26.6 | 66 | Doping ZnO with Al | Jia et al. (2019) |
| ITO/ZnO/PbS-PbX$_2$/PbS-EDT/Au | 7.6 | 0.55 | 35.2 | 39.7 | ZnO fabricated via atomic layered deposition | Jo et al. (2021) |
| ITO/PbS-PbX$_2$/PbS-EDT/Au | 10.56 | 0.59 | 27.74 | 65 | ETL-free solar cell | Jia et al. (2021) |
| ITO/In$_2$O$_3$/ZnO/PbS-PbX$_2$/PbS-EDT/Au | 10.7 | 0.63 | 27.33 | 61.9 | In$_2$O$_3$ as buffer layer | Kirmani et al. (2020) |
| ITO/ZnO/PbS-PbX$_2$(1250 nm)/PbS-PbX$_2$+BTA(950 nm)/PbS-EDT/Au | 12.3 | 0.63 | 28.8 | 68 | Graded bandgap structure | Kim et al. (2018) |
| ITO/ZnO/PbS-PbX$_2$/PbS-EDT+S/Au | 10.35 | 0.58 | 26.51 | 65 | Doping PbS-EDT with sulfur-infuse | Chiu et al. (2020) |
| ITO/ZnO/PbS-PbX$_2$+CPT/PbS-EDT+NH$_4$SCN/Au | 10.5 | 0.6 | 30.2 | 58 | NH$_4$SCN rinsed PbS-EDT layer | Sharma et al. (2021) |
| ITO/ZnO/PbS-MAPbI$_3$/PbS-EDT+Mo(tfd-COCF3)$_3$/Au | 9 | 0.61 | 21 | 68.8 | Doping PbS-EDT with organic | Kirmani et al. (2017) |
| ITO/ZnO/PbS-PbX2/PbS-EDT(PRN)/Au | 13.3 | 0.65 | 29.2 | 69.9 | Propionitrile(PRN) as solvent | Biondi et al. (2021) |
| ITO/MZO/PbS-PbX$_2$/PbS-EDT+MPA/Au | 10.4 | 0.633 | 25.3 | 66.8 | EDT + MPA hybrid exchange | Teh et al. (2020) |
| ITO/PbS-MPA/PbS-PbX$_2$/ZnO/Al | 9.04 | 0.57 | 26.22 | 61 | MPA PTLE ink as p-type material and n-i-p junction | Goossens et al. (2021) |
| FTO/ZnO/PCBM/PbS-FAPbI$_3$/PMMA-GO/PbS-EDT/Au | 15.45 | 0.66 | 31.5 | 74.3 | Morphology enhancement via buffer layer | Ding et al. (2022) |

*(Continued)*

**Table 15.2** (Continued)

| CQDSC structure | PCE (%)[a] | $V_{oc}$ (V) | $J_{sc}$ (mA/cm²) | FF (%) | Comment | References |
|---|---|---|---|---|---|---|
| ITO/ZnO/PbS-PbX₂/PbS-MA/Au | 13 | 0.64 | 29.1 | 70 | PbS-MA as p-type material | Biondi et al. (2020) |
| ITO/ZnO/PbS-PbX₂/PbS-ABT/Au | 12.2 | 0.64 | 28.8 | 66 | PbS-ABT as p-type material | Chen et al. (2021) |
| ITO/ZnO/PbS-PbX₂(1200 nm)/PbS-BA/Au[b] | 1.43 | 0.438 | 5.34 | 61.3 | PbS-BA ink as p-type material | Lee et al. (2020) |
| ITO/ZnO/PbS-PbX₂/PTB7/MoO3/Au | 9.6 | 0.57 | 27.92 | 60 | One of the first organic HTL report | Aqoma et al. (2018) |
| ITO/ZnO/PbS-PbX₂/PBDTTPD-HT/PTB₇/MoO₃/Au | 11.53 | 0.626 | 27.43 | 0.671 | π-conjugated organic HTL | Mubarok et al. (2020) |
| ITO/ZnO/PbS-PbX₂ (1200 nm)/PBDTTT-E-T/MoO₃/Ag[b] | 1.34 | 0.43 | 5.6 | 56 | Organic HTL and patterned electrode | Baek et al. (2019) |
| ITO/ZnO/PbS-PbX₂/PbS-EDT/PBDTTT-E-T/MoO₃/Au | 13.74 | 0.67 | 32.44 | 63.14 | Inserting buffer layer before organic HTL | Kim et al. (2022) |
| ITO/ZnO/PbS-PbX₂/PD2FCT-29DBPP/MoO₃/Au | 14 | 0.66 | 30.3 | 70 | D-A alternating organic HTL | Kim et al. (2020) |
| ITO/ZnO/PbS-PbX₂/PTB7:PCBM/MoO₃/Au | 12.02 | 0.65 | 27.933 | 66.2 | Organic BHJ as HTL | Zhang et al. (2020) |
| ITO/ZnO/PbS-PbX₂/TIPS-TPD/MoO₃/Au | 13.03 | 0.66 | 28.8 | 0.69 | π-conjugated organic HTL | Mubarok et al. (2020) |
| ITO/ZnO/PbS-PbX₂/PTAA:C₆₀F₄₈/Au | 10.41 | 0.583 | 29.6 | 60.4 | Organic BHJ as HTL | Becker-Koch et al. (2021) |

a) Device parameters listed here pertain to the highest performing device reported in the single article when tested with solar irradiation at AM 1.5.
b) The PCE of this work refers to the IRPCE, where a long-pass c-Si filter on 1100 nm is applied in front of the device.

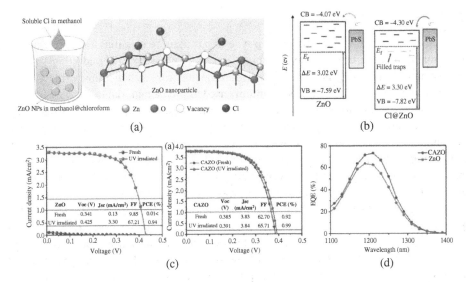

**Figure 15.11** (a) Schematic illustration of the Cl-passivation mechanism of ZnO (Cl@ZnO) nanoparticles by elimination of oxygen vacancies. (b) Schematic band alignment diagram of ZnO and Cl@ZnO. Source: Choi et al. (2017)/John Wiley & Sons. (c) Current density–voltage (J–V) characteristics of IR CQDs PV devices prepared using ZnO NCs ETL (left) and Al/Cl hybrid doping of ZnO sol-gel (CAZO) ETL (right). (d) EQE spectra of IR CQDs PV prepared with both ETL. Source: Choi et al. (2018)/John Wiley & Sons.

level splitting and increase the free carrier density, thus improving the built-in voltage and depletion region, which is beneficial to carrier extraction and $V_{oc}$ improvement. The CQDSC with this ETL was demonstrated to achieve a PCE of 11.63%, with improved $V_{oc}$ and fill factor.

Further studies showed that Al/Cl-doping of ZnO NCs can improve the band alignment with PbS CQDs and the passivation of the nanoparticles, and thus ensure good charge carrier transport (Choi et al. 2018). As shown in Figure 15.11c,d, the surface passivation through Al/Cl doping is improved to the extent that UV activation is no longer required, which largely simplifies the fabrication process.

The interface between the ETL and the active layer was also improved by several authors through the insertion of buffer layers. Both $In_2O_3$ (Kirmani et al. 2020) and CdSe CQD (Zhao et al. 2016) were found to be appropriate buffer layers that are compatible with the PbS CQDs. The role of these buffer layers is multiple. From one side, they can improve the alignment with the conduction band of the active layer, reducing eventual barriers with the ETL, but they can also affect the morphology via surface energy engineering and reduce the trap states at the interface, improving the photovoltaic performance of the device in all aspects. It has also been reported that the fullerene derivative PCBM can effectively block the corrosion of the ETL film during fabrication while passivating defect states at the ZnO interface (Ding et al. 2022). However, it is generally reported a trade-off when growing buffer layers between the passivation capacity and the limitation in transport, which affects both $J_{sc}$ and FF of the solar cells.

### 15.7.2 Hole Transport Layer

In this subsection, we will discuss both p-type CQDs that are essential for the formation of p-n junction at the heart of the absorbing layer in CQDs solar cells and other types of p-type materials that are generically indicated in the literature as hole transport layers.

As mentioned earlier, PbS-EDT is commonly used as p-type material in CQDs solar cells (Chuang et al. 2014). However, similarly to the other organic capped CQDs using SSLE, the PbS-EDT suffers from severe carrier recombination due to nonoptimal surface passivation, and it does not provide a good electron blocking layer, as its bandgap is too narrow for suppressing electron transport (Becker-Koch et al. 2020). It is reported that the behavior of PbS-EDT is largely improved by oxygen doping via dry air exposure. This oxygen-doping leads to an increase in the p-doping level but was also reported to give rise to a size reduction of the dot and the filling of surface defects(Kirmani et al. 2018). Interestingly, with the optimization of the exposure time, the air-doped solar cell can achieve a PCE of 11% and a $V_{oc}$ 35% higher than that of the as-prepared ones. Other doping sources, such as elemental sulfur doping, were also proposed as options for reducing stoichiometric imbalance and ensuring sufficient p-type behavior of the p-type PbS (Chiu et al. 2020; Speirs et al. 2017). Biondi et al. recently reported that it is possible to tune the reactivity of EDT ligand on the surface of PbS by substituting the conventionally used acetonitrile (ACN), with propionitrile (PRN) (Biondi et al. 2021). This less volatile solvent can regulate the reactivity of the SSLE and thus have positive effects on the CQDs' order and defect passivation by preventing random agglomeration of the CQDs. The resulting device showed a PCE of 13.3%, while a further slowdown of the reactivity utilizing butyronitrile (BTN) gave rise to similar results (Biondi et al. 2021).

Recently, a hybrid organic ligand strategy utilizing both EDT and MPA has been proposed for obtaining p-type PbS (Teh et al. 2020). This approach takes advantage of the main characteristics provided by the two ligands, namely the high $V_{oc}$ from the EDT ligand, attributed to the oxidation of the surface of the CQDs, and the high $J_{sc}$ provided by PbS treated with MPA, which is attributed to the better charge carrier transport. The resulting device shows a PCE of 10.4%.

Biondi et al. have also revealed that during the deposition of PbS-EDT, the high-reactivity of EDT in ACN solution penetrates the HTL layer and harms the surface chemistry of the underneath n-type layer, resulting in weakened charge collection efficiency (Biondi et al. 2020). The mechanism and the analytics showing the effect of EDT are illustrated in Figure 15.12a,b. Therefore, the authors decided to use, instead of PbS-EDT, an alternative ligand, namely malonic acid (MA), which shows a low reactivity and is chemically orthogonal to the n-type layer and has only negligible impact on its surface chemistry (Biondi et al. 2020). The solar cell fabricated with the PbS-MA as p-type material shows complete interfacial charge collection and achieves a PCE of 13.0%, with a high FF of 70%. Further investigation of chemically orthogonal ligands also found that 4-aminobenzenethiol (ABT), can also show good photovoltaic performance (Chen et al. 2021). The device using PbS-ABT as p-type CQDs shows very good operational stability, which is attributed to the thiol passivation of the CQD.

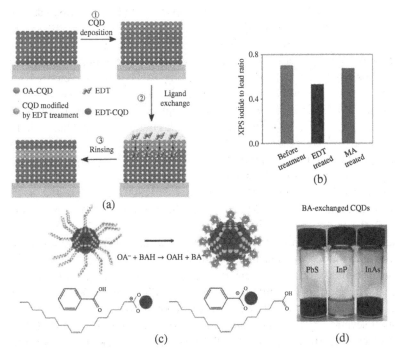

**Figure 15.12** (a) Schematic of the EDT HTL fabrication procedure. (b) Iodide-to-lead ratio from XPS measurement before HTL treatment and after treatment with EDT and MA solution. Source: Biondi et al. (2020)/John Wiley & Sons. (c) Schematic illustrations of the benzoic acid (BA) PTLE procedure. (d) Photograph showing the dispersibility of the BA-exchanged CQDs in chlorobenzene. Source: Lee et al. (2020)/Springer Nature/CC BY 4.0.

Till now, we have discussed the p-type CQDs layers all deposited by using the SSLE process. These are extremely time-consuming and far from being a technique used in the industrial environment.

However, PbS-MPA can also be obtained through the PTLE method using butylamine (BA) and water as a solvent (Goossens et al. 2021). The p-type ink shows sufficient colloidal stability and substrate wettability to be deposited by blade coating. Due to the nature of the solvent, the n–i–p structure could not be used, but all solution-processed devices were fabricated using the p–i–n structure. A power conversion efficiency of up to 9% was obtained (Goossens et al. 2021).

Lee et al. also achieved the enhancement of charge collection by using p-type CQDs obtained by PTLE. The highly polarizable benzoic acid (BZ) ligands were used to obtain ink in chlorobenzene (see Figure 15.12d) (Lee et al. 2020). The ink shows higher hole mobility, lower defect density, and slightly deeper energy levels compared to that of PbS-EDT. The corresponding device using CQDs with 1300 nm excitonic peak, in the n–i–p device configuration, shows an IRPCE of 1.43%, the present record IRPCE with large dot-size CQD.

In 2016, a device was reported where poly(3-hexylthiophene) (P3HT) was first used as organic HTLs in CQDs solar cells where an n-type iodine-capped PbS was used as the absorber (Neo et al. 2016). With respect to EDT-capped PbS, the organic HTL had the advantage of a tunable energy level low-temperature solution-processability, giving rise to homogeneous films. However, this first example using P3HT was characterized by limited $V_{oc}$ due to the non-optimal band alignment and the low polymer hole mobility. Devices using polythieno(3,4-*b*)-thiophene-*co*-benzodith-iophene (PTB7) as HTL exhibited superior performance to that using P3HT, owing to favorable energy levels and higher hole mobility (Aqoma et al. 2018). Though the PCE (9.6%) of solar cells based on the PTB7 HTL is much higher than that of the P3HT, it is still lower than that with the p-type CQDs due to the low charge collection and short diffusion length (Chen et al. 2017).

Recently, Kim et al. used a D-A alternating copolymer (PD2FCT-29DPP) as the HTL, attaching the cyclopentadithiophene (CPDT) and diketopyrrolopyrrole (DPP) units (Kim et al. 2020). The CPDT component is an electron-rich moiety with a pseudo-planar structure that has excellent charge transport (Lee et al. 2014), while the DPP acceptor with bithiophene is known to induce π-conjugation (Uddin et al. 2015). The resulting polymer matrix has high hole mobility $\sim 2.36 \times 10^{-3}$ cm$^2$/Vs and smooth morphology owing to the π-π stacking. The device fabricated using this polymer as HTL reached a PCE of 14.0%, one of the highest for CQDs solar cells. Moreover, recent reports show the CQDs absorbers are compatible with several organic HTLs, including asy-ranPBTBDT (Kim et al. 2020), TIPS-TPD (Mubarok et al. 2020), and PBDTTT-E-T (Kim et al. 2022). The PCE using some of these organics is even higher than that using the p-type CQDs. While the single-step solution-based deposition of this organic HTL is very favorable. It should also be noted that several of the organic molecules recently reported having complicated structures, which reflects negatively on the cost and scalability of the synthesis of the materials. It is also important to underline that the hole mobility of the best organic HTL introduced in CQD solar cells is still lower than that of CQD p-type layers.

## 15.8 Conclusion

In the last ten years, CQDs solar cells have matured enormously, going from 5% efficiency devices fabricated with a wasteful layer-by-layer deposition process to devices with efficiency above 15% fabricated with a minimal number of steps thanks to the developments of stable short ligands inks. This power conversion efficiency value, together with the large tunability of the bandgap, makes CQDs solar cells extremely interesting for multijunction devices, where they may get the role of absorbing the photons of lower energy.

Still, a few challenges are waiting for a solution, among them are important to mention the still not-optimal physical properties (mobility and doping level) and processability of the p-type CQDs. More work in the next years will be necessary to make the p-type inks achieve similar performances as the n-type inks. Furthermore, the lack of focus in the past on the selection of solvents and ligands compatible with industrial processes would need a clear improvement. As last, we would like to underline that the work described, while we have been focusing on solar cells, is also of enormous relevance for the development of IR photodetectors with tunable spectral sensitivity.

## Author Biographies

**Han Wang** obtained a B.Sc. degree majoring in Opto-Electronic Information in 2019 from the South China Normal University, China. In 2020, he gained his M.Sc. degree with distinction in sensor & imaging systems from the University of Glasgow and the University of Edinburgh jointly in the UK. He is currently a Ph.D. candidate investigating the physical properties of semiconducting quantum dots and the fabrication of quantum dot opto-electronic devices in the Photophysics and Optoelectronics group of Prof. Maria Antonietta Loi at the University of Groningen.

**Maria Antonietta Loi** studied physics at the University of Cagliari in Italy, where she received her PhD in 2001. In the same year, she joined the Linz Institute for Organic Solar Cells, of the University of Linz, Austria, as a postdoctoral fellow. Later she worked as researcher at the Institute for Nanostructured Materials of the Italian National Research Council in Bologna, Italy. In 2006, she became assistant professor and Rosalind Franklin Fellow at the Zernike Institute for Advanced Materials of the University of Groningen, The Netherlands. She is now full professor in the same institution and chair of the Photophysics and OptoElectronics group.

She has published more than 280 peer-reviewed articles on photophysics and optoelectronics of different types of materials. In 2013, she has received an ERC Starting Grant and in 2022 an ERC Advanced Grant from the European Research Council. She currently serves as deputy editor-in-chief of Applied Physics Letters, and she is member of the international advisory board of several international journals in physics and materials physics. In 2018, she received the Physicaprijs from the Dutch Physics Association. In 2020, she became a fellow of the American Physical Society. In 2022, she was elected fellow of the Dutch Academy of Science (KNAW). In the same year, she became fellow of the European Academy of Science (EURASC) and of the Royal Society of Chemistry.

## References

Ahmad, W., He, J., Liu, Z. et al. (2019). Lead selenide (PbSe) colloidal quantum dot solar cells with >10% efficiency. *Advanced Materials* 31 (33): 1900593.

Albaladejo-Siguan, M., Baird, E.C., Becker-Koch, D. et al. (2021). Stability of quantum dot solar cells: a matter of (life)time. *Advanced Energy Materials* 11 (12): 2003457.

Aqoma, H. and Jang, S.Y. (2018). Solid-state-ligand-exchange free quantum dot ink-based solar cells with an efficiency of 10.9%. *Energy and Environmental Science* 11 (6): 1603–1609. Royal Society of Chemistry.

Aqoma, H., Mubarok, M.A., Lee, W. et al. (2018). Improved processability and efficiency of colloidal quantum dot solar cells based on organic hole transport layers. *Advanced Energy Materials* 8 (23): 1800572.

Azmi, R., Aqoma, H., Hadmojo, W.T. et al. (2016). Low-temperature-processed 9% colloidal quantum dot photovoltaic devices through interfacial management of p-n heterojunction. *Advanced Energy Materials* 6 (8): 1502146.

Azmi, R., Sinaga, S., Aqoma, H. et al. (2017). Highly efficient air-stable colloidal quantum dot solar cells by improved surface trap passivation. *Nano Energy* 39 (March): 86–94. Elsevier Ltd. Retrieved from http://dx.doi.org/10.1016/j.nanoen.2017.06.040.

Baek, S.W., Molet, P., Choi, M.J. et al. (2019). Nanostructured back reflectors for efficient colloidal quantum-dot infrared optoelectronics. *Advanced Materials* 31 (33): 1901745.

Balazs, D.M. and Loi, M.A. (2018). Lead-chalcogenide colloidal-quantum-dot solids: novel assembly methods, electronic structure control, and application prospects. *Advanced Materials* 30 (33): 1800082.

Balazs, D.M., Nugraha, M.I., Bisri, S.Z. et al. (2014). Reducing charge trapping in PbS colloidal quantum dot solids. *Applied Physics Letters* 104 (11): 112104.

Balazs, D.M., Dirin, D.N., Fang, H.H. et al. (2015). Counterion-mediated ligand exchange for PbS colloidal quantum dot superlattices. *ACS Nano* 9 (12): 11951–11959.

Balazs, D.M., Bijlsma, K.I., Fang, H.H. et al. (2017). Stoichiometric control of the density of states in PbS colloidal quantum dot solids. *Science Advances* 3 (9): eaao1558.

Balazs, D.M., Rizkia, N., Fang, H.H. et al. (2018). Colloidal quantum dot inks for single-step-fabricated field-effect transistors: the importance of postdeposition ligand removal. *ACS Applied Materials and Interfaces* 10 (6): 5626–5632.

Barkhouse, D.A.R., Pattantyus-Abraham, A.G., Levina, L., and Sargent, E.H. (2008). Thiols passivate recombination centers in colloidal quantum dots leading to enhanced photovoltaic device efficiency. *ACS Nano* 2 (11): 2356–2362.

Bashir, R., Bilal, M.K., Bashir, A. et al. (2021). A low-temperature solution-processed indium incorporated zinc oxide electron transport layer for high-efficiency lead sulfide colloidal quantum dot solar cells. *Nanoscale* 13 (30): 12991–12999. Royal Society of Chemistry.

Beard, M.C., Luther, J.M., and Nozik, A.J. (2010). Semiconductor quantum dots and quantum dot arrays and applications of multiple exciton generation to third-generation photovoltaic solar cells. *Colloidal Quantum Dot Optoelectronics and Photovoltaics* 110 (11): 6873–6890.

Beard, M.C., Luther, J.M., Semonin, O.E., and Nozik, A.J. (2013). Third generation photovoltaics based on multiple exciton generation in quantum confined semiconductors. *Accounts of Chemical Research* 46 (6): 1252–1260.

Becker-Koch, D., Albaladejo-Siguan, M., Lami, V. et al. (2020). Ligand dependent oxidation dictates the performance evolution of high efficiency PbS quantum dot solar cells. *Sustainable Energy and Fuels* 4 (1): 108–115. Royal Society of Chemistry.

Becker-Koch, D., Albaladejo-Siguan, M., Hofstetter, Y.J. et al. (2021). Doped organic hole extraction layers in efficient PbS and AgBiS2 Quantum dot solar cells. *ACS Applied Materials and Interfaces* 13 (16): 18750–18757.

Bederak, D., Sukharevska, N., Kahmann, S. et al. (2020). On the colloidal stability of PbS quantum dots capped with methylammonium lead iodide ligands. *ACS Applied Materials and Interfaces* 12 (47): 52959–52966.

Bi, Y., Pradhan, S., Gupta, S. et al. (2018). Infrared solution-processed quantum dot solar cells reaching external quantum efficiency of 80% at 1.35 μm and Jsc in excess of 34 mA cm⁻². *Advanced Materials* 30 (7): 1704928.

Biondi, M., Choi, M.J., Ouellette, O. et al. (2020). A chemically orthogonal hole transport layer for efficient colloidal quantum dot solar cells. *Advanced Materials* 32 (17): 1906199.

Biondi, M., Choi, M.J., Lee, S. et al. (2021). Control over ligand exchange reactivity in hole transport layer enables high-efficiency colloidal quantum dot solar cells. *ACS Energy Letters* 6 (2): 468–476.

Bisri, S.Z., Piliego, C., Yarema, M. et al. (2013). Low driving voltage and high mobility ambipolar field-effect transistors with PbS colloidal nanocrystals. *Advanced Materials* 25 (31): 4309–4314.

Brown, A.S. and Green, M.A. (2002). Detailed balance limit for the series constrained two terminal tandem solar cell. *Physica E: Low-dimensional Systems and Nanostructures* 14: 96–100. Retrieved from papers2://publication/uuid/A979552C-0DF1-484E-9DF9-B7138A7117B0.

Brown, P.R., Kim, D., Lunt, R.R. et al. (2014). Energy level modification in lead sulfide quantum dot thin films through ligand exchange. *ACS Nano* 8 (6): 5863–5872.

Cao, Y., Stavrinadis, A., Lasanta, T. et al. (2016). The role of surface passivation for efficient and photostable PbS quantum dot solar cells. *Nature Energy* 1 (4): 16035.

Chatterjee, S. and Maitra, U. (2016). A novel strategy towards designing a CdSe quantum dot-metallohydrogel composite material. *Nanoscale* 8 (32): 14979–14985. Royal Society of Chemistry.

Chen, Z., Du, X., Zeng, Q., and Yang, B. (2017). Recent development and understanding of polymer-nanocrystal hybrid solar cells. *Materials Chemistry Frontiers* 1 (8): 1502–1513. Royal Society of Chemistry.

Chen, J., Zheng, S., Jia, D. et al. (2021). Regulating thiol ligands of p-type colloidal quantum dots for efficient infrared solar cells. *ACS Energy Letters* 6 (5): 1970–1979.

Chiu, A., Rong, E., Bambini, C. et al. (2020). Sulfur-infused hole transport materials to overcome performance-limiting transport in colloidal quantum dot solar cells. *ACS Energy Letters* 5 (9): 2897–2904.

Choi, H., Ko, J.H., Kim, Y.H., and Jeong, S. (2013). Steric-hindrance-driven shape transition in PbS quantum dots: understanding size-dependent stability. *Journal of the American Chemical Society* 135 (14): 5278–5281.

Choi, H., Lee, J.G., Mai, X.D. et al. (2017). Supersonically spray-coated colloidal quantum dot ink solar cells. *Scientific Reports* 7 (1): 622. Springer US. Retrieved from http://dx.doi.org/10.1038/s41598-017-00669-9.

Choi, J., Kim, Y., Jo, J.W. et al. (2017). Chloride passivation of ZnO electrodes improves charge extraction in colloidal quantum dot photovoltaics. *Advanced Materials* 29 (33): 1702350.

Choi, J., Jo, J.W., de Arquer, F.P.G. et al. (2018). Activated electron-transport layers for infrared quantum dot optoelectronics. *Advanced Materials* 30 (29): 1801720.

Choi, J., Choi, M.J., Kim, J. et al. (2020). Stabilizing surface passivation enables stable operation of colloidal quantum dot photovoltaic devices at maximum power point in an air ambient. *Advanced Materials* 32 (7): 1906497.

Choi, M.J., Baek, S.W., Lee, S. et al. (2020). Colloidal quantum dot bulk heterojunction solids with near-unity charge extraction efficiency. *Advanced Science* 7 (15): 2000894.

Choi, M.J., García de Arquer, F.P., Proppe, A.H. et al. (2020). Cascade surface modification of colloidal quantum dot inks enables efficient bulk homojunction photovoltaics. *Nature Communications* 11 (1): 103.

Chuang, C.H.M., Brown, P.R., Bulović, V., and Bawendi, M.G. (2014). Improved performance and stability in quantum dot solar cells through band alignment engineering. *Nature Materials* 13 (8): 796–801.

Clifford, J.P., Konstantatos, G., Johnston, K.W. et al. (2009). Fast, sensitive and spectrally tuneable colloidal-quantum-dot photodetectors. *Nature Nanotechnology* 4 (1): 40–44.

Cunningham, P.D., Boercker, J.E., Foos, E.E. et al. (2011). Enhanced multiple exciton generation in quasi-one-dimensional semiconductors. *Nano Letters* 11 (8): 3476–3481.

Ding, C., Liu, F., Zhang, Y. et al. (2020). Passivation strategy of reducing both electron and hole trap states for achieving high-efficiency PbS quantum-dot solar cells with power conversion efficiency over 12%. *ACS Energy Letters* 5 (10): 3224–3236.

Ding, C., Wang, D., Liu, D. et al. (2022). Over 15% efficiency PbS quantum-dot solar cells by synergistic effects of three interface engineering: reducing nonradiative recombination and balancing charge carrier extraction. *Advanced Energy Materials* 12 (15): 2201676.

Dirin, D.N., Dreyfuss, S., Bodnarchuk, M.I. et al. (2014). Lead halide perovskites and other metal halide complexes as inorganic capping ligands for colloidal nanocrystals. *Journal of the American Chemical Society* 136 (18): 6550–6553.

Domanski, K., Alharbi, E.A., Hagfeldt, A. et al. (2018). Systematic investigation of the impact of operation conditions on the degradation behaviour of perovskite solar cells. *Nature Energy* 3 (1): 61–67. Springer US. Retrieved from http://dx.doi.org/10.1038/s41560-017-0060-5.

Fan, J.Z., Liu, M., Voznyy, O. et al. (2017). Halide re-shelled quantum dot inks for infrared photovoltaics. *ACS Applied Materials and Interfaces* 9 (43): 37536–37541.

Fan, J.Z., Andersen, N.T., Biondi, M. et al. (2019). Mixed lead halide passivation of quantum dots. *Advanced Materials* 31 (48): 1904304.

Fischer, A., Rollny, L., Pan, J. et al. (2013). Directly deposited quantum dot solids using a colloidally stable nanoparticle ink. *Advanced Materials* 25 (40): 5742–5749.

Ge, C., Yang, E., Zhao, X. et al. (2022). Efficient near-infrared PbS quantum dot solar cells employing hydrogenated $In_2O_3$ transparent electrode. *Small* 18 (44): 2203677.

Gong, X., Yang, Z., Walters, G. et al. (2016). Highly efficient quantum dot near-infrared light-emitting diodes. *Nature Photonics* 10 (4): 253–257.

Goodwin, H., Jellicoe, T.C., Davis, N.J.L.K., and Böhm, M.L. (2018). Multiple exciton generation in quantum dot-based solar cells. *Nanophotonics* 7 (1): 111–126.

Goossens, V.M., Sukharevska, N.v., Dirin, D.N. et al. (2021). Scalable fabrication of efficient p-n junction lead sulfide quantum dot solar cells. *Cell Reports Physical Science* 2 (12): 100655. The Authors. Retrieved from https://doi.org/10.1016/j.xcrp.2021.100655.

Green, M., Dunlop, E., Hohl-Ebinger, J. et al. (2021). Solar cell efficiency tables (version 57). *Progress in Photovoltaics: Research and Applications* 29 (1): 3–15.

Gu, M., Wang, Y., Yang, F. et al. (2019). Stable PbS quantum dot ink for efficient solar cells by solution-phase ligand engineering. *Journal of Materials Chemistry A* 7 (26): 15951–15959. Royal Society of Chemistry.

Guyot-Sionnest, P. (2012). Electrical transport in colloidal quantum dot films. *Journal of Physical Chemistry Letters* 3 (9): 1169–1175.

Halim, M.A. (2013). Harnessing sun's energy with quantum dots based next generation solar cell. *Nanomaterials* 3 (1): 22–47.

Hoye, R.L.Z., Muñoz-Rojas, D., Nelson, S.F. et al. (2015). Research Update: Atmospheric pressure spatial atomic layer deposition of ZnO thin films: reactors, doping, and devices. *APL Materials* 3 (4): 040701. Retrieved from http://dx.doi.org/10.1063/1.4916525.

Hu, L., Lei, Q., Guan, X. et al. (2021). Optimizing surface chemistry of PbS colloidal quantum dot for highly efficient and stable solar cells via chemical binding. *Advanced Science* 8 (2): 2003138.

Huang, J., Yuan, Y., Shao, Y., and Yan, Y. (2017). Understanding the physical properties of hybrid perovskites for photovoltaic applications. *Nature Reviews Materials* 2: 17042.

Ip, A.H., Thon, S.M., Hoogland, S. et al. (2012). Hybrid passivated colloidal quantum dot solids. *Nature Nanotechnology* 7 (9): 577–582.

Ip, A.H., Kiani, A., Kramer, I.J. et al. (2015). Infrared colloidal quantum dot photovoltaics via coupling enhancement and agglomeration suppression. *ACS Nano* 9 (9): 8833–8842.

Jaeger, H.M., Fischer, S., and Prezhdo, O.V. (2012). Decoherence-induced surface hopping. *Journal of Chemical Physics* 137 (22): 22A545.

Jaeger, H.M., Hyeon-Deuk, K., and Prezhdo, O.V. (2013). Exciton multiplication from first principles. *Accounts of Chemical Research* 46 (6): 1280–1289.

Jagadamma, L.K., Abdelsamie, M., el Labban, A. et al. (2014). Efficient inverted bulk-heterojunction solar cells from low-temperature processing of amorphous ZnO buffer layers. *Journal of Materials Chemistry A* 2 (33): 13321–13331. Royal Society of Chemistry.

Jia, D., Chen, J., Zheng, S. et al. (2019). Highly stabilized quantum dot ink for efficient infrared light absorbing solar cells. *Advanced Energy Materials* 9 (44): 1902809.

Jia, Y., Wang, H., Wang, Y. et al. (2021). High-performance electron-transport-layer-free quantum junction solar cells with improved efficiency exceeding 10%. *ACS Energy Letters* 6 (2): 493–500.

Jiang, X., Li, H., Shang, Y. et al. (2019). Bi-inorganic-ligand coordinated colloidal quantum dot ink. *Chemical Communications* 55 (64): 9483–9486. Royal Society of Chemistry.

Jiao, X., Ye, L., and Ade, H. (2017). Quantitative morphology–performance correlations in organic solar cells: insights from soft X-ray scattering. *Advanced Energy Materials* 7 (18): 1700084.

Jo, J.W., Kim, Y., Choi, J. et al. (2017). Enhanced open-circuit voltage in colloidal quantum dot photovoltaics via reactivity-controlled solution-phase ligand exchange. *Advanced Materials* 29 (43): 1703627.

Jo, J.W., Choi, J., García De Arquer, F.P. et al. (2018). Acid-assisted ligand exchange enhances coupling in colloidal quantum dot solids. *Nano Letters* 18 (7): 4417–4423.

Jo, H., Kim, J.K., Kim, J. et al. (2021). Unprecedentedly large photocurrents in colloidal PbS quantum-dot solar cells enabled by atomic layer deposition of zinc oxide electron buffer layer. *ACS Applied Energy Materials* 4 (12): 13776–13784.

Johnston, K.W., Pattantyus-Abraham, A.G., Clifford, J.P. et al. (2008). Schottky-quantum dot photovoltaics for efficient infrared power conversion. *Applied Physics Letters* 92 (15): 151115.

Jung, B.K., Woo, H.K., Shin, C. et al. (2022). Suppressing the dark current in quantum dot infrared photodetectors by controlling carrier statistics. *Advanced Optical Materials* 10 (2): 2101611.

Kahmann, S. and Loi, M.A. (2020). Trap states in lead chalcogenide colloidal quantum dots – origin, impact, and remedies. *Applied Physics Reviews* 7 (4): 041305. AIP Publishing LLC.

Kedikova, S., Pavlova, E., and Ivanov, S. (2011). Efficient photodiodes from interpenetrating polymer networks. *Akusherstvo i Ginekologiia* .

Khan, J., Yang, X., Qiao, K. et al. (2017). Lowerature-processed SnO2-Cl for efficient PbS quantum-dot solar cells: via defect passivation. *Journal of Materials Chemistry A* 5 (33): 17240–17247.

Kiani, A., Sutherland, B.R., Kim, Y. et al. (2016). Single-step colloidal quantum dot films for infrared solar harvesting. *Applied Physics Letters* 109 (18): Retrieved from http://dx.doi.org/10.1063/1.4966217.

Kim, H. II, Baek, S.W., Cheon, H.J. et al. (2020). A tuned alternating D–A copolymer hole-transport layer enables colloidal quantum dot solar cells with superior fill factor and efficiency. *Advanced Materials* 32 (48): 2004985.

Kim, H. II, Lee, J., Choi, M.J. et al. (2020). Efficient and stable colloidal quantum dot solar cells with a green-solvent hole-transport layer. *Advanced Energy Materials* 10 (39): 2002084.

Kim, D., Kim, D.H., Lee, J.H., and Grossman, J.C. (2013). Impact of stoichiometry on the electronic structure of PbS quantum dots. *Physical Review Letters* 110 (19): 196802.

Kim, G.H., García De Arquer, F.P., Yoon, Y.J. et al. (2015). High-efficiency colloidal quantum dot photovoltaics via robust self-assembled monolayers. *Nano Letters* 15 (11): 7691–7696.

Kim, J., Ouellette, O., Voznyy, O. et al. (2018). Butylamine-catalyzed synthesis of nanocrystal inks enables efficient infrared CQD solar cells. *Advanced Materials* 30 (45): 1803830.

Kim, Y., Che, F., Jo, J.W. et al. (2019). A facet-specific quantum dot passivation strategy for colloid management and efficient infrared photovoltaics. *Advanced Materials* 31 (17): 1805580.

Kim, B., Baek, S.W., Kim, C. et al. (2022). Mediating colloidal quantum dot/organic semiconductor interfaces for efficient hybrid solar cells. *Advanced Energy Materials* 12 (2): 2102689.

Kirmani, A.R., García De Arquer, F.P., Fan, J.Z. et al. (2017). Molecular doping of the hole-transporting layer for efficient, single-step-deposited colloidal quantum dot photovoltaics. *ACS Energy Letters* 2 (9): 1952–1959.

Kirmani, A.R., Sheikh, A.D., Niazi, M.R. et al. (2018). Overcoming the ambient manufacturability-scalability-performance bottleneck in colloidal quantum dot photovoltaics. *Advanced Materials* 30 (35): 1801661.

Kirmani, A.R., Eisner, F., Mansour, A.E. et al. (2020). Colloidal quantum dot photovoltaics using ultrathin, solution-processed bilayer $In_2O_3/ZnO$ electron transport layers with improved stability. *ACS Applied Energy Materials* 3 (6): 5135–5141.

Klem, E.J.D., Shukla, H., Hinds, S. et al. (2008). Impact of dithiol treatment and air annealing on the conductivity, mobility, and hole density in PbS colloidal quantum dot solids. *Applied Physics Letters* 92 (21): 212105.

Konstantatos, G., Howard, I., Fischer, A. et al. (2006). Ultrasensitive solution-cast quantum dot photodetectors. *Nature* 442 (7099): 180–183.

Konstantatos, G., Clifford, J., Levina, L., and Sargent, E.H. (2007). Sensitive solution-processed visible-wavelength photodetectors. *Nature Photonics* 1 (9): 531–534.

Kroupa, D.M., Vörös, M., Brawand, N.P. et al. (2017). Tuning colloidal quantum dot band edge positions through solution-phase surface chemistry modification. *Nature Communications* 8 (May): 2–9.

Kumar, M., Vezzoli, S., Wang, Z. et al. (2016). Hot exciton cooling and multiple exciton generation in PbSe quantum dots. *Physical Chemistry Chemical Physics* 18 (45): 31107–31114. Royal Society of Chemistry.

Lan, X., Voznyy, O., García De Arquer, F.P. et al. (2016). 10.6% certified colloidal quantum dot solar cells via solvent-polarity-engineered halide passivation. *Nano Letters* 16 (7): 4630–4634.

Law, M., Luther, J.M., Song, Q. et al. (2008). Structural, optical, and electrical properties of PbSe nanocrystal solids treated thermally or with simple amines. *Journal of the American Chemical Society* 130 (18): 5974–5985.

Lee, J., Jang, M., Myeon Lee, S. et al. (2014). Fluorinated benzothiadiazole (BT) groups as a powerful unit for high-performance electron-transporting polymers. *ACS Applied Materials and Interfaces* 6 (22): 20390–20399.

Lee, S., Choi, M.J., Sharma, G. et al. (2020). Orthogonal colloidal quantum dot inks enable efficient multilayer optoelectronic devices. *Nature Communications* 11 (1): 4814. Springer US. Retrieved from http://dx.doi.org/10.1038/s41467-020-18655-7.

Li, H., Chen, D., Li, L. et al. (2010). Size- and shape-controlled synthesis of PbSe and PbS nanocrystals via a facile method. *CrystEngComm* 12 (4): 1127–1133.

Li, Y., Yang, F., Wang, Y. et al. (2020). Magnetron sputtered $SnO_2$ constituting double electron transport layers for efficient PbS quantum dot solar cells. *Solar RRL* 4 (7): 2000218.

Li, M., Chen, S., Zhao, X. et al. (2022). Matching charge extraction contact for infrared PbS colloidal quantum dot solar cells. *Small* 18 (1): 2105495.

Liu, Y., Gibbs, M., Puthussery, J. et al. (2010). Dependence of carrier mobility on nanocrystal size and ligand length in pbse nanocrystal solids. *Nano Letters* 10 (5): 1960–1969.

Liu, H., Tang, J., Kramer, I.J. et al. (2011). Electron acceptor materials engineering in colloidal quantum dot solar cells. *Advanced Materials* 23 (33): 3832–3837.

Liu, Y., Gibbs, M., Perkins, C.L. et al. (2011). Robust, functional nanocrystal solids by infilling with atomic layer deposition. *Nano Letters* 11 (12): 5349–5355.

Liu, M., de Arquer, F.P.G., Li, Y. et al. (2016). Double-sided junctions enable high-performance colloidal-quantum-dot photovoltaics. *Advanced Materials* 28 (21): 4142–4148.

Liu, M., Voznyy, O., Sabatini, R. et al. (2017). Hybrid organic-inorganic inks flatten the energy landscape in colloidal quantum dot solids. *Nature Materials* 16 (2): 258–263.

Liu, M., Chen, Y., Tan, C.S. et al. (2019). Lattice anchoring stabilizes solution-processed semiconductors. *Nature* 570 (7759): 96–101. Springer US. Retrieved from http://dx.doi.org/10.1038/s41586-019-1239-7.

Liu, J., Xian, K., Ye, L., and Zhou, Z. (2021). Open-circuit voltage loss in lead chalcogenide quantum dot solar cells. *Advanced Materials* 33 (29): 2008115.

Liu, X., Li, J., Wang, X., and Yang, D. (2022). Inorganic lead-based halide perovskites: from fundamental properties to photovoltaic applications. *Materials Today* 61 (December): 191–217. Elsevier Ltd. Retrieved from https://doi.org/10.1016/j.mattod.2022.11.002.

Lu, K., Wang, Y., Liu, Z. et al. (2018). High-efficiency PbS quantum-dot solar cells with greatly simplified fabrication processing via "solvent-curing". *Advanced Materials* 30 (25): 1707572.

Luther, J.M., Law, M., Beard, M.C. et al. (2008). Schottky solar cells based on colloidal nanocrystal films. *Nano Letters* 8 (10): 3488–3492.

Luther, J.M., Law, M., Song, Q. et al. (2008). Structural, optical, and electrical properties of self-assembled films of PbSe nanocrystals treated with 1,2-ethanedithiol. *ACS Nano* 2 (2): 271–280.

Mandal, D., Goswami, P.N., and Rath, A.K. (2019). Thiol and halometallate, mutually passivated quantum dot ink for photovoltaic application. *ACS Applied Materials and Interfaces* 11 (29): 26100–26108.

Mandal, D., Dambhare, N.V., and Rath, A.K. (2021). Reduction of hydroxyl traps and improved coupling for efficient and stable quantum dot solar cells. *ACS Applied Materials and Interfaces* 13 (39): 46549–46557.

McGuire, J.A., Sykora, M., Joo, J. et al. (2010). Apparent versus true carrier multiplication yields in semiconductor nanocrystals. *Nano Letters* 10 (6): 2049–2057.

Midgett, A.G., Luther, J.M., Stewart, J.T. et al. (2013). Size and composition dependent multiple exciton generation efficiency in PbS, PbSe, and PbSxSe1-x alloyed quantum dots. *Nano Letters* 13 (7): 3078–3085.

Moreels, I., Lambert, K., Smeets, D. et al. (2009). Size-dependent optical properties of colloidal PbS quantum dots. *ACS Nano* 3 (10): 3023–3030.

Moreels, I., Justo, Y., de Geyter, B. et al. (2012). Size-tunable, bright, and stable PbS quantum dots: a surface chemistry study. *ACS Nano* 5 (3): 2004–2012.

Mubarok, M.A.L., Aqoma, H., Wibowo, F.T.A. et al. (2020). Molecular engineering in hole transport $\pi$-conjugated polymers to enable high efficiency colloidal quantum dot solar cells. *Advanced Energy Materials* 10 (8): 1902933.

Mubarok, M.A.L., Wibowo, F.T.A., Aqoma, H. et al. (2020). PbS-based quantum dot solar cells with engineered I-conjugated polymers achieve 13% efficiency. *ACS Energy Letters* 5 (11): 3452–3460.

Murray, C.B., Kagan, R., and Bawendi, M.G. (2006). Synthesis and characterization of monodisperse nanocrystals and close-packed nanocrystal assemblies. *Annual Review of Materials Science* 30 (June): 545–610.

Nag, A., Kovalenko, M.V., Lee, J.S. et al. (2011). Metal-free inorganic ligands for colloidal nanocrystals: $S^{2-}$, $HS^-$, $Se^{2-}$, $HSe^-$, $Te^{2-}$, $HTe^-$, $TeS_3^{2-}$, $OH^-$, and $NH_2^-$ as surface ligands. *Journal of the American Chemical Society* 133 (27): 10612–10620.

Neo, D.C.J., Zhang, N., Tazawa, Y. et al. (2016). Poly(3-hexylthiophene-2,5-diyl) as a hole transport layer for colloidal quantum dot solar cells. *ACS Applied Materials and Interfaces* 8 (19): 12101–12108.

Ning, Z., Ren, Y., Hoogland, S. et al. (2012). All-inorganic colloidal quantum dot photovoltaics employing solution-phase halide passivation. *Advanced Materials* 24 (47): 6295–6299.

Ning, Z., Zhitomirsky, D., Adinolfi, V. et al. (2013). Graded doping for enhanced colloidal quantum dot photovoltaics. *Advanced Materials* 25 (12): 1719–1723.

Ning, Z., Dong, H., Zhang, Q. et al. (2014). Solar cells based on inks of n-type colloidal quantum dots. *ACS Nano* 8 (10): 10321–10327.

Ning, Z., Voznyy, O., Pan, J. et al. (2014). Air-stable n-type colloidal quantum dot solids. *Nature Materials* 13 (8): 822–828.

Ning, Z., Gong, X., Comin, R. et al. (2015). Quantum-dot-in-perovskite solids. *Nature* 523 (7560): 324–328.

Ono, L.K., Liu, S., and Qi, Y. (2020). Reducing detrimental defects for high-performance metal halide perovskite solar cells. *Angewandte Chemie – International Edition* 59 (17): 6676–6698.

Park, D., Azmi, R., Cho, Y. et al. (2019). Improved passivation of PbS quantum dots for solar cells using triethylamine hydroiodide. *ACS Sustainable Chemistry and Engineering* 7 (12): 10784–10791.

Pattantyus-Abraham, A.G., Kramer, I.J., Barkhouse, A.R. et al. (2010). Depleted-heterojunction colloidal quantum dot solar cells. *ACS Nano* 4 (6): 3374–3380.

Peng, L., Tang, J., and Zhu, M. (2012). Recent development in colloidal quantum dots photovoltaics. *Frontiers of Optoelectronics* 5 (4): 358–370.

Piliego, C., Protesescu, L., Bisri, S.Z. et al. (2013). 5.2% efficient PbS nanocrystal Schottky solar cells. *Energy and Environmental Science* 6 (10): 3054–3059.

Pinna, J., Mehrabi, K.R., Gavhane, D.S. et al. (2022). Approaching bulk mobility in PbSe colloidal quantum dots 3D superlattices. *Advanced Materials* 35 (8): 2207364.

Pitaro, M., Tekelenburg, E.K., Shao, S., and Loi, M.A. (2022). Tin halide perovskites: from fundamental properties to solar cells. *Advanced Materials* 34 (1): 2105844.

Proppe, A.H., Xu, J., Sabatini, R.P. et al. (2018). Picosecond charge transfer and long carrier diffusion lengths in colloidal quantum dot solids. *Nano Letters* 18 (11): 7052–7059.

Roberge, A., Dunlap, J.H., Ahmed, F., and Greytak, A.B. (2020). Size-dependent PbS quantum dot surface chemistry investigated via gel permeation chromatography. *Chemistry of Materials* 32 (15): 6588–6594.

Rong, Y., Hu, Y., Mei, A. et al. (2018). Challenges for commercializing perovskite solar cells. *Science* 361 (6408): eeat8235.

Sambur, J.B., Novet, T., and Parkinson, B.A. (2010). Multiple exciton collection in a sensitized photovoltaic system. *Science* 330 (6000): 63–66.

Sandberg, R.L., Padilha, L.A., Qazilbash, M.M. et al. (2012). Multiexciton dynamics in infrared-emitting colloidal nanostructures probed by a superconducting nanowire single-photon detector. *ACS Nano* 6 (11): 9532–9540.

Sayevich, V., Guhrenz, C., Sin, M. et al. (2016). Chloride and indium-chloride-complex inorganic ligands for efficient stabilization of nanocrystals in solution and doping of nanocrystal solids. *Advanced Functional Materials* 26 (13): 2163–2175.

Sayevich, V., Guhrenz, C., Dzhagan, V.M. et al. (2017). Hybrid N-butylamine-based ligands for switching the colloidal solubility and regimentation of inorganic-capped nanocrystals. *ACS Nano* 11 (2): 1559–1571.

Schaller, R.D. and Klimov, V.I. (2004). High efficiency carrier multiplication in PbSe nanocrystals: implications for solar energy conversion. *Physical Review Letters* 92 (18): 186601.

Service, R.F. (2008). Can the upstarts top silicon? *Science* 319 (5864): 718–720.

Shabaev, A., Hellberg, C.S., and Efros, A.L. (2013). Efficiency of multiexciton generation in colloidal nanostructures. *Accounts of Chemical Research* 46 (6): 1242–1251.

Shaheen, S.E., Brabec, C.J., Sariciftci, N.S. et al. (2001). 2.5% efficient organic plastic solar cells. *Applied Physics Letters* 78 (6): 841–843.

Sharma, A., Mahajan, C., and Rath, A.K. (2020). Reduction of trap and polydispersity in mutually passivated quantum dot solar cells. *ACS Applied Energy Materials* 3 (9): 8903–8911.

Sharma, A., Dambhare, N.v., Bera, J. et al. (2021). Crack-free conjugated PbS quantum dot-hole transport layers for solar cells. *ACS Applied Nano Materials* 4 (4): 4016–4025.

Shrestha, A., Jin, B., Kee, T.W. et al. (2016). A robust strategy for "living" growth of lead sulfide quantum dots. *ChemNanoMat* 2 (1): 49–53.

Shrestha, A., Spooner, N.A., Qiao, S.Z., and Dai, S. (2016). Mechanistic insight into the nucleation and growth of oleic acid capped lead sulphide quantum dots. *Physical Chemistry Chemical Physics* 18 (20): 14055–14062. Royal Society of Chemistry.

Shrestha, A., Batmunkh, M., Tricoli, A. et al. (2019). Near-infrared active lead chalcogenide quantum dots: preparation, post-synthesis ligand exchange, and applications in solar cells. *Angewandte Chemie – International Edition* 58 (16): 5202–5224.

Shulga, A.G., Kahmann, S., Dirin, D.N. et al. (2018). Electroluminescence generation in PbS quantum dot light-emitting field-effect transistors with solid-state gating. *ACS Nano* 12 (12): 12805–12813.

Sliz, R., Lejay, M., Fan, J.Z. et al. (2019). Stable colloidal quantum dot inks enable inkjet-printed high-sensitivity infrared photodetectors. *ACS Nano* 13 (10): 11988–11995.

Song, K., Yuan, J., Shen, T. et al. (2019). Spray coated colloidal quantum dot films for broadband photodetectors. *Nanomaterials* 9 (12): 1738.

Speirs, M.J., Balazs, D.M., Fang, H.H. et al. (2015). Origin of the increased open circuit voltage in PbS-CdS core-shell quantum dot solar cells. *Journal of Materials Chemistry A* 3 (4): 1450–1457. Royal Society of Chemistry.

Speirs, M.J., Dirin, D.N., Abdu-Aguye, M. et al. (2016). Temperature dependent behaviour of lead sulfide quantum dot solar cells and films. *Energy and Environmental Science* 9 (9): 2916–2924. Royal Society of Chemistry.

Speirs, M.J., Balazs, D.M., Dirin, D.N. et al. (2017). Increased efficiency in pn-junction PbS QD solar cells via NaHS treatment of the p-type layer. *Applied Physics Letters* 110 (10): 103904. Retrieved from http://dx.doi.org/10.1063/1.4978444.

Sukharevska, N., Bederak, D., Goossens, V.M. et al. (2021). Scalable PbS quantum dot solar cell production by blade coating from stable inks. *ACS Applied Materials and Interfaces* 13 (4): 5195–5207.

Sun, B., Voznyy, O., Tan, H. et al. (2017). Pseudohalide-exchanged quantum dot solids achieve record quantum efficiency in infrared photovoltaics. *Advanced Materials* 29 (27): 1700749.

Sun, B., Johnston, A., Xu, C. et al. (2020). Monolayer perovskite bridges enable strong quantum dot coupling for efficient solar cells. *Joule* 4 (7): 1542–1556.

Sze, S.M. and Ng, K.K. (2006). Photodetectors and solar cells. In: *Physics of Semiconductor Devices*, 663–742. John Wiley & Sons, Ltd. Retrieved from https://onlinelibrary.wiley.com/doi/abs/10.1002/9780470068328.ch13.

Szendrei, K., Gomulya, W., Yarema, M. et al. (2010). PbS nanocrystal solar cells with high efficiency and fill factor. *Applied Physics Letters* 97 (20): 203501.

Talapin, D.v. and Murray, C.B. (2005). Applied physics: PbSe nanocrystal solids for n- and p-channel thin film field-effect transistors. *Science* 310 (5745): 86–89.

Tang, C.W. (1986). Two-layer organic photovoltaic cell. *Applied Physics Letters* 48 (2): 183–185.

Tang, J., Kemp, K.W., Hoogland, S. et al. (2011). Colloidal-quantum-dot photovoltaics using atomic-ligand passivation. *Nature Materials* 10 (10): 765–771. Nature Publishing Group.

Teh, Z.L., Hu, L., Zhang, Z. et al. (2020). Enhanced power conversion efficiency via hybrid ligand exchange treatment of p-type PbS quantum dots. *ACS Applied Materials and Interfaces* 12 (20): 22751–22759.

Trinh, M.T., Polak, L., Schins, J.M. et al. (2011). Anomalous independence of multiple exciton generation on different Group IV-VI quantum dot architectures. *Nano Letters* 11 (4): 1623–1629.

Turner, G.M., Beard, M.C., and Schmuttenmaer, C.A. (2002). Carrier localization and cooling in dye-sensitized nanocrystalline titanium dioxide. *Journal of Physical Chemistry B* 106 (45): 11716–11719.

Uddin, M.A., Lee, T.H., Xu, S. et al. (2015). Interplay of intramolecular noncovalent Coulomb interactions for semicrystalline photovoltaic polymers. *Chemistry of Materials* 27 (17): 5997–6007.

Vo, N.T., Ngo, H.D., Do Thi, N.P. et al. (2016). Stability investigation of Ligand-Exchanged CdSe/ZnS-Y (Y =3-mercaptopropionic acid or Mercaptosuccinic Acid) through zeta potential measurements. *Journal of Nanomaterials* 2016: 8562648.

Wang, Y., Lu, K., Han, L. et al. (2018). In situ passivation for efficient PbS quantum dot solar cells by precursor engineering. *Advanced Materials* 30 (16): 1704871.

Xi, J. and Loi, M.A. (2021). The fascinating properties of tin-alloyed halide perovskites. *ACS Energy Letters* 6 (5): 1803–1810.

Xu, J., Voznyy, O., Liu, M. et al. (2018). 2D matrix engineering for homogeneous quantum dot coupling in photovoltaic solids. *Nature Nanotechnology* 13 (6): 456–462. Springer US. Retrieved from http://dx.doi.org/10.1038/s41565-018-0117-z.

Xu, K., Zhou, W., and Ning, Z. (2020). Integrated structure and device engineering for high performance and scalable quantum dot infrared photodetectors. *Small* 16 (47): 2003397.

Yang, F. (2021). Size effect on the bandgap change of quantum dots: thermomechanical deformation. *Physics Letters, Section A: General, Atomic and Solid State Physics* 401: 127346. Elsevier B.V. Retrieved from https://doi.org/10.1016/j.physleta.2021.127346.

Yang, Z., Janmohamed, A., Lan, X. et al. (2015). Colloidal quantum dot photovoltaics enhanced by perovskite shelling. *Nano Letters* 15 (11): 7539–7543.

Yang, Z., Fan, J.Z., Proppe, A.H. et al. (2017). Mixed-quantum-dot solar cells. *Nature Communications* 8 (1): 1325. Springer US. Retrieved from http://dx.doi.org/10.1038/s41467-017-01362-1.

Yin, S., Ho, C.H.Y., Ding, S. et al. (2022). Enhanced surface passivation of lead sulfide quantum dots for short-wavelength photodetectors. *Chemistry of Materials*, 34(12), 5433–5442.

Yuan, M., Kemp, K.W., Thon, S.M. et al. (2014). High-performance quantum-dot solids via elemental sulfur synthesis. *Advanced Materials* 26 (21): 3513–3519.

Yuan, M., Wang, X., Chen, X. et al. (2022). Phase-transfer exchange lead chalcogenide colloidal quantum dots: ink preparation, film assembly, and solar cell construction. *Small* 18 (2): 2102340.

Zarghami, M.H., Liu, Y., Gibbs, M. et al. (2010). P-type PbSe and PbS quantum dot solids prepared with short-chain acids and diacids. *ACS Nano* 4 (4): 2475–2485.

Zhang, J., Gao, J., Miller, E.M. et al. (2014). Diffusion-controlled synthesis of PbS and PbSe quantum dots with in situ halide passivation for quantum dot solar cells. *ACS Nano* 8 (1): 614–622.

Zhang, Y., Ding, C., Wu, G. et al. (2016). Air stable PbSe colloidal quantum dot heterojunction solar cells: ligand-dependent exciton dissociation, recombination, photovoltaic property, and stability. *Journal of Physical Chemistry C* 120 (50): 28509–28518.

Zhang, X., Zhang, J., Phuyal, D. et al. (2018). Inorganic CsPbI3 perovskite coating on PbS quantum dot for highly efficient and stable infrared light converting solar cells. *Advanced Energy Materials* 8 (6): 1702049.

Zhang, X., Cappel, U.B., Jia, D. et al. (2019). Probing and controlling surface passivation of PbS quantum dot solid for improved performance of infrared absorbing solar cells. *Chemistry of Materials* 31 (11): 4081–4091.

Zhang, Y., Kan, Y., Gao, K. et al. (2020). Hybrid quantum dot/organic heterojunction: a route to improve open-circuit voltage in PbS colloidal quantum dot solar cells. *ACS Energy Letters* 5 (7): 2335–2342.

Zhao, T., Goodwin, E.D., Guo, J. et al. (2016). Advanced architecture for colloidal PbS quantum dot solar cells exploiting a CdSe quantum dot buffer layer. *ACS Nano* 10 (10): 9267–9273.

Zhitomirsky, D., Furukawa, M., Tang, J. et al. (2012). N-type colloidal-quantum-dot solids for photovoltaics. *Advanced Materials* 24 (46): 6181–6185.

Zhitomirsky, D., Kramer, I.J., Labelle, A.J. et al. (2012). Colloidal quantum dot photovoltaics: the effect of polydispersity. *Nano Letters* 12 (2): 1007–1012.

Zhu, L., Zhang, M., Xu, J. et al. (2022). Single-junction organic solar cells with over 19% efficiency enabled by a refined double-fibril network morphology. *Nature Materials* 21 (6): 656–663. Springer US.

# 16

# Singlet Fission for Solar Cells

*Timothy W. Schmidt[1] and Murad J. Y. Tayebjee[2]*

[1] School of Chemistry and Chief Investigator of the ARC Centre of Excellence in Exciton Science, University of New South Wales, Sydney, NSW, Australia
[2] School of Photovoltaic and Renewable Energy Engineering, University of New South Wales, Sydney, NSW, Australia

## 16.1 Introduction

Single-threshold solar cells are limited to an energy conversion efficiency of 33.7 % (the Shockley–Queisser limit) (Shockley and Queisser 1961; Tayebjee et al. 2012). The two major sources of efficiency loss are the inability to absorb photons with energy below the threshold, and the thermalization of carriers formed by photons with energy in excess of the threshold (Hirst and Ekins-Daukes 2011). One strategy among many to ameliorate this second efficiency loss is singlet fission, a form of multiple exciton generation that occurs in organic molecules (Tayebjee et al. 2015). By harnessing singlet fission, the electrical current may be doubled for photons absorbed above a second energy threshold.

Singlet fission was first discovered in anthracene molecules in the 1960s (Singh et al. 1965), and its application to photovoltiacs was first proposed by Dexter (1979). In the push toward what was then called "third-generation photovoltaics," singlet fission saw renewed interest in the context of renewable energy from the mid-2000s (Paci et al. 2006).

The success of silicon photovoltaics has seen a shift in strategies that seek to overcome the Shockley–Queisser limit away from thin-film photovoltaics and back toward silicon. The compatibility of singlet fission with silicon has spurred research into this application, and recently, devices have started being reported (MacQueen et al. 2018; Einzinger et al. 2019). As such, it is an exciting time in singlet fission photovoltaics, especially as it is applied to silicon.

This chapter is arranged as a part tutorial and mini-review. We introduce the electronic structure and photophysics of organic chromophores and then how the excited states interact in solids. We then undergo a deep dive into the theory of singlet fission while providing the reader solid references from which to seek further clarification. We then describe singlet fission solar cells and their mode of operation and discuss singlet fission materials before reviewing a selection of reported devices. We round out the chapter with a perspective on the current challenges that remain as barriers to successful implementation of singlet fission photovoltaics.

*Photovoltaic Solar Energy: From Fundamentals to Applications, Volume 2*, First Edition.
Edited by Wilfried van Sark, Bram Hoex, Angèle Reinders, Pierre Verlinden, and Nicholas J. Ekins-Daukes.
© 2024 John Wiley & Sons Ltd. Published 2024 by John Wiley & Sons Ltd.
Companion website: www.wiley.com/go/PVsolarenergy

## 16.2 Molecular Electronic Structure and Photophysics

The chromophores considered in this section are exclusively planar aromatic molecules. Their electronic structure can be understood in terms of $\sigma$-bonding orbitals localized between carbon atoms, or between carbons and hydrogens. Each $\sigma$-orbital is occupied by two electrons and is photophysically inert. In aromatic molecules, carbon atoms bond to three other atoms in a plane and contribute one valence electron to each $\sigma$-bond. This leaves one valence electron per carbon atom to contribute to the molecular orbitals, which come about through the interaction of the unhybridized atomic p-orbitals perpendicular to the molecular plane (Figure 16.1). These so-called $\pi$-orbitals are antisymmetric with respect to reflection through the molecular plane.

The energy structure of the $\pi$-orbitals can be understood within the framework of Hückel theory. Assuming a basis of atomic p-orbitals to be orthonormal, one obtains a Hamiltonian matrix with diagonal elements $\alpha$, representing the energy of a lone p-orbital, and off-diagonal elements $\beta$, representing the resonance integrals for adjacent p-orbitals in the basis. For e.g. benzene,

$$\mathbf{H} = \begin{pmatrix} \alpha & \beta & 0 & 0 & 0 & \beta \\ \beta & \alpha & \beta & 0 & 0 & 0 \\ 0 & \beta & \alpha & \beta & 0 & 0 \\ 0 & 0 & \beta & \alpha & \beta & 0 \\ 0 & 0 & 0 & \beta & \alpha & \beta \\ \beta & 0 & 0 & 0 & \beta & \alpha \end{pmatrix} \tag{16.1}$$

The $\pi$-electronic structures of linear acenes in their ground states are shown in Figure 16.2. The excited states of aromatic molecules are brought about by excitations from the occupied orbitals into the unoccupied orbitals. For chromophores larger than naphthalene, the lowest energy excited singlet state is the short-axis polarized $L_a$ state, which is characterized by excitation from the highest-occupied molecular orbital (HOMO) to the lowest unoccupied molecular orbital (LUMO), as shown in Figure 16.2. This state is designated $S_1$, with the "S" symbolizing that it has singlet spin multiplicity (no net spin, $S = 0$). The subscript "1" denotes the 1st excited state. The energy of the $S_1$ state decreases as the chromophore is grown in extent from naphthalene to pentacene.

There exists a lower-energy electronic state with $S = 1$. This is a so-called triplet state which, in high magnetic fields, splits into components with $M_S = -1, 0, +1$. To understand the energy difference between $S_1$ and $T_1$, we consider a two-electron system. The triplet

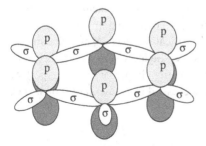

**Figure 16.1** A cartoon of the electronic structure of benzene. Each carbon atom at the vertex of the hexagon makes three $\sigma$-bonds to two carbons and one hydrogen (not all are visible). The carbon atoms contribute three of their four valence electrons to these bonds. The fourth electron resides in the $\pi$-system, which is brought about through the interaction of unhybridized p-orbitals perpendicular to the molecular plane.

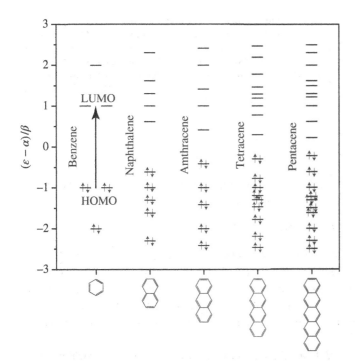

**Figure 16.2** The $\pi$-electronic energy levels of benzene and the acene series as calculated by Hückel theory. The HOMO–LUMO gap diminishes with increased chromophore size. In reality, the optical gap in tetracene is about half that of benzene.

state is written

$$|T_1\rangle = \frac{\phi_H(1)\,\phi_L(2) - \phi_L(1)\,\phi_H(2)}{\sqrt{2}} \times \begin{cases} \alpha(1)\alpha(2) \\ (\alpha(1)\,\beta(2) + \beta(1)\alpha(2))/\sqrt{2} \\ \beta(1)\,\beta(2) \end{cases} \qquad (16.2)$$

where $\phi_H$ and $\phi_L$ are spatial wave functions for the HOMO and LUMO and the number in parentheses indicates the coordinates of the electron with that index. The functions $\alpha$ and $\beta$ here represent spin functions for $m_s = \pm\frac{1}{2}$, respectively. Due to the antisymmetry requirement for fermions, here the spatial wave function is antisymmetric because the spin function must be symmetric. The state with an antisymmetric spin function is the singlet state, $S_1$.

$$|S_1\rangle = \frac{\phi_H(1)\,\phi_L(2) + \phi_L(1)\,\phi_H(2)}{\sqrt{2}} \times \frac{\alpha(1)\,\beta(2) - \beta(1)\alpha(2)}{\sqrt{2}} \qquad (16.3)$$

The energy of $|T_1\rangle$ is

$$E_{T_1} = \langle T_1 |H| T_1 \rangle = \langle \phi_H\Phi_L |H| \phi_H\Phi_L\rangle - \langle \phi_H\Phi_L |H| \phi_L\phi_H\rangle \qquad (16.4)$$

and the energy of the singlet state is

$$E_{S_1} = \langle S_1 |H| S_1 \rangle = \langle \phi_H\Phi_L |H| \phi_H\Phi_L\rangle + \langle \phi_H\Phi_L |H| \phi_L\phi_H\rangle \qquad (16.5)$$

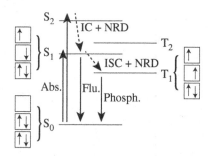

**Figure 16.3** Photophysical processes of single chromophores. Abs. – absorption, Flu. – fluorescence, Phosph. – phosphorescence, IC + NRD – internal conversion and non-radiative decay, ISC + NRD – intersystem crossing and non-radiative decay. The arrows in boxes are cartoon representations of leading electronic configurations.

where the notation is compressed and the electron coordinates are implied by the order of the spatial orbitals in the bra or ket. It follows that

$$\Delta E_{ST} = E_{S_1} - E_{T_1} = 2\langle \phi_H \Phi_L | H | \phi_L \phi_H \rangle \tag{16.6}$$

This matrix element, which exchanges two electrons, is called the exchange integral. The only operator in the Hamiltonian that includes the coordinates of two electrons and can, therefore, operate between wave functions differing by two spin-orbitals is the Coulomb operator, which accounts for the repulsion between electrons. As such, the exchange integral is a positive quantity and the $T_1$ state nearly always lies lower than $S_1$.

We can now construct the Jablonski diagram, which indicates photophysical processes occurring in single chromophores (Figure 16.3). The absorption spectrum of an organic chromophore is due to excitations from the ground state ($S_0$) to the excited states $S_n$. Usually, states above $S_1$ are so short-lived that their fate can be considered entirely due to internal conversion to the lowest excited state, $S_1$. The excess vibrational energy is transferred to the surroundings.

The $S_1$ state can undergo fluorescence on a nanosecond time scale to return the molecule to $S_0$, or it can undergo a spin-flip to decay to the lowest triplet state, $T_1$. In some cases, $T_2$ is near degenerate with $S_1$ and this internal conversion process is accelerated. Once in $T_1$, the molecule can undergo phosphorescence over a long time scale ($\gtrsim \mu s$) or undergo intersystem crossing back to $S_0$.

## 16.3 Davydov Splitting

Molecular crystals that have more than one site per unit cell display a phenomenon called Davydov splitting – the splitting of electronic spectra, usually observed in absorption spectroscopy. In the following, we formalize the electronic states that may give rise to Davydov splitting.

The electronic states of crystals of acenes such as those depicted in Figure 16.2 can be understood in terms of a basis of chromophore states. For parallel chromophores arranged in a line, the ground state is

$$|g\rangle = \prod_i |S_0^i\rangle \tag{16.7}$$

where the index runs over the individual chromophores. A basis of excited states may be written

$$|e^j\rangle = |S_1^j\rangle \prod_{i \neq j} |S_0^i\rangle \tag{16.8}$$

These states interact via the $1/r$ Coulomb operator. The interaction is strongest for neighboring chromophores (in atomic units),

$$\left\langle e^j \left| \frac{1}{r} \right| e^{j+1} \right\rangle \approx \left\langle S_1^j \left| \left\langle S_0^{j+1} \left| \frac{1}{r} \right| S_1^{j+1} \right\rangle \right| S_0^j \right\rangle \tag{16.9}$$

$$\approx \int \phi_H^j \phi_L^j \phi_H^{j+1} \phi_H^{j+1} \frac{1}{r} \phi_H^j \phi_H^j \phi_H^{j+1} \phi_L^{j+1} d\tau \tag{16.10}$$

$$\approx \int \phi_L^j(2) \phi_H^j(2) \frac{1}{r_{24}} \phi_H^{j+1}(4) \phi_L^{j+1}(4) d\tau \tag{16.11}$$

where for the terms chosen, electrons 2 and 4 change state. The product of $\phi_H \phi_L$ for the coordinates of one electron is the transition density, which we can approximate as a transition dipole, $\mu$. The Coulomb operator thus acts on two dipoles, which have an interaction energy

$$V = \frac{\mu_j \cdot \mu_{j+1}}{4\pi\varepsilon_0 R^3}, \tag{16.12}$$

where $R$ is the interchromophore distance for parallel dipoles perpendicular to the line joining them. The Hamiltonian has a structure reminiscent of that derived in Hückel theory. For six chromophores arranged linearly,

$$\mathbf{H} = \begin{pmatrix} E_{S_1} & V & 0 & 0 & 0 & 0 \\ V & E_{S_1} & V & 0 & 0 & 0 \\ 0 & V & E_{S_1} & V & 0 & 0 \\ 0 & 0 & V & E_{S_1} & V & 0 \\ 0 & 0 & 0 & V & E_{S_1} & V \\ 0 & 0 & 0 & 0 & V & E_{S_1} \end{pmatrix} \tag{16.13}$$

The states disperse into a band, with the lowest energy state being dark for side-on dipoles and the highest energy state carrying the majority of the oscillator strength (Kasha 1963). For a system with dipoles that are arranged head-to-tail, the lowest-energy state carries the majority oscillator strength. In systems such as tetracene, the herringbone structure brings about two exciton branches that differ in the phase of the transition dipole on pairs of non-equivalent chromophores, resulting in a splitting of the 0–0 band. Triplet states do not experience the same Coulombic interactions as singlets and are thus much less dispersed in organic crystals.

## 16.4 Singlet Fission

A pair of chromophores may, as an ensemble, possess a triplet pair state with energy $2E_{T1} \approx E_{S_1}$. In this case, a transition from the state with a single $S_1$ excitation to the triplet pair can occur if

$$\langle S_0 S_1 | H |^1(T_1 T_1)\rangle \neq 0 \tag{16.14}$$

To conserve spin, this is held to occur as an intermolecular two-electron transfer. The transition can occur via a charge transfer "intermediate" as two sequential one-electron transfers. It should be stressed that these considerations hold in the diabatic representation of the electronic wave functions, but the matrix elements manifest as a coupling between the $S_0 S_1$ and $^1(TT)$ surfaces in the adiabatic representation.

We introduce the notation for spin-orbitals where a bar indicates a down-spin electron and an unbarred orbital is assumed to be up-spin. In this notation, we write the states as combinations of Slater determinants.

$$|S_1 S_0\rangle = N_{S_0 S_1} \left( \left| \phi_H \overline{\phi}_L \phi'_H \overline{\phi'_H} \right| - \left| \overline{\phi}_H \phi_L \phi'_H \overline{\phi'_H} \right| \right) \tag{16.15}$$

where the terms $N$ are normalization factors and the primes distinguish the two chromophores involved. The state $|S_0 S_1\rangle$ can be written analogously. Similarly, we introduce the triplet-pair state

$$|^1(T_1 T_1)\rangle = \frac{N_{T_1 T_1}}{\sqrt{3}} \left( \left| \phi_H \phi_L \overline{\phi'_H} \overline{\phi'_L} \right| + \left| \overline{\phi}_H \overline{\phi}_L \phi'_H \phi'_L \right| \right.$$
$$- \frac{1}{2} \left( \left| \phi_H \overline{\phi}_L \phi'_H \overline{\phi'_L} \right| + \left| \phi_H \overline{\phi}_L \overline{\phi'_H} \phi'_L \right| \right.$$
$$\left. + \left| \overline{\phi}_H \phi_L \phi'_H \overline{\phi'_L} \right| + \left| \overline{\phi}_H \phi_L \overline{\phi'_H} \phi'_L \right| \right) \right) \tag{16.16}$$

and the charge–transfer state

$$|^1 CA\rangle = \frac{N_{^1 CA}}{\sqrt{2}} \left( \left| \phi_H \overline{\phi'_L} \phi'_H \overline{\phi'_H} \right| - \left| \overline{\phi}_H \phi'_L \phi'_H \overline{\phi'_H} \right| \right) \tag{16.17}$$

with $|^1 AC\rangle$ defined analogously, and C and A indicate the radical cation and anion states of the individual chromophores.

The following treatment of singlet fission is taken from Havlas and Michl (2016), which is based on work in Harcourt et al. (1994). Because the basis states in this treatment are not orthogonal (due to overlap between the $\phi$ and $\phi'$), the matrix element of the Hamiltonian, which leads from reactants to products, is

$$T_{RP} = H_{RP} - S_{RP} E_R \tag{16.18}$$

where $S_{RP}$ is the overlap between the states and $E_R$ is the energy of the reactant state. In singlet fission, we consider that the reactant is $|S_1 S_0\rangle$ and the product is $|^1(T_1 T_1)\rangle$. So the matrix element for singlet fission becomes

$$T_{SF^d} = \langle ^1(T_1 T_1)|H|S_1 S_0\rangle - \langle ^1(T_1 T_1)|S_1 S_0\rangle\langle S_1 S_0|H|S_1 S_0\rangle \tag{16.19}$$

This term is the "direct" term, which corresponds to the concerted transfer of two electrons (Figure 16.4). It is typically on the order of a few meV. To address the indirect, two-step transfer, we first consider the mixing of the charge transfer states into the reactant state. Applying perturbation theory,

$$|R\rangle = |S_1 S_0\rangle - \frac{T_{CT}}{\Delta E_{CT}} |^1 CA\rangle - \frac{T_{CT'}}{\Delta E_{CT'}} |^1 AC\rangle \tag{16.20}$$

**Figure 16.4** Cartoon showing the interactions that lead to singlet fission.

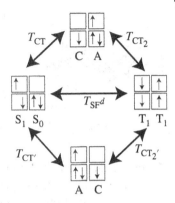

where the energy differences are generally positive quantities, e.g. $\Delta E_{CT} = E_{{}^1CA} - E_{S_1S_0}$, and the $T$ terms are derived analogously to Eq. (16.18). Invoking this state with charge transfer character as the reactant state, the matrix element of interest is

$$T_{SF} = T_{SF^d} - \left( T_{CT_2} \frac{T_{CT}}{\Delta E_{CT}} + T_{CT_2'} \frac{T_{CT'}}{\Delta E_{CT'}} \right) \tag{16.21}$$

where the $T_{CT_2}$ terms pertain to charge transfer to or from the triplet pair state (Figure 16.4).

The term $T_{SF}$ represents the overall *electronic* coupling from the excited singlet state to the triplet pair. Taking into account the coupling to the charge transfer states, called the "mediated" pathway, this matrix element now expands to some hundreds of meV. Application of the Fermi golden rule would then make the rate of singlet fission

$$W_{SF} = \frac{2\pi}{\hbar} T_{SF}^2 \rho(E_R) \tag{16.22}$$

where $\rho$ is the density of states of the products at the energy of the reactant.

Singlet fission may also be treated in terms of Marcus non-adiabatic theory (Yost et al. 2014). In this case the expression is

$$W_{SF} \approx \frac{2\pi}{\hbar} T_{SF}^2 \frac{1}{\sqrt{4\pi \lambda k_B T}} \exp\left( -\frac{(\Delta G + \lambda)^2}{4\pi \lambda k_B T} \right) \tag{16.23}$$

where $\lambda$ is the reorganization energy. Deviation from this model is seen for couplings exceeding a few tens of meV, but the considerations of which factors accelerate and attenuate singlet fission still apply. Where $T_{SF}$ becomes large, the problem must be treated from the standpoint of an adiabatic potential energy surface as plotted in Figure 16.5.

An instructive model system to explore the adiabatic singlet fission surface is a pair of ethylene molecules. These are calculated, at a suitable level of theory, to exhibit a triplet pair state in the vicinity of the excited singlet. At the most symmetric geometry with the ethylene planes disposed vertically, $T_{SF}$ vanishes and the intersection between the $^1(TT)$ and $S_1S_0$ surfaces manifests as a conical intersection (Schmidt 2019). At the canonical $\pi$-stacking distance of 3.4 Å, the coupling $T_{SF}$ was found to exceed 100 meV.

It is instructive to consider the adiabatic singlet fission potential energy surface in terms of an *intermolecular* coordinate, that modulates the coupling, and an *intramolecular* coordinate, that links the ground state to the triplet pair. A plot of this surface is shown in

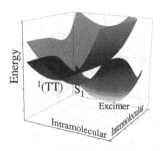

**Figure 16.5** A computed adiabatic potential energy surface for model chromophores as a function of the intramolecular coordinate that connects the optimized geometries of the involved states, and an intermolecular coordinate that modulates the coupling between the states. At close range, the surface collapses into an excimeric well. Data obtained from Schmidt (2019).

Figure 16.5 (Schmidt 2019; Dover et al. 2018). Because the triplet pair state contains two "excitons" but the excited singlet contains only one, the triplet pair is displaced approximately twice as far from the ground state as the excited singlet. This intramolecular coordinate thus serves as the Franck–Condon coordinate as well as the Marcus–Hush reaction coordinate (Schmidt 2019). The intermolecular coordinate modulates the coupling between $S_1$ and $^1$(TT). As can be seen in Figure 16.5, the Marcus–Hush double-well morphs into a deep potential minimum – the excimer. The word "excimer" is a portmanteau of excited dimer, and represents a strongly coupled state where the excitation is spread across both moieties. It is considered to comprise excited singlet state as well as charge–transfer character. As may be expected, the excimer has been identified as a trap in singlet fission dynamics (Dover et al. 2018).

Experimentally observed singlet fission rates are observed to be on the order of that predicted by the above considerations. Pentacene is the "drosophila" of singlet fission studies, and is known to transit to a triplet pair state in under 100 fs (Wilson et al. 2011). Tetracene, which is mildly endothermic, will darken in an amorphous film in $\sim$ 80 ps, regardless of the temperature (Burdett et al. 2011; Tayebjee et al. 2013). However, in crystalline tetracene, singlet fission occurs on the several 100 ps time-scale (Piland and Bardeen 2015).

## 16.5 The Potential Benefits of Singlet Fission

While the initial interest in singlet fission arose in the 1960s to explain spectroscopic measurements of acenes (Singh et al. 1965), the topic has experienced a resurgence in light of their potential to augment the current of solar cells, as shown in Figure 16.6. To understand this motivation, it is useful to consider the detailed balance limit of a singlet fission solar cell.

Detailed balance calculations of solar cells were first proposed by Shockley and Queisser (1961). In this canonical work the limiting efficiency of a p–n junction solar cell was reported, by assuming

1. All recombination is radiative
2. Every photon with energy greater than the bandgap of the solar cell, $E_g$, is absorbed
3. There is infinite carrier mobility
4. All absorption occurs at the surface of the cell. That is, the extinction coefficient approaches infinity. We can equivalently say that the cell has an absorbance of 1.

**Figure 16.6** Schematic of a potential architecture of singlet fission (SF) tandem solar cell, where a singlet fission material is deposited on an interdigitated back-contact inorganic cell. Singlet fission generates two excitons from high energy (bluer) light that are transferred into the inorganic photovoltaic (PV) device, augmenting the current. The inorganic PV can also generate current through direct absorption of the lower energy (redder) light.

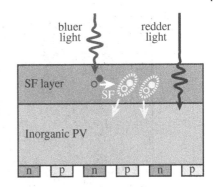

While these assumptions may appear to be unreasonable at first, real-world silicon solar cells are actually approaching this limit! Indeed, the current world record is 26.7% (Yoshikawa et al. 2017; Green et al. 2022) as compared to the detailed balance limiting efficiency of 33% under standard test conditions (STC, AM1.5G illumination and an operating temperature of 298 K). Considering that the first practical silicon solar cell efficiency was 6% in 1954 (Chapin et al. 1954), this is a clear demonstration of the effect of decades of research and development.

As such, if only for aspirational reasons, it is useful to compare the detailed balance limit of a p–n junction solar cell with that of a solar cell augmented by singlet fission. The first calculation of this limit was performed by Hanna and Nozik in 2006, where they showed a limit of 44% (Hanna and Nozik 2006). These calculations were motivated, in part, by multiple exciton generation in quantum dot structures. While this is similar to a singlet fission formalism, it also contained some simplifications (Pusch et al. 2021). Moreover, only exothermic multiple exciton generation in quantum dots has been observed, and the authors therefore limited their calculations to isoergic multiple exciton generation.

By contrast, endothermic singlet fission has been observed in a range of materials, and therefore should be considered in detailed balance calculations. In 2012, we demonstrated that, if one accounts for endothermic fission, the detailed balance limit is 45.9% under AM1.5G radiation and an operating temperature of 300 K (Tayebjee et al. 2012). This is illustrated in Figure 16.7. Here, the x-axis is the bandgap of the overall device, and the y-axis shows the change in enthalpy of the singlet fission process.

Interestingly, a limiting efficiency of 45% was calculated for a bandgap of 1.1 eV under STC (Tayebjee et al. 2015); that is, the bandgap of silicon. This will become important as we discuss the direction of photovoltaic research in Section 16.6.

While detailed balance calculations are useful, several attempts have been made to predict more reasonable limits given current singlet fission material properties. In 2014, a semi-empirical approach to the limiting efficiencies of the two most commonly studied singlet fission molecules, tetracene, and pentacene, was undertaken (Tayebjee et al. 2014). The calculations assumed that the singlet fission layer was a p-type material coupled to an idealized n-type semiconductor. For tetracene this resulted in a limiting efficiency of 35.8%. The inclusion of a free-energy sacrifice at the interface to separate the bound polaron at the interface was also considered; an energy-sacrifice of ≈ 0.4 eV negated the increased efficiency from singlet fission. In 2021 Ehrler and coworkers published another efficiency limit paper, arriving at a maximum efficiency of 34.6% if triplet energy is transferred via

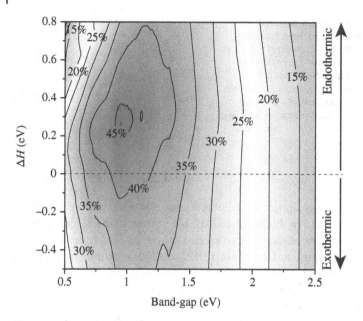

**Figure 16.7** Limiting efficiency as a function of bandgap and change in enthalpy of singlet fission. Here $\Delta H$ is defined as the energy of two $T_1$ excitons less the energy of a $S_1$ exciton.

charge transfer (Daiber et al. 2021). Also in 2021, the lifespan of singlet fission silicon tandem devices was considered in light of the reduced operating temperature that arises from limiting thermalization losses (Jiang et al. 2021). Moreover, this work considered the optical properties of the degradation products of tetracene to demonstrate that a "boost" from singlet fission was possible prior to degradation; due to the transparency of the degradation products the underlying silicon cell would operate as normal. Considering all of these "realistic" approaches, it seems clear that singlet fission has the potential to be used in solar cells, which exceed 30% efficiency, and could be a significant player in the photovoltaic market.

## 16.6 Materials for Singlet Fission

There are two broad approaches to the identification of singlet fission materials: empirical and rational. The empirical approach comes from accumulated knowledge of which systems undergo singlet fission: mostly acenes. Singlet fission was discovered in anthracene (Singh et al. 1965), and is known to occur in tetracene and pentacene. Only in pentacene is it an efficient exothermic process, but the triplet energy is rather too low to be technologically relevant. Much focuss has been placed on modified tetracenes.

While tetracene itself is an endothermic system with $\Delta G \sim 200\,\text{meV}$, rubrene (tetraphenyltetracene) is less drastically so, and is found to be capable of both singlet fission and triplet–triplet annihilation (TTA). The substituted triisopropylsiloethynyl tetracene (TIPS-tetracene) has been found to exhibit exothermic singlet fission, despite assumptions to the contrary (Dover et al. 2018). In many acene-like compounds, singlet fission, and TTA are both found to occur. Indeed, any structural motif known to exhibit delayed fluorescence by TTA is a possible singlet fission candidate. To this end the photophysically robust perylene diimides have been intensely studied. A further category which should be mentioned are the polyenes, which include carotenoids. While these are unlikely to be of technological relevance, they serve as model chemical systems for fundamental studies.

The requirement for a large singlet-triplet gap such that $2E_{T_1} \approx E_{S_1}$ is difficult to engineer from the standpoint of the exchange integral alone (Eq. (16.6)). Michl proposed another approach by considering diradicaloid systems. If one first considers $H_2$, then at large internuclear distances, the lowest triplet state is degenerate with the ground state. There will be some excited singlet state at a higher energy. As one brings the two atoms together, the triplet surface is repulsive while the ground state is attractive and forms a bond. As the coupling between the two radical sites is increased, the triplet state increases in energy and at some point will lie at half the energy of the excited state with respect to the ground state. Zethrenes form a biradicaloid class of compounds, which are of present interest as singlet fission candidates (Lukman et al. 2017; Liu et al. 2020).

An attractive class of compounds for technological application are those used as automotive pigments. These compounds are demonstrated to be photostable. The diketopyrrolopyrroles are photostable dyes used in organic photovoltaics (OPVs) and are closely related to the pigment known as Ferrari Red. They are under investigation as singlet fission materials (Levine et al. 2021; Mauck et al. 2017). Futhermore, as mentioned above, the pereylene diimide (PDI) class of compounds also shows promise. These are also used as automotive pigments. A selection of compounds known to undergo singlet fission is displayed in Figure 16.8.

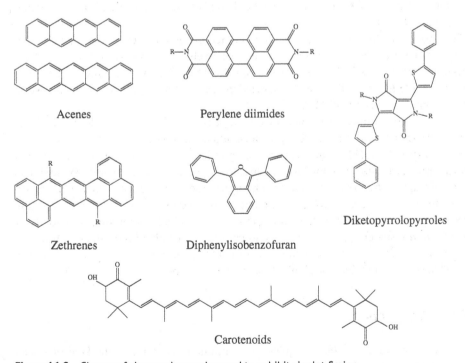

Acenes

Perylene diimides

Diketopyrrolopyrroles

Zethrenes

Diphenylisobenzofuran

Carotenoids

**Figure 16.8** Classes of chromophores observed to exhibit singlet fission.

## 16.7 Devices Reported to Date

There have been several reviews of singlet fission photovoltaics in recent years (Xia et al. 2017; Ehrler 2021; Baldacchino et al. 2022). It is not the aim of this section to revisit these in detail. Instead we will briefly outline the types of devices being considered, followed by an examination of the most promising avenue of research and development: a singlet fission silicon tandem solar cell.

As outlined in our recent review (Baldacchino et al. 2022), it is useful to classify singlet fission solar cells by the triplet acceptor material with which it is being used. The earliest demonstrations were in organic photovoltaics, with the notable milestone of an external quantum efficiency exceeding 100% (Congreve et al. 2013). Other structures include the use of inorganic quantum dots as triplet acceptors (Ehrler et al. 2012).

However, as alluded to, we believe the most promising avenue is the implementation of singlet fission on crystalline silicon solar cells. The reason this is so attractive is that it is clear that silicon technologies will continue to dominate the photovoltaic marketplace for decades to come. However, the 2022 International Technology Roadmap for Photovoltaic suggests that silicon tandems will comprise about 5% of market share by 2030. While issues such as stability and low efficiencies remain, there are good reasons for singlet fission to be considered alongside competing perovskite and III–V technologies.

There exist three conceptual designs for singlet fission silicon tandems. The first demonstration was a four-terminal parallel tandem device, which was an organic cell stacked upon a silicon cell (Pazos et al. 2017). Impressively, the authors showed an external quantum efficiency (EQE) in excess of 100%, although the overall efficiency was low (4.9%). This tandem architecture is the simplest to understand and is conceptually identical to any other four-terminal structure.

The second, by MacQueen et al. used a p–n junction approach where the singlet fission layer (tetracene) and crystalline silicon were respectively the p- and n-type materials (MacQueen et al. 2018). This device has the important difference that there are only two electrical contacts, while retaining limiting efficiencies in excess of 45%. The device in this work, however, had an efficiency of 11%.

Finally, in 2019, a direct triplet injection device was reported by a group at MIT (Einzinger et al. 2019). Here the singlet fission layer transports triplet *excitons* into a silicon device, as shown in Figure 16.6. This is the most promising architecture because no electrical current passes through the singlet fission layer. Instead, only excitons are transported to the silicon layer. This architecture requires minimal changes to the underlying silicon device. Importantly, this removes current/voltage-matching requirements for the tandem architecture since energy is transferred to silicon rather than current. Finally, if the singlet fission layer degrades to transparent products there is no parasitic absorption (as occurs in tetracene), the underlying silicon cell can operate as normal (Jiang et al. 2021). As such, there is a "boost" to the current while the singlet fission layer is active.

## 16.8 Prospects

The authors are confident that progress will continue to be made in the field of singlet fission photvoltaics, especially as applied to silicon, in the coming decade. There are presently two main camps working in the field. In one camp, researchers aim to harvest triplet excitons using semiconductor nanocrystals which then radiate energy into the photovoltaic silicon. The specific challenges in this implementation include optimizing the proportion of emitted light that can be harvested. This will be in excess of 50%, as it must be, due to the high refractive index of silicon as well as the singlet fission material, as compared to air. The emission quantum yield of the semiconductor nanocrystals is another concern. The usual way to increase this well above 50% is to coat the nanocrystal with a giant shell. And yet, this would create an insurmountable barrier to triplet injection from the singlet fission material. As such, researchers must find a way to make the electronic structure of the semiconductor nanocrystal accessible to the singlet fission material while also passivated with respect to surface traps. This is not an impossible task.

The other camp requires access to the electronic structure of silicon. For exciton injection into or electron transfer to silicon, the singlet fission material must necessarily interact with the electron states of silicon. And yet, high efficiencies in silicon have been achieved by judiciously passivating the surface of silicon. As such, progress in this area demands the discovery or design of a material that can simultaneously passivate silicon while mediating exciton transfer from the photovoltaic material. This is also not an impossible task.

## Author Biographies

**A/Prof. Murad Tayebjee** is an ARC Future Fellow at the School of Photovoltaic and Renewable Energy Engineering at UNSW and Deputy Leader of OMEGA Silicon. He completed a B. Sc. (Advanced) Hons I majoring in both physics and chemistry in 2009. He was awarded his PhD in physical chemistry from The University of Sydney in 2012. He then undertook two consecutive fellowships: an ARENA Fellowship at UNSW and a Marie Sklodowska-Curie Fellowship at The University of Cambridge.

A/Prof. Tayebjee leads the Spectre Research Group (SPECTroscopy, Renewables & Education), studying methods of efficient solar energy harvesting and renewable energy education. His work has been recently recognised with a NSW Young Tall Poppy Award and the 2023 Royal Australian Chemistry Institute Physical Chemistry Lectureship.

**Professor Timothy Schmidt** was educated at The University of Sydney, winning the University Medal for Theoretical Chemistry in 1997. He undertook his PhD at The University of Cambridge in the field of femtosecond spectroscopy under the supervision of the late Dr Gareth Roberts. He was a postdoctoral research associate of Prof. Dr John Paul Maier, FRS in Basel, Switzerland, where he researched highly unsaturated hydrocarbon molecules of astrophysical relevance. Tim returned to Australia in 2003 to take up a position at CSIRO researching artificial photosynthesis. He was appointed as a lecturer in the

School of Chemistry at The University of Sydney in 2004 and there rose to Associate Professor before moving to UNSW in 2014 as Professor and ARC Future Fellow. At UNSW he is Head of the School of Chemistry and Chief Investigator of the ARC Centre of Excellence in Exciton Science. Professor Schmidt has been the recipient of a number of awards for his research including the Coblentz Award (2010) for contributions to the science of molecular spectroscopy. He is a Fellow of the Royal Australian Chemical Institute, The Royal Society of Chemistry and the Royal Society of New South Wales.

## References

Baldacchino, A.J., Collins, M.I., Nielsen, M.P. et al. (2022). Singlet fission photovoltaics: progress and promising pathways. *Chemical Physics Reviews* 3: 021304

Burdett, J.J., Gosztola, D., and Bardeen, C.J. (2011). The dependence of singlet exciton relaxation on excitation density and temperature in polycrystalline tetracene thin films: kinetic evidence for a dark intermediate state and implications for singlet fission. *The Journal of Chemical Physics* 135: 214508

Chapin, D.M., Fuller, C.S., and Pearson, G.L. (1954). A new silicon *p-n* junction photocell for converting solar radiation into electrical power. *Journal of Applied Physics* 25: 676–677.

Congreve, D.N., Lee, J., Thompson, N.J. et al. (2013). External quantum efficiency above 100 in a singlet-exciton-fission-based organic photovoltaic cell. *Science* 340: 334–337.

Daiber, B., van den Hoven, K., Futscher, M.H., and Ehrler, B. (2021). Realistic efficiency limits for singlet-fission silicon solar cells. *ACS Energy Letters* 6: 2800–2808.

Dexter, D. (1979). Two ideas on energy transfer phenomena: ion-pair effects involving the OH stretching mode, and sensitization of photovoltaic cells. *Journal of Luminescence* 18-19: 779–784.

Dover, C.B., Gallaher, J.K., Frazer, L. et al. (2018). Endothermic singlet fission is hindered by excimer formation. *Nature Chemistry* 10: 305–310.

Ehrler, B. (2021). Emerging strategies to reduce transmission and thermalization losses in solar cells redefining the limits of solar power conversion efficiency. In: *Singlet Fission Solar Cells*, Chapter 15 (ed. J.S. Lissau and M. Madsen), 313–341. Sonderborg, Denmark: Springer Nature.

Ehrler, B., Wilson, M.W., Rao, A. et al. (2012). Singlet exciton fission-sensitized infrared quantum dot solar cells. *Nano Letters* 12: 1053–1057.

Einzinger, M., Wu, T., Kompalla, J.F. et al. (2019). Sensitization of silicon by singlet exciton fission in tetracene. *Nature* 571: 90–94.

Green, M.A., Dunlop, E.D., Hohl-Ebinger, J. et al. (2022). Solar cell efficiency tables (Version 60). *Progress in Photovoltaics: Research and Applications* 30: 687–701.

Hanna, M. and Nozik, A. (2006). Solar conversion efficiency of photovoltaic and photoelectrolysis cells with carrier multiplication absorbers. *Journal of Applied Physics* 100: 74510

Harcourt, R.D., Scholes, G.D., and Ghiggino, K.P. (1994). Rate expressions for excitation transfer. II. Electronic considerations of direct and through–configuration exciton resonance interactions. *The Journal of Chemical Physics* 101: 10521–10525.

Havlas, Z. and Michl, J. (2016). Guidance for mutual disposition of chromophores for singlet fission. *Israel Journal of Chemistry* 56: 96–106.

Hirst, L.C. and Ekins-Daukes, N.J. (2011). Fundamental losses in solar cells. *Progress in Photovoltaics: Research and Applications* 19: 286–293.

Jiang, Y., Nielsen, M.P., Baldacchino, A.J. et al. (2021). Singlet fission and tandem solar cells reduce thermal degradation and enhance lifespan. *Progress in Photovoltaics: Research and Applications* 29: 899–906.

Kasha, M. (1963). Energy transfer mechanisms and the molecular exciton model for molecular aggregates. *Radiation Research* 20: 55–70.

Levine, A.M., He, G., Bu, G. et al. (2021). Efficient free triplet generation follows singlet fission in diketopyrrolopyrrole polymorphs with goldilocks coupling. *The Journal of Physical Chemistry C* 125: 12207–12213.

Liu, X., Tom, R., Gao, S., and Marom, N. (2020). Assessing zethrene derivatives as singlet fission candidates based on multiple descriptors. *The Journal of Physical Chemistry C* 124: 26134–26143.

Lukman, S., Richter, J.M., Yang, L. et al. (2017). Efficient singlet fission and triplet-pair emission in a family of zethrene diradicaloids. *Journal of the American Chemical Society* 139: 18376–18385.

MacQueen, R.W., Liebhaber, M., Niederhausen, J. et al. (2018). Crystalline silicon solar cells with tetracene interlayers: the path to silicon-singlet fission heterojunction devices. *Materials Horizons* 5: 1065–1075.

Mauck, C.M., Hartnett, P.E., Wu, Y.-L. et al. (2017). Singlet fission within diketopyrrolopyrrole nanoparticles in water. *Chemistry of Materials* 29: 6810–6817.

Paci, I., Johnson, J.C., Chen, X. et al. (2006). Singlet fission for dye-sensitized solar cells: can a suitable sensitizer be found? *Journal of the American Chemical Society* 128: 16546–16553.

Pazos, L.M., Lee, J.M., Kirch, A. et al. (2017). A silicon-singlet fission tandem solar cell exceeding 100 external quantum efficiency with high spectral stability. *ACS Energy Letters* 2: 476–480.

Piland, G.B. and Bardeen, C.J. (2015). How morphology affects singlet fission in crystalline tetracene. *Journal of Physical Chemistry Letters* 6: 1841–1846.

Pusch, A., Bremner, S.P., Tayebjee, M.J.Y., and Daukes, N.J.E. (2021). Microscopic reversibility demands lower open circuit voltage in multiple exciton generation solar cells. *Applied Physics Letters* 118: 151103

Schmidt, T.W. (2019). A Marcus-Hush perspective on adiabatic singlet fission. *The Journal of Chemical Physics* 151: 054305

Shockley, W. and Queisser, H.J. (1961). Detailed balance limit of efficiency of p–n junction solar cells. *Journal of Applied Physics* 32: 510–519.

Singh, S., Jones, W.J., Siebrand, W. et al. (1965). Laser generation of excitons and fluorescence in anthracene crystals. *The Journal of Chemical Physics* 42: 330–342.

Tayebjee, M.J.Y., Gray-Weale, A.A., and Schmidt, T.W. (2012). Thermodynamic limit of exciton fission solar cell efficiency. *The Journal of Physical Chemistry Letters* 3: 2749–2754.

Tayebjee, M.J.Y., Clady, R.G.C.R., and Schmidt, T.W. (2013). The exciton dynamics in tetracene thin films. *Physical Chemistry Chemical Physics* 15: 14797–14805.

Tayebjee, M.J.Y., Mahboubi Soufiani, A., and Conibeer, G.J. (2014). Semi-empirical limiting efficiency of singlet-fission-capable polyacene/inorganic hybrid solar cells. *The Journal of Physical Chemistry C* 118: 2298–2305.

Tayebjee, M.J.Y., McCamey, D.R., and Schmidt, T.W. (2015). Beyond Shockley–Queisser: molecular approaches to high-efficiency photovoltaics. *The Journal of Physical Chemistry Letters* 6: 2367–2378.

Wilson, M.W., Rao, A., Clark, J. et al. (2011). Ultrafast dynamics of exciton fission in polycrystalline pentacene. *Journal of the American Chemical Society* 133: 11830–11833.

Xia, J., Sanders, S.N., Cheng, W. et al. (2017). Singlet fission: progress and prospects in solar cells. *Advanced Materials* 29: 1601652

Yoshikawa, K., Kawasaki, H., Yoshida, W. et al. (2017). Silicon heterojunction solar cell with interdigitated back contacts for a photoconversion efficiency over 26. *Nature Energy* 2: 17032

Yost, S.R., Lee, J., Wilson, M.W.B. et al. (2014). A transferable model for singlet-fission kinetics. *Nature Chemistry* 6: 492–497.

**Part Seven**

**Characterization and Measurements Methods**

# 17

# Temperature-Dependent Lifetime and Photoluminescence Measurements

*Ziv Hameiri and Yan Zhu*

*The University of New South Wales, Sydney, NSW, Australia*

## 17.1 Temperature-Dependent Lifetime Spectroscopy

Identification of defects in silicon wafers and cells is essential for their elimination (Graff 2013). Determination of their electrical properties is also required for evaluating the defects' impact on device performance (Coletti 2011; Schmidt et al. 2012). However, as the quality of silicon continues to improve, the concentration of bulk defects significantly decreases, making their detection very challenging. Nevertheless, even in a dilute concentration, some defects still have a substantial impact on solar cells' operation (Rein 2006). Lifetime spectroscopy refers to characterization techniques that extract defect parameters from charge carrier lifetime measurements (Rein 2006; Zhu and Hameiri 2021). If the recombination induced by a defect dominates the overall charge carrier recombination, the defect can be detected by lifetime spectroscopy regardless of its absolute concentration. This unique characteristic makes lifetime spectroscopy a powerful method for the identification of defects and the determination of their properties.

### 17.1.1 Lifetime Spectroscopy

The term "lifetime spectroscopy" describes several types of measurement techniques (Rein 2006; Zhu and Hameiri 2021). Temperature-dependent lifetime spectroscopy (TDLS) is performed at low injection at various temperatures (Hayamizu et al. 1991; Kirino et al. 1990), while injection-dependent lifetime spectroscopy (IDLS) is done by measurements across a range of excess carrier concentrations (Ferenczi et al. 1991; Horányi 1996; Walz et al. 1996). IDLS with temperature variation is referred to as T-IDLS (Birkholz et al. 2005; Schmidt 2003), whereas IDLS with sample doping variation is referred to as $N_{dop}$-IDLS (Macdonald and Cuevas 2001; Murphy et al. 2012; Sun et al. 2014). The first applications of lifetime spectroscopy were primarily based on TDLS (Hayamizu et al. 1991; Kaniava et al. 1995; Shimura 1986). IDLS grew to be the common lifetime spectroscopy with the development of the quasi-steady state photoconductance (QSS-PC) lifetime measurement technique (Sinton et al. 1996). However, it has been shown that IDLS is ambiguous for the parameterization of a Shockley–Read–Hall (SRH) defect (Hall 1952;

Shockley and Read 1952). An infinite number of sets of defect parameters can fit data obtained by IDLS equally well. These infinite sets can only be confined to a solution surface (Rein 2006; Rein et al. 2002). It was demonstrated that only T-IDLS and $N_{dop}$-IDLS can reduce the number of solutions to two, the true solution and a false solution in the opposite halves of the bandgap (Rein 2006). The full potential of TIDLS to characterize defects was further highlighted by the systematic study of (Rein et al. 2002). TIDLS was soon used to investigate the electrical properties of various defects in silicon (Inglese et al. 2016; Niewelt et al. 2015; Roth et al. 2008; Vargas et al. 2017).

### 17.1.2 Analysis Methods

Using the SRH expression, the injection-depended lifetime of a single-level defect can be presented as (Hall 1952; Shockley and Read 1952):

$$\tau(\Delta n) = \frac{\dfrac{p_0 + p_1 + \Delta n}{N_t \sigma_e v_e} + \dfrac{n_0 + n_1 + \Delta n}{N_t \sigma_h v_h}}{n_0 + p_0 + \Delta n} \tag{17.1}$$

where $n_0$ ($p_0$) is the electron (hole) concentration at thermal equilibrium, $N_t$ is the defect concentration, $\sigma_e$ ($\sigma_h$) is the electron (hole) capture cross-section, $v_e$ ($v_h$) is the electron (hole) thermal velocity, $n_1$ ($p_1$) is the electron (hole) concentration when the Fermi-level is at the defect energy level $E_t$. Thus, $n_1 = N_C \times \exp[-(E_C - E_t)/k_B \times T]$ and $p_1 = N_V \times \exp[-(E_t - E_V)/k_B \times T]$, where $N_C$ and $N_V$ are, respectively, the density of states in the conduction and valence band, $E_C$ is the conduction band edge, $E_V$ is the valence band edge, $k_B$ is the Boltzmann constant, and $T$ is the temperature.

Hence, the lifetime associated with the defect can be determined if the defect parameters ($E_t$, $\sigma_e$, and $\sigma_h$) and concentration ($N_t$) are known. Lifetime spectroscopy is essentially the reverse process where the defect parameters are *extracted* from the measured lifetime.

#### 17.1.2.1 Defect Parameterization Solution Surface

The common method to analyze lifetime measurements is the defect parameterization solution surface (DPSS) (Rein 2006). Below, its principle is demonstrated with a set of simulated lifetime curves of a single-level defect in a p-type silicon wafer at three temperatures. The defect and bulk parameters, as well as the used models, are included in the figure's caption. During the DPSS process, Eq. (17.1) is used to fit each injection-dependent lifetime curve using three independent defect parameters: $E_t$, $N_t \times \sigma_e$ and the ratio of capture cross sections $k = \sigma_e/\sigma_h$. However, instead of fitting the three parameters simultaneously, $E_t$ is fixed, and the values of $N_t \times \sigma_e$ and $k$ are obtained from the optimal fit. This process is repeated while the value of $E_t$ is swept within the bandgap. The resulting DPSS curves plot the obtained $N_t \times \sigma_e$ and $k$ as a function of $E_t$. Figure 17.2 presents the DPSS curves obtained from the simulated lifetime of Figure 17.1. Note that $\sigma_e$ and $\sigma_h$ are coupled with $N_t$ in Eq. (17.1). Hence, it is not possible to extract the absolute value of each of these parameters from lifetime spectroscopy (only their product can be extracted).

One important feature of the DPSS curves is that the quality of the fit is *identical* along the curve. Thus, any combination of ($E_t$, $N_t \times \sigma_e$, $k$) along the curves provides a

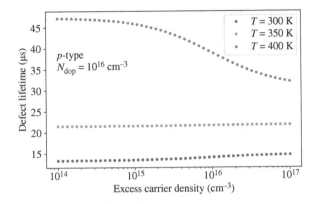

**Figure 17.1** Simulated injection-dependent lifetime associated with a defect in a p-type silicon wafer ($N_{dop} = 10^{16}$ cm$^{-3}$) at three temperatures. Models: The density of states is calculated using the model by (Couderc et al. 2014), while the thermal velocity is calculated using the model by (Green 1990) and the bandgap is calculated with the model of (Pässler 2002) and the bandgap narrowing with the model by (Schenk 1998). Defect parameters: $E_t = E_c - 0.8$ eV, $\sigma_e = \sigma_{e0} \times T^{-3}$ with $\sigma_{e0} = 10^{-7}$ cm$^2$, $\sigma_h = \sigma_{h0} \times T^{-3}$ with $\sigma_{h0} = 10^{-6}$ cm$^2$, $N_t = 10^{12}$ cm$^{-3}$.

similar fit to the measured injection-dependent lifetime. Therefore, it is impossible to determine ($E_t$, $N_t \times \sigma_e$, $k$) from an injection-dependent lifetime measurement at a single temperature.

To reduce ambiguity in the determination of defect parameters, IDLS measurements at various temperatures (or various doping concentrations) are required. Here, we use the temperature variation. As can be seen (Figure 17.2), the three DPSS curves for $k$ intersect at one point that corresponds to the simulated values of $E_t$ and $k$. However, the DPSS curves of $N_t \times \sigma_e$ do not intersect at one point, highlighting a significant limitation of the DPSS method. As the simulated $\sigma_e$ is temperature-dependent, a sharp intersect cannot be obtained. However, this is not the case for $k$, which is temperature-independent in this example. Nevertheless, the value of $N_t \times \sigma_e$ can still be extracted at each temperature after the determination of $E_t$ from the DPSS curves for $k$. In the upper half of the bandgap (i.e. in Figure 17.2, for low values of $E_c - E_t$), the DPSS-$k$ curves have a set of diffused intersections, referred to as the "false solution." In T-IDLS, it is possible to distinguish the "false solution" from the "true solution" if the asymmetry in capture cross sections is large ($k > 10$ or $k < 0.1$) (Rein 2006). In real measurements, it is quite difficult to distinguish the two solutions if $0.1 < k < 10$, and it is impossible when $N_{dop}$-IDLS is used.

As can be seen in Figure 17.2, the DPSS curves at 300 and 350 K are continuous curves covering the mid-gap, whereas the DPSS curve at 400 K is split between two sections that are near the two band edges. To reduce measurement uncertainty, it is preferred to have both types of curves ("continuous" and "split"), as it is often challenging to identify the intersection if only one type of curve is obtained (Rein 2006). Note that the transition temperature between the two types can be easily calculated for a given set of defect parameters.

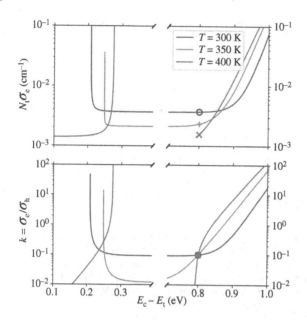

**Figure 17.2** DPSS curves obtained from simulated injection-dependent lifetime curves at three different temperatures. The symbols ("O", "+", and "×") indicate the parameters of the simulated defect.

### 17.1.2.2 Defect Parameter Contour Map

A different method of visualization for the defect parameter solution surface has been proposed by (Bernardini et al. 2017, 2018; Narland et al. 2018), named the defect parameter contour map (DPCM). In this method, a discrete space of $(E_t, k)$ combinations is created. For each $(E_t, k)$ combination, the value of $N_t \times \sigma_e$ is obtained from the optimal fitting of all the injection-dependent lifetime measurements. The resulting map of residuals in the $(E_t, k)$ space provides a visualization of possible solutions for the defect parameters. Figure 17.3 presents the DPCM of the simulated lifetime using the same defect parameters as in Figure 17.1, except for the capture cross sections that are now defined as temperature

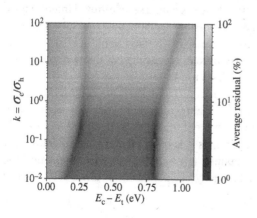

**Figure 17.3** Defect parameter contour map for a set of simulated T-IDLS data. The red dot indicates the point with the lowest average residual.

independent ($\sigma_e = 10^{-14}$ cm$^2$ and $\sigma_h = 10^{-13}$ cm$^2$). Clear visualization of possible ($E_t$, $k$) combinations is obtained with two identified bands of low residual, including the point with the lowest residual (marked with a red dot) that well agrees with the ground truth values.

One major difference between DPSS and DPCM is that, in DPSS, each set of IDLS data is fitted independently, whereas in DPCM, all the IDLS data are fitted simultaneously. In DPSS analysis, the defect parameters can be obtained even if $N_t \times \sigma_e$ changes with temperature as long as $k$ is constant at all measurement conditions. In contrast, DPCM analysis requires both $N_t \times \sigma_e$ and $N_t \times \sigma_h$ to be constant as well. However, this assumption is often not valid: the capture cross sections are often temperature dependent (T-IDLS), or $N_t$ is different in different samples ($N_{dop}$-IDLS). Bernardini et al. tried to address this issue by assigning an error to the extracted $k$, however, this requires knowledge regarding temperature dependency of capture cross sections (Bernardini et al. 2018).

### 17.1.2.3 Linearization-Based Methods

Apart from DPSS and DPCM, other alternative methods for defect parameterization have been proposed (Morishige et al. 2016; Murphy et al. 2012; Zhu et al. 2017). These methods are based on linearization of the SRH lifetime (Murphy et al. 2012; Voronkov et al. 2010, 2011). For p-type silicon fulfilling conditions of $n_0 \ll p_0$ and $n_0 \ll \Delta n$, the SRH lifetime in Eq. (17.1) can be transformed into a linear form by introducing the variable $X = n/p = (n_0 + \Delta n)/(p_0 + \Delta n)$, where $n(p)$ is the total free electron (hole) concentration:

$$\tau(X) = \frac{\left(1 + \dfrac{kn_1 v_e}{p_0 v_h} + \dfrac{p_1}{p_0}\right) + \left(\dfrac{k v_e}{v_h} - \dfrac{kn_1 v_e}{p_0 v_h} - \dfrac{p_1}{p_0}\right) X}{N_t \sigma_e v_e} \qquad (17.2)$$

An illustration of this linearization process is made in Figure 17.4, where the simulated lifetime of Figure 17.1 is plotted against $X$. Using this method, each set of IDLS data is simply

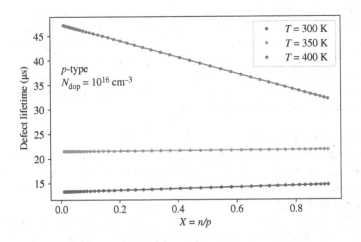

**Figure 17.4** Simulated injection-dependent defect lifetime plotted as a function of $X$ at three different temperatures. The solid lines indicate the linear fitting of the data.

fitted as a straight line, with a slope $m$ and intercept $b$, both functions of $(E_t, N_t \times \sigma_e, k)$:

$$m = \left( \frac{k}{v_h} - \frac{kn_1}{p_0 v_h} - \frac{p_1}{p_0 v_e} \right) / (N_t \sigma_e)$$

$$b = \left( \frac{1}{v_e} + \frac{kn_1}{p_0 v_h} + \frac{p_1}{p_0 v_e} \right) / (N_t \sigma_e) \tag{17.3}$$

(Morishige et al. 2016) combined the linearization and the DPSS concept and transferred Eq. (17.3) into:

$$k = \frac{\dfrac{m}{m+b} + \dfrac{p_1}{p_0}}{1 - \dfrac{n_1}{p_0} - \dfrac{m}{m+b}} \times \frac{v_h}{v_e}$$

$$N_t \sigma_e = \frac{\left[ \left( \dfrac{1}{v_e} + \dfrac{1}{p_0} \right) \left( \dfrac{kn_1 + \dfrac{p_1 v_h}{v_e}}{\dfrac{v_h}{v_e}} \right) \right]}{b} \tag{17.4}$$

As can be seen, $k$ is now expressed as a function of $m$, $b$, $E_t$. Therefore, when $m$ and $b$ are obtained from the linear fit, $k$ can be calculated for any given $E_t$. After the determination of $k$, $N_t \times \sigma_e$ can also be easily obtained as a function of $E_t$. This method greatly simplifies the original DPSS method as there is no need to fit the lifetime measurement for each $E_t$. Instead, only one fit of the linearized lifetime is required, and all DPSS curves are obtained from simple calculations using the extracted slope and intercept.

#### 17.1.2.4 The Newton–Raphson Method

Another approach to extract $(E_t, N_t \times \sigma_e, k)$ is to consider Eq. (17.3) as a system of two equations of three unknown parameters. Since the number of unknowns is larger than the number of equations, there is an infinite number of solutions. This provides a mathematical explanation for the ambiguity of a single IDLS. By combining IDLS from two temperatures or two doping conditions, the number of equations becomes four and exceeds the number of unknowns. Therefore, a unique solution can be obtained using, for example, the Newton–Raphson method (Raphson 1702), as demonstrated by Zhu et al. (2017). Similar to other methods, two sets of $(E_t, N_t \times \sigma_e, k)$ are obtained. Since this approach attempts to find a single set of $(E_t, N_t \times \sigma_e, k)$ for all the measurements, it is limited to only cases where both $N_t \times \sigma_e$ and $N_t \times \sigma_h$ are constant (similar to DPCM).

#### 17.1.2.5 Machine Learning-Based Methods

Recently, Buratti et al. (2020) proposed a new method for extracting the defect parameters using machine learning. By training machine learning algorithms with a large set of simulated SRH lifetime data, the method can predict the defect parameters with similar accuracy as the other methods. Furthermore, the machine learning approach can also predict the bandgap half (Buratti et al. 2019, 2020) and seems to handle cases where k has temperature dependency well – both capabilities are a significant improvement compared to the methods reviewed.

### 17.1.3 Challenges

#### 17.1.3.1 Extraction of the Defect-Associated Lifetime

The accuracy of any of the above-mentioned analysis methods strongly depends on the quality of the defect-associated lifetime that is extracted from the measured *effective* lifetime. The accuracy of the extracted defect lifetime strongly depends on the dominancy of the defect-associated recombination. If the defect-associated recombination is responsible for only a small portion of the overall recombination in the sample, the uncertainty associated with the extracted defect lifetime significantly increases. Other main challenges that may impact the extraction of the defect-associated lifetime are nonuniform distribution of the defect within the detection area, metastable defects (especially during temperature-dependent measurements), and the stability of surface passivation quality during measurements.

#### 17.1.3.2 Temperature Dependencies of Capture Cross Sections

Temperature dependency of capture cross sections poses a significant challenge for the T-IDLS analysis. Although the above demonstration of the DPSS method indicates that defect parameters can be accurately extracted even if the capture cross sections are temperature dependent, it should be noted that in our example, $\sigma_e$ and $\sigma_h$ have the same temperature dependency and therefore, $k$ is temperature independent. However, this is not the case for most defects (Graff 2013; Juhl et al. 2018; Rougieux et al. 2018). Compared to T-IDLS, $N_{dop}$-IDLS does not suffer from this limitation, however, it requires samples with different doping concentrations and the same dominant defect.

One possible approach to address the challenge of temperature dependencies of capture cross-sections is to combine T-IDLS with other measurement techniques; for example, obtaining the defect energy level or the capture cross-section of the majority carriers from deep-level transient spectroscopy (DLTS). In this case, IDLS can be used to extract the information of the minority carrier capture cross section (Zhu et al. 2021). As discussed, it appears the machine learning method may, at least partly, overcome this challenge.

#### 17.1.3.3 Two-Level (or More) Defects

All the methods discussed above use the SRH recombination statistics, assuming defects with a single energy level within the bandgap. However, most of the characterized defects have two or more energy levels (Rougieux et al. 2018) that follow the Sah-Shockley recombination statistics (Sah and Shockley 1958). Thus, it is challenging for any of the reviewed methods to accurately extract the parameters of two (or more)-level defects.

It should be noted that, in many cases, the measured injection-dependent lifetime curves cannot be fitted with a single SRH defect (Bothe and Schmidt 2006; Macdonald et al. 2002; Rougieux et al. 2015; Warta 2006). In these cases, a secondary SRH defect is assumed to be present in the sample. Fitting of the lifetime with two SRH defects usually greatly improves fitting quality. However, fitting quality can also be improved by using a two-level defect. Although, in some cases, it is possible to distinguish two-level defects from two single-level defects in IDLS, there are cases that it is impossible. This is a severe limitation, as it has been demonstrated by (Zhu et al. 2019; Zhu et al. 2020) that misinterpreting a two-level defect as two single-level defects can result in a significant error in the extracted defect parameters. A proper parameterization process for two-level defects in IDLS has been proposed by Zhu et al. (2020).

### 17.1.4 T-IDLS Like Measurements

#### 17.1.4.1 Suns-$V_{oc}$(T)

Commonly, photoconductance (PC)-based measurements are used for lifetime measurements (Schroder 2006; Sinton et al. 1996). However, samples with a significant amount of metal on the surface (i.e. solar cells) cannot be measured using this method. Recently, a method based on temperature-dependent Suns-$V_{oc}$ [Suns-$V_{oc}$(T)] was suggested for the purpose of extracting the defect parameters of metalized samples (Jafari et al. 2022). A Suns-$V_{oc}$ measures the open-circuit voltage ($V_{oc}$) at different illumination intensities that are easily converted to current. Importantly, lifetime curves can also be extracted from Suns-$V_{oc}$ measurements, and unlike PC-based lifetime measurements, the obtained lifetime is not overestimated due to artifacts that impact low injection levels – such as minority carrier traps (Macdonald and Cuevas 1999) and depletion region modulation (DRM) (Cousins et al. 2004). The Suns-$V_{oc}$(T) method has already been employed to investigate light-induced degradation (LID) in gallium (Ga)-doped cells and, for the first time, to extract using DPSS the parameters of defect associated with this degradation in Ga-doped wafers (Jafari et al. 2022).

#### 17.1.4.2 Investigation of Surface Passivation

The sections above discussed the use of TDLS to investigate bulk defects. However, lifetime spectroscopy can also be applied to investigate surface passivation properties. Bernardini and Bertoni utilized T-IDLS to study the surface recombination velocity (SRV) at the interface between silicon and amorphous silicon (Bernardini and Bertoni 2019). They concluded that the passivation provided by amorphous silicon degrades due to the effusion of hydrogen from the passivation layer. Nie et al. investigated the recombination at the silicon–silicon dioxide ($SiO_2$) interface using T-IDLS (Nie et al. 2021). Using the extended SRH recombination model, they extracted the silicon-$SiO_2$ interface defects' parameters and determined the value and the temperature-dependency of the capture cross-sections at this interface (Nie et al. 2021).

## 17.2 Temperature-Dependent Photoluminescence Imaging

### 17.2.1 Photoluminescence Imaging

Photoluminescence (PL)-based methods have been extensively used for decades to characterize a wide variety of photovoltaic (PV) materials. However, the development of PL imaging techniques in 2005 (Trupke et al. 2006) has completely revolutionized the characterization of silicon-based PV across the entire value chain. Today, PL imaging is a standard inspection method in silicon PV from the starting point of ingots and bricks (Mitchell et al. 2011) through to wafers (Trupke et al. 2006), cells (Glatthaar et al. 2010; Hameiri and

Chaturvedi 2013), and modules (Bhoopathy et al. 2018; Zafirovska et al. 2016). PL images capture a wide variety of essential electrical properties (Michl et al. 2014). Most obviously, as a result of the direct correlation between PL emission and the separation of the quasi-Fermi levels (Würfel et al. 1995), PL can image one of the most critical parameters of a PV device, namely its $V_{oc}$. PL imaging measurements are also used to obtain the distribution of shunt resistance (Augarten et al. 2013), saturation currents (Glatthaar et al. 2009; Hameiri and Chaturvedi 2013), local ideality factor (Hameiri et al. 2013), and many more (Michl et al. 2014; Trupke et al. 2012). However, to date, only rarely have PL images been taken at different temperatures.

### 17.2.2 T-IDLS Using Photoluminescence Imaging

Spatially resolved temperature- and injection-dependent lifetime spectroscopy (T-IDLS) has been attempted by measuring PL images at various excitation light intensities and at various temperatures (Hameiri et al. 2015; Haug et al. 2019; Mundt et al. 2015). Since PL intensity is related to the separation of quasi-Fermi levels, which is directly related to the excess carrier concentration, PL images can be converted into carrier concentration images (Kiliani et al. 2011; Sio et al. 2014). With knowledge of the excitation light intensity and the sample's absorption, generation rate in PL imaging measurements can be obtained. This can further convert the carrier concentration images into lifetime images (Giesecke et al. 2011; Herlufsen et al. 2008; Trupke et al. 2006). By measuring PL images at various excitation light intensities, the injection-dependent lifetime at each pixel can be obtained. By further measuring at various temperatures, T-IDLS analysis can be done at each pixel of the resulting images. This spatially resolved T-IDLS analysis is particularly useful when the defect exhibits a nonuniform distribution in the measured samples, such as the ring defects in n-type Cz silicon (Coletti et al. 2014).

However, one needs to carefully evaluate the impact of lateral carrier flow on the accuracy of injection-dependent lifetime data extracted from an individual pixel (Phang et al. 2016; Sio et al. 2014; Zhu et al. 2018). As mentioned previously, inaccurate lifetime data will lead to inaccurate defect parameterization from T-IDLS.

### 17.2.3 Spatially Resolved Temperature Coefficients

The temperature has a critical impact on the performance of solar cells (Dupré et al. 2015, 2017; Green 2003) mainly due to the temperature dependence of the bandgap energy ($E_g$) (Steinkemper et al. 2017). In silicon, $E_g$ reduces with increasing temperature. As a result, the short-circuit current increases while the $V_{oc}$ reduces at higher temperatures. The temperature dependency of the electrical parameters of solar cells is often described using temperature coefficients (TCs) (Dupré et al. 2017). These coefficients are commonly used to predict energy production under field conditions. For silicon solar cells, $V_{oc}$ variation dominates the temperature sensitivity of cell efficiency (Green 2003). The absolute TC of the $V_{oc}$ is expressed as (Green 1998):

$$TC(V_{oc}) = \frac{dV_{oc}}{dT} = -\frac{\dfrac{E_{g0}}{q} - V_{oc} + \dfrac{\gamma k_B T}{q}}{T} \tag{17.5}$$

where $E_{g0}$ is $E_g$ extrapolated to 0 K, $q$ is the elementary charge, and $\gamma$ represents the temperature dependency of the diode saturation current density $J_0$ (Dupré et al. 2017; Green 1998) and contains information regarding the dominant recombination mechanisms in the material (Dupré et al. 2017).

Temperature dependence of solar cells is normally reported as an *average* value for the entire cell (Berthod et al. 2019; Dupré et al. 2015; Steinkemper et al. 2017). However, since many other solar cell parameters vary across the cell area, it can reasonably be expected that TC is not uniform in many cases. There has been growing but as-yet limited, research into spatially resolved temperature dependence across wafers and solar cells (Eberle et al. 2019, 2018; Haug et al. 2017). Existing studies show a large variation in temperature sensitivity across wafers and cells. It has been shown that temperature sensitivity increases, particularly for $V_{oc}$, in cell areas with a high concentration of impurities (Eberle et al. 2018). Surprisingly, a reduction in temperature sensitivity of $V_{oc}$ has been identified in regions containing dislocation clusters (Eberle et al. 2019). Nie et al. investigated the impact of position along silicon bricks and dislocation densities on the temperature sensitivity of wafers and cells (Nie et al. 2020). They demonstrated that dislocated areas show both high and low-temperature sensitivity and that $\gamma$ exhibits very low values (even negative) in dislocated regions with low-temperature sensitivity. They also found that TC($V_{oc}$) is less negative with increasing brick height (Nie et al. 2020).

Figure 17.5 shows a schematic of a temperature-dependent calibrated PL imaging system. The system consists of a camera [often a silicon charge-coupled device (CCD)] and an illumination source – in this case, a laser. Optical filters are placed in front of the camera to avoid detection of any reflected excitation light. A temperature-controlled stage is used to heat the wafers and cells to higher temperatures.

In this example, the calibration of PL images is achieved using the QSS-PL front detection method (Dumbrell et al. 2018a, 2018b). PL emission from a selected region is focused on a PL detector via a lens. A filter set is attached to the lens to ensure detection of only

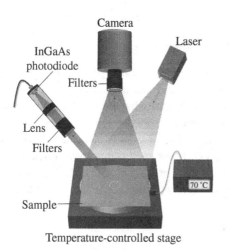

**Figure 17.5** Schematic of a temperature-dependent calibrated PL imaging system (not to scale).

Camera

Laser

InGaAs photodiode

Filters

Lens

Filters

70 °C

Sample

Temperature-controlled stage

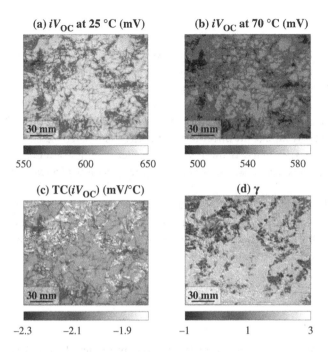

**Figure 17.6** Spatially resolved $iV_{oc}$ at 25 °C (a), $iV_{oc}$ at 70 °C (b), TC($iV_{oc}$) (c), and $\gamma$ (d) of a $1.8 \pm 0.1\,\Omega$ cm p-type multicrystalline wafer under 0.5 sun illumination.

band-to-band PL emission from the samples. A second photodiode is used to monitor the incident photon flux during measurements. Both photodiodes are connected to low-noise pre-amplifiers. The main advantage of this QSS-PL front detection system is that it can calibrate both metallized and non-metallized samples without the impact of trapping or DRM. This method provides both local and global temperature characteristics of wafers and cells.

Figure 17.6 depicts an example of the various parameters obtained by temperature-dependent PL imaging. Figure. 17.6a presents implied $V_{oc}$ ($iV_{oc}$) map at 25 °C of a $1.8 \pm 0.1\,\Omega$ cm p-type multicrystalline wafer, while Figure 17.6b shows the $iV_{oc}$ map of the same wafer at 70 °C, and the calculated TC($iV_{oc}$) is displayed in Figure 17.6c. A large variation of TC($iV_{oc}$) across the wafer is observed. It seems that TC($iV_{oc}$) is more negative in grain boundaries and some dislocation clusters, indicating a larger reduction of $iV_{oc}$ with increasing temperature in these regions compared to other areas across the wafer. The majority of dislocated regions show lower temperature sensitivity, which contradicts the common belief that temperature sensitivity is expected to increase with decreasing $iV_{oc}$ (Green and Emery 1982; Dupré et al. 2017). Figure 17.6d presents a $\gamma$ map calculated by applying Eq. (17.5) to each pixel. Lower $\gamma$ values are observed at dislocation clusters.

Figure 17.6 clearly highlights the importance of performing spatially resolved analysis to assess the TC of silicon wafers. This provides more detailed information about the specific samples' electrical properties in comparison to global single-value measurements.

## 17.3 Summary

Temperature-dependent measurements provide a wide range of information that cannot be obtained by the more common room-temperature measurements.

T-IDLS is a powerful technique to extract critical parameters of recombination active defects. As discussed, standard IDLS measurements can provide only a "surface" of solutions, whereas T-ILDS can greatly reduce this ambiguity. Nevertheless, when using T-IDLS, special care should be given to (a) evaluate the uncertainties in *defect*-associated recombination lifetime extracted from measured *effective* lifetime, (b) consider the impact of temperature dependencies of capture cross sections, and (c) consider the possibility of a two-level (or more) defect.

Temperature-dependent PL imaging is a useful method for various applications, one of which is imaging of the $iV_{oc}$ TC of metallized and non-metallized samples. This spatially resolved approach enables assessment of both local and global temperature characteristics of wafers and cells and provides more information regarding the material properties than conventional global measurements.

## Author Biographies

**Ziv Hameiri** was awarded his PhD in 2011 from the University of New South Wales (UNSW). He then joined the National University of Singapore (NUS) for three years before returning to UNSW as a Research Fellow. In the last decade, his research has focused on the development of novel characterization methods for photovoltaic devices and applying them to investigate loss mechanisms in silicon and non-silicon cells. He and his group developed a unique temperature- and injection-dependent lifetime spectroscopy (TIDLS) that has been used by researchers across the world for defects in silicon wafers and cells.

**Yan Zhu** has been working on the characterization of solar cells for more than five years. He received his Master's degree in Renewable Energy Science and Technology from École Polytechnique in France in 2015. He then started his PhD at the School of Photovoltaic and Renewable Energy Engineering of the University of New South Wales (UNSW). His PhD thesis focuses on the characterization of defects in silicon solar cells. Yan is now an ACAP (Australian Centre for Advanced Photovoltaics) postdoctoral fellow at UNSW. He continues to work on developing new characterization techniques of both silicon and non-silicon semiconductors for photovoltaic applications.

## References

Augarten, Y., Trupke, T., Lenio, M. et al. (2013). Calculation of quantitative shunt values using photoluminescence imaging. *Progress in Photovoltaics: Research and Applications* 21 (5): 933–941.

Bernardini, S. and Bertoni, M.I. (2019). Insights into the degradation of amorphous silicon passivation layer for heterojunction solar cells. *Physica Status Solidi A-Applications and Materials Science* 216 (4): 1800705.

Bernardini, S., Naerland, T.U., Blum, A.L. et al. (2017). Unraveling bulk defects in high-quality c-Si material via TIDLS. *Progress in Photovoltaics: Research and Applications* 25 (3): 209–217.

Bernardini, S., Naerland, T.U., Coletti, G., and Bertoni, M.I. (2018). Defect parameters contour mapping: a Powerful tool for lifetime spectroscopy data analysis. *Physica Status Solidi B* 255 (8): 1800082.

Berthod, C., Kristensen, S.T., Strandberg, R. et al. (2019). Temperature sensitivity of multicrystalline silicon solar cells. *IEEE Journal of Photovoltaics* 9: 957.

Bhoopathy, R., Kunz, O., Juhl, M. et al. (2018). Outdoor photoluminescence imaging of photovoltaic modules with sunlight excitation. *Progress in Photovoltaics: Research and Applications* 26: 69–73.

Birkholz, J.E., Bothe, K., Macdonald, D., and Schmidt, J. (2005). Electronic properties of iron-boron pairs in crystalline silicon by temperature-and injection-level-dependent lifetime measurements. *Journal of Applied Physics* 97 (10): 103708.

Bothe, K. and Schmidt, J. (2006). Electronically activated boron-oxygen-related recombination centers in crystalline silicon. *Journal of Applied Physics* 99 (1): 013701.

Buratti, Y., Dick, J., Gia, Q. L., & Hameiri, Z. (2019). A machine learning approach to defect parameters extraction: using random forests to inverse the Shockley Read Hall equation. Paper Presented at 46th IEEE Photovoltaic Specialists Conference.

Buratti, Y., Le Gia, Q.T., Dick, J. et al. (2020). Extracting bulk defect parameters in silicon wafers using machine learning models. *NPJ Computational Materials* 6 (1): 1–8.

Coletti, G. (2011). *Impurities in Silicon and their Impact on Solar Cell Performance*. Eurosolare SPA.

Coletti, G., Manshanden, P., Bernardini, S. et al. (2014). Removing the effect of striations in n-type silicon solar cells. *Solar Energy Materials and Solar Cells* 130: 647–651.

Couderc, R., Amara, M., and Lemiti, M. (2014). Reassessment of the intrinsic carrier density temperature dependence in crystalline silicon. *Journal of Applied Physics* 115 (9): 093705.

Cousins, P., Neuhaus, D., and Cotter, J. (2004). Experimental verification of the effect of depletion-region modulation on photoconductance lifetime measurements. *Journal of Applied Physics* 95 (4): 1854–1858.

Dumbrell, R., Juhl, M.K., Trupke, T., and Hameiri, Z. (2018a). Extracting metal contact recombination parameters from effective lifetime data. *IEEE Journal of Photovoltaics* 8 (6): 1413–1420.

Dumbrell, R., Juhl, M. K., Trupke, T., & Hameiri, Z. (2018b). Extracting surface saturation current density from lifetime measurements of samples with metallized surfaces. Paper Presented at 7th World Conference on Photovoltaic Energy Conversion.

Dupré, O., Vaillon, R., and Green, M.A. (2015). Physics of the temperature coefficients of solar cells. *Solar Energy Materials and Solar Cells* 140: 92–100.

Dupré, O., Vaillon, R., and Green, M.A. (2017). *Thermal Behaviour of Photovoltaic Devices*. Springer.

Eberle, R., Haag, S.T., Geisemeyer, I. et al. (2018). Temperature coefficient imaging for silicon solar cells. *IEEE Journal of Photovoltaics* 8 (4): 930–936.

Eberle, R., Fell, A., Mägdefessel, S. et al. (2019). Prediction of local temperature-dependent performance of silicon solar cells. *Progress in Photovoltaics: Research and Applications* 27 (11): 999–1006.

Ferenczi, G., Pavelka, T., and Tüttô, P. (1991). Injection level spectroscopy: A novel non-contact contamination analysis technique in silicon. *Japanese Journal of Applied Physics* 30: 3630–3633.

Giesecke, J.A., Schubert, M.C., Michl, B. et al. (2011). Minority carrier lifetime imaging of silicon wafers calibrated by quasi-steady-state photoluminescence. *Solar Energy Materials and Solar Cells* 95 (3): 1011–1018.

Glatthaar, M., Giesecke, J., Kasemann, M. et al. (2009). Spatially resolved determination of the dark saturation current of silicon solar cells from electroluminescence images. *Journal of Applied Physics* 105 (11): 113110.

Glatthaar, M., Haunschild, J., Zeidler, R. et al. (2010). Evaluating luminescence based voltage images of silicon solar cells. *Journal of Applied Physics* 108 (1): 014501.

Graff, K. (2013). *Metal Impurities in Silicon-Device Fabrication*, vol. vol. 24. Springer Science and Business Media.

Green, M.A. (1990). Intrinsic concentration, effective densities of states, and effective mass in silicon. *Journal of Applied Physics* 67 (6): 2944–2954.

Green, M.A. (1998). *Solar Cells: Operating Principles, Technology and System Application.* Sydeny: University of New South Wales.

Green, M.A. (2003). General temperature dependence of solar cell performance and implications for device modelling. *Progress in Photovoltaics: Research and Applications* 11 (5): 333–340.

Green, M. and Emery, K.A. (1982). Silicon solar cells with reduced temperature sensitivity. *Electronics Letters* 18 (2): 97–98.

Hall, R.N. (1952). Electron-hole recombination in germanium. *Physical Review* 87 (2): 387–387.

Hameiri, Z. and Chaturvedi, P. (2013). Spatially resolved electrical parameters of silicon wafers and solar cells by contactless photoluminescence imaging. *Applied Physics Letters* 102 (7): 073502.

Hameiri, Z., Chaturvedi, P., and McIntosh, K. (2013). Imaging the local ideality factor by contactless photoluminescence measurement. *Applied Physics Letters* 103 (2): 023501.

Hameiri, Z., Juhl, M. K., Carlaw, R., & Trupke, T. (2015). Spatially resolved lifetime spectroscopy from temperature-dependent photoluminescence imaging. Paper presented at 42nd IEEE Photovoltaic Specialist Conference.

Haug, H., Søndenå, R., Wiig, M.S., and Marstein, E.S. (2017). Temperature dependent photoluminescence imaging calibrated by photoconductance measurements. *Energy Procedia* 124: 47–52.

Haug, H., Søndenå, R., Berg, A., and Wiig, M.S. (2019). Lifetime spectroscopy with high spatial resolution based on temperature- and injection dependent photoluminescence imaging. *Solar Energy Materials and Solar Cells* 200: 109994.

Hayamizu, Y., Hamaguchi, T., Ushio, S. et al. (1991). Temperature dependence of minority-carrier lifetime in iron-diffused p-type silicon wafers. *Journal of Applied Physics* 69 (5): 3077–3081.

Herlufsen, S., Schmidt, J., Hinken, D. et al. (2008). Photoconductance-calibrated photoluminescence lifetime imaging of crystalline silicon. *Physica Status Solidi – Rapid Research Letters* 2 (6): 245–247.

Horányi, T.S. (1996). Identification possibility of metallic impurities in p-type silicon by lifetime measurement. *Journal of The Electrochemical Society* 143 (1): 216.

Inglese, A., Lindroos, J., Vahlman, H., and Savin, H. (2016). Recombination activity of light-activated copper defects in p-type silicon studied by injection- and temperature-dependent lifetime spectroscopy. *Journal of Applied Physics* 120 (12): 125703.

Jafari, S., Abbott, M., Zhang, D. et al. (2022). Bulk defect characterization in metalized solar cells using temperature-dependent Suns-$V_{oc}$ measurements. *Solar Energy Materials and Solar Cells* .

Juhl, M. K., Heinz, F. D., Coletti, G., Macdonald, D., Rougieux, F. E., Schindle, F., Niewelt, T., Schubert, M. C., & Ise, F. (2018). An open source-based repository for defects in silicon. 7th World Conference on Photovoltaic Energy Conversion, 0328-0332.

Kaniava, A., Rotondaro, A.L.P., Vanhellemont, J. et al. (1995). Recombination activity of iron-related complexes in silicon studied by temperature dependent carrier lifetime measurements. *Applied Physics Letters* 67 (26): 3930.

Kiliani, D., Micard, G., Steuer, B. et al. (2011). Minority charge carrier lifetime mapping of crystalline silicon wafers by time-resolved photoluminescence imaging. *Journal of Applied Physics* 110 (5): 54508.

Kirino, Y., Buczkowski, A., Radzimski, Z.J. et al. (1990). Noncontact energy level analysis of metallic impurities in silicon crystals. *Applied Physics Letters* 57 (26): 2832–2834.

Macdonald, D. and Cuevas, A. (1999). Trapping of minority carriers in multicrystalline silicon. *Applied Physics Letters* 74 (12): 1710–1712.

Macdonald, D., & Cuevas, A. (2001). *Lifetime spectroscopy of FeB pairs in silicon.* 11th Workshop of Crystalline Silicon Solar Cell Materials and Processes, 24–31.

Macdonald, D., Brendle, W., Cuevas, A., & Istratov, A. A. (2002). Injection-dependent lifetime studies of copper precipitates in silicon. 12th Workshop on Crystalline Silicon Solar Cell Materials and Processes.

Michl, B., Padilla, M., Geisemeyer, I. et al. (2014). Imaging techniques for quantitative silicon material and solar cell analysis. *IEEE Journal of Photovoltaics* 4 (6): 1502–1510.

Mitchell, B., Trupke, T., Weber, J., and Nyhus, J. (2011). Bulk minority carrier lifetimes and doping of silicon bricks from photoluminescence intensity ratios. *Journal of Applied Physics* 109 (8): 083111.

Morishige, A.E., Jensen, M.A., Needleman, D.B. et al. (2016). Lifetime spectroscopy investigation of light-induced degradation in p-type multicrystalline silicon PERC. *IEEE Journal of Photovoltaics* 6 (6): 1466–1472.

Mundt, L.E., Schubert, M.C., Schon, J. et al. (2015). Spatially resolved impurity identification via temperature- and injection-dependent photoluminescence imaging. *IEEE Journal of Photovoltaics* 5 (5): 1503–1509.

Murphy, J.D., Bothe, K., Krain, R. et al. (2012). Parameterisation of injection-dependent lifetime measurements in semiconductors in terms of Shockley-Read-Hall statistics: an application to oxide precipitates in silicon. *Journal of Applied Physics* 111 (11): 113709.

Narland, T.U., Bernardini, S., Wiig, M.S., and Bertoni, M.I. (2018). Is it possible to unambiguously assess the presence of two defects by temperature-and injection-dependent lifetime spectroscopy? *IEEE Journal of Photovoltaics* 8 (2): 465–472.

Nie, S., Kristensen, S.T., Gu, A. et al. (2020). Photoluminescence-based spatially resolved temperature coefficient maps of silicon wafers and solar cells. *IEEE Journal of Photovoltaics* 10: 585–594.

Nie, S., Bonilla, R.S., and Hameiri, Z. (2021). Unravelling the silicon-silicon dioxide interface under different operating conditions. *Solar Energy Materials and Solar Cells* 224: 111021.

Niewelt, T., Schön, J., Broisch, J. et al. (2015). Electrical characterization of the slow boron oxygen defect component in Czochralski silicon. *Physica Status Solidi – Rapid Research Letters* 9 (12): 692–696.

Pässler, R. (2002). Dispersion-related description of temperature dependencies of band gaps in semiconductors. *Physical Review B* 66 (8): 85201.

Phang, S.P., Sio, H.C., and Macdonald, D. (2016). Applications of carrier de-smearing of photoluminescence images on silicon wafers. *Progress in Photovoltaics: Research and Applications* 24 (12): 1547–1553.

Raphson, J. (1702). *Analysis Aequationum Universalis*. Typis Tho. Braddyll.

Rein, S. (2006). *Lifetime Spectroscopy: A Method of Defect Characterization in Silicon for Photovoltaic Applications*, vol. vol. 85. Springer Science and Business Media.

Rein, S., Rehrl, T., Warta, W., and Glunz, S.W. (2002). Lifetime spectroscopy for defect characterization: systematic analysis of the possibilities and restrictions. *Journal of Applied Physics* 91 (3): 2059–2070.

Roth, T., Rüdiger, M., Warta, W., and Glunz, S.W. (2008). Electronic properties of titanium in boron-doped silicon analyzed by temperature-dependent photoluminescence and injection-dependent photoconductance lifetime spectroscopy. *Journal of Applied Physics* 104 (7): 074510.

Rougieux, F.E., Grant, N.E., Barugkin, C. et al. (2015). Influence of annealing and bulk hydrogenation on lifetime-limiting defects in nitrogen-doped floating zone silicon. *IEEE Journal of Photovoltaics* 5 (2): 495–498.

Rougieux, F.E., Sun, C., and Macdonald, D. (2018). Determining the charge states and capture mechanisms of defects in silicon through accurate recombination analyses: a review. *Solar Energy Materials and Solar Cells* 187: 263–272.

Sah, C. and Shockley, W. (1958). Electron-hole recombination statistics in semiconductors through flaws with many charge conditions. *Physical Review* 109 (4): 1103–1115.

Schenk, A. (1998). Finite-temperature full random-phase approximation model of band gap narrowing for silicon device simulation. *Journal of Applied Physics* 84 (7): 3684–3695.

Schmidt, J. (2003). Temperature- and injection-dependent lifetime spectroscopy for the characterization of defect centers in semiconductors. *Applied Physics Letters* 82 (13): 2178–2180.

Schmidt, J., Lim, B., Walter, D., Bothe, K., Gatz, S., Dullweber, T., & Altermatt, P. P. (2012). Impurity-related limitations of next-generation industrial silicon solar cells. 38th IEEE Photovoltaic Specialists Conference.

Schroder, D.K. (2006). *Semiconductor Material and Device Characterization*. Wiley-IEEE Press.

Shimura, F. (1986). Carbon enhancement effect on oxygen precipitation in Czochralski silicon. *Journal of Applied Physics* 59 (9): 3251–3254.

Shockley, W. and Read, W.T. (1952). Statistics of the recombination of holes and electrons. *Physical Review* 87 (46): 835–842.

Sinton, R. A., Cuevas, A., & Stuckings, M. (1996). Quasi-steady-state photoconductance, a new method for solar cell material and device characterization. 25th IEEE Photovoltaic Specialists Conference, 457–460.

Sio, H.C., Phang, S.P., Trupke, T., and Macdonald, D. (2014). An accurate method for calibrating photoluminescence-based lifetime images on multi-crystalline silicon wafers. *Solar Energy Materials and Solar Cells* 131: 77–84.

Steinkemper, H., Geisemeyer, I., Schubert, M.C. et al. (2017). Temperature-dependent modeling of silicon solar cells: $E_g$, $n_i$, recombination, and $V_{oc}$. *IEEE Journal of Photovoltaics* 7 (2): 450–457.

Sun, C., Rougieux, F.E., and Macdonald, D. (2014). Reassessment of the recombination parameters of chromium in n- and p-type crystalline silicon and chromium-boron pairs in p-type crystalline silicon. *Journal of Applied Physics* 115 (21): 214907.

Trupke, T., Bardos, R.A., Schubert, M.C., and Warta, W. (2006). Photoluminescence imaging of silicon wafers. *Applied Physics Letters* 89 (4): 44107.

Trupke, T., Mitchell, B., Weber, J.W. et al. (2012). Photoluminescence imaging for photovoltaic applications. *Energy Procedia* 15: 135–146.

Vargas, C., Zhu, Y., Coletti, G. et al. (2017). Recombination parameters of lifetime-limiting carrier-induced defects in multicrystalline silicon for solar cells. *Applied Physics Letters* 110 (9): 092106.

Voronkov, V.V., Falster, R.J., Schmidt, J. et al. (2010). Lifetime degradation in boron doped czochralski silicon. *ECS Transactions* 33 (11): 103–112.

Voronkov, V.V., Falster, R., Bothe, K. et al. (2011). Lifetime-degrading boron-oxygen centres in p-type and n-type compensated silicon. *Journal of Applied Physics* 110 (6): 063515.

Walz, D., Joly, J.P., and Kamarinos, G. (1996). On the recombination behaviour of iron in moderately boron-doped p-type silicon. *Applied Physics A: Materials Science and Processing* 62 (4): 345–353.

Warta, W. (2006). Advanced defect and impurity diagnostics in silicon based on carrier lifetime measurements. *Physica Status Solidi A* 203 (4): 732–746.

Würfel, P., Finkbeiner, S., and Daub, E. (1995). Generalized 'Planck's radiation law for luminescence via indirect transitions. *Applied Physics A* 60 (1): 67–70.

Zafirovska, I., Juhl, M. K., Weber, J. W., Kunz, O., & Trupke, T. (2016). Module inspection using linescanning photoluminescence imaging. Paper Presented at 32nd European Photovoltaic Solar Energy Conference and Exhibition.

Zhu, Y. and Hameiri, Z. (2021). Review of injection dependent charge carrier lifetime spectroscopy. *Progress in Energy* 3 (1): 012001.

Zhu, Y., Le Gia, Q.T., Juhl, M.K. et al. (2017). Application of the Newton–Raphson method to lifetime spectroscopy for extraction of defect parameters. *IEEE Journal of Photovoltaics* 7 (4): 1092–1097.

Zhu, Y., Heinz, F.D., Juhl, M. et al. (2018). Photoluminescence imaging at uniform excess carrier density using adaptive nonuniform excitation. *IEEE Journal of Photovoltaics* 8 (6): 1787–1792.

Zhu, Y., Coletti, G., & Hameiri, Z. (2019). Injection dependent lifetime spectroscopy for two-level defects in silicon. 46th IEEE Photovoltaic Specialists Conference, 0829–0832.

Zhu, Y., Sun, C., Niewelt, T. et al. (2020). Investigation of two-level defects in injection dependent lifetime spectroscopy. *Solar Energy Materials and Solar Cells* 216: 110692.

Zhu, Y., Rougieux, F., Grant, N.E. et al. (2021). Electrical characterization of thermally activated defects in n-type float-zone silicon. *IEEE Journal of Photovoltaics* 11 (1): 26–35.

# 18

## Advanced Flash Testing in High-Volume Manufacturing

*Karoline Dapprich and Ronald A. Sinton*

Sinton Instruments, Boulder, CO, USA

## 18.1 Capacitive Devices

The first challenge to measuring modern, high-efficiency cells is to measure the steady state characteristic at line speed, despite the slow time response of the cell. When a cell or module does not respond immediately to changes in applied illumination, voltage, or current, the error introduced is called capacitive error. It is caused by the departure from steady-state conditions due to charging or discharging of the sample as carriers diffuse during the measurement. The result is an over- or under-estimation of the device's true steady-state power, depending on the direction of the voltage sweep. $J$–$V$ measurements taken with a sweep from $J_{sc}$ to $V_{oc}$ will tend to underestimate module power, while measurements taken in the opposite direction will tend to overestimate it (Figure 18.1).

The charging or discharging current preventing the device from reaching steady state is called the transient current and is given by the following set of equations:

$$\frac{dQ}{dt} = \frac{dQ}{dV_j}\frac{dV_j}{dt} \tag{18.1}$$

$$\frac{dQ}{dV_j} = \frac{q}{kT}\frac{qWn_i^2\,\exp\left(qV_j/kT\right)}{\sqrt{N_{A,D}^2 + 4n_i^2\,\exp\left(qV_j/kT\right)}} \tag{18.2}$$

$$V_j = V + JR_s \tag{18.3}$$

in which $dV_j/dt$ is the junction voltage sweep rate and $dQ/dV_j$ is the differential capacitance (Blum et al. 2018). The differential capacitance is entirely dependent on cell or module characteristics, such as thickness, doping, and operating voltage. In contrast, the measured junction voltage depends on tester parameters such as voltage ramp rate (under steady-state illumination) and sweep direction.

The magnitude of error is also dependent on the sample being tested (Figure 18.2). For PERC and Al-BSF cells, the error in measured power produced by short flashes remains below 0.1% for ramp rate less than 14 V/s, which corresponds to a linearly ramped measurement of about 50 ms. For HIT cells, a ramp rate of 10 V/s would result in an error >5% (Blum et al. 2018).

*Photovoltaic Solar Energy: From Fundamentals to Applications, Volume 2*, First Edition.
Edited by Wilfried van Sark, Bram Hoex, Angèle Reinders, Pierre Verlinden, and Nicholas J. Ekins-Daukes.
© 2024 John Wiley & Sons Ltd. Published 2024 by John Wiley & Sons Ltd.
Companion website: www.wiley.com/go/PVsolarenergy

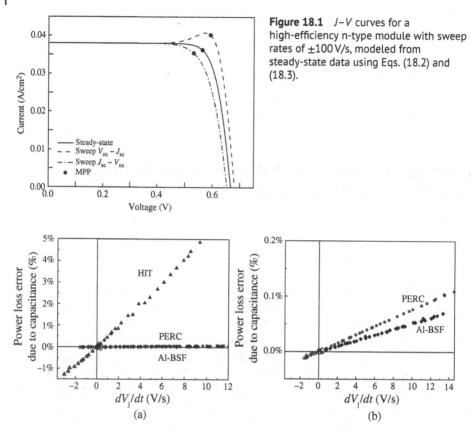

**Figure 18.1** *J–V* curves for a high-efficiency n-type module with sweep rates of ±100 V/s, modeled from steady-state data using Eqs. (18.2) and (18.3).

**Figure 18.2** (a) Error in measured power due to capacitive effects for three cell types. (b) Close-up of (a).

**Table 18.1** Capacitance of cell types and junction voltage sweep rate required for <0.1% error in power.

| Sample | Capacitance ($\mu F/cm^2$) | $dV_j/dt$ (V/s) |
|---|---|---|
| p-type Al-BSF | 1.76 | 19.2 |
| p-type mono PERC | 2.97 | 13.2 |
| n-type HIT | 144 | 0.189 |

Table 18.1 shows the capacitance at $V_{mp}$ of various cell types along with the minimum required junction voltage sweep rate to achieve less than 0.1% error in measured power (Sinton et al. 2017). Of the cell types listed, the most capacitive are HIT, whose junction voltage changes slowly enough to require over 3.8 s per flash to remain below 0.1% error. This makes high-speed production testing using a conventional linear voltage ramp impossible.

To keep measurements short and error low, the junction voltage sweep rate, which depends on tester parameters, can be used strategically to minimize the transient current.

By enforcing $dV_j/dt = 0$ in Eq. (18.1), the transient current $dQ/dt$ is also forced to 0. This can be done by modulating the applied voltage during the flash (Sinton 2006). The degree of modulation can then be validated by checking that the current and illumination signals are in phase – if not, it means there is a nonzero transient current and capacitive effects will be present. The voltage modulation can then be adjusted until the current and illumination are in phase. By compensating for the source of the capacitance in this way, conveniently short flashes can be used without sending the device out of steady state. An alternative method is to sweep the cell in both directions, $J_{sc}$–$V_{oc}$ and then $V_{oc}$–$J_{sc}$, at constant illumination intensity. The resulting hysteresis loop can be analyzed to determine the steady-state $J$–$V$ curve using Eqs. (18.1)–(18.3). This self consistently determines $R_s$ and the junction voltage ramp rates and current at each junction voltage, which enables the construction of the steady-state $J$–$V$ curve (Meixner et al. 2014).

## 18.2 Bifacial Devices

Bifacial cells make use of light incident on both sides of the cell and are gaining an increasingly large share of the world market, with 20 GW cumulatively installed at the end of 2020 (Kopecek and Libal 2021). The question of how to test them centers on whether to reproduce the dual-sided illumination they see in the field or to use a simpler single-sided illumination method that mimics this bifacial boost. The IEC has defined standards for both methods (IEC TS 60904-1-2 2019) and recognizes them as equivalent, which has been confirmed by independent studies (Rauer et al. 2019; Sabater et al. 1999). In the first method, dual illumination, both sides of the cell are simultaneously illuminated, with 1 sun on the front and 0.2 suns on the rear. This is an intuitive representation of the operating conditions a bifacial module is expected to encounter, but it can be difficult to implement in practice. The second method, the equivalent intensity method, first requires measuring a subset of cells on each side individually at 1 sun. Their average bifaciality $\Phi$ is calculated as the percentage of $I_{sc\_back}$ to $I_{sc\_front}$. Once the nominal bifaciality has been measured, the equivalent intensity (in $W/m^2$) at which to illuminate is calculated according to:

$$G_{equivalent} = 1000 + \Phi \cdot G_{rear} \tag{18.4}$$

with $G_{rear}$ equal to 0.2 suns. Then every cell in the lot is measured on the front side only at this equivalent intensity, which is intended to replicate the power increase due to bifaciality that the cell would see in the field. As an example, cells with 80% bifaciality would be illuminated at an equivalent intensity of 1.16 suns.

Although each cell in a lot is assumed to be identically bifacial, the uncertainty introduced by this is very small (Dapprich et al. 2020) and does not affect the resulting module power (Sabater et al. 1999). Aside from the extra step to measure a subset of cells on both sides, it is very straightforward to implement. In contrast, illuminating a cell on both sides simultaneously introduces light management complications. Reflections and light leaked from one side of the cell to the other increase measurement error and cannot be completely avoided while holding the cell with minimal shadowing on both sides. A complex correction is therefore required to bring the measurement back into agreement with both individual single-sided and equivalent intensity measurements (Krieg et al. 2018).

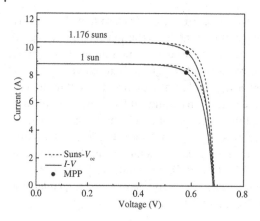

**Figure 18.3** $I-V$ and Suns-$V_{oc}$ curves for a bifacial cell at 1 and 1.176 suns.

**Table 18.2** Bifacial cell parameters measured at 1 and 1.176 suns.

| Intensity (suns) | 1.176 suns | 1 sun |
|---|---|---|
| $I_{sc}$ (A) | 10.4 | 8.823 |
| $V_{oc}$ (V) | 0.6884 | 0.6847 |
| $I_{mp}$ (A) | 9.675 | 8.215 |
| $V_{mp}$ (V) | 0.5814 | 0.5786 |
| $P_{mp}$ (W) | 5.626 | 4.753 |
| $R_{sh}$ ($\Omega$) | 70.23 | 70.23 |
| $R_{s}$ ($\Omega$) | 0.002946 | 0.002946 |
| Efficiency (%) | 19.58 | 19.45 |
| FF (%) | 78.62 | 78.68 |

The equivalent intensity method, though less intuitive, requires only one side of the cell to be illuminated, so there is no leaked light or correction for the same to contend with, and the tester is mechanically simpler to build. It has the additional benefit of enabling high-speed measurements at both 1 sun and at the higher equivalent intensity, for binning cells for use in both monofacial and bifacial modules (Dapprich et al. 2020). Figure 18.3 shows $I-V$ curves and Table 18.2 shows numerical results for a bifacial cell with 88% bifaciality, measured at 1 sun and at its 1.176 sun equivalent intensity.

## 18.3 Aggregate $J_{0s}$

The recombination in a cell or module is monitored throughout production to assess passivation and substrate quality, ideally with the bulk recombination distinguished from surface recombination. Without this distinction, it is not clear which region dominates the recombination and, therefore, which element of the cell needs improvement or tighter process control. The most widely used method for identifying specific recombination losses, developed by Kane and Swanson (1985), uses a fit to inverse lifetime versus carrier density data, and works well for lightly doped n-type cells. However, the strategy does not work well for all cell types, namely those that are heavily doped. The reason for this, as well as a generalization of the method suitable for heavily doped cells, is discussed here.

Kane and Swanson's method extracts $J_{0s}$ and $\tau_{bulk}$ from a linear fit to inverse lifetime versus carrier density data according to the equation:

$$\left(\frac{1}{\tau_{measured}}\right) - \left(\frac{1}{\tau_{Auger}}\right) = \left(\frac{1}{\tau_{bulk}}\right) + \frac{(J_{0\,front} + J_{0\,back})}{qWn_i^2}(N_A + \Delta n) \qquad (18.5)$$

The success of this method, which was originally developed for undoped materials and then generalized to include doping, is dependent on the existence of a high-injection regime where the bulk lifetime is constant and inverse lifetime increases linearly with carrier density (Figure 18.4a). This means it fails for many heavily doped p-type samples where 1 sun illumination is not enough to create high-injection conditions ($\Delta n \gg$ doping) (Figure 18.4b).

The Kane and Swanson fit relies on $\Delta n$ varying during the measurement of a single cell, but the independent variable in this equation is really ($\Delta n +$ doping). If doping level is measured for every cell, it is often possible to use the natural variation in doping across a production lot to get the range required for a fit. Instead of relying on the entire inverse lifetime curve of a single cell, single-point measurements at $V_{oc}$ for every cell are compiled into one plot (Figure 18.5a). This is fitted to extract one "aggregate $J_0$" and bulk lifetime for the lot (Dapprich et al. 2021).

Each point in Figure 18.5a corresponds to an inverse lifetime curve (Figure 18.5b). The points at $V_{oc}$ and $V_{mp}$ are both straightforward to identify, but the points at $V_{oc}$ are better for the fit because they fall in the nominally linear portion of the inverse lifetime curve, where bulk lifetime is constant, and the fit assumptions are met.

For sample sets where the power loss in a single cell due to recombination cannot be separated into surface and bulk, they can often be separated for the batch by looking at

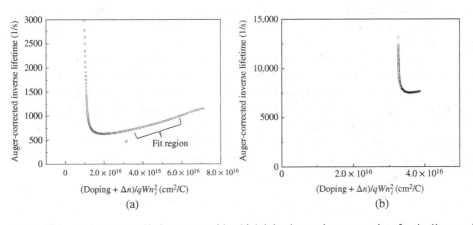

**Figure 18.4** (a) An inverse lifetime curve with a high injection region appropriate for the Kane and Swanson fit. The x-axis is chosen so the slope is $J_0$. The y-intercept is $1/\tau_{bulk}$. (b) Heavily doped cell for which the fit is not possible.

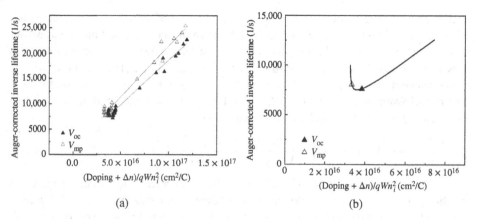

**Figure 18.5** (a) Inverse lifetime at $V_{oc}$ and $V_{mp}$ for a set of 32 cells. (b) The inverse lifetime curve of one cell. The slopes of the two plots both provide a way to extract a $J_0$.

the aggregate $J_{0s}$. This requires some spread in doping across a cell lot (which is common due to depth-dependent doping variation in ingots) and a consistent bulk lifetime between cells but is successful for doping levels above $1 \times 10^{16}$ cm$^{-3}$. It is conveniently validated by placing some samples under high-intensity illumination to create high injection data, and applying the Kane and Swanson fit.

## 18.4 Power Loss Analysis

An analysis of sources of power loss is one of the most fundamental functions of an $I–V$ tester, allowing the user to identify where they can improve process control or cell technology to get the most efficiency gains. In addition to measuring parameters such as bulk recombination and series resistance, the extent to which these mechanisms reduce the cell's power can also be quantified. The technique to do this requires defining an upper power limit and then identifying how much each mechanism drops the maximum power (Wilterdink et al. 2021).

The upper limit to power includes intrinsic recombination only – the Auger recombination (radiative recombination is neglected due to its small contribution to silicon). This upper limit depends on cell doping and carrier density, so takes cell parameters into account and does not apply a universal limit to every cell. To calculate it, the parameterization of Richter et al. (2012) is used to determine $\tau_{Auger}$ as a function of $\Delta n$ and convert it to $J_{Auger}$. Following Eq. (18.6), $J_{Auger}$ is subtracted from $J_{sc}$ for the total current with only intrinsic recombination present ("Auger limit" in Figure 18.6) (Sinton and Swanson, 1990). This is the "ideal" scenario to which all other implied $I–V$ curves are compared.

$$J = J_{sc} - \frac{V_j}{R_{sh}} - [J_{0SRH} + J_{0s} + J_{0Auger}]e^{qV/kT} \tag{18.6}$$

The other implied $I–V$ curves are similarly constructed with certain loss mechanisms removed. For example, the current including only intrinsic and surface recombination,

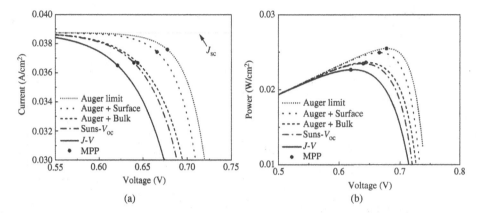

**Figure 18.6** (a) Light *J–V* and Suns-$V_{oc}$ curve compared to implied *J–V* curves with different loss mechanisms removed. The Auger limit curve is the idealized scenario with only intrinsic recombination present. (b) Power curves corresponding to *J–V* curves in (a).

where losses due to series resistance and bulk recombination are set to zero, is equal to $J_{sc} - (J_{0Auger} + J_{0s})e^{qV/kT}$ ("Auger + Surface" in Figure 18.6a).

If the bulk and surface recombination can be separated, unique curves for the bulk and surface recombination can be constructed. If they cannot be separated, only one implied *I–V* curve can be made including both types of recombination.

The series resistance losses are removed from the light *I–V* curve with the familiar Suns-$V_{oc}$ method (Sinton and Cuevas 2000), which represents the performance of the cell if it had no series resistance since it is measured under open-circuit conditions.

The power-voltage characteristic can be calculated for each of these implied *I–V* curves in the same way (Figure 18.6b). However, comparison between them is tricky because comparing different curves at a constant voltage does not accurately represent the power gains between them. Comparing points on the Auger and the bulk curves at 0.7 V, one might say the Auger power is what would exist if all bulk recombination were eliminated. However, this is not exactly true because by removing bulk recombination, the operating voltage of the cell has been shifted up, thereby increasing $\Delta n$ and changing the amount of recombination caused by the *other* mechanisms as well. Ultimately, by removing one source of current, two things occur – the terminal current at the original maximum power point is increased, and the maximum power point is shifted to a higher voltage. The result is a nonlinear increase in power compared to the increase in current.

In Figure 18.7, this nonlinear increase in power is broken down into parts of the total recombination loss. The total percentage is the difference in maximum power between the Auger and *I–V* curves. The components are in the same proportion as they are at the light *I–V* curve $V_{mp}$, in order to best inform about which loss is greatest in the operating regime of the cell. The change in power per decrease in total $J_0$ at the measured maximum power point is also given.

For large datasets, such as are produced daily in production environments, it is useful to investigate the distribution of causes of power loss. For example, the effect of series resistance variation on efficiency across the lot can be examined by comparing pseudo efficiency to efficiency (Sinton et al. 2021). At a given pseudo efficiency, cells with nominally equal

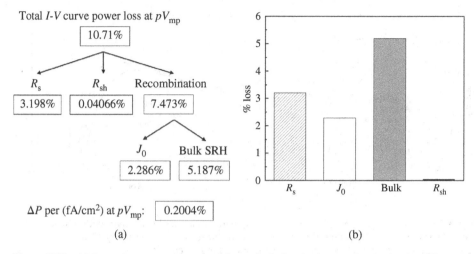

**Figure 18.7** (a) Power loss as percentage of Auger limit $P_{mp}$, broken out by mechanism. The percentages are in the same proportion as at $V_{mp}$, scaled to reach the Auger limit $P_{mp}$. (b) Losses by mechanism.

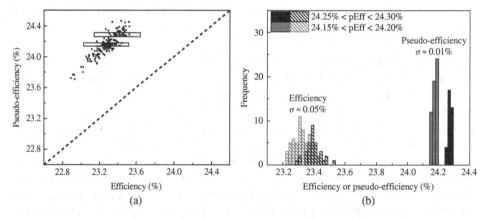

**Figure 18.8** (a) Pseudo efficiency versus efficiency for 234 cells. Boxes indicate 0.05%-wide slices in pseudo efficiency shown in (b). (b) Histograms of pseudo efficiency and efficiency. The variation due to $R_s$ is 0.05% and the mean loss is 0.87%.

$V_{mp}$ values will have equal amounts of recombination. This is because the pseudo efficiency is calculated from the Suns-$V_{oc}$ curve, which does not contain any series resistance effects, so the only difference between efficiency and pseudo efficiency is due to series resistance (shunt losses are equally present in both $I$–$V$ and Suns-$V_{oc}$ curves).

As a result, a set of cells with a very narrow range in pseudo efficiency will have a much greater range in efficiency (Figure 18.8). The distribution of efficiency and pseudo efficiency is shown in Figure 18.8 for two pseudo efficiency slices that are 0.05% wide; the flattened distribution of the efficiency is due to $R_s$ variation. With very large data sets, the slices in pseudo efficiency can be made vanishingly thin so that series resistance is truly the only

**Figure 18.9** Distribution of power loss due to series resistance, surface recombination, and bulk recombination for a set of 235 cells.

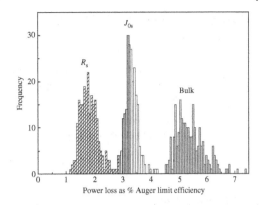

thing affecting the efficiency. In practice, process variations can be compared using this approach with the goal of reducing both series resistance and its variation.

Comparing the magnitude of different losses is also informative. For the sample lot shown in Figure 18.9, the largest source of loss is bulk recombination. It is also the most widely varying source of loss. This provides a clear indication of what cell parameters should be improved for the most gain, as well as which processing steps need to be more tightly controlled. For heavily doped cells where the Kane and Swanson method for finding $J_0$ does not work, the aggregate $J_0$ can be extracted from the entire lot, providing more detail to the power loss analysis.

With the power loss broken down by mechanism, manufacturers can identify the biggest sources of efficiency loss and target improvements in those areas. It is also possible to translate "power loss per type of recombination current" to "power loss per cost", and focus on those areas which achieve the most cost-effective power increase.

## 18.5 Conclusion

The $I$–$V$ test station is a fundamental tool for both research and production, enabling clear identification of efficiency loss in cells or modules through a thorough device physics analysis. A comprehensive measurement system is required to take full advantage, with attention paid to the specific requirements of the sample type under test. High-capacitance devices require special attention to counteract and quantify the capacitive effects in the measurement. Measurements of bifacial samples must include the bifacial boost without introducing uncertainty from leaked light. P- and n-type samples can also require different treatments, such as for separating bulk and surface recombination. Large data sets can provide more information than single-cell or single-module results, identifying actionable sources of power loss due to series resistance and transport as well as recombination. Along with increasingly prevalent luminescence imaging, such $I$–$V$ test information provides a valuable tool for production managers in understanding the quality of and routes to improvement for their cells and modules.

## Abbreviations and Acronyms

| | |
|---|---|
| BSF | back surface field |
| HIT | heterojunction with intrinsic thin layer |
| *I–V* | current-voltage characteristics |
| MPP | maximum power point |
| PERC | passivated emitter and rear cell |
| PV | photovoltaic or photovoltaics |
| SRH | Shockley-Read-Hall |

## Nomenclature

| | |
|---|---|
| FF | fill factor |
| $G_{equivalent}$ | equivalent light intensity of combined front and rear illumination |
| $G_{rear}$ | rear side light intensity (W/m$^2$) |
| $I_{sc}$ | short-circuit current |
| $I_{mp}$ | current at the maximum power point |
| $J_0$ | saturation current density |
| $J_{0front}$ | saturation current density due to the front recombination |
| $J_{0back}$ | saturation current density due to the rear recombination |
| $J_{Auger}$ | recombination current density due to Auger recombination |
| $J_{0Auger}$ | equivalent Auger saturation current density |
| $J_{0s}$ | saturation current density due to surface recombination |
| $J_{0SRH}$ | saturation current density due to bulk SRH recombination |
| $k$ | Boltzmann constant |
| $n_i$ | intrinsic carrier concentration |
| $N_A$ or $N_D$ | doping level of acceptor or donor impurities |
| $P_{mp}$ | maximum power output |
| $Q$ | charge (generated or injected in the base of the solar cell) |
| $q$ | charge of electron |
| $R_s$ | series resistance |
| $R_{sh}$ | shunt resistance |
| $T$ | absolute temperature |
| $V_j$ | junction voltage |
| $V_{mp}$ | voltage at the maximum power point |
| $V_{oc}$ | open-circuit voltage |
| $w$ | thickness of the solar cell |
| $\Delta n$ | excess carrier density |
| $\tau_{measured}$ | effective measured carrier lifetime |
| $\tau_{Auger}$ | Auger recombination lifetime |
| $\tau_{bulk}$ | carrier lifetime in the bulk of the device |

## Author Biographies

**Karoline Dapprich** received a B.A. in Physics from the University of Colorado and joined Sinton Instruments, where she specializes in characterization and data interpretation for silicon solar cells using $I$–$V$, Suns-$V_{oc}$, lifetime, and luminescence measurements.

**Ronald A. Sinton** received a Ph.D. in Applied Physics at Stanford University following a B.S. in Engineering Physics from the University of Colorado. He has specialized in the device physics and characterization of high-efficiency silicon solar cells, founding Sinton Instruments in 1992 to supply test and measurement equipment for both R&D and manufacturing lines. He received the IEEE PVSC Cherry Award in 2014.

## References

Blum, A. L., Sinton, R. A., & Wilterdink, H. W. (2018). Determining the accuracy of solar cell and module measurements on high-capacitance devices. 2018 IEEE 7th World Conference on Photovoltaic Energy Conversion, WCPEC 2018 – A Joint Conference of 45th IEEE PVSC, 28th PVSEC and 34th EU PVSEC, 3603–3606.

Dapprich, K., Dinger, J., Sinton, R., & Wilterdink, H. (2020). Production-line binning of bifacial cells using Suns-$V_{oc}$ analysis. 47th IEEE PVSC.

Dapprich, K., Sinton, R., Wilterdink, H., Dobson, W., Sainsbury, C., & Dinger, J. (2021). Using doping variation to determine $J_{0s}$ at $I$–$V$ test.Silicon PV *AIP Conference Proceedings* 2487, 030003.

International Electrochemical Commission. (2019). IEC TS 60904-1-2: Photovoltaic devices-Part 1–2: Measurement of current-voltage characteristics of bifacial photovoltaic (PV) devices. Retrieved from www.iec.ch.

Kane, D. E., & Swanson, R. M. (1985). Measurement of the emitter saturation current by a contactless decay method. 18th IEEE Photovoltaic Specialists Conference, 578–583.

Kopecek, R. and Libal, J. (2021). Bifacial photovoltaics 2021: status, opportunities and challenges. *Energies* 14 (8).

Krieg, A., Greulich, J., Ramspeck, K., Dzafic, D., Wöhrle, N., Rauer, M., & Rein, S. (2018). IV-measurements of bifacial solar cells in an inline solar simulator with double-sided illumination. 35th European Photovoltaic Solar Energy Conference and Exhibition.

Meixner, M., Metz, A., Komp, L., Schenk, S., & Ramspeck, K. (2014). Accurate efficiency measurements on very high efficiency silicon solar cells using pulsed light sources. 29th European Photovoltaic Solar Energy Conference and Exhibition, 1253–1256.

Rauer, M., Guo, F., & Hohl-Ebinger, J. (2019). Accurate measurement of bifacial solar cells with single- and both-sided illumination. 36th European PV Solar Energy Conference and Exhibition. Marseille.

Richter, A., Glunz, S.W., Werner, F. et al. (2012). Improved quantitative description of Auger recombination in crystalline silicon. *Physical Review B – Condensed Matter and Materials Physics* 86 (16): 1–14.

Sabater, A.A., Wöhrle, N., Greulich, J.M. et al. (1999). Impact of bifacial illumination and sorting criteria of bifacial solar cells on module power. In: *AIP Conference Proceedings*. American Institute of Physics Inc.

Sinton, R.A. (2006). A solution to the problem of accurate flash-testing of high-efficiency modules. In: *21st EU PVSEC. Dresden*, 634–638.

Sinton, R.A. and Cuevas, A. (2000). A quasi-steady-state open-circuit voltage method for solar cell characterization. In: *16th European Photovoltaic Solar Energy Conference*, 1–4. Glasgow.

Sinton, R.A. and Swanson, R.M. (1990). High efficiency silicon solar cells. In: *Advances in Solar Cells* (ed. K. Boer), 427–485. Plenum Press.

Sinton, R.A., Wilterdink, H.W., and Blum, A.L. (2017). Assessing transient measurement errors for high-efficiency silicon solar cells and modules. *IEEE Journal of Photovoltaics* 7 (6): 1591–1595.

Sinton, R., Wilterdink, H., Sainsbury, C. et al. (2021). *Characterization of a Cell Process from IV Cell-Test Data*, 1–5. Silicon PV.

Wilterdink, H., Sinton, R., Blum, A., and Dapprich, K. (2021). *Power Loss Analysis for Silicon PV Cells and Modules using the Richter Recombination Limit*. PVRW.

# 19

# Machine Learning for Photovoltaic Applications

*Priya Dwivedi and Ziv Hameiri*

*The University of New South Wales, Sydney, NSW, Australia*

## 19.1 Machine Learning

### 19.1.1 Types of Machine Learning

Machine learning (ML) algorithms can be classified into three broad categories depending on the available input and feedback – supervised, unsupervised, and reinforcement learning.

#### 19.1.1.1 Supervised Learning

Supervised algorithms learn from labeled examples (Russell and Norvig 2002). Figure 19.1a shows an example of supervised learning. Toys of various shapes and colors are labeled as either "Yes" ("1") or "No" ("0"). The input features and corresponding output labels are included in Figure 19.1b. Supervised learning aims to develop a model to predict the unseen toys' labels. First, the model predicts an output for each example. This output is then compared with the given label and a feedback is used to tune the model. Once the model sees enough examples, it learns to predict the output even for unseen cases. In essence, supervised learning algorithms learn to decipher the pattern between shown examples and their labels. Based on the type of prediction, they are divided into two types – classification and regression. In classification, the aim is to predict the label of an input. The predicted label is a discrete variable, which can be binary (only two classes as the example of Figure 19.1) or multiclass (three or more classes). In "exclusive" classification, an input can be assigned to only one class (label) at the same time, while in "multilabel" classification an input can be assigned multiple labels. In regression, the output label is a continuous variable (such as the price of a toy in Figure 19.1 or the efficiency of a solar cell).

#### 19.1.1.2 Unsupervised Learning

In unsupervised learning [also named "knowledge discovery" (Murphy 2012)], the labels of the examples are unknown (Hinton and Sejnowski 1999). The algorithms identify patterns hidden in the input data, which function as a springboard for conducting further analysis. They are used, for example, to cluster data before applying supervised algorithms. As obtaining labeled datasets is expensive (Murphy 2012), unsupervised learning is more

*Photovoltaic Solar Energy: From Fundamentals to Applications*, *Volume 2*, First Edition.
Edited by Wilfried van Sark, Bram Hoex, Angèle Reinders, Pierre Verlinden, and Nicholas J. Ekins-Daukes.
© 2024 John Wiley & Sons Ltd. Published 2024 by John Wiley & Sons Ltd.
Companion website: www.wiley.com/go/PVsolarenergy

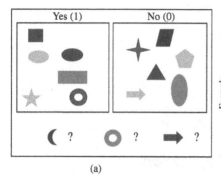

(a)

(b)

**Figure 19.1** Supervised algorithm example based on Leslie Kaelbling. Source: Adapted from Murphy (2012). (a) The colored toys are labeled with "Yes" ("1") or "No" ("0") along with three test cases. (b) Training dataset with rows and columns representing the examples and features, respectively. The last column is the example's label.

**Figure 19.2** In iteration 1, three centroids are initialized, and data points are assigned to each cluster based on their distance from the centroids. In iterations 2 and 3, the centroids are updated, and new clusters are formed.

Iteration 1    Iteration 2    Iteration 3

widely applicable than supervised learning. However, assessing this type of learning is challenging as the output of the corresponding input variables is not known. Two types of unsupervised learning are detailed below – clustering and latent factors.

***Clustering*** The objective of clustering algorithms is to group the data based on similarity. It is frequently used in high-dimension datasets. In Figure 19.2, a *k*-mean algorithm (Forgy 1965) is used to cluster the data. It first assumes the number of clusters (three in this example). It then initializes the centroid of each of the clusters and calculates the distance between each data point and each centroid. A data point will belong to the cluster of the closest centroid distance. The centroids are recalculated based on the formed clusters and updated. These steps are repeated until the average distance between the data points and their respective cluster's centroid does not improve.

***Latent Factors (Unsupervised Transformation)*** Latent factors transform high dimensional datasets consisting of numerous input features into a new data representation that is easier to understand for humans and machines. Frequently, not all the features in the high dimensional dataset are essential for predictive accuracy. There might be a lower dimensional representation that still contains all the necessary information for better predictive accuracy. Filtering the inessential information makes the model computationally inexpensive. Moreover, low dimensional data can be better visualized compared to high dimensional data.

Principal component analysis (PCA) (Hotelling 1933; Pearson 1901) is the most common approach to reduce the dimensionality of datasets and increase interpretability while minimizing information loss. Uncorrelated variables (or latent variables) are created from

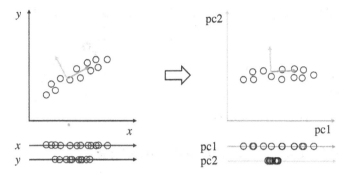

**Figure 19.3** Left: Data points and their variance along the *x*- and *y*-axes are shown in the cartesian coordinate. Right: Data points are transformed in a way that one dimension has the minimum variance. Hence, the dimension with minimum variance can be removed with only negligible information loss. Source: Adapted from Powell (2015).

observed variables (input features) by solving an eigenvalue/eigenvector problem that successively maximizes variance (Murphy 2012). In Figure 19.3, 2D data points are shown in cartesian coordinates and their principal components (PC) in color. The PC are created such that one dimension has the least variance so that this dimension can be dropped with minimum information loss.

### 19.1.1.3 Reinforcement Learning
Reinforcement learning is one of the challenging categories of ML. It is learning by trial and error. The model is presented with a vast number of examples (input) without labels. However, unlike unsupervised learning, each example contains a reward or penalty. Based on the model prediction, the model receives feedback that is used to improve its predictive accuracy. The best solution is the one that collects the highest rewards (Sutton and Barto 2018).

An example of reinforcement learning is training a model to play chess (Sutton and Barto 2018). During the training, a large number of scenarios are presented to the model. The model then receives a reward or penalty based on its decision for the next move. Hence, the model learns the game rules and the winning strategies by trial and error with no prior knowledge about the game.

## 19.1.2 Machine Learning-based Process Optimization

Another common application of ML is process optimization. A general framework of process optimization is shown in Figure 19.4. Firstly, a dataset of process parameters (inputs) and corresponding device parameters (outputs) is generated either by simulations or experiments ("Data generation"). Then, the generated labeled dataset is used to train ML models ("Regression"). The trained models are then used as objective functions for other optimization algorithms ("Optimization"). The combination of the trained ML models and the optimization algorithms generates new sets of process parameters (inputs) that will produce an optimum device. These recommended parameters are then validated either by simulation or experiment.

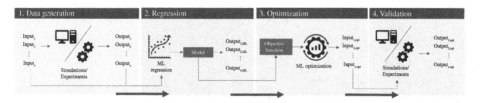

**Figure 19.4** General framework for process optimization. Source: Credit: Gaia Maria Javier (UNSW).

## 19.2 Applications of Machine Learning in Photovoltaics

### 19.2.1 Ingot

Casting mono-like silicon (Si) ingots by directional solidification (DS) offers low-cost solar cells. However, the ingots produced suffer from highly dislocated top regions (Gu et al. 2012) and highly contaminated bottom regions ("red zone") introduced by the impurities present by the crucible walls (Song and Yu 2021) and a high concentration of oxygen precipitates (Huang et al. 2021). A large effort has been made to optimize the solidification process to reduce the dislocation concentration (Schubert et al. 2021; Takahashi et al. 2014). An ML-based processing optimization approach was recently proposed (Liu et al. 2022). First, a crystal growth simulator [CGSim (Artemyev et al. 2019)] was used to generate $150 \times 51 \times 51$ sets of measurements (where $51 \times 51$ are the mesh coordinates on the ingot). An artificial neural network (ANN) algorithm was trained where the input features were the control parameters, and the outputs were the residual stress, the dislocation density, and the solidification time. The trained network was used as an optimization function for a genetic algorithm that predicted the optimized solidification conditions. This approach speeds up the evaluation process by 27,000 times compared to computational fluid dynamics (CFD) (Liu et al. 2022).

Interstitial oxygen ($O_i$) concentration is a critical parameter, especially for Czochralski (Cz)-grown ingots (Borghesi et al. 1995). An ML-based method was developed to predict the $O_i$ concentration based on the growth parameters (Kutsukake et al. 2020). Using the data of 450 ingots with $O_i$ concentrations between $0.7 \times 10^{18}$ and $1.7 \times 10^{18}$ atoms/cm$^3$ that were grown in the same furnace, an algorithm was developed to determine the $O_i$ concentration with a root mean square error (RMSE) and mean absolute error (MAE) of ~$4.2 \times 10^{16}$ and $2.7 \times 10^{16}$ atoms/cm$^3$, respectively.

### 19.2.2 Wafer

#### 19.2.2.1 Process Optimization

ML has been used to optimize the diamond wire sawing (DWS) process considering the surface roughness uniformity (Kayabasi et al. 2017). Based on various process parameters (spool speed, $z$-axis speed, and coolant oil) an ANN model was developed to predict and then minimize the surface roughness. Using 28 experimental datasets, the model was trained to predict the roughness (with an RMSE of 1.43%). By sweeping the process parameters within a defined range and using the model to predict the roughness, an optimum set of process parameters was identified.

Another study focused on the lapping process that is used to reduce the thickness variations subsequent to the DWS process (Ozturk et al. 2018). An ANN algorithm was trained to determine the roughness variations based on the rotation speed, lapping duration, and lapping pressure. After achieving a relative error of below 1%, the model was used to optimize the process. They found that reducing the rotation speed and lapping pressure reduces the roughness while also lowering the energy and material consumption.

### 19.2.2.2 Defect Inspection

Identification of cracks in wafers is an important task as it prevents further expensive processes of defective wafers. As the currently used identification techniques are often expensive and inefficient (Han et al. 2020), ML algorithms have been developed for crack segmentation based on photoluminescence (PL) images (Han et al. 2020). First, a region proposal network (RPN) (Ren et al. 2015) was used to generate bounding boxes around the defects, then a modified U-Net (Ronneberger et al. 2015) was used for segmentation. Initial results indicated an intersection over union (IoU) – the ratio between the overlapped and union areas between the predicted and ground-truth bounding boxes (Tanimoto 1958) – of 0.635 and $F_1$-score – the harmonic mean of precision and recall (Murphy 2012) – of 0.79.

### 19.2.2.3 Quality Rating

Early identification of wafers that may produce low-efficiency solar cells can significantly reduce waste of resources and costs. A few studies have developed ML algorithms to predict the electrical parameters of the completed solar cells using the PL images of the as-cut wafers. For example, DenseNet-based convolutional neural network (CNN) (Huang et al. 2017) was trained by Demant et al. using 7,300 wafers from 74 bricks (fabricated by ten manufacturers) for training and testing (Demant et al. 2019a). The trained model predicted open-circuit voltage ($V_{oc}$) and efficiency with a MAE of 2.05 mV and 0.11%, respectively.

In a subsequent study (Demant et al. 2019b), two visualization techniques were proposed to identify what had been learned by deep learning. First, the final layer of the model was adjusted to resolve the spatial quality of the image. This is noted as a regression activation map (Wang and Yang 2018). The image was then compared with an image of the recombination saturation current ($I_0$) of the corresponding solar cell. It was found that the CNN learned to assign low $V_{oc}$ to contaminated and dislocated regions. Second, the last layer of the trained model (feature vector) was mapped into low-dimensional space to analyze similarities and anomalies in the dataset. Unsupervised technique, like t-distributed stochastic neighborhood embedding (t-SNE) (Van der Maaten and Hinton 2008), was used to visualize the high-dimension dataset in 2D. It was shown that this transformation helps to identify faults and material anomalies.

## 19.2.3 Solar Cells

### 19.2.3.1 Process Optimization

Increasing the efficiency of solar cells is one of the key paths to reducing the cost of PV-generated electricity even further (Green 2019). Optimization of PV production lines is critical to increasing efficiency. Traditionally, optimization methods have been expensive and difficult because of the enormous number of experiments required (Myers et al. 2016).

ML, sometimes together with numerical device modelling software, has been identified as a promising method to simplify the optimization process.

A recent study demonstrated the capability of ML and a genetic algorithm to optimize PV production lines (Buratti et al. 2020). The database was generated using a virtual production line of aluminum-back surface field (Al-BSF) solar cells and PC1D simulations (Haug and Greulich 2016). Several ML models were trained to predict the efficiency of the fabricated cells using 400,000 combinations of process parameters ("recipes"). It was shown that ANN outperforms the other ML models by achieving an $R^2$ of 99.6% and RMSE of 0.019%. The ANN was then used as an objective function for a genetic algorithm to optimize the production line and obtain an increase in relative maximum energy efficiency by 2%.

Another study investigated the impact of material and design properties on solar cell performance (Wagner-Mohnsen and Altermatt 2020). Since, in most cases and for obvious cost reason, individual wafers are not tracked in production lines, it is challenging to determine the causes for the inconsistency in the cell performance. Using Sentaurus, a common technology computer-aided design (TCAD) software (Sentaurus Device 2019), the electrical parameters of passivated emitter and rear contact (PERC) cells were simulated with given material parameters (such as wafer resistivity and quality) and design (such as diffusion profile, grid dimensions, and passivation quality). The generated dataset was then used to train a supervised ML algorithm to predict electrical parameters. The trained algorithm was also used to generate a billion digital twins by sweeping the material and design properties. The measured electrical parameters of manufactured ~15,700 PERC cells were then compared with the generated dataset to identify their digital twins. This allowed tracing of their material properties, hence, significantly improving understanding of the impact of material properties on the efficiency of solar cells from different quality bins.

The same group also optimized the $POCl_3$ (phosphoryl chloride) diffusion process using numerical simulation, ML, and a genetic algorithm (Wagner-Mohnsen et al. 2021). Sentaurus was used to generate a dataset of 2,400 solar cells with different emitters and their corresponding electrical parameters. The dataset was then used to train supervised ML algorithms such as Gaussian process regression (Rasmussen and Williams 2006). Based on the process parameters (diffusion time, process temperature, and etch depth), the ML model predicts the obtained cell efficiency with an RMSE of 0.085%. To identify the optimum process parameters, the trained ML model was used as a cost function for a genetic algorithm achieving maximum efficiency of 23.4%.

A recent study used Quokka2 (Fell 2013) to generate a database of PERC cells that was later used to train an ML model [based on a multilayer perception (MLP)] to determine the impact of wafer and passivation quality on the obtained efficiency (Zhu et al. 2022).

### 19.2.3.2 Defect Inspection

*Defect Classification* Defect classification is often based on PL or electroluminescence (EL) images. Considering the target, the number of classes ("labels") changes.

A recent study used a visual geometry group (VGG) network (Simonyan and Zisserman 2015) to classify EL images from the ELPV dataset, a collection of 2,654 EL images (Buerhop-Lutz et al. 2018), as functional or defective cells, achieving an accuracy of 88.4% (Deitsch et al. 2019). A later study utilizing the same dataset used multiple classes (no defect, micro defect, large defect, and low-resolution defect), attaining a prediction accuracy of 73.8% (Acharya et al. 2020). Using a larger dataset of 13,835 images of multicrystalline Si (mc-Si) cells extracted from EL images of modules, an accuracy of 93% for a multiple classification problem was demonstrated (Ying et al. 2018). To overcome the limited size of the ELPV dataset, data augmentation (Shorten and Khoshgoftaar 2019) was suggested (Akram et al. 2019). In this process, artificial images are created from the dataset itself. The network was simplified by reducing the number of learnable parameters, for example, the number of layers. They also implemented regularization techniques (Kukačka et al. 2017) to avoid overfitting – dropout (Srivastava et al. 2014) and batch normalization (Ioffe and Szegedy 2015). These techniques enhanced the model's generalizability and achieved an accuracy of >93%.

**Defect Detection**  Defect detection is localization of a defect in an image. While classification predicts the class of the defects, detection identifies where the defect is located (see Figure 19.5b).

In one of the first studies, defects were detected using a deep learning model (Liu et al. 2019). The model was trained on 5,000 EL images of mc-Si cells. They modified a faster R-CNN model (Ren et al. 2015) and improved the defect detection mAP (mean average precision) – a commonly used metric for detection that compares the ground-truth and predicted bounding boxes of the defects (Zhu 2011) – from 83.96% (using the Faster R-CNN) to 94.62% (modified). Another study performed data processing by denoising and binarization (Xu, Wu, and Fan 2021). The researchers used a modified single shot

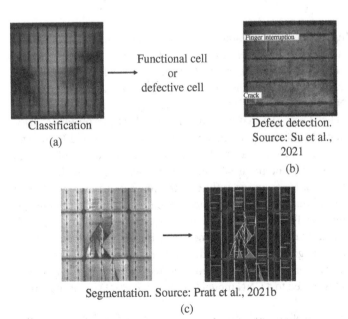

Classification
(a)

Functional cell
or
defective cell

Defect detection.
Source: Su et al.,
2021
(b)

Segmentation. Source: Pratt et al., 2021b
(c)

**Figure 19.5**  Defect inspection using EL images: (a) Classification – An EL image is classified as a functional or defective cell. (b) Defect detection – The defects are localized by a bounding box. (c) Segmentation. Source: Pratt et al. (2021a)/with permission of Elsevier – Each pixel is colored based on its group. For example, pixels belonging to cracks and busbar are marked with white and green, respectively.

multibox detector (SSD) algorithm (Liu et al. 2016) and achieved a mAP of 73.4%, outperforming the classic object detection algorithms such as fast R-CNN (Girshick 2015) and YOLOv3 (Redmon and Farhadi 2018) that used as baselines. Other studies (Su et al. 2021a, 2021b) utilized the recently developed deep learning technique [attention networks (Vaswani et al. 2017)] to improve the performance of defect detection, achieving an mAP >87.3%.

*Defect Segmentation*  A different method to classify an image is segmentation (Shapiro and Stockman 2001). In segmentation, a pixel level classification is performed. An example is given in Figure 19.5c that presents a segmented EL image. Pixels with the same color correspond to a similar class enabling a detailed analysis of the defect types and the use of augmented computer vision techniques.

A segmentation of the ELPV dataset was done using three algorithms (Tian et al. 2021) – a fully convolutional network (FCN) (Long et al. 2015), U-Net, and a combination of modified VGGNet (Simonyan and Zisserman 2015), and U-Net$^{++}$ (Zhou et al. 2018). It was found that U-Net$^{++}$ outperforms the other models achieving an IoU of 0.955. Another study (Su et al. 2021c) presented a relatively new segmentation approach based on generative adversarial network (GAN) (Goodfellow et al. 2014) – a CyclicGAN (Zhu et al. 2017) was used to translate an image with defects into a defect-free image. Then, the image difference between the image with a defect and the defect-free image was used to segment a defect achieving an $F_1$-score of 90.34%.

### 19.2.3.3  Quality Rating

The possibility of extracting the electrical parameters of a solar cell from its luminescence image using ML, potentially replacing the expensive current–voltage (I–V) testers, was recently demonstrated (Buratti et al. 2021). As I–V testers are used for (i) detecting underperforming cells and (ii) sorting the cells based on their quality, a two-step process was suggested. First, a CNN-based network is trained to classify cells based on their EL image to detect rejected cells. Then, the feature extraction block of the trained CNN is fed into a regressor to predict the efficiency. Using a dataset of ~20,000 paired EL images with the corresponding I–V parameters, the method achieved an accuracy of 96% for efficiency binning, while the regressor predicted the cell efficiency with an $R^2$ of 0.93 and an RMSE ~0.10% (absolute). More importantly, it was shown that the power loss in modules fabricated from cells binned by the new method is identical to the loss obtained by the traditional (I–V) binning. In a follow up study, the group developed an automated efficiency loss analysis method based on GAN and luminescence images (Buratti et al. 2022). First, a defective region in an image is identified (Abdullah-Vetter et al. 2021). Then, GAN is used to in-paint the defective region to generate a defect-free image of the cell. Images with and without the defect were fed to a CNN model (Buratti et al. 2021) to estimate their efficiencies.

Point defects can significantly limit the performance of PV devices. They are often investigated using spectroscopy techniques (Lang 1974; Rein et al. 2002) on wafers. However, it is challenging to study these defects in completed solar cells. A method based on I–V measurements of cells and Bayesian parameter estimation (BPE) (Kurchin et al. 2019) was proposed to extract the Shockley–Read–Hall (SRH) parameters of point defects

(Kurchin et al. 2020). The extracted parameters were found to be comparable to those obtained by the temperature- and injection-dependent lifetime spectroscopy (TIDLS) and deep-level transient spectroscopy (DLTS) methods.

### 19.2.4 Module

Classifying defects in the modules based on their I–V measurements and ML was proposed (Kumar and Maheshwari 2021). The I–V measurements and the corresponding EL images of >6,000 modules were investigated. The defects were identified in the images and correlated with their I–V measurements. First, unsupervised methods such as t-SNE (Van der Maaten and Hinton 2008) and uniform manifold approximation and projection (UMAP) (McInnes et al. 2018) were used to reduce the dimensionality of the dataset. Then, feature selection methods were applied to maintain the highly correlated features for prediction. Lastly, random forest (Breiman 2001) and XGboost (Chen and Guestrin 2016) were applied to predict the defect type based only on the I–V measurements, aiming to remove the need for EL imaging.

An approach for module defect detection was developed using YOLO-PV, utilizing the state-of-the-art object detection networks YOLOv4 (Bochkovskiy et al. 2020). It was shown that YOLO-PV is faster and provides higher average precision compared to YOLOv4 (Meng et al. 2022).

Segmentation of module EL images was also developed (Pratt et al. 2021b). A U-Net-based network was trained on 120 labeled EL images at the end of production lines and after thermal and mechanical stress testing. Several types of defects were segmented, demonstrating outstanding capabilities despite having a very small dataset.

An interesting deep learning-based approach to predict the power of the modules using their luminescence images was recently developed (Hoffmann et al. 2021). In this study, a dataset of almost 720 EL images of three types of modules were used to train a ResNet network (He et al. 2016). Their prediction achieved a MAE of $7.3 \pm 6.5\,W_P$. Moreover, the developed approach can predict the power loss at the cell level and analyze the impact of defect severity as shown in Figure 19.6. It was concluded that the module power loss due to fractures is significant, whereas power loss due to cracks is relatively minor.

### 19.2.5 New Materials

The development of new materials for PV applications has always been a time-consuming task, mainly due to the slow and repetitive nature of the fabrication and the following device characterization. A method based on BPE was demonstrated to significantly accelerate the process (Brandt et al. 2017). It predicts four critical device parameters (minority carrier lifetime, mobility, surface recombination velocity, and conduction band offset) with comparable accuracy to state-of-the-art spectroscopic techniques. Its main advantage is its speed; in less than 24 hours, 576,000 discrete device simulations were performed.

A combination of ML and the density functional theory (DFT) was used to screen 5,158 previously unexplored hybrid organic–inorganic perovskite materials (Lu et al. 2018). Multiple ML algorithms, including gradient boosting regression (GBR) (Friedman 2001), were trained on a small database of 212 organic–inorganic perovskite materials that

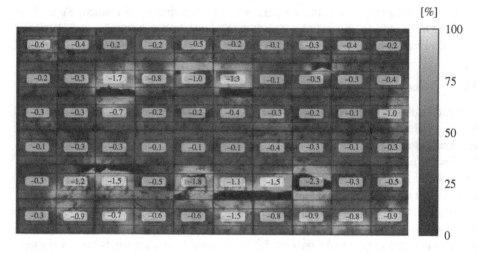

**Figure 19.6** The relative power loss at the cell level with the color scheme indicating the magnitude of the loss. Source: Hoffmann et al. (2021)/John Wiley & Sons.

were earlier identified by high throughput first-principles calculations (Nakajima and Sawada 2017). Six new thermally stable lead-free hybrid organic–inorganic perovskite materials with suitable bandgaps were discovered. The combination of DFT and ML was also used to screen 19,841 potential compositions of ferroelectrics PV (FPV) materials (Lu et al. 2019). The aim was to find new materials with good stability, spontaneous polarization, and narrow bandgap. First, all-inorganic perovskites were identified by a classification model. From this selected group, perovskites with a polar structure were predicted using regression. Regression was also used to determine the bandgap. They identified 151 new FPV compositions, all are thermally stable and have the required bandgap and large polarization. The results were further validated with DFT calculations. Another approach for finding new high-performing PV materials is by establishing the relationship between the chemical structures and electrical properties of a material. Supervised ML was used to determine this relationship using a dataset of 1,700 organic donor materials (Sun et al. 2019). The reliability of the ML-based approach was validated using ten newly designed donor materials. A good agreement between the ML predictions and experimental results was found.

Li et al. presented an ML-based approach for replacing DFT. The approach is divided into two steps. First, the algorithm predicts the bandgap of the material from its compositions. Then, it predicts the material electrical parameters and the expected power conversion efficiency (Li et al. 2019).

## 19.3 Conclusions

ML impacts various aspects of our life including science and engineering. It enables automated analysis of large datasets, which otherwise is a tedious or even impossible task, to fuel R&D efforts. Although the full potential of ML has not been captured by the PV industry yet, the enormous amount of data that is generated by its exponential growth provides an outstanding opportunity to utilize ML for significant improvement in PV. We are sure that the ML applications that are discussed in this chapter are only the beginning of an ML-driven revolution in the PV sector.

## Author Biographies

**Ziv Hameiri** was awarded his PhD in 2011 from the University of New South Wales (UNSW). He then joined the National University of Singapore (NUS) for three years before returning to UNSW as a research fellow. In the last decade, his research focuses on the development of novel characterization methods for photovoltaic devices and applying them to investigate loss mechanisms in silicon and nonsilicon cells. Recently, he and his group have developed several machine learning algorithms for photovoltaic applications.

**Priya Dwivedi** is a postdoctoral research fellow at the School of Photovoltaic and Renewable Energy Engineering, UNSW, Australia, with an academic background in image processing, imaging systems, and machine learning. Her current research entails devising novel and faster characterization techniques based on machine learning algorithms for the photovoltaic industry. She completed her doctorate in lensless imaging systems, computational imaging in 2019. She has over two years of industrial experience in automating the process to reduce the cost of operations using machine learning, and data analytics.

## References

Abdullah-Vetter, Z., Buratti, Y., Dwivedi, P., Sowmya, A., Trupke, T., & Hameiri, Z. (2021). Localization of defects in solar cells using luminescence images and deep learning. IEEE 48th Photovoltaic Specialists Conference (PVSC), 0745–0749.

Acharya, A.K., Sahu, P.K., and Jena, S.R. (2020). Deep neural network based approach for detection of defective solar cell. *Materials Today: Proceedings* 39: 2009–2014.

Akram, M.W., Li, G., Jin, Y. et al. (2019). CNN based automatic detection of photovoltaic cell defects in electroluminescence images. *Energy* 189: 116319.

Artemyev, V., Smirnov, A., Kalaev, V. et al. (2019). Use of computer modeling for defect engineering in Czochralski silicon growth. *Journal of Power Technologies* 99 (2).

Bail, C.A. (2014). The cultural environment: measuring culture with big data. *Theory and Society* 43 (3–4): 465–482.

Bochkovskiy, A., Wang, C.-Y., & Liao, H.-Y. M. (2020). YOLOv4: optimal speed and accuracy of object detection. ArXiv:2004.10934 [Cs, Eess].

Borghesi, A., Pivac, B., Sassella, A., and Stella, A. (1995). Oxygen precipitation in silicon. *Journal of Applied Physics* 77 (9): 4169–4244.

Brandt, R.E., Kurchin, R.C., Steinmann, V. et al. (2017). Rapid photovoltaic device characterization through Bayesian parameter estimation. *Joule* 1 (4): 843–856.

Breiman, L. (2001). Random forests. *Machine Learning* 45 (1): 5–32.

Buerhop-Lutz, C., Deitsch, S., Maier, A., Gallwitz, F., Berger, S., Doll, B., Hauch, J., Camus, C., & Brabec, C. J. (2018). A benchmark for visual identification of defective solar cells in electroluminescence imagery. 35th European Photovoltaic Solar Energy Conference and Exhibition, 1287–1289.

Buratti, Y., Eijkens, C., and Hameiri, Z. (2020). Optimization of solar cell production lines using neural networks and genetic algorithms. *ACS Applied Energy Materials* 3 (11): 10317–10322.

Buratti, Y., Sowmya, A., Evans, R. et al. (2021). Half and full solar cell efficiency binning by deep learning on electroluminescence images. *Progress in Photovoltaics: Research and Applications* 1–12.

Buratti, Y., Sowmya, A., Dumbrell, R. et al. (2022). Automated efficiency loss analysis by luminescence image reconstruction using generative adversarial networks. *Joule* 6 (6): 1320–1332.

Chen, T., & Guestrin, C. (2016). Xgboost: a scalable tree boosting system. Proceedings of the 22nd Acm Sigkdd International Conference on Knowledge Discovery and Data Mining. 785–794.

Deitsch, S., Christlein, V., Berger, S. et al. (2019). Automatic classification of defective photovoltaic module cells in electroluminescence images. *Solar Energy* 185: 455–468.

Demant, M., Virtue, P., Kovvali, A. et al. (2019a). Learning quality rating of as-cut mc-Si wafers via convolutional regression networks. *IEEE Journal of Photovoltaics* 9 (4): 1064–1072.

Demant, M., Virtue, P., Kovvali, A. et al. (2019b). Visualizing material quality and similarity of mc-Si wafers learned by convolutional regression networks. *IEEE Journal of Photovoltaics* 9 (4): 1073–1080.

Fell, A. (2013). A free and fast three-dimensional/two-dimensional solar cell simulator featuring conductive boundary and quasi-neutrality approximations. *IEEE Transactions on Electron Devices* 60 (2): 733–738.

Forgy, E.W. (1965). Cluster analysis of multivariate data: efficiency versus interpretability of classifications. *Biometrics* 21: 768–769.

Friedman, J.H. (2001). Greedy function approximation: a gradient boosting machine. *Annals of Statistics* 1189–1232.

Girshick, R. (2015). Fast r-cnn. Proceedings of the IEEE International Conference on Computer Vision, 1440–1448.

Goodfellow, I., Pouget-Abadie, J., Mirza, M. et al. (2014). Generative adversarial nets. *Advances in Neural Information Processing Systems* 27.

Green, M.A. (2019). How did solar cells get so cheap? *Joule* 3 (3): 631–633.

Gu, X., Yu, X., Guo, K. et al. (2012). Seed-assisted cast quasi-single crystalline silicon for photovoltaic application: towards high efficiency and low cost silicon solar cells. *Solar Energy Materials and Solar Cells* 101: 95–101.

Han, H., Gao, C., Zhao, Y. et al. (2020). Polycrystalline silicon wafer defect segmentation based on deep convolutional neural networks. *Pattern Recognition Letters* 130: 234–241.

Haug, H. and Greulich, J. (2016). PC1Dmod 6.2 – Improved simulation of c-Si devices with updates on device physics and user interface. *Energy Procedia* 92: 60–68.

He, K., Zhang, X., Ren, S., & Sun, J. (2016). Deep residual learning for image recognition. Proceedings of the IEEE Conference on Computer Vision and Pattern Recognition, 770–778.

Hinton, G. and Sejnowski, T.J. (1999). *Unsupervised Learning: Foundations of Neural Computation.* MIT Press.

Hoffmann, M., Buerhop-Lutz, C., Reeb, L. et al. (2021). Deep-learning-based pipeline for module power prediction from electroluminescense measurements. *Progress in Photovoltaics: Research and Applications* 1–16.

Hotelling, H. (1933). Analysis of a complex of statistical variables into principal components. *Journal of Educational Psychology* 24 (6): 417.

Huang, C., Wu, P., Wang, L., and Yang, D. (2021). Effect of oxygen concentration on minority carrier lifetime at the bottom of quasi-single crystalline silicon. *Materials Science in Semiconductor Processing* 123: 105497.

Huang, G., Liu, Z., van der Maaten, L., & Weinberger, K. Q. (2017). Densely connected convolutional networks. Proceedings of the IEEE Conference on Computer Vision and Pattern Recognition, 4700–4708.

IEA-PVPS-Trends-Report (2021). IEA-PVPS-Trends-report-2021-1.pdf. Retrieved February 21, 2022, from: https://iea-pvps.org/wp-content/uploads/2022/01/IEA-PVPS-Trends-report-2021-1.pdf.

Ioffe, S., & Szegedy, C. (2015). Batch normalization: accelerating deep network training by reducing internal covariate shift, Proceedings of the 32nd International Conference on Machine Learning. 448–456.

Kayabasi, E., Ozturk, S., Celik, E., and Kurt, H. (2017). Determination of cutting parameters for silicon wafer with a diamond wire saw using an artificial neural network. *Solar Energy* 149: 285–293.

Kukačka, J., Golkov, V., & Cremers, D. (2017). Regularization for deep learning: a taxonomy. ArXiv:1710.10686 [Cs, Stat].

Kumar, V. and Maheshwari, P. (2021). Advanced analytics on IV curves and electroluminescence images of photovoltaic modules using machine learning algorithms. *Progress in Photovoltaics: Research and Applications* 30 (8): 880–888.

Kurchin, R., Romano, G., and Buonassisi, T. (2019). Bayesim: a tool for adaptive grid model fitting with Bayesian inference. *Computer Physics Communications* 239: 161–165.

Kurchin, R.C., Poindexter, J.R., Vähänissi, V. et al. (2020). How much physics is in a current–voltage curve? Inferring defect properties from photovoltaic device measurements. *IEEE Journal of Photovoltaics* 10 (6): 1532–1537.

Kutsukake, K., Nagai, Y., Horikawa, T., and Banba, H. (2020). Real-time prediction of interstitial oxygen concentration in Czochralski silicon using machine learning. *Applied Physics Express* 13 (12): 125502.

Lang, D.V. (1974). Deep-level transient spectroscopy: a new method to characterize traps in semiconductors. *Journal of Applied Physics* 45 (7): 3023–3032.

Li, J., Pradhan, B., Gaur, S., and Thomas, J. (2019). Predictions and strategies learned from machine learning to develop high-performing perovskite solar cells. *Advanced Energy Materials* 9 (46): 1901891.

Liu, L., Zhu, Y., Ur Rahman, M. R., Zhao, P., & Chen, H. (2019). Surface defect detection of solar cells based on feature pyramid network and GA-Faster-RCNN. China Symposium on Cognitive Computing and Hybrid Intelligence (CCHI), 292–297.

Liu, W., Anguelov, D., Erhan, D. et al. (2016). SSD: single shot multibox detector. In: *Computer Vision – ECCV 2016* (ed. B. Leibe, J. Matas, N. Sebe, and M. Welling), 21–37. Springer International Publishing.

Liu, X., Dang, Y., Tanaka, H. et al. (2022). Data-driven optimization and experimental validation for the lab-scale mono-like silicon ingot growth by directional solidification. *ACS Omega* 7 (8): 6665–6673.

Long, J., Shelhamer, E., & Darrell, T. (2015). Fully Convolutional Networks for Semantic Segmentation. https://arxiv.org/abs/1411.4038. arXiv:1411.4038v2 3431–3440.

Lu, S., Zhou, Q., Ouyang, Y. et al. (2018). Accelerated discovery of stable lead-free hybrid organic-inorganic perovskites via machine learning. *Nature Communications* 9 (1): 3405.

Lu, S., Zhou, Q., Ma, L. et al. (2019). Rapid discovery of ferroelectric photovoltaic perovskites and material descriptors via machine learning. *Small Methods* 3 (11): 1900360. https://doi.org/10.1002/smtd.201900360.

Marr, B. (2018). *How Much Data Do We Create Every Day? The Mind-Blowing Stats Everyone Should Read*. Forbes.

McCarthy, J. (2004). What Is Artificial Intelligence?. Computer Science Department Stanford University, 2007 (revised version). http://jmc.stanford.edu/articles/whatisai/whatisai.pdf.

McInnes, L., Healy, J., & Melville, J. (2018). Umap: uniform manifold approximation and projection for dimension reduction. ArXiv Preprint ArXiv:1802.03426.

Meng, Z., Xu, S., Wang, L. et al. (2022). Defect object detection algorithm for electroluminescence image defects of photovoltaic modules based on deep learning. *Energy Science & Engineering* 10 (3).

Murphy, K.P. (2012). *Machine Learning: A Probabilistic Prespective*. MIT Press.

Myers, R.H., Montgomery, D.C., and Anderson-Cook, C.M. (2016). *Response Surface Methodology: Process and Product Optimization Using Designed Experiments*. John Wiley & Sons.

Nakajima, T. and Sawada, K. (2017). Discovery of Pb-free perovskite solar cells via high-throughput simulation on the K computer. *The Journal of Physical Chemistry Letters* 8 (19): 4826–4831.

Ozturk, S., Kayabasi, E., Celik, E., and Kurt, H. (2018). Determination of lapping parameters for silicon wafer using an artificial neural network. *Journal of Materials Science: Materials in Electronics* 29 (1): 260–270.

Pearson, K. (1901). LIII. On lines and planes of closest fit to systems of points in space. *The London, Edinburgh, and Dublin Philosophical Magazine and Journal of Science* 2 (11): 559–572.

Powell, V. (2015). Principal Component Analysis explained visually. Explained Visually. https://setosa.io/ev/principal-component-analysis/.

Pratt, L., Govender, D., and Klein, R. (2021a). Defect detection and quantification in electroluminescence images of solar PV modules using U-net semantic segmentation. *Renewable Energy* 178: 1211–1222.

Pratt, L., Govender, D., and Klein, R. (2021b). Defect detection and quantification in electroluminescence images of solar PV modules using U-net semantic segmentation. *Renewable Energy* 178: 1211–1222.

Rasmussen, C.E. and Williams, C.K.I. (2006). *Gaussian Processes for Machine Learning*. MIT Press.

Redmon, J., & Farhadi, A. (2018). YOLOv3: an incremental improvement. ArXiv:1804.02767 [Cs].

Rein, S., Rehrl, T., Warta, W., and Glunz, S.W. (2002). Lifetime spectroscopy for defect characterization: systematic analysis of the possibilities and restrictions. *Journal of Applied Physics* 91 (4): 2059–2070.

Ren, S., He, K., Girshick, R., and Sun, J. (2015). Faster R-CNN: towards real-time object detection with region proposal networks. *Advances in Neural Information Processing Systems* 28.

Ronneberger, O., Fischer, P., and Brox, T. (2015). U-Net: convolutional networks for biomedical image segmentation. *Medical Image Computing and Computer-Assisted Intervention – MICCAI* 2015: 234–241.

Russell, S., & Norvig, P. (2002). *Artificial Intelligence: A Modern Approach.* Pearson, London, England, https://www.pearson.com/en-us/subject-catalog/p/artificial-intelligence-a-modern-approach/P200000003500/9780137505135.

Samuel, A.L. (1959). Some studies in machine learning using the game of checkers. *IBM Journal of Research and Development* 3 (3): 210–229. https://doi.org/10.1147/rd.33.0210.

Schubert, M.C., Schindler, F., Benick, J. et al. (2021). The potential of cast silicon. *Solar Energy Materials and Solar Cells* 219: 110789.

Sentaurus Device. (2019). Synopsys Inc. https://www.synopsys.com/manufacturing/tcad/device-simulation/sentaurus-device.html

Shapiro, L.G. and Stockman, G.C. (2001). *Computer Vision*, vol. vol. 3. New Jersey: Prentice Hall.

Shorten, C. and Khoshgoftaar, T.M. (2019). A survey on image data augmentation for deep learning. *Journal of Big Data* 6 (1): 60.

Simonyan, K., & Zisserman, A. (2015). Very deep convolutional networks for large-scale image recognition. ArXiv:1409.1556.

Song, L. and Yu, X. (2021). Defect engineering in cast mono-like silicon: a review. *Progress in Photovoltaics: Research and Applications* 29 (3): 294–314.

Srivastava, N., Hinton, G., Krizhevsky, A., Sutskever, I., & Salakhutdinov, R. (2014). Dropout: A Simple Way to Prevent Neural Networks from Overfitting. 30.

Su, B., Chen, H., and Zhou, Z. (2021a). BAF-Detector: an efficient CNN-based detector for photovoltaic cell defect detection. *IEEE Transactions on Industrial Electronics* 1–1.

Su, B., Chen, H., Chen, P. et al. (2021b). Deep learning-based solar-cell manufacturing defect detection with complementary attention network. *IEEE Transactions on Industrial Informatics* 17 (6): 4084–4095.

Su, B., Zhou, Z., Chen, H., & Cao, X. (2021c). SIGAN: a novel image generation method for solar cell defect segmentation and augmentation. ArXiv:2104.04953 [Cs, Eess].

Sun, W., Zheng, Y., Yang, K. et al. (2019). Machine learning-assisted molecular design and efficiency prediction for high-performance organic photovoltaic materials. *Science Advances* 5 (11): eaay4275. https://doi.org/10.1126/sciadv.aay4275.

Sutton, R.S. and Barto, A.G. (2018). *Reinforcement learning: An introduction.* MIT press.

Takahashi, I., Joonwichien, S., Kentaro, K., Matsushima, S., Yonenaga, I., & Usami, N. (2014). Improvement of annealing procedure to suppress defect generation during impurity gettering in multicrystalline silicon for solar cells. 2014 IEEE 40th Photovoltaic Specialist Conference (PVSC), 3017–3020.

Tanimoto, T.T. (1958). *Elementary Mathematical Theory of Classification and Prediction.* International Business Machines Corp.

Tian, S., Li, W., Li, S., Tian, G., Sun, L., & Ning, X. (2021). Image defect detection and segmentation algorithm of solar cell based on convolutional neural network. International Conference on Intelligent Computing and Signal Processing (ICSP), 154–157.

Van der Maaten, L. and Hinton, G. (2008). Visualizing data using t-SNE. *Journal of Machine Learning Research* 9 (11).

Vaswani, A., Shazeer, N., Parmar, N. et al. (2017). Attention is all you need. *Advances in Neural Information Processing Systems* 30.

Wagner-Mohnsen, H. and Altermatt, P.P. (2020). A combined numerical modeling and machine learning approach for optimization of mass-produced industrial solar cells. *IEEE Journal of Photovoltaics* 10 (5): 1441–1447.

Wagner-Mohnsen, H., Esefelder, S., Klöter, B., Mitchell, B., Schinke, C., Bredemeier, D., Jäger, P., & Brendel, R. (2021). Combining numerical simulations, machine learning and genetic algorithms for optimizing a POCl3 diffusion process. IEEE 48th Photovoltaic Specialists Conference (PVSC), 0528–0531.

Wang, Z., & Yang, J. (2018). Diabetic retinopathy detection via deep convolutional networks for discriminative localization and visual explanation. Workshops at the Thirty-Second AAAI Conference on Artificial Intelligence.

Xu, Z., Wu, Z., and Fan, W. (2021). Improved SSD-assisted algorithm for surface defect detection of electromagnetic luminescence. *Proceedings of the Institution of Mechanical Engineers, Part O: Journal of Risk and Reliability* 235 (5): 761–768.

Ying, Z., Li, M., Tong, W., & Haiyong, C. (2018). Automatic detection of photovoltaic module cells using multi-channel convolutional neural network. Chinese Automation Congress (CAC), 3571–3576.

Zhou, Z., Rahman Siddiquee, M. M., Tajbakhsh, N., & Liang, J. (2018). UNet++: a nested U-Net architecture for medical image segmentation. Deep Learning in Medical Image Analysis and Multimodal Learning for Clinical Decision Support, 3–11.

Zhu, H., Yan, W., Liu, Y. et al. (2022). Design investigation on 100 µm-thickness thin silicon PERC solar cells with assistance of machine learning. *Materials Science in Semiconductor Processing* 137: 106198.

Zhu, J.-Y., Park, T., Isola, P., & Efros, A. A. (2017). Unpaired image-to-image translation using cycle-consistent adversarial networks. Proceedings of the IEEE International Conference on Computer Vision, 2223–2232.

Zhu, M. (2011). Recall, Precision and Average Precision. Wayback Machine. https://web.archive.org/web/20110504130953/http://sas.uwaterloo.ca/stats_navigation/techreports/04WorkingPapers/2004-09.pdf

# 20

## Non-Destructive, Spatially Resolved, Contactless Current Measurement

*Kai Kaufmann, Tino Band, and Dominik Lausch*

DENKweit GmbH, Halle (Saale), Germany

## 20.1 Introduction

The introduction of automated soldering processes and the increasing usage of novel contacting methods to save costs can cause harmful defects, for example, missing and defective contacts, and thus, affect the electrical performance and longevity of solar modules after production or in-field usage. Furthermore, the trend toward completely new soldering technologies and solar module schemes, i.e. shingled solar cells, opens unknown physical and technical challenges. Established methods such as electroluminescence (EL) or thermography are mostly sensitive to voltage drops at the p-n transition (and effective lifetime according to Fuyuki approximation) (Fuyuki et al. 2005) or lack spatial resolution, speed, and sensitivity for industrial applications, respectively (Jahn et al. 2018). For the analysis of the contacting quality, electrical defects, and modern module schemes, it is necessary to obtain detailed information about the direction and strength of the electrical currents, ideally in high lateral resolution. One method that can be used to analyze electrical currents in a spatially resolved, noncontact, real-time manner is magnetic field imaging (MFI). In this chapter, the method is presented by means of different applications. In addition, the problem of quantifying the data by solving the inverse magnetic problem for the idealized conductor structure of a solar cell is assessed.

## 20.2 Theory and Practical Application to PV Modules

Any electrical current generates a magnetic field, which is described by a fundamental relationship in magnetostatics, the Biot–Savart law.

$$d\vec{B}(\vec{r}) = \frac{\mu_0}{4\pi} I d\vec{l} \times \frac{\vec{r} - \vec{r}'}{|\vec{r} - \vec{r}'|3}$$

where $\vec{B}$ is the magnetic flux density, $I$ is the current strength, $d\vec{l}$ is an element of the conductor, the constant $\mu_0$ is the permeability of free space, and $\vec{r}'$ and $\vec{r}$ are the location of the conductor and sensor, respectively.

*Photovoltaic Solar Energy: From Fundamentals to Applications*, *Volume 2*, First Edition.
Edited by Wilfried van Sark, Bram Hoex, Angèle Reinders, Pierre Verlinden, and Nicholas J. Ekins-Daukes.
© 2024 John Wiley & Sons Ltd. Published 2024 by John Wiley & Sons Ltd.
Companion website: www.wiley.com/go/PVsolarenergy

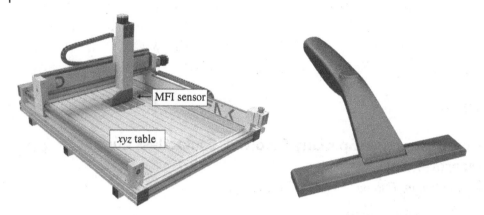

**Figure 20.1** The DENKWeit B-Lab device (left) consisting of a *xyz* table with an attached B-Tech sensor and a handheld device for field applications (right).

The method itself is not new. In 1989, it was shown that 2D current distributions can be investigated by magnetic field measurements (Roth et al. 1989). In the following years, several publications with applications in the field of photovoltaics were published (Paduthol et al. 2019; Kaufmann et al. 2021; Lausch et al. 2018).

Typical currents in solar applications range from a few mA to several A. This results in magnetic flux densities in the µT range at a few mm. The DENKweit magnetic field sensor B-TECH covers the measuring range from 1 µT up to 1.5 mT and is therefore well suited for solar applications. It is integrated into laboratory systems as well as in handheld devices (see Figure 20.1) developed by DENKweit. Currents can be generated either by illuminating the module or by applying an electric current to the module in the forward direction. The required current depends on the type of sensor used. For sensors with a measuring range in the µT range, currents that are comparable to STC conditions are suitable. For example, a current of 1 A at a measurement distance of 3 mm produces a flux density of 66 µT. The measurement itself is done by a rastering method, either by moving a point sensor in two directions or by moving a line sensor in only one direction. The Earth's magnetic field is subtracted from all measurements shown. With the B-TECH sensor, all three spatial components (Bx, By, Bz) are measured simultaneously.

The measurement distance is critical for the measurement. For an ideal line conductor, the magnetic flux density decreases reciprocally with the distance. The spatial resolution also decreases since sharply defined signals of line conductors become wider with increasing distance. For the examination of solar modules it, therefore, makes a difference if the measurement is performed from the glass side (sunny side) or through the back side foil if present. As an example, the evaluation of individual solder joints requires a measurement from the rear side since in most cases the spatial resolution is not sufficient for a front-side measurement through the glass. Broken cross-connections or cell connectors can also be detected through the glass-protected sunny side, as these defects result in large differences in the magnetic field distribution.

## 20.3 Application to Solar Modules

### 20.3.1 Cross and Interconnectors

Critical points for failure in the module are the cross-connectors and cell connectors. The cross-connectors are connected to the ribbons of the adjacent cell by solder joints. This connection can break due to thermomechanical loads, or the contact may not be well-formed from the start due to a nonoptimal soldering process. The absence of these electrical connections can lead to performance loss (Colvin et al. 2021). Missing electrical connections are often not easy to detect visually or even via EL imaging. Using MFI a quantitative evaluation of the solder quality can be performed.

Figure 20.2 shows the magnetic field components Bx of a solar cell at the bottom of a full module. A current of 8 A was applied to the module in forward direction. Bx is mainly dominated by currents flowing parallel to the busbars. Busbar 4 has no contact with the cross-connector, so no current can flow through this connection from the cell to the cross-connector and vice-versa. The current strength at the point of failure is 0 A, leading to an absent magnetic field after correction for the Earth's magnetic field. Towards the point of failure, the magnetic field decreases linearly, which can also be seen well in the 3D representation. Since no current can flow due to the missing connection at busbar 4, the current is redistributed to the other busbars, preferably to the nearest neighbors, in this case, busbar 3, since it is the nearest of the other busbars. The increased current flow is then reflected in a higher magnetic flux density. Furthermore, wave-like structures can be seen on the busbars in 3D view. These are caused by the solder joints on the back side of the module because the current changes direction here. The relationships are explained in detail in (Kaufmann et al. 2021). MFI can also visualize the current that contributes to this lateral redistribution. Figure 20.3 shows the By-component of the magnetic flux density of the same solar cell, which is mainly dominated by currents perpendicular to the busbars. Here, current flows from busbar 3 to busbar 4, which is shown by a spatially extended effect in the magnetic flux density. The fracture of the busbar here leads to cross-flowing currents

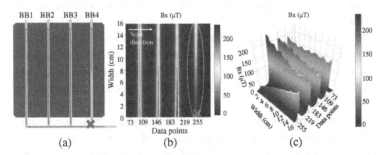

**Figure 20.2** (a) Sketch of a solar cell with one missing connection between the busbar and the cross connector. (b) Measured MFI of the Bx component of the flux density of solar cell is reduced near the missing connection. Besides this, the flux density increased in the neighboring busbar. (c) 3D representation of the same measurement. The wave-like structures on top of the busbar signals are caused by the individual soldering joints of the backside soldering.

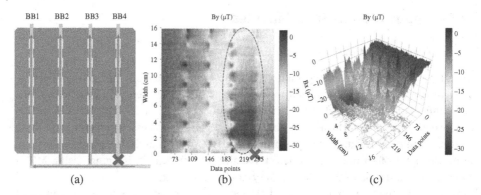

BB1  BB2  BB3  BB4

(a)

(b)

(c)

**Figure 20.3** (a) Sketch of a solar cell, with one missing connection between the busbar, is not connected to the cross-connector. (b) Measured MFI of the By-component, the solder joints become visible in this component because the current changes its direction when it passes the solder joint and also flows in the *x* and *z* directions. In busbar 4, no solder joints are visible, as the rear side busbar is separated from the cross-connector. The current is redistributing across the grid fingers which is visible as increased flux density between busbar 3 and busbar 4. (c) 3D representation of the same measurement.

over the grid fingers, which then become visible in the By-component. In addition, the individual solder connections on the back are visible in this component, which will be discussed later in more detail.

### 20.3.2 Soldering Joints

On both monofacial and bifacial Al-BSF or PERC solar cells, the ribbons on the rear side are usually contacted to the bulk silicon via a silver contact through a silicon nitride (SiN) passivation layer. The SiN layer has small openings to enable the contact between the metal contacts and the silicon. To save silver and reduce backside recombination at the poorer interfaces at the silver contacts, usually, only 10–15 individually isolated solder pads are used per ribbon. Therefore, there is only a small area where the electric current flows from the cell to the rear soldering pads of the solar cell or vice versa (depending on whether the current was injected electrically or generated by illumination) leading to the increased signal in By-component which can be seen in Figure 20.2. Due to the point-like connection, the electric current has to flow laterally to these single solder joints. This leads locally to an increased signal caused by two lateral currents from left and right toward the solder joints.

The solder joints could fail during the lifespan of the module, especially under thermo-mechanical stress. In Figure 20.4, the solder pads of a solar cell after zero, one hundred temperature cycles (TC100) from 40 to 85 °C, and after four hundred cycles (TC400) are compared. As a reference, the initial MFI image before thermal cycle (TC) testing is also shown in Figure 20.4. In this representation, the *z*-component of the flux density was chosen, and the derivative in the *y*-direction was calculated to show the effect of the current flow at the solder joints more clearly. At TC0, the individual soldering joints are barely visible. The press contact between the solder ribbon and the full-surface aluminum backside seems to be good enough for an electrical conduction. With further cycles, the connections

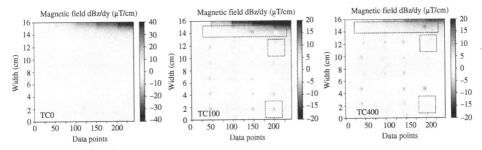

**Figure 20.4** Magnetic field of a solar cell after (a) 0, (b) 100, and (c) 400 TC cycles. For this representation, the z-component was derived in the y-direction. This facilitates the visibility of the solder pads. The soldering joints in the dotted areas vanish or get significantly weaker from TC100 to TC400. The strong signal on the top of the cell is caused by the cross-connector, which connects the individual bus bars.

become visible as the press contact weakens. One reason for this could be the formation of an oxide layer. However, the cause for this was not investigated in detail in this example.

From TC100 to TC400, several solder joints become invisible in the magnetic field image. The solder joints of interest are marked with a dotted box. The regions at the edge of the cell are most often affected, as this is where the mechanical stresses are greatest. This effect was observed frequently with MFI. The more thermal cycles a solar module undergoes, the more solder pads fail. A similar result is shown in an earlier study (Urban et al. 2019).

### 20.3.3 Shingled Modules

MFI can be used to optimize production methods and processes for new cell concepts. Shingled modules, for example, have higher power densities and their market share is expected to increase in the future due to a strongly simplified production process and reduced costs. Here, conductive adhesives are often used to connect the shingles. This can lead to new defects that are only relevant for this type of module. MFI provides access to local current distribution, which facilitates the evaluation of new contacting options and the analysis of defects.

Adjacent shingles are connected by a line of conductive adhesive. The adhesive ensures that current can flow from one shingle to the next across the entire cell width. In the fault-free state, the MFI scan shows no signals in the Bx component, since no currents flow in the y-direction, and a homogeneous signal in By is caused by the uniform current flow through the module in the x-direction. Various faults, such as short circuits or missing electrical contact, can occur at the contact line. Figure 20.5a shows a section of a shingled solar module. The module has a missing contact between two adjacent shingles at the marked location. In the EL image (Figure 20.5b), only a slightly darker area is visible. Figure 20.5c shows the Bx signal of an MFI scan of the region in question. In the region of reduced contact, an increased flux density parallel to contact line can be seen, which is caused by currents flowing inside the conductive adhesive. In the neighboring shingle, currents can be measured in the opposite direction, as the current flow here returns, it is redistributed over the full shingle area. Figure 20.5d shows the By-component. Clearly visible are regions of reduced flux density (and thus reduced current) in the area of the

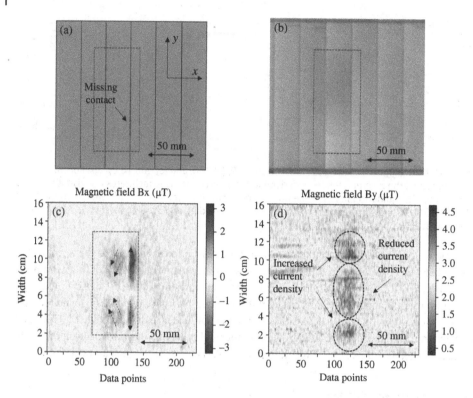

**Figure 20.5** (a) Shingled modules may have areas of missing contact (red dotted line). (b) Section of an EL image of a shingled solar module. The lack of contact only results in a slightly darker EL signal near the defect. (c) In the magnetic field, the currents along the contact line become visible. (d) In the area of the missing contact, the current density is reduced, and next to it, it increases.

missing contact. Above and below this, regions of increased flux density appear. These effects show that the current flows around the area without contact.

### 20.3.4 Diode Analysis in Junction Boxes

Defective bypass diodes in junction boxes often are frequent causes of module failures. For example, diodes can fail, i.e. be short-circuited or no longer conduct electricity in the forward direction. In addition, the junction boxes are usually potted in the field, so direct examination of the electrical components is not readily possible. Here, MFI offers the possibility to examine the function of the diodes in the field during operation. Short-circuited diodes already show up by a simple scan. In the case of a short circuit, current can flow in the reverse direction across the diode, which is then visible in the magnetic field image. Even faulty diodes that only partially block the current under normal operation can be easily detected.

Whether diodes work correctly in the case of shading can be shown in the field with a simple experiment. One cell of a string is shaded at a time, and the magnetic field is measured outside over the junction box. Figure 20.6 shows results from a non-shaded module

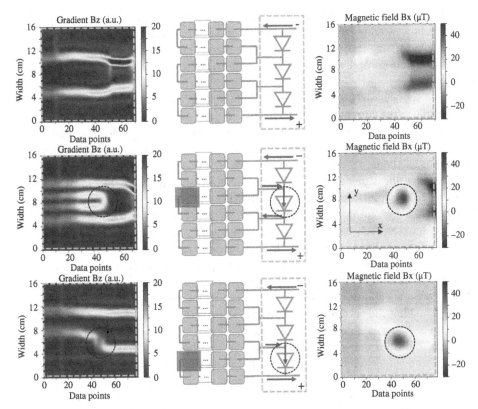

**Figure 20.6** Behavior of bypass diodes under different shading conditions with current path (left), equivalent circuit (middle), and the diode lighting up in the MFI (Bx-component), if a string is shaded.

and from two shading scenarios (string outside left, string in the middle) measured during daylight in a solar field. In the results with shading, the changed current path can be clearly seen. For better understanding, the equivalent circuit diagrams are shown. The current flow is marked with arrows for three different cases.

Without shading, you can see the ribbons leading in and out of the box. There is no current flowing through the diodes themselves. If a cell in the middle string is shaded, no current flows through the cells of the middle string. In this case, the current has to flow through the bypass diode. An additional current path corresponding to the equivalent circuit is visible in the Bz component. The position of the bypass diode is marked with a circle. Since here, the current flows within the diode in y-direction, a signal becomes visible in the Bx component. An analogous behavior shows up with shadowing of the lower string. In this case, the lower diode becomes visible in the Bx component.

### 20.3.5 Solder Joint Analysis in Junction Boxes

One common cause of solar module failures in the first year is the junction box solder contacts. The lack of redundancy in the system often leads to a total failure, which negatively affects the performance of the entire string. This means that the solder contacts could be subjected to detailed quality control to detect potential failures in the field. Currently, however, only random mechanical checks are carried out manually or by means of a pull-off test. Magnetic field analysis has significant potential for the future, paving the way for complete control, although further improvements are needed to fully realize this potential.

Soldered connections in junction boxes can be very different due to different box types. This also affects the current flow around the solder joint, which must be taken into account when analyzing the magnetic field images. As an example for the following results, two ribbons were soldered at a 90-degree angle. While this is only an approximation of a solder joint inside the junction box, it still shows the potential of magnetic field analysis. Representative measurements for the three B-field components Bx, By, and Bz of this 90-degree solder joint are shown in Figure 20.7a–c, respectively. The vertical sensor distance while measuring was approximately 1.5 mm. A current of 4 A was injected into the lower busbar. Each busbar has a width of 6 mm. Therefore, the solder joint has a quadratic shape and within that area, the current direction changes from "bottom-to-top" to "left-to-right". Those types of solder joints are typically for junction boxes. The related analysis and interpretation of those boxes are more complex in contrast to 180-degree solder joints since all three b-field components carry information about the current flow. Bx and By exhibit the currents in each busbar, namely the currents in $y$- and $x$-direction, respectively. The $z$-component approximately shows the mean path of the current. Obviously, the transition of the current flow inside the solder joint is sensitive to the quality of the solder joint itself. Consequently, the induced magnetic field is likewise influenced.

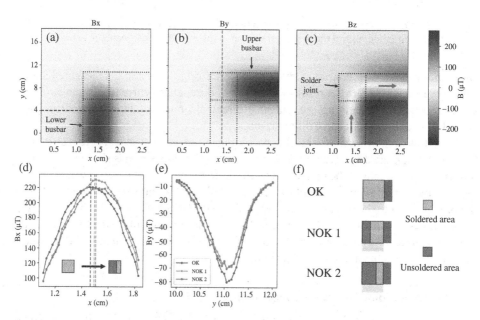

**Figure 20.7** Analysis of a 90-degree solder joint via MFI. (a)–(c) Bx, By, and Bz-component with implied solder joint (dotted lines) and current flow (red arrows). (d) Bx(x) profile at $y = 10.5$ cm (dashed blue line in MFI) and (e) By(y) profile at $x = 1.4$ cm (dashed red line) of OK and not OK (NOK) samples with (d) related legend.

The related Bx profile of the horizontal dashed blue line in Figure 20.7a is separately shown in Figure 20.7d (blue curve). Additionally, the same profiles for two solder joints with prepared errors are included. Sketches for the error characteristics are shown in Figure 20.7f. The successfully soldered area of the joint is homogenously reduced from left to right. The maximum of the Bx(x) curve shifts rightwards with increasing defect area, indicating a shifted current flow to the remaining successfully soldered area. Absolute measurement values differ by 10–20 μT, which is clearly above the measurement accuracy. Simultaneously, the By profiles close to the unsoldered area at $x = 1.4$ cm (vertical dashed red line in Figure 20.7b) indicate an increased or closer current flow in the OK solder joint at that position. That is also plausible since the current in the NOK-joints only flows, if any, in the lower busbar and not in the upper busbar due to the missing contact.

For industrial applications, such an intensive analysis of the current flow within a high-resolution MFI is not applicable, mainly due to time constraints, since the method requires a precise scan of the solder joint. Here, the primary importance is a quality control due to an error classification based on a fast and reliable measurement. This requires intelligent algorithms to deal with the big data from in-line measuring. Figure 20.8d shows a quality rating for solder joints based on a $4 \times 4$ grid sensor with a measurement area of $6 \times 6$ mm$^2$. According to that, each MFI has only 16 measurement values. The quality rating factor

$$q(B) = \frac{\sum B_i \cdot B_{\mathrm{OK},i}}{\sqrt{\sum B_i^2 \cdot \sum B_{\mathrm{OK},i}^2}}$$

of a solder joint corresponds to the mean cross-correlation of the b-field components to an OK solder joint with the values $B_{\mathrm{OK}}$. With that, a distinction between OK and NOK solder joints is possible by a threshold value (dashed red line). If the rating factor is above the threshold, the quality of the solder joint is good, else bad. Furthermore, a hierarchical clustering with the same metric also finds different error classes as indicated by the dashed black lines between the solder sketches. For the same clusters, the derived rating factor scales with the size of the defect area.

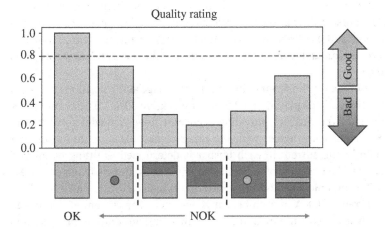

**Figure 20.8** Quality rating of different solder joints with specific unsoldered areas (red) based on a correlation analysis.

### 20.3.6 Inverse Problem, Quantitative Current Analysis

When comparing MFI of different cell types, investigating efficiencies, photocurrent, or insulation problems, a quantitative measurement of the current is mandatory, since the magnetic flux also depends on the distance between the sensor and the current source. To deduce the current from the MFI, one must solve the so-called inverse magnetic field problem. Knowing the current distribution in space makes it straightforward to calculate the magnetic flux at a certain position. However, the inverse way is ill-posed and has, in general, infinite solutions. Assumptions about the specific measurement setup are unavoidable to take care of that. In this way, a two-dimensional current distribution can be estimated from the discretized Biot–Savart law (Brauchle et al. 2021). An alternative for a system of conductors with simple geometries is assuming infinitely long wires. The corresponding in-plane field perpendicular to the current direction of such a wire can be calculated from Biot–Savart by:

$$B_{xy}(x, I, z, x_c) = \frac{\mu_0 I}{2\pi} \frac{z}{(x - x_c)^2 + z^2}$$

where $x$ is the $x$-position of the sensor and $z$ is the vertical distance between the sensor and the conductor, which is at the position $x = x_c$. This function has an extreme point at the position of the conductor, and the full width at half maximum is $2 \cdot z$, so that the width of the curve is not influenced by the current. According to that, the parameters $I$, $z$, and $x_c$ can be estimated via nonlinear regression of a profile from a spatially resolved MFI.

Figure 20.9a shows a sketch of a 2-half-cell mini-module with a broken cell connector at busbar 3 (marked by a cross). In Figure 20.9b,c are the corresponding magnetic field components Bx and By. A current of 4 A was applied to the module in the laboratory. The current flows into the mini-module via the top cross-connector and then splits according to the number of busbars. According to the proportion of the current flowing into each busbar, the current in the cross-connector decreases, resulting in a step-like decrease in flux density. The Bx- and the By-components contain almost exclusively the busbar and the cross-connector currents, respectively, and currents in $z$-direction can be ignored. Therefore, the solar cell can be described by seven simple conductors – five busbars and two cross-connectors. Of course, this is also a simplification of the cell since a busbar consists of two conductors – one at the bottom and one at the top of the semiconductor. Those two conductors cannot be separated due to the small thickness of the semiconductor compared to the sensor distance.

The related profiles of the two horizontal lines in the Bx image at the positions $y = 14$ cm and $y = 20$ cm are additionally displayed in a 2d diagram (Figure 20.9d). The profiles obey peaks at the positions of the busbars providing a current flow. A model consisting of five or rather four straight wires agrees with the experimental data, as the fitting lines point out. Consequently, the current in each busbar is determined as fitting parameter. The so-obtained spatial dependency of each busbar current is shown in Figure 20.8e. The broken cell exhibits inhomogeneous busbar currents, which contrasts with an undamaged cell, where a current of 0.8 A in each busbar is expected. The current of the broken busbar 3 drops linearly from 0.6 A to zero at the cell connector. The adjacent busbars 2 and 4 carry significantly more current due to cross currents from the broken busbar as already discussed. Each current drops gradually to zero at the edges of the busbars, where the

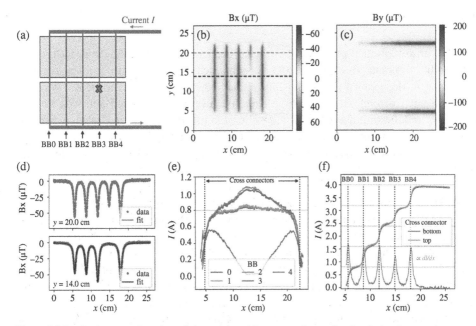

**Figure 20.9** Basic example solar mini-module with current determination in busbars and cross connectors. (a) Sketch of solar mini-module, (b) Bx- and (c) By-components from magnetic field imaging, (d) Bx-profiles with nonlinear regression fitting, (e) spatial dependence of busbar currents, and (f) currents in each cross-connector.

cross-connectors are. This transition is due to the insufficient model describing the end of conductors since infinitely long wires are assumed. Nevertheless, the estimated currents in the cross-connectors also show in good approximation the expected stepwise behavior along with the absolute values (Figure 20.9f). The step height at busbar 3 is noticeably smaller than the others, indicating the broken cell connector due to higher resistance in this busbar.

## 20.3.7 Conclusion

MFI, as a measurement technique, offers a new approach to the visualization of electric currents flowing in solar cells and modules generated by light or injected under production conditions. MFI differs significantly from EL, where the measurement signal depends on the voltage drop at the pn junction, while the measurement signal of MFI is directly proportional to the flowing current. By evaluating the individual spatial components of the magnetic flux density, conclusions can also be drawn about the direction of the current, which contributes to a detailed understanding of defects and their effects on module efficiency. The method offers added value in quality control and process optimization in module production. We showed examples of inspection of cross-connectors, back-side solder joints, and solder joints in junction boxes. Moreover, the thorough investigation of novel techniques and concepts, exemplified by the analysis of shingled solar modules, can yield valuable insights using the highly reproducible and quantitative measurement provided by MFI. This opens up the possibility of tracing back electric currents, making it a powerful tool for further research and development.

## Author Biographies

**Kai Kaufmann** studied physics at Martin Luther University in Halle (Saale), Germany, earning his diploma in 2011. In 2018, he obtained his Ph.D. with a focus on laser processing in the field of CIGS thin film photovoltaics. During this time, Kai was part of PV research projects at Hochschule Anhalt University of Applied Sciences Köthen, Germany, and Fraunhofer CSP. In 2018, he cofounded DENKweit GmbH, where he contributed significantly to the development of magnetic field measurements of solar modules. Now he serves as the company's COO, driving innovative solutions in renewable energy.

**Tino Band** received his master's degree in physics from Martin Luther University in Halle (Saale) in 2014, followed by his PhD in 2023. During this time, he specialized in polymer-based dielectric materials for energy storage. As early as 2020, he joined the distribution grid operator, where he was involved in research projects and led an internal project on future control of the low-voltage grid. At the end of 2021, he moved to Anhalt University in Köthen, where he closely collaborated with DENKweit GmbH on a magnetic field measurement project. Since 2023, Tino has been an integral member of DENKweit, overseeing the application site of the magnetic field technology including sensor development, simulation, and data analysis.

**Dominik Lausch** received a diploma degree in physics from the University of Leipzig, Germany, in 2009. During his studies, he worked with the company Q-Cells SE on various subjects, including crack detection and pre-breakdown luminescence. In 2012, he received a Ph.D. degree from the University of Halle, Halle (Saale), Germany, in cooperation with the Fraunhofer Center for Silicon Photovoltaics CSP, Halle, and Q-Cells SE, Thalheim, Germany. His dissertation explored the subject of the "Influence of recombination active defects on the electrical properties of recombination active defects in silicon solar cells," which received the PVSEC Student Award for outstanding scientific research in 2012. After finishing his Ph.D.Ph.D. work, he built up a team at Fraunhofer CSP dealing with plasma texturing and hydrogen passivation topics. In 2018, he founded the company DENKweit GmbH where he now works as CEO.

## References

Brauchle, F., Grimsmann, F., Kessel, O.v., and Birke, K.P. (2021). Direct measurement of current distribution in lithium-ion cells by magnetic field imaging. *Journal of Power Sources* 507: 230292. 0378-7753.

Colvin, D.J., Schneller, E.J., and Davis, K.O. (2021). Impact of interconnection failure on photovoltaic module performance. *Progress in Photovoltaics: Research and Applications* 29: 524–532.

Fuyuki, T., Kondo, H., Yamazaki, T. et al. (2005). Photographic surveying of minority carrier diffusion length in polycrystalline silicon solar cells by electroluminescence. *Applied Physics Letters* 86: 262108.

Jahn, U., Herz, M., Köntges, M. et al. (2018). Review on infrared and electroluminescence imaging for PV field applications. In: *International Energy Agency Photovoltaic Power Systems Programme, IEA PVPS Task 13, Subtask 3.3, Report IEA-PVPS T13-12*. Paris: International Energy Agency.

Kaufmann, K., Lausch, D., Lin, C. et al. (2021). Evaluation of the quality of solder joints within silicon solar modules using magnetic field imaging. *Physica Status Solidi A: Applications and Materials Science* 218: 2000292.

Lausch, D., Patzold, M., Rudolph, M. et al. (2018). *Proc. of the EU-PVSEC, Brussels*, 1060–1064.

Paduthol, A., Kunz, O., Kaufmann, K. et al. (2019). Magnetic field imaging: strengths and limitations in characterizing solar cells. In: *2019 IEEE 46th Photovoltaic Specialists Conference (PVSC)*, 0822–0824.

Roth, B.J., Sepulveda, N.G., and Wikswo, J.P. (1989). Using a magnetometer to image a two-dimensional current distribution. *Journal of Applied Physics* 65: 361.

Urban, T., Mette, A., Müller, M., and Heitmann, J. (2019). Front and rear pressure contact degradation in solar modules due to thermal cycle treatment. *IEEE Journal of Photovoltaics* 9 (5): 1360–1365.

# Part Eight

# PV Modules

# 21

# Advanced Industrial High-Efficiency Silicon PV Module Design

*Shu Zhang[1,2], Xue Chen[1], Jianmei Xu[1], and Pierre J. Verlinden[1,3,4,5]*

[1] *R&D Department, State Key Laboratory of Photovoltaic Science and Technology, Trina Solar, Changzhou, Jiangsu, 213031, China*
[2] *College of Materials Science and Technology, Nanjing University of Aeronautics and Astronautics, Nanjing, Jiangsu, 211106, China*
[3] *AMROCK Pty Ltd, McLaren Vale, SA 5171, Australia*
[4] *School of Photovoltaic and Renewable Energy Engineering, UNSW, Sydney, Australia*
[5] *Yangtze Institute of Solar Technology, Jiangyin, China*

## 21.1 Introduction

With continuously increasing requirements for cost reduction and efficiency increase to bring photovoltaic (PV) power generation costs down to be competitive with any other energy source, the PV market demand for larger and higher power modules with high-efficiency cells continues to rise. The improvement in module conversion efficiency is mainly realized by improving cell efficiency, reducing optical losses in encapsulation materials, and optimizing module layout design. The factors affecting module efficiency are (i) losses due to non-generating areas, such as frame, gaps between cells, gaps between cells and frame; (ii) front cover glass reflection; (iii) front cover glass absorption; (iv) reflection at the encapsulation material glass interface due to a refractive index mismatch; (v) absorption in the encapsulation material; (vi) shading by ribbons or wires; (vii) lack of total internal reflection (TIR) for the light reflected by fingers, busbars, ribbons, wires or backsheet; (viii) interconnection losses (cells and strings); (ix) electrical mismatch between cells; and (x) junction box, wire, and connector losses. Most factors lead to a decline in module efficiency compared to the cell efficiency, but a good optical design of the module could also result in a module power output higher than the sum of the power output of the cells (for example, when the gap between cells is large and the module benefits from TIR for the light reflected by a white backsheet) and, sometimes but rarely, in a higher module efficiency than the average cell efficiency, mostly due to good TIR. Over the years, module and material manufacturers have deeply studied the factors mentioned above to improve the power and efficiency of PV modules through various technologies and have recently achieved tremendous improvements in module technology. At the present time, a modern PV module usually incorporates the following improvements which were implemented over the last five years: (i) large silicon wafer technology, (ii) multi-busbar (MBB) technology with wire interconnection, or Smart Wire Connection Technology

*Photovoltaic Solar Energy: From Fundamentals to Applications, Volume 2*, First Edition.
Edited by Wilfried van Sark, Bram Hoex, Angèle Reinders, Pierre Verlinden, and Nicholas J. Ekins-Daukes.

(SWCT™) technology, (iii) advanced layout design: half cut and trisection cut, even multi-cut technology, (iv) non-destructive cutting (NDC) technology, (v) high-density encapsulation technology: small spacing, overlap soldering technology or shingled module technology, (vi) bifacial module design, and (vii) reduction of encapsulation loss through the improvement of encapsulation materials. This chapter will review the most recent technological improvements in crystalline silicon PV module technology.

## 21.2 Large Silicon Wafers

The wafer size in the PV industry has been growing in the same way that the wafer size has increased in the semiconductor industry. Partially driven by "Moore's Law," which observed back in 1965 that the number of gates per integrated circuit roughly doubles every two years, a "law" that still appears valid today and which has been used to guide long-term planning, the wafer size in the semiconductor industry has grown from 2″ in 1960s to 12″ (300 mm) in 2005 (Mack 2011). Similarly, in the PV industry, the standard wafers for manufacturing crystalline silicon solar cells have grown from 3″ in diameter in the 1970s to 210 mm square wafers, corresponding to an ingot diameter of 295 mm in 2021. The evolution of wafer size is shown in Figure 21.1. The 5-inch, 6-inch, and 8-inch round wafers for the semiconductor industry correspond to 100, 125, and 156 mm edge lengths, respectively, for square or quasi-square wafers for the PV industry. Due to the need to reduce costs and increase the module power output, PV wafer manufacturers have fine-tuned the wafer size several times based on 8-inch wafer technology from 2010 to 2020, which has allowed them to increase the module power output without fundamentally changing the manufacturing tools. As a result, various sizes of silicon wafers, such as M2 (156.75 mm), M4 (161.75 mm), G1 (158.75 mm), and M6 (166 mm), have appeared, almost simultaneously, in the market

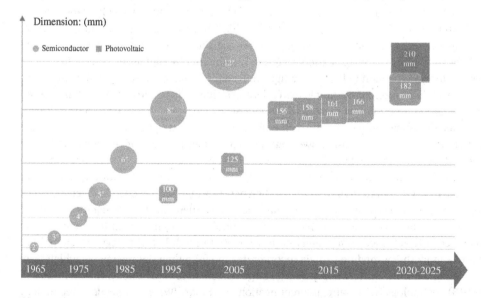

**Figure 21.1** Development of silicon wafer size in the semiconductor and PV industry.

and have created a bit of "chaos" in PV manufacturing. The variety of wafer sizes leads to a wide range of module powers and sizes, causing many challenges for downstream system design and applications. The lack of a uniform standard for power and size, including the position of mounting holes for the modules, and the quick change in specifications from one year to the next have surprised customers and complicated the work of PV system designers.

In 2019, Tianjin Zhonghuan Semiconductor Co. Ltd announced that 12-inch Cz-Si technology would be applied to the solar industry and launched the 210 mm PV square monocrystalline silicon wafer, or G12 wafer, based on 295 mm ingot (Kopecek and Libal 2021). About at the same time, Longi announced the launch of 182 mm quasi-square monocrystalline silicon wafer or M10 wafer. Since then, these two wafer standards have dominated the PV industry and forced the other standards to disappear gradually from the market. It is expected that by 2023, only these two standards, M10 and G12, will remain in the market. So far, the possible next-generation of 18-inch wafers has not been able to replace the 12-inch wafers (M10 and G12) for mainstream size due to issues such as the hot-field size in ingot puller, crystal pulling process, breakage rate in slicing and cell manufacturing, and high cost. However, having at present two wafer-size standards is not good for the PV industry, as it slows down the overall standardization and cost reduction. The M10 wafer standard allows for keeping existing manufacturing tools with some obvious adaptations and seems to be a transition standard, while the G12 standard requires completely new factories and is expected to be the wafer standard for at least the next decade, which will help to standardize the design of PV systems.

The PV industry will continue to reduce the levelized cost of electricity (LCOE) per kilowatt hour to satisfy customer needs, and advanced PV module technology development is an important part of this process. Ultra-high-power modules will become the mainstream product of the PV industry, at least for utility-scale applications, and modules made with large cells (M10 and G12) will have the largest market share. As shown in Figures 21.2 and 21.3 (Lin et al. 2022), the manufacturing capacities of the large-size cells and modules

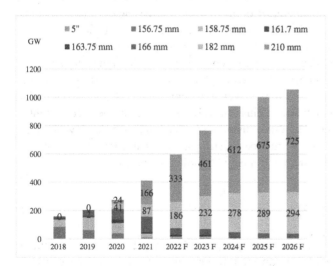

**Figure 21.2** Current and projected cell manufacturing capacity per wafer size.

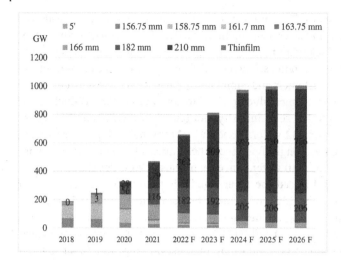

**Figure 21.3** Current and projected module manufacturing capacity per wafer size.

will gradually increase in the next several years, far exceeding the capacities of other sizes, which will are expected to disappear in a relatively short period of time.

Typically, 60 or 72 solar cells are encapsulated in series to form a PV module, which are connected in series to form strings. Note that, with half-cut cells, sub-strings are also connected in parallel within a module, which has a significant advantage in reducing the cell temperature in the case of partial shading. The strings are then connected in parallel to form an array, and the arrays then form a PV power plant. For a given power plant capacity, the higher the string power, the fewer the number of strings, which reduces the cost per watt of the various components, such as the mounting system, foundation piles, cables, and cable tray, thus reducing the whole system cost. However, in a utility-scale power plant, since the maximum string voltage is usually fixed, limited by the maximum DC voltage of the inverter and the minimum temperature of the location, increasing the string power by increasing the voltage of the module does not help. One can only increase the string power by increasing the module's current, hence the advantage of large wafer sizes.

Recently, it became clear to developers that using modules with large wafer sizes can greatly reduce the balance of system (BOS) cost of a power plant, thanks to its high current, low voltage (because of large wafers), high power, and allowing for a better layout design with higher current and power per string. This results in a reduction in the overall number of modules, while the number of modules (using a smaller number of large wafers) per string is increased, a reduction of the number of strings, the cable length, the amount of steel used in mounting systems, and the number of foundation piles.

## 21.3  Multi-Busbar (MBB)

Originally developed by the German company Schmid-Group, MBB has become (Braun et al. 2013) the mainstream cell interconnection technology in the industry since 2017 as a technical solution to achieve the best balance between reducing electrical losses and improving optical utilization (Liu et al. 2020).

The application of MBB technology significantly shortens the path of current flowing through the fingers to the busbar, effectively reducing the power loss in the metal fingers and improving the current collection capability (Papargyri et al. 2020). Moreover, the sensitivity of the module power to micro-cracks is significantly reduced, as well as the cell and interconnection series resistance, thereby increasing the module power by 1–1.5% relative compared to the five-busbar (5BB) design, as shown in Figure 21.4a,c. MBB is also a necessary technology for large-size solar cells. For example, if a 5BB design would be used in large wafers such as 210 mm cells, the current lateral collection pathlength would increase by more than 30% compared with the 158 mm cells, and the finger series resistance would increase by more than 75%.

Another advantage of MBB technology is its increased optical efficiency. The use of solder-coated round wire, instead of flat ribbons, enables 75% of the incident light that is reflected by the wires to be reflected again by TIR at the front glass-air interface and directed back to the cell surface, and this independently of the incidence angle. In comparison, with flat ribbon technology that is applied in traditional 5BB modules, only about 5% of the light reflected by the ribbon is reflected by TIR back to the cell surface, as shown in Figure 21.4b. The improved optical utilization of the module leads to a 1–1.5% power increase.

SWCT™ developed by Meyer Burger, is an alternative to the standard MBB technology (Papet et al. 2015). SWCT™ offers the same advantages as the standard MBB technology, however, they are not identical. The main innovation of SWCT™ is that the cells are busbar-less, and the multi-wires are connected to the fingers during the module lamination process instead of soldered before lamination, as in the case of the MBB. After PECVD, the fingers of the cell are screen printed, but without busbars. The SWCT™ wires, made of solder-coated copper wires, are attached to a polymer film that is laminated onto the cell during the stringing step (Pieters et al. 2019). The surface of the copper wires embedded in the polymer film is coated with a solder alloy with a low melting point, typically Sn–Bi or Sn–Bi–Ag. The soldering process happens during the module lamination step at a relatively low temperature (~140 °C), the pressure and heat in the laminator helping the copper wires and the screen-printed fingers to form an ohmic contact (Figure 21.5).

The copper wires at the end of the string are also connected together to a wide ribbon, in the same lamination step. SWCT™ has similar advantages as MBB with, in addition, the advantage of reducing the amount of screen-printed silver paste, but the disadvantage of a significant bismuth consumption.

Similar to the principle of SWCT™, the innovative TWILL technology developed by IMEC is based on a weaved-style fabric structure with electrically conductive solder-coated wires weaved into a non-conductive polymer fabric (Govaert 2020). It was originally designed to interconnect Interdigitated Back Contact cells, but recently was applied to Building Integrated PV modules. Unlike SWCT™, TWILL ribbons are soldered to the cells before lamination.

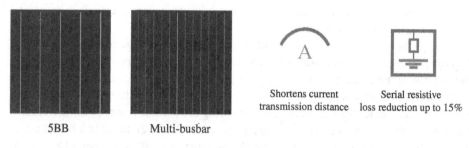

5BB        Multi-busbar

Shortens current transmission distance    Serial resistive loss reduction up to 15%

Shortens current transmission distance on the finger
by more than 50%, reducing power loss due to internal resistance

(a)

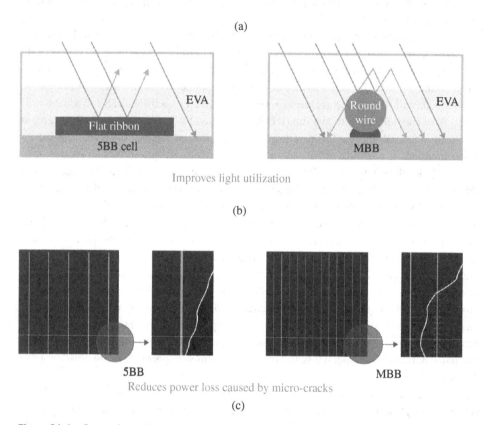

Improves light utilization

(b)

Reduces power loss caused by micro-cracks

(c)

**Figure 21.4** Comparison of multi-busbar (MBB) and 5BB cells in terms of (a) current transfer path, (b) light utilization, and (c) resistance to micro-cracks.

**Figure 21.5** Smart wire connection technology (Pieters et al. 2019).

## 21.4  Diversified Layout Design

Traditional PV modules generally use a $6 \times 10$ array of cells, for light-weight rooftop modules, or a $6 \times 12$ layout, for more cost-effective utility-application modules, where 6 represents the number of cell columns, or sub-string, of a PV module and 10 or 12 represents the number of series-connected cells within each cell column. Nowadays, with the increase in silicon wafer size, modules with a different layout, e.g. $5 \times 8$, $5 \times 10$, $5 \times 11$, $6 \times 10$, and $6 \times 11$, are becoming available to provide customers with more options for different applications. These layouts are designed to balance the modules' electrical performances, optimize the area and weight, maximize the number of modules per shipping container, and improve the installation compatibility while avoiding additional costs. In addition, parallel interconnections within the modules allow for a better tolerance to partial shading and, therefore, improve reliability.

The change in layout design comes simultaneously with the adoption of multi-cut technology. In addition to the resistive losses in the cells, the electrical power loss within the module includes losses in ribbons or wires, busbars, cables and connectors, in which the ribbons or wires account for the largest proportion of the loss. Therefore, cutting the cells in half, effectively reduces the resistive losses in the current interconnection scheme and improves the module power output. There are two ways to reduce the power loss in the ribbons or wires: (i) to increase the cross-sectional area of the ribbons or wires or (ii) to reduce the current flowing through the ribbon. Since most of the power loss is due to the ribbon or wire interconnects, the technology of cutting the cell into half or trisection, as shown in Figures 21.6 and 21.7, allows to reduce the current in each ribbon

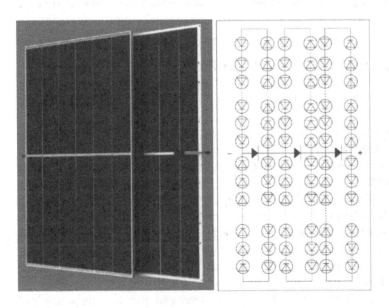

**Figure 21.6**  Half-cut module and circuit diagram.

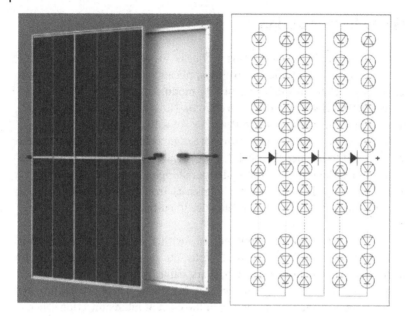

**Figure 21.7** Trisection-cut module and circuit diagram.

or wire by a factor of 2, 3 and reduce the interconnection resistive loss in the ribbons or wires by a factor of 4 or 9 (Zhang et al. 2020). The resistance loss of ribbon in the half-cut module can be reduced by 75%, and the loss of cutting into trisection or multi-section can be further reduced. At present, half-cut technology has become the mainstream of the industry.

## 21.5 Non-Destructive Cutting (NDC) Technology

With the development of multi-cut technology and the observation of non-negligible efficiency loss after traditional cutting, NDC technology became a critical element in the development of the recent large modules made with very large cells. The NDC technology was also found able to reduce the potential risk of the formation of micro-cracks due to mechanical stresses of the modules (wind, snow, handling, transport) observed with large wafer sizes (Wang and Yang 2021).

Traditional cell cutting is divided into two steps. First, the cell surface is ablated and grooved by a laser, which can locally create a very high temperature (over 1500 °C). After the creation of the groove, the cell is separated along the laser line using mechanical stress in a step called cleavage. This technique will generally introduce micro-cracks on the cutting edge, as shown in Figure 21.8b, which eventually affects the efficiency, as well as the mechanical strength of the cell.

To address this problem, the NDC technology was developed. Based on the principle of thermal expansion and contraction instead of ablation of silicon, the cells are naturally

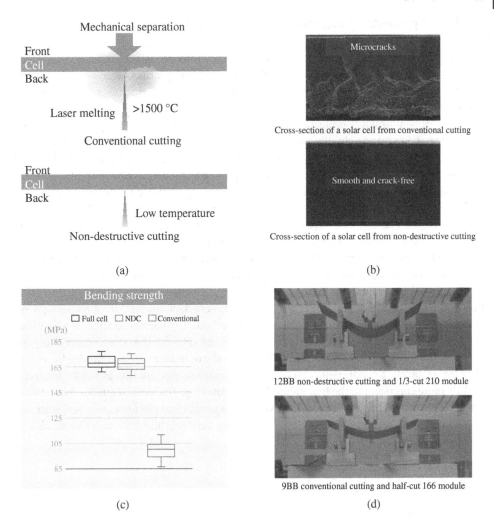

**Figure 21.8** Comparison of traditional and non-destructive cutting techniques: (a) schematic diagram, (b) cross-section, (c) three-point bending test, and (d) 210 mm non-destructive cutting (NDC) versus 166 mm traditional cutting.

separated due to thermal stress without the formation of a groove using low-temperature laser technology. As shown in Figure 21.8b, the cutting edge is very smooth, without any micro-cracks. The mechanical strength of the cell after NDC is comparable to that of the full cell and is much higher than that produced using traditional cutting techniques, as shown in Figure 21.8c (Zhang et al. 2020).

Mechanical three-point bending tests show that the 210 mm non-destructive cut cells have a bending strength of about 165 MPa or about 1.5 times that of traditional cut cells (about 100 MPa), and the cell deformation limit is 1.5–2.5 times higher than that of 166 mm half cells with traditional cutting, as shown in Figure 21.8d.

## 21.6 High-Density Assembly

The high-density assembly techniques can be divided into two types: the cells connected with ribbons or wires with solder or the cells connected without ribbons or wires, in which case the interconnection technology requires an electrically conductive adhesive (ECA) (Beaucarne et al. 2015).

For the ribbon- or wire-connection type, as shown in Figure 21.8a, the adjacent cell gap within one string was originally, and until very recently, around 2–3 mm. To increase the module efficiency, this gap must be reduced, and more cells need to be placed in the limited area of the module. With breakthroughs in technologies related to ribbon treatment and soldering processes, high-density encapsulation technology has been developed.

The high-density encapsulation techniques with cells connected with ribbons or wires can also be divided into two types, as shown in Figure 21.9. The first method consists of flattening the wires in the area between the two adjacent cells. The gap between cells is reduced to around 0.5 mm, which reduces the module size by about 1.5% for a typical 72-cell half-cut module and effectively improves the module efficiency by the same relative amount. In this method, there is still a small gap between cells. However, the yield risk in the production process and the risk of micro-cracks forming during handling and operation are reduced compared to the second method. The second method of assembly consists of overlapping a small part (0.2–0.5 mm) of one cell with the adjacent cell. The gap between cells is completely eliminated. This method can reduce the module size by about 2.3% for a typical 72-cell module and increase the module efficiency by about 2% relative, slightly more compared to the first method. Due to overlap, this method creates more damage and breakage in the production process, and the production yield is somewhat lower. In addition, the risk of micro-cracks in the cells during handling and operation (wind, snow) is increased.

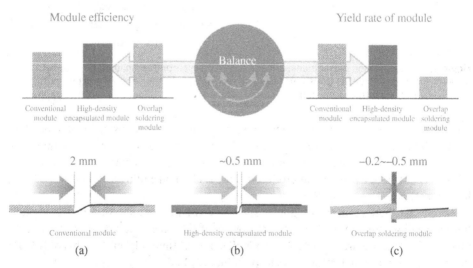

**Figure 21.9** Schematic diagram of the soldering structure of traditional, high-density encapsulated and overlap soldering modules and comparison of their efficiencies and yields.

The techniques described above can achieve high-density encapsulation, even with ribbons or wires and soldering, but the highest-density encapsulation technology was developed earlier. It does not use ribbons or wires and uses ECA instead of soldering. The module design consists of shingled cells, like the tiles on the roof of a house. This technology was first developed in the 1960s by D. Dickson et al. of Hoffman Electronics Corporation (US Patent 2,938,938) and also by 1970s by the US government (US Patent 3,769,091). The technology of solar cell shingling was later re-developed for modern PV modules by companies like Sun Power, Solaria, Seraphim and PI Berlin (Patent application US2016/0163902). The technology uses solar cells processed through a multi-cut laser tool and cut into strips of 1/5 or 1/6 of the original cell. The edges of the small cell strips are overlapped, covering the busbar along the long strip edge and edge bonded together with ECA. By overlapping the cells in the module and hiding the busbars, the number of cells and active area of the module is increased.

As shown in Figure 21.10, the string of shingled cells, using ECA as the bonding agent, is flexible and relatively easy to handle before lamination. Parallel connections of the long strings of shingled cell strips show an excellent tolerance to partial shading. On the other hand, the cutting losses increase with the number of cuts. Also, the electrical loss by series resistance in the long fingers running perpendicular to the busbar and the shingle edge is

**Figure 21.10** Shingled cell strips connected by ECA with excellent flexibility.

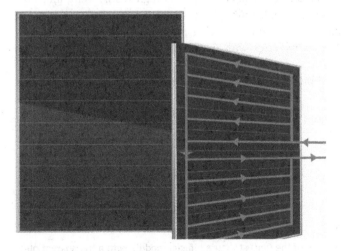

**Figure 21.11** Shingled module and the layout.

much greater than for an MBB or SWCT™ interconnection. Together, these two additional losses largely lower the gain brought by the shingle technology, featuring a significant reduction of the inactive areas of the module. In addition, the product's reliability needs to be further verified, especially the reliability of the long-term outdoor performance. One example of a shingled module and its layout is shown in Figure 21.11.

## 21.7 Bifacial Modules

Because the back side of the cell can absorb the light reflected by the ground surrounding the modules, as well as diffused sunlight (for example, coming from the clouds), bifacial modules potentially have higher power generation than mono-facial modules (Figure 21.12). In good conditions, the fraction of the light reflected by the ground, called albedo, can be significant, for example, with snow-covered ground or above white gravel or concrete. In these conditions, the annual energy generation gain can be increased by about 5% with a grass ground cover up to about 30% when white gravel is used. The specific energy gain is highly influenced by the ground reflectivity, the array height above ground, the spacing between rows, the rack design, the tracking array or fixed tilt and the surrounding environment. Over the recent years, the overall market share of bifacial modules has increased to around 40%.

Bifacial modules have two main encapsulation structures: (i) double glass structure and (ii) glass integrated with a transparent backsheet structure (Zhou et al. 2020; Hua 2020). The double-glass modules are traditionally significantly heavier than glass-backsheet modules, which brought some difficulties in the installation process and restricted their usage to ground-mounted PV systems. Nowadays, double-glass modules, with a thickness of two glass panels as low as 1.6 mm, have the same weight as standard glass-backsheet modules with 3.2 mm glass. Double-glass modules also have much higher reliability and durability than their glass-backsheet counterparts. Double-glass modules have demonstrated a better suitability to harsh environments, such as high temperature and high UV exposure (for example, desert), high humidity (for example, tropical, maritime or floating PV applications) and high altitude (showing no creation of micro-cracks due to high wind or snow). Bifacial modules with a glass-transparent backsheet or with a double-1.6 mm-glass structure can effectively keep the weight of bifacial module in line with standard mono-facial modules. This option has attracted the attention of major

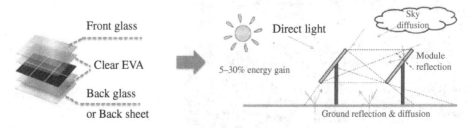

**Figure 21.12** Schematic illustration of the (left) structure bifacial module with a transparent glass or backsheet at the rear and (right) pathways for sunlight to reach the back of bifacial modules.

module manufacturers in recent years. It is expected that the production capacity of bifacial modules will continue to increase gradually in the next few years. Both double-glass and glass-transparent backsheet will continue to coexist.

## 21.8    Advanced Materials

Module and material factories have also been committed to developing and using advanced materials to reduce the power loss due to encapsulation and improve the efficiency cell-to-module (CTM) ratio. We list below the different improvements that are either still in development or already applied to commercial products.

(1)  Application of an antireflective film on the front glass:

Due to the difference in refractive index with air, a small part of the light is reflected at the front air-glass interface. The reflection is given by the Fresnel equation, which is, for example, at normal incidence for glass ($n_{glass} = 1.45$) and air ($n_{air} = 1$):

$$R = \left[ \frac{n_{glass} - n_{air}}{n_{glass} + n_{air}} \right]^2 \sim 4\%$$

By using an antireflection coating (ARC) with a refractive index $n_{ARC} = \sqrt{n_{glass}} = 1.2$ with an optical thickness equal to quarter-wavelength of the target wavelength of minimal reflection (in this case, 600 nm), it is possible to reduce the reflection by about half to 2%. However, materials with a low refractive index are usually very soft and not durable. Harder and more durable materials generally have a higher refractive index and consequently, the optical gain is not as high. Therefore, a compromise must be found between optical efficiency and durability. Typical ARCs on glass consist of porous film (Grosjean et al. 2018). Practical antireflective film-coated glass can improve the light transmittance of glass by about 1% relative, with even greater improvement for large angles of incidence of the light. This is widely used in existing commercial modules. The power of PV modules with double antireflective film-coated glass can be further increased by about 0.4% relative.

(2)  Potential induced degradation and high-performance encapsulation materials: ethylene vinyl acetate (EVA), polyolefin elastomer (POE) or EPE:

Potential-induced degradation (PID) can be avoided by improving the insulation performance of the encapsulation material. The advantages of POE's high resistivity and water vapor barrier performance can effectively inhibit the PID problem and power degradation (Lechner et al. 2020; Ben et al. 2020). But POE has issues such as high cost, slippage and lower adhesion to the glass. Due to the low friction between glass and POE, POE and cell surface, relative slippage is easy to occur between materials during transportation on the manufacturing line. For example, when laminated parts enter the laminating machine, due to slippage, problems such as unacceptable cell spacing and improper creepage distance between cells will occur after lamination. A new encapsulant material, called EPE (EVA–POE–EVA), not only can effectively improve the anti-PID performance of the module compared to pure EVA, but also can effectively prevent slippage during the lamination process. EPE encapsulant has become widely accepted by module manufacturers.

Several recent studies (Swanson et al. 2005; Hara et al. 2015) have found that some new cell technologies, such as TOPCon technology, have PID problems mainly caused by polarization, which is called PID-p. It refers to the fact that dust deposits or water condensation on the glass surface essentially form a polarization plate at the same potential as the grounded frame but with a high voltage difference (positive or negative) with respect to the cell voltage. Moisture ingress will also lower the resistivity of typical encapsulants like EVA. The large electric field created between the cells and the top surface of the glass made conductive by the presence of dust or water, or between the encapsulant with lower resistivity and the cell surface would induce a significant amount of charges in the top dielectric of the solar cells. Depending on the type of cell and its passivating dielectric layer, and depending on the direction of the electric field, the induced charges will enhance or worsen the passivation effect of the dielectric layer. Such polarization effect may lower the performance of the module over time, but the degradation is 100% reversible. The initial power output can be restored when the modules are illuminated. An updated edition of the IEC 61215 series (Terrestrial PV modules – Design qualification and type approval) was published in Feb. 2021. The document mentions that after PID test (that takes place in the dark), the PV modules shall be exposed to an irradiation dose of $(2.0 \pm 0.2)$ kWh/m$^2$ for less than 24 h at a temperature above 40 °C.

High-efficiency TOPCon cells can present a minor degradation in performance after a PID test with EVA encapsulation, and often, POE encapsulant is preferred because it maintains a high resistivity even after moisture ingress. However, the PID-test results can vary significantly depending on the type of EVA encapsulant. For example, Trina Solar researchers have demonstrated (Zhang et al. 2022) that the power of their TOPCon modules with EVA encapsulation films is as stable as that of TOPCon modules with POE encapsulation films. These modules have, for example, been in operation in a PV plant on water in Changzhou, China, for over five years without showing degradation.

(3) UV down-conversion

UV down-conversion encapsulation film can be added to the module optical design. In such structure, a UV down-conversion dye is used to convert high-energy UV photons, which usually do not generate any photocurrent because of absorption in the cell ARC or due to high surface recombination in the front emitter of the cell, into lower energy visible photons. The addition of the down-conversion film protects the encapsulant film from UV damage and improves the optical utilization of the sun spectrum, thus increasing the power generation of the module. UV down-conversion encapsulation film has been primarily applied in HJT modules as they have a particularly low quantum efficiency for short-wavelength photons. Compared to highly UV-transparent encapsulation films, UV down-conversion films achieve not only an equivalent initial module power but a much better module reliability.

(4) Triangular ribbons:

Triangular ribbons can potentially make efficient use of almost all incident light at normal incidence and non-normal incidence, as shown in Figure 21.13. Currently, some

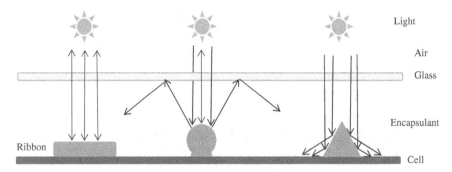

**Figure 21.13** Comparison of light utilization of three shapes of ribbon.

manufacturers use triangular ribbons on the front of the cell and flat ribbons on the back. The soldering process is relatively complex and incompatible with most existing production lines. There may be reliability risks at the overlap between a triangular and flat ribbon, which needs further verification. At the same time, the maturity of mass production equipment to perform such stringing process also light needs to be improved.

(5) Highly reflective back encapsulation material:

The incident light falling between cells within a sub-string or falling between adjacent sub-strings is reflected, usually in a non-specular Lambertian way, toward the front glass-air interface where it can be internally reflected toward the cell surface, and in particular totally reflected if the light ray is outside of the TIR cone, or has an incidence angle greater than 43.6°. The reflectivity of the back surface of the module is, therefore, critical to the efficiency of the module. The white backsheet historically provided good reflectivity. This is not possible with simple double-glass module with clear encapsulant. Therefore, the backside reflection must be provided by an ultra-white EVA encapsulant between the cells and the rear glass, or by a white ceramic coating on the rear glass, in a grid pattern, providing a good light reflection between cells and still a good transparency for bifacial modules.

The rear encapsulation material of the module (the light receiving surface is on the top) may be made of white EVA, which can effectively improve the reflection of light instead of using the white backsheet to provide the light reflection from the back. This material is widely used in double-glass modules. Even with backsheet modules, the efficiency of single glass module can be improved by about 0.5–0.7% by using white EVA between the cells and the backsheet. In addition, white EVA can effectively block ultraviolet rays and protect the backsheet, so as to improve the reliability of modules.

Similarly, for bifacial modules, to avoid the loss of light falling between cells, one can apply a white ceramic coating on the rear glass between cells or a white coating on the transparent backsheet (of course, also only between cells). This technique effectively improves the reflection of light falling between cells and strings, but the light reflection and the power gain are not as good as with white EVA.

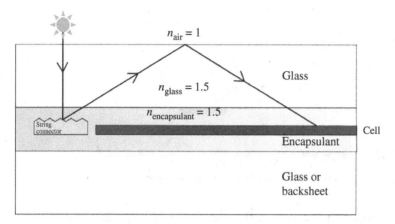

**Figure 21.14** Light utilization improvement of string connector.

Modules for distributed energy are generally used for household or industrial and commercial roofs, and the aesthetics of the modules are important. In order to get the overall black appearance effect of modules, which some customers prefer, a black backsheet could be used despite the loss in efficiency. As a blackbody-like material, the ordinary black backsheet basically does not reflect much light back to the cell. Recently, "high-reflection black" backsheets have shown a greater reflection for IR light but still look black to the human eye. It can slightly improve the reflection of infrared light so that the light can still be used by the cell, and the power is improved by about 0.5 % relative compared to the ordinary black backsheet.

(6) Reflective string connector:

Within a module, the string connectors are made of large Cu ribbons, coated with solder, and occupy a significant area of the module. It is possible to partially capture the light reflected by these string connector ribbons and redirect it toward the cell surface to increase the photogenerated current. This is done by creating a series of V-shaped groves on the ribbon surface that can reflect the light toward the cell while using TIR at the glass–air interface. The specific principle is shown in Figure 21.14. The string connectors can also be covered by a piece of white backsheet material. The effect on the module power output is, however, very small because it only improves the current of the cells close to the string ribbons, which creates a current mismatch within the module.

## 21.9 Summary

The improvements in performance of PV systems are not only due to cell efficiency improvements. Recent developments in module technology have also played a significant role. The technology of modules with half-cut cells has rapidly become a standard throughout the PV industry due to their better performance under high solar intensity or partial shading. Larger cells such as M10 (182 mm, quasi-square) and G12 (210 mm square) have been adopted by the industry, and smaller-size PV wafer standards have already almost disappeared. Modules with larger cells, having larger current and lower voltage than their predecessors, allow for reductions in cost of BOS. The technology of cutting cells, usually in half for most modules, but also in 1/3 or even 1/5 for modules with shingled cells, has pushed the industry to develop a NDC technology to reduce the edge defects, the cut-related efficiency losses and improve the bending capability of the cell. MBB technology has also been widely adopted in the industry, even for HJT modules that use MBB with a low-temperature solder based on Sn–Bi alloys. Some HJT modules are, however, still using either ECA with ribbon or SWCT™. Inactive areas of the modules (gap between adjacent cells) are reduced as much as possible, and reflectivity of the gap between cells is often enhanced by using more reflective white EVA between cells and backsheet or between the cells and the rear glass panel for double-glass non-bifacial modules. Bifacial modules are increasing their market share. Their weight with transparent backsheet or with 1.6mm/1.6mm double-glass structure is comparable to the standard non-bifacial modules. The performance of bifacial modules is enhanced thanks to the printing of white reflective strips printed on the rear glass internal surface and located between cells. Viewed from the front, the bifacial module looks like any standard white backsheet module and has equivalent internal light capture, while maintaining good bifacial properties. ARC are nowadays standard on the front glass surface of modules and provide about 1% relative gain. This is a trade-off between AR properties and durability of the ARC. Several other important developments have been implemented in module manufacturing to improve performance or reliability, such as POE or EVA-POE-EVA (also called EPE) encapsulants for the prevention of PID, UV down-conversion and structured string connectors.

## Abbreviations and Acronyms

| | |
|---|---|
| ARC | antireflection coating |
| BOS | balance of system |
| CTM | cell-to-module ratio (power or efficiency) |
| Cz | Czochralski |
| DC | direct current |
| ECA | electrically conductive adhesive |
| EPE | EVA-POE-EVA encapsulant |
| EVA | ethylene vinyl acetate |
| HJT | heterojunction with thin layer |
| LCOE | levelized cost of electricity |
| MBB | multi-busbar |
| NDC | non-destructive cutting |
| PECVD | plasma-enhanced chemical vapor deposition |

| PID | potential-induced degradation |
|---|---|
| POE | polyolefin elastomer |
| PV | photovoltaic or photovoltaics |
| SWCT™ | Smart Wire Connection Technology |
| TIR | total internal reflection |
| TOPCon | tunnel oxide polysilicon contact |
| UV | ultra-violet |
| 5BB | 5 busbar |

## Nomenclature

| $R$ | reflectance |
|---|---|
| $n$ | refractive index |
| $n_{air}$ | refractive index of air |
| $n_{glass}$ | refractive index of glass |

## Author Biographies

**Shu Zhang** was born in China in 1984. She received her B.S. degree in Material Science and Engineering from Nanjing University of Aeronautics and Astronautics, Nanjing, Jiangsu Province, China, in 2005. She got the M.S. degree in Material Processing from the same university in 2008. She has worked at Trina Solar for over 15 years. She is the Leader of the Advanced Module Team and one of the core technical staff of the State Key Laboratory of Photovoltaic Science and Technology. Her research interests include theoretical analysis of power loss from solar cells to photovoltaic (PV) modules, electrics of PV Module, optics of PV Module, advanced cell assembling process, light trapping technology, laser cutting process of solar cell, cell/module matching technology, advanced module design, module reliability, module performance, building integrated photovoltaics, as well as industrial module technologies development. She holds 64 authorized patents. As the project leader, she created 10 world records for large-area module power and aperture efficiency. She led the development of multi-busbar technology, which is leading in the PV industry. Moreover, she developed a series of high-efficiency PV module products based on 210 mm large-size silicon wafers, leading the era of 600W+ products. She is the winner of the 2022 Red Dot Award.

**Xue Chen** was born in China in 1982. She received her B.S. degree in Material Science and Engineering from Suzhou University, Suzhou, Jiangsu Province, China, in 2004. She got the M.S. degree in Material Science from the same university in 2007. She has worked at Trina Solar for over 16 years. She is the principal engineer of the Advanced Module Team at Trina Solar. She has worked for the State Key Lab of PV Science and Technology, which belongs to Trina Solar, for seven years. She has several research achievements on high-performance crystal growth, thermal field design, module reliability study and characterization, advanced module design, module performance simulation, power generation capability improvement, as well as related industrial technologies development. She has holds 27 authorized patents.

**Xu Jianmei**, born in China in 1974, received her M.S. degree in PV Materials Science and Engineering from Sun Yat-sen University in 2011. She has been a member of Trina Solar since 2005, currently titled as chief engineer of photovoltaic module technology. She has both solid theoretical knowledge and practical experience in module technology & process, material and reliability. Under her leadership, the team developed several advanced PV module products like double-glass modules, multi-busbar (MBB) module technology, and new frame and installation structures, which have all been put into mass production and leading the PV module industry. She owns 38 authorized patents as the first author. Also, she participated in the writing of crystalline silicon PV module as the second author.

**Pierre J. Verlinden** was born in D.R. Congo in 1957 and received his M.S. and PhD. degrees in Electrical Engineering from the Catholic University of Louvain, Belgium, in 1979 and 1985, respectively. He is currently an Adjunct Professor at the University of New South Wales, Sydney, Australia, and Chief Scientist of the Yangtze Institute for Solar Technology (YIST), Jiangyin, China. Dr. Verlinden has been working in the field of photovoltaics for 43 years. From 2012 to 2018, Dr. Verlinden served as Chief Scientist, Vice-President and Vice-Chair of the State Key Laboratory of PV Science and Technology at Trina Solar, China. Previously, he has managed the R&D department or served as

Chief Scientist in several PV companies in the USA and Australia, including SunPower, Origin Energy and Solar Systems. Dr. Verlinden is the recipient of the 2016 IEEE William Cherry Award, and the 2019 Becquerel Prize, for his dedication over the past four decades at the forefront of PV technology and his overall leadership of key R&D organizations. Dr. Verlinden is also the recipient of the 2017 Chinese Government Friendship Award, the highest award given by China to foreigners.

## References

Beaucarne et al. (2015). Innovative cell interconnection based on ribbon bonding of busbar-less cells using silicone-based electrically conductive adhesives. *Energy Procedia* 67: 185–193.

Ben, X., Yu, J., Lv, R. et al. (2020). Durability of polyolefin encapsulation based modules: a cross-comparison of commercially available solutions. In: *37th European Photovoltaic Solar Energy Conference and Exhibition*, 1065–1067.

Braun, S., Hahn, G., Nissler, R. et al. (2013). The multi-busbar design: an overview. *Energy Procedia* 43: 86–92.

Grosjean, A., Soum-Glaude, A., Neveu, P., and Thomas, L. (2018). Comprehensive simulation and optimization of porous $SiO_2$ antireflective coating to improve glass solar transmittance for solar energy applications. *Solar Energy Materials and Solar Cells* 182: 166–177.

Govaert (2020). Encapsulant-integrated interconnection of bifacial solar cells for BIPV applications: latest results in the TWILL-BIPV project. In: *37th European Photovoltaic Solar Energy Conference and Exhibition*, 33–37.

Hara, K., Jonai, S., and Masuda, A. (2015). Potential-induced degradation in photovoltaic modules based on n-type single crystalline Si solar cells. *Solar Energy Materials & Solar Cells* 140: 361–365.

Hua, F. (2020). Zhongtian T3 fluorine film transparent backplane provides the best solution for double-sided module packaging. *Modern Transmission* 1 (2).

Kopecek, R. and Libal, J. (2021). Bifacial photovoltaics 2021: status, opportunities and challenges. *Energies* 14 (8).

Lechner, P., Schnepf, J., Hummel, S. et al. (2020). Extreme testing of PID resistive c-Si PV modules with 1500 V system voltage. In: *37th European Photovoltaic Solar Energy Conference and Exhibition*, 792–795.

Lin, Y. R., et al. (2022) *INFOLINK PV Technology Trend Report*.

Liu, S., He, S., Shan, W. et al. (2020). Performance analysis of multi-busbar photovoltaic modules. *Solar Energy* 4 (2).

Mack, C.A. (2011). Fifty years of Moore's law. *IEEE Transactions on Semiconductor Manufacturing* 24 (2): 202–207.

Papargyri, L., Theristis, M., Kubicek, B. et al. (2020). Modelling and experimental investigations of microcracks in crystalline silicon photovoltaics: a review. *Renewable Energy* 145: 2387–2408.

Papet, P., Andreetta, L., Lachenal, D. et al. (2015). New cell metallization patterns for heterojunction solar cells interconnected by the smart wire connection technology. *Energy Procedia* 67: 203–209.

Pieters, P., Govaerts, J., Borgers, T. et al. (2019). Sillion-PV module technology: advanced interconnection approaches for two-side and back contacted solar cells. In: *15th China SoG Silicon and PV Power Conference*.

Swanson, R., Cudzinovic, M., Deceuster, D. et al. (2005). The surface polarization effect in high-efficiency silicon solar cells. In: *15th International Photovoltaic Science & Engineering Conference*.

Wang, H.J. and Yang, T. (2021). A review on laser drilling and cutting of silicon. *Journal of the European Ceramic Society* 41 (10): 4997–5015.

Zhang, S., Wang, L., and Chen, X. (2020). Research progress of 210 high power module performance. In: *16th China SoG Silicon and PV Power Conference, Wuxi, Jiangsu*.

Zhang, S., Wang, L., Chen, X., and Chen, D.M. (2022). Research on Vertex N module performance. In: *18th China SoG Silicon and PV Power Conference, Taiyuan, Shanxi*.

Zhou, X., Li, M., Zeng, M., and Mei, J. (2020). Performance analysis of backplane glass materials for double glass modules. *Glass* 47 (5): 5.

# 22

## Smart Modules for Shade Resilience

*Wilfried van Sark*

*Copernicus Institute of Sustainable Development, Utrecht University, Utrecht, The Netherlands*

## 22.1 Introduction

The growth in photovoltaic (PV) system capacity has led to total installed capacity of 1 TWp in early 2022 (Weaver 2022). These installations are grid-connected, either large-scale centralized or smaller-scale decentralized. About 40% of the PV systems can be found in residential areas (Masson and Kaizuka 2021). PV systems in the urban environment are typically prone to lower yields than large-scale centralized systems due to (partial) shading from various objects (dormer, tree, chimney, neighboring buildings) (Moraitis et al. 2018), see Figure 22.1a. In PV systems, typically, several PV modules are connected in series to central inverters. Consequently, shading-induced power losses at the (single) module level will lower the maximum attainable power compared to when shading would be absent. Figure 22.1b shows an example of the effect of shading on time series of power on a sunny and overcast day. This particular single PV panel suffers from reduced power (about 2/3rd compared to the unshaded situation) in the morning on a sunny day. Shade effects are not discernable on an overcast day (Tsafarakis et al. 2019).

Shading-induced loss is mainly due to the fact that irradiance levels differ per cell (as a result of shading) in the same series connection within one PV module, and the shaded cells can be forced to support current levels exceeding their characteristic short-circuit current values. This may push the shaded cells into reverse voltage regimes where they may act as rectifying diodes. As a consequence, thermal power dissipation by those cells causes excessive power losses and the formation of localized "hot spots," which may lead to possible permanent cell damage (Sinapis et al. 2016). The use of bypass diodes will limit this by bypassing shaded substrings of the solar module (Hasyim et al. 1986). Typically, three so-called bypass diodes are used in antiparallel connection to the cells for a 60-cell module for each of three 20-cell substrings (Silvestre et al. 2009). Upon partial shading, power loss will be limited to 1/3rd of 2/3rd of the maximum power. Adding more by-pass diodes, thus

*Photovoltaic Solar Energy: From Fundamentals to Applications, Volume 2*, First Edition.
Edited by Wilfried van Sark, Bram Hoex, Angèle Reinders, Pierre Verlinden, and Nicholas J. Ekins-Daukes.
© 2024 John Wiley & Sons Ltd. Published 2024 by John Wiley & Sons Ltd.
Companion website: www.wiley.com/go/PVsolarenergy

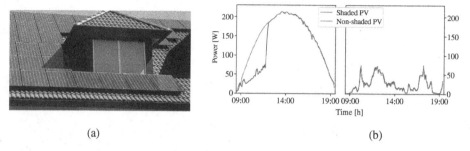

(a)                                                                                    (b)

**Figure 22.1** (a) PV system on a roof with dormer with shading; (b) power as a function of time for one panel in a PV system on a sunny day (left), and at an overcast data (right) illustrating the effect of shading. Source: Tsafarakis et al. (2019)/MDPI/CC by 4.0.

increasing granularity of cell groups, can increase performance, especially under partial shading conditions (Pannebakker et al. 2017; Sinapis et al. 2021).

PV system design for optimal performance should take potential shading effects into account. This involves the design of string connections and use of central inverters, or module-level power electronics, that aim to optimize the maximum power of individual PV modules (Sinapis et al. 2016). Both microinverters (MIs) (Çelik et al. 2018) and power optimizers (MacAlpine et al. 2013) have been available commercially for some years now, and system designers are challenged to find an optimum between additional cost and additional yield, which will depend on the expected shading loss.

Mitigation of shading effects on PV performance can be classified into two groups: circuit-based topologies and modified MPPT (maximum power point tracker) based techniques (Das et al. 2017). These can be applied on a per-module level. For example, increasing the number of bypass diodes, or so-called total cross tied (TCT) topologies, which can be fixed connections between cells in a module, or adaptive connections, within a module. These approaches are also used for (reconfigurable) connections between modules in a system. The TCT topology requires microcontrollers that control a switching matrix and, indeed, mitigate shading, but this is quite complex by design (Nguyen and Lehman 2008; Velasco-Quesada et al. 2009; Bidram et al. 2012; Krishna and Moger 2019; Baka et al. 2019; Ajmal et al. 2020; Yang et al. 2021; Calcabrini et al. 2021).

In this chapter, some examples of shade mitigation strategies are described. More examples can be found in Littwin et al. (2020, 2021).

## 22.2 Shade Mitigation Strategies

### 22.2.1 Module Level

#### 22.2.1.1 Increasing Module Granularity

One strategy for mitigating nonlinear shading effects is to organize the cells in a PV module into groups of less than the traditional number of 20 cells, while at the same time deployinga so-called active bypass diode circuits for each group of cells instead of passive bypass diodes (Golroodbari et al. 2018). An active by-pass diode is actually an electrical circuit consisting of Metal-Oxide-Semiconductor Field-Effect Transistors (MOSFETs), which can be controlled via a duty cycle as a control signal. These module-integrated electronics (MIEs) can be categorized as follows: (i) conventional systems, consisting of three bypass diodes per module and a central converter to change the output voltage level, (ii) buck converters, to which normal PV modules are connected and thus the output current of the shaded module is to be controlled, (iii) buck-boost converters, in which both current and voltage are controlled, and (iv) voltage equalizers, which are a combination of different converters or even bidirectional converters to equalize the voltage by power processing (Schmidt et al. 2010; Olalla et al. 2013; Uno and Kukita 2017).

Conventionally, cells are series-connected in three groups and to each group, one bypass diode is connected. Power is optimized by means of the MPPT in the inverter, in the case of a central inverter, or in a local optimizer, in case of module-level power electronics. MIEs using buck or buck-boost converters are potentially interesting for controlling the current or voltage and current of a shaded group. While in the ultimate case of one by-pass diode per cell, shade mitigation is optimal, this is certainly not economically optimal (Pannebakker et al. 2017). Hence, a compromise must be found between the number of groups and bypass diodes.

Figure 22.2a shows the principle design of a module with $n_g$ solar cells in $N_G$ groups of cells in a module (Golroodbari et al. 2018), thus employing distributed MPP tracking. Using 60 monocrystalline silicon solar cells (open circuit voltage ($V_{OC}$) of 613 mV, short circuit current ($I_{SC}$) of 7.92 A, maximum power ($P_{max}$) of 3.7 W, efficiency ($\eta$) of 15.4%) and a number of buck converters, in this case Linear Technology LTM4611, the optimum configuration in terms of converter efficiency has been determined. Figure 22.1b shows the efficiency contour plots of the converter, using converter efficiency graphs from the LTM4611 specification sheet (see Golroodbari et al. 2018), for input and output current. For a module with 10 groups of 6 cells, specifications per group are $V_{OC} = 3.67$ V, $I_{SC} = 7.92$ A, and maximum power ($P_{max}$) of 22.18 W. Because of the system topology, the output current flow of each converter is equal as all converters are connected in series. This strategy extracts as much power as each group of cells can generate, even though some groups may be very heavily shaded. In the smart module architecture, all cells, even shaded ones, are thus producing power efficiently, and none of the cells is bypassed.

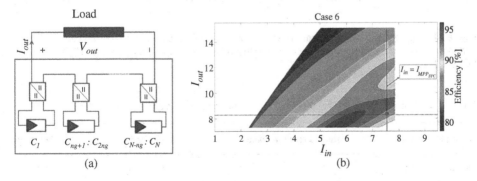

**Figure 22.2** (a) Smart module architecture with $n_g$ cells in $N_G$ groups with electronic circuits, (b) contour plots of efficiency as a function of input ($I_{in}$) and output current ($I_{out}$) of the converter for a smart module with 10 groups of 6 cells. Source: Golroodbari et al. (2018)/MDPI/CC by 4.0.

Four different module topologies have been modeled to assess and compare the shading mitigation potential: (1) smart module with 10 groups of 6 cells (as in Figure 22.2); (2) a module with three parallel-connected strings that consists of each string consisting of 20 cells connected in series and a blocking diode (normally this topology is implemented for strings of PV modules instead of PV cells); (3) a standard reference module with series-connected strings consists of three series groups of 20 cells, where each group is equipped with one by-pass diode; and (4) an ideal module, in which for each cell a DC-DC converter is responsible for leveling up the current for shaded cells. Figure 22.3 shows the calculated output power for two different shading patterns. The power of the ideal module (second column in the figure) clearly reflects the shade that is cast on the module, i.e. most clearly in Figure 22.3a. The current and voltage of the series/parallel module are considerably affected by the shade. Figure 22.3b, in which only the edge of the module is shaded, reveals that two groups of cells in the smart module are affected, and 1/3rd of the series/parallel module. Module power is nearly equal when there is no shade, illustrating some loss in the smart module, see also Table 22.1.

A smart module with ten groups of six cells has been manufactured and tested (Golroodbari et al. 2019). At the backside of the module developed earlier by

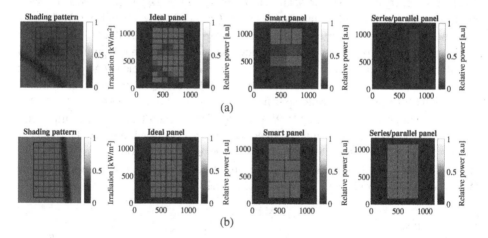

**Figure 22.3** Effect of different shading patterns on different module topologies, (a) combined pole and random shading pattern, (b) pole shading pattern. Source: Golroodbari et al. (2018)/MDPI/ CC by 4.0.

**Table 22.1** Output power (W) for three different shading patterns and four architectures.

| Architecture | Pole and random shade | Pole shade | No shade |
| --- | --- | --- | --- |
| Smart | 18 | 69 | 109 |
| Parallel | 5 | 63 | 113 |
| Series | 1 | 31 | 112 |
| Ideal | 48 | 84 | 117 |

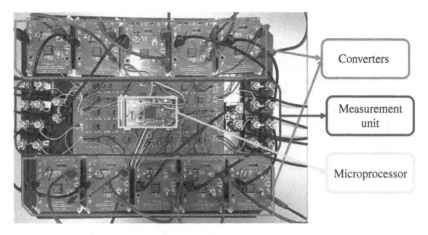

**Figure 22.4** Prototype electronics installed at the back of the module, showing 10 converters, *I–V* measurement unit and microprocessor (for details, see Golroodbari et al. 2019/John Wiley & Sons, Inc.).

Pannebakker et al. (2017) an electronics prototype was attached, see Figure 22.4. This has been tested with several shading patterns.

An example of the measurements is shown in Figure 22.5a. This shows a static shade over the bottom part of the module and a moving pole shade over the top of the module for three different timesteps for a period of 60 minutes. Also, a tree with some leaves was moving due to wind in front of the module as well. Figure 22.5b shows that cell groups 6–10 do not generate power, while the other groups show fluctuating power, which clearly corresponds to the shading pattern. The average output power in the smart module is higher than the expected average power of a conventional module with three bypass diodes. The smart module generated an average of 8.13 W under a very restricted shading condition with three different shade types at a maximum irradiation of 350 W/m$^2$. Based on the outputs from groups with severe shading, a conventional panel under these testing conditions would generate only about 1.64 W. This is due to the activation of the bypass diodes. The smart module generates five times more power under these conditions.

### 22.2.1.2 Increasing Cell Granularity

The so-called Tessera solar module concept (Figure 22.6), named by TNO after "tesserae" being used as individual tiles in ancient mosaics, makes use of small cells that are cut by laser scribing from standard-size (156 × 156 mm$^2$) silicon cells (Carr et al. 2015). The used cells are 6 in. Metal Wrap Through (MWT) c-Si cells, which have been cut to produce 16 mini cells. These mini cells are subsequently interconnected in series and connected to a bypass diode that is laminated in the back sheet (see Figure 22.6a top-left). Being back-contact solar cells, interconnection schemes can be designed using a conductive backsheet mimicking a printed circuit board (PCB), which allows the integration of diodes. Blocks of 64 mini cells (Figure 22.6 top-right) can then be connected in parallel, forming modules with custom size and rated power (Figure 22.6 bottom). These module building blocks show $V_{OC}$ of around 40 V and $I_{SC}$ of 0.6 A, under standard test conditions (Carr et al. 2015). A standard module size of 1.6 m$^2$ comprises 15 of these module building blocks of 64 mini cells that are connected in parallel, thus yielding $I_{SC}$ of 8 A, and module

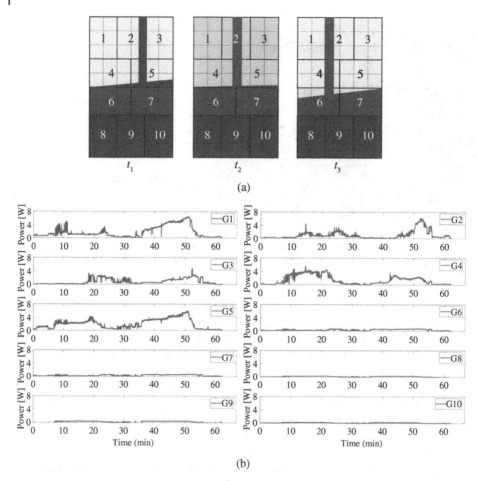

**Figure 22.5** (a) Moving shading pattern during the experiments for different time frames ($0 < t_1 < 20$ minutes; $20 < t_2 < 40$ minutes; $40 < t_3 < 60$ minutes), (b) output power of the 10 cell groups. Source: Golroodbari et al. (2019)/The Institution of Engineering and Technology/CC BY 3.0.

rating of 265 Wp. Module manufacturing equipment is the same as for standard back contact modules, albeit with a small modification in the production line.

The design of the blocks is optimized for shade resilience. The subcells can be regarded as pixels of about $40 \times 40$ mm$^2$ in size, and only a shaded pixel will be affected by the shade, and not 1/3rd of a (standard) module, because of a somewhat different output compared to a standard module, a dedicated inverter may be required. However, the voltage of the module is almost independent of shading conditions, which would make the inverter design less complex (Slooff et al. 2017).

The performance of the Tessera module has been experimentally tested and compared to a conventional reference module. The shade performance was defined as the ratio of maximum power of the Tessera module and a standard 3-string module, applying (artificial, i.e. pole, dormer, tree) shading conditions in a solar simulator. Table 22.2 shows that the reduction in performance of the Tessera module is considerably less compared to the reference

**Figure 22.6** (top-left) Tessera module concept cell and (top-right) building block, consisting of four four-size MWT cells, (bottom) full modules of 1.6 m² consisting of 960 mini cells (Sinapis et al. 2021/John Wiley & Sons, Inc.).

**Table 22.2** Effect of shading conditions on the shade performance and shade linearity of the Tessera module compared to a standard module.

|  |  | Pole shade | Dormer shade | Tree shade |
|---|---|---|---|---|
| Shade performance (%) | Tessera | 67 | 64 | 73 |
|  | Reference | 17 | 26 | 53 |
| Linearity (%) | Tessera | 76 | 92 | 92 |
|  | Reference | 20 | 38 | 66 |

Source: Adapted from Carr et al. (2015).

module. Clear differences are also seen in shade linearity. Shade linearity is 100% if an $x$% sized shade (related to the area of a module) will lead to the same $x$% of power loss induced by that shading (Carr et al. 2015). Table 22.2 shows that shade linearity is very high for the Tessera module compared to the reference module for all shading conditions.

A field test with six Tessera modules has been performed comparing their performance to two reference systems, all with the same orientation and tilt: one with six standard modules, also of 265 Wp capacity each, connected in series to a string inverter (SI), and one with

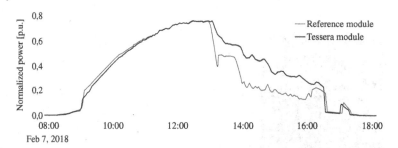

**Figure 22.7** Normalized power output of a Tessera and reference module under the same partial shading conditions. Source: Sinapis et al. (2021)/John Wiley & Sons/CC BY 4.0.

six standard modules each equipped with a MI (Sinapis et al. 2018). The Tessera modules are also connected to individual MIs. A pole shade has been used for assessment of partial shading effects. Figure 22.7 compares the normalized power outputs of the Tessera and reference module under identical shading conditions of a day in winter. The pole affects the performance of the modules between 1 p.m. and 4:30 p.m., while in the morning, power generation is identical. For this particular day, the Tessera module generates 15% more energy than the reference module. Interestingly, micro- inverter efficiency is somewhat higher in the case of the Tessera module due to the more stable voltage output of the module. The shade leads to voltage reduction of about 1/3rd for the reference module, while this was not observed for the Tessera module (Sinapis et al. 2021).

A longer outdoor test for the months of February to September showed the effect of solar elevation (shadow length) on shade resilience. In summer months, the performance ratio (Reich et al. 2012) of the Tessera module coupled to the MI and reference module coupled to either a string or MI is similar, while in other months, the differences are clear. This especially holds for the reference module coupled to a SI. Based on these results, it is estimated that annual shading loss in residential systems with Tessera modules and MIs would be similar to a module with one bypass diode per cell. Compared to standard modules with three bypass diodes, annual shading loss would be decreased from 9.3% to 3.3% (Sinapis et al. 2017). A more recent estimate, based on a 3D simulation of the shade effects for a typical row house from the Dutch building stock with dormer, exhaust pipe and chimney, showed a 6.3% annual loss for a standard module with three bypass diodes. The Tessera module showed a 3.2% loss, and the standard module with one bypass diode per cell would lead to a 2.8% loss (Table 22.3).

**Table 22.3** AC performance ratio for the three module systems.

| Module/inverter | Feb | Mar | Apr | May | Jun | Jul | Aug | Sep |
|---|---|---|---|---|---|---|---|---|
| | | | | Months | | | | |
| Reference/SI | 0.76 | 0.84 | 0.81 | 0.77 | 0.84 | 0.78 | 0.68 | 0.78 |
| Reference/MI | 0.88 | 0.87 | 0.83 | 0.80 | 0.85 | 0.81 | 0.73 | 0.82 |
| Tessera/MI | 0.91 | 0.88 | 0.85 | 0.84 | 0.83 | 0.81 | 0.76 | 0.84 |

### 22.2.2 System Level

Another way to address shading effects is to change the electrical configuration of the PV system itself, depending on the daily and seasonal variation of the shading pattern. Different interconnections of individual PV modules are suggested typically categorized in (i) series-parallel (SP), (ii) TCT, (iii) bridge-linked (BL), and (iv) honeycomb (HC) configurations (Bidram et al. 2012; Karatepe and Hiyama 2010; Krishna and Moger 2019; Yang et al. 2021), see also Figure 22.8. In a series-connected system, module current is equal in all modules and module voltages are added. In a parallel system, current is added while voltage is equal in all modules. The SP approach combines this in connecting some modules in series in separate strings, and these strings are then connected in parallel. System design can be optimized considering shading patterns. As this is quite simple, the SP approach is most commonly used.

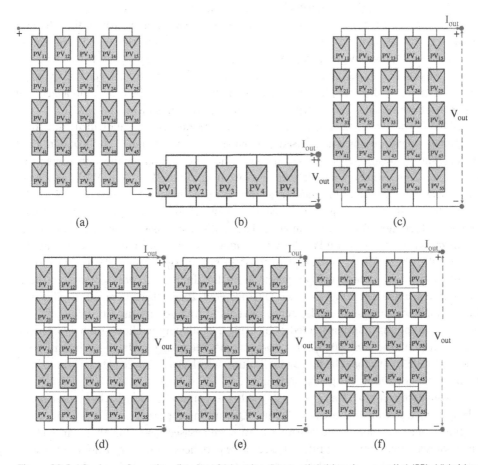

**Figure 22.8** Basic configuration circuits of (a) series, (b) parallel, (c) series–parallel (SP), (d) bridge-link (BL), (e) total-cross tied (TCT) and (f) honey-comb (HC), Source: Ajmal et al. (2020)/with permission of Elsevier.

In a TCT configuration, solar modules are connected in parallel, after which they are connected in series as well. This requires much wiring, but BL and HC configurations are designed with lower amount of wiring. TCT configurations perform better under partial shading conditions than SP or standard series/parallel configurations. Analyses of shading patterns to be expected are needed to design an optimized connection layout of modules in the system. This can involve optimization techniques such as SuDoKu (Rani et al. 2013), but others are suggested, e.g. (Pillai et al. 2018; Nihanth et al. 2019; Venkateswari and Rajasekar 2020). Obviously, such static optimization only makes sense if the shading conditions do not change, such as neighboring buildings, while it may also involve customized connection schemes that may be labor-intensive and thus expensive. However, this may be compensated by attaining higher yields.

The above methods are all static: once designed, wiring schemes are not changed. In contrast, dynamic methods have also been suggested. Dynamic methods encompass switches that can be automatically modified to change the connection scheme between modules due to changes in shading (La Manna et al. 2014). For example, Serna-Garcés et al. (2016) proposed an electrical array reconfiguration, which uses a switching matrix to find the best configuration. This method operates at the module level and is implemented to maximize the available DC power by grouping modules with similar shading patterns. In one example, it is shown that 25% more energy can be gained. A critical point is that, in general, dynamic reconfiguration methods implemented on the module level may be very complicated and may perform at a sluggish pace, depending on the optimization algorithm used.

Following Gao et al. (2023), dynamic reconfiguration can be subdivided into three categories: (1) performance data-based, (2) mathematics-based, and (3) full configuration search. The performance data-based method finds the optimal connection scheme by training it using actual operational performance data of the PV system. With the advent of artificial intelligence techniques, these are also used here, e.g. Karakose et al. (2014, 2016) employ fuzzy techniques and neural networks. Downside is the amount of data needed, which not always be available. The mathematics-based method uses classic optimization algorithms to find the best-reconfigured interconnection scheme by solving a mathematical cost function with certain constraints, see, e.g. Shams El-Dein et al. (2013). A full configuration search uses all potential module interconnection schemes and tests them until a constraint is met (Velasco-Quesada et al. 2009; Obane et al. 2012). This may lead to high computational loads; hence, various techniques are used to reduce this by limiting the options to be considered or by using evolutionary methods, such as a genetic algorithm (Deshkar et al. 2015). Many others have been suggested as well recently (see Gao et al. (2023)). It should be noted that limiting search space may lead to lock-in to local minima and not finding the global optimum.

A recent approach from Gao et al. (2023) addresses some of the drawbacks of these methods by using a so-called Divide and Conquer Q-Learning approach, using a TCT interconnection scheme of modules. Q-learning is an unsupervised algorithm that learns about an unknown environment or future based on experiences. However, it does not need a lot of input data. Also, computational load is lowered. An example is provided showing an 11% increased annual yield for system with partial shading conditions compared to a non-reconfigurable TCT system.

## 22.3   Commercial Applications

Smart shade-resilient modules are offered by several companies today (fall 2022). Examples include A–E Solar, Sunchip, Maxim Integrated, and Taylor. The German company AE Solar markets hot-spot-free modules that are shade-resistant by using one small bypass diode per cell located in between cells. They report that additional cost is paid back by increased yields (AE Solar 2022). Maxim Integrated has developed a cell optimizer chip that can replace a standard bypass diode in a module; hence, three of these are needed per module. This optimizer is said to boost the current of the weak (shadowed) cells to match those of the stronger (unshaded) (Maxim Integrated 2022). The chips are used by several major module manufacturers in their shade-resilient module types. The Dutch company Taylor Solar also uses integrated electronics in a similar manner. Their cell string optimizer (CSO) is marketed to have 20% higher yields compared to standard modules (Taylor Solar 2022). Another Dutch company, Sunchip, has also designed a CSO termed nanovoltage optimizer, which can be used for 2 to up to 20 cells in a module to step up voltage in order to optimize power and claim a 12% higher yield (Sunchip 2022).

While this overview is not complete, it can be expected that smart modules will become omnipresent in the next years. As PV systems keep getting cheaper, they will be installed at locations that are not ideal, with potential shading, soiling-induced mismatch, and combinations of tilt and orientations, which requires a per-module optimization.

## 22.4   Conclusion

In this chapter, examples of smart modules and systems are presented. In the residential environment, more and more PV systems will be installed that will suffer in some way from the shadow effect, leading to lower annual yields. Smart shade-resilient modules could be an option to mitigate this, however, this depends clearly on the type of shade and the size of the shading effect. Note that soiling could also be seen as a type of shade, as it will decrease the intensity of light on the module, and it may case mismatch conditions due to differing levels of soiling spread out over the module surface. Technological options are available, also commercially, that can limit shade effects. The question always remains if these, sometimes complex, options that come at extra cost in the end lead to increased financial revenues for the PV system owner. Hence, a case-by-case analysis is always needed.

## Acknowledgments

We would like to sincerely thank Boudewijn Pannebakker, Panos Moraitis, Sara Golroodbari, Kostas Sinapis, Odysseas Tsafarakis, Anne de Waal, Geert Litjens, Tom Rooijakkers (UU), Anna Carr, Lenneke Slooff-Hoek (TNO), and many participants in the IEA-PVPS Task 13 for various contributions and fruitful discussions in the past years working on smart shade resilient module development. This work was partly financially supported by the Netherlands Enterprise Agency (RVO) within the framework of the Dutch Topsector Energy (project Scalable Shade Tolerant Modules, SSTM).

Finally, we would like to thank the jury of the Science2Business Challenge Award 2010, who awarded us this award for intelligent solar power development, which sparked the funding obtained from RVO (DUB 2022).

### Author Biography

**Wilfried van Sark** is a full professor "Integration of Photovoltaics" at the Copernicus Institute of Sustainable Development. He is an experimental physicist by training (M.Sc./Ph.D.) and has 40 years of experience in the field of photovoltaics (PVs). He worked on various material systems, such as crystalline and thin-film silicon and III–V solar cells, both experimentally and theoretically. His current activities focus on employing spectrum conversion to increase solar cell conversion efficiency for next-generation PV energy converters such as luminescent solar concentrators, as well as performance analysis of building-integrated and standard PV systems in the field. This, in particular, links to the integration of PV systems in smart grids in the built environment, in which Electrical Vehicles, demand response, self-consumption and self-sufficiency play a major role. He is the author of some 250 publications, cited more than 12,000 times, and various (text)books on Photovoltaics. He is associate editor of Solar Energy and Frontiers in Energy Research and a member of the editorial boards of Renewable Energy, Energies, and Materials. He is member of the International Solar Energy Society, and senior member of the Institute of Electrical and Electronic Engineers (IEEE) and a member of scientific committees of various EU and IEEE PV conferences. He presently is national representative in the International Energy Agency (IEA) PVPS Task 13 (PV Performance, Operation and Reliability of Photovoltaic Systems) and Task 16 (Solar Resource for High Penetration and Large Scale Applications).

## Abbreviations and Acronyms

| | |
|---|---|
| BL | Bridge Linked |
| HC | Honeycomb |
| MI | Microinverter |
| MIE | Module Integrated Electronics |
| MOSFET | Metal-Oxide-Semiconductor Field-Effect |
| MPPT | Maximum Power Point Tracker |
| PCB | Printed Circuit Board |
| PV | Photovoltaics |
| SI | String inverter |
| SP | Series-Parallel |
| TCT | Total Cross Tied |

## Nomenclature

| | |
|---|---|
| $\eta$ | efficiency (%) |
| $I_{SC}$ | short circuit current (Ampere) |

| $n_g$ | number of solar cells per group (-) |
| $N_G$ | number of groups of solar cells in a module (-) |
| $P_{max}$ | maximum power (Watt) |
| $t_1, t_2, t_3$ | time periods 1, 2, 3 (s) |
| $V_{OC}$ | open circuit voltage (Volt) |

## References

AE Solar, 2022 https://ae-solar.com/wp-content/uploads/2018/10/AE-Solar-HSF-Introduction-light.pdf (last accessed 18 September 2022)

Ajmal, A.M., Babu, T.S., Ramachandaramurthy, V.K. et al. (2020). Static and dynamic reconfiguration approaches for mitigation of partial shading influence in photovoltaic arrays. *Sustainable Energy Technologies and Assessments* 40: 100738.

Baka, M., Manganiello, P., Soudris, D., and Catthoor, F. (2019). A cost-benefit analysis for reconfigurable PV modules under shading. *Solar Energy* 178: 69–78.

Bidram, A., Davoudi, A., and Balog, R.S. (2012). Control and circuit techniques to mitigate partial shading effects in photovoltaic arrays. *IEEE Journal of Photovoltaics* 2: 532–546.

Calcabrini, A., Muttillo, M., Weegink, R. et al. (2021). A fully reconfigurable series-parallel photovoltaic module for higher energy yields in urban environments. *Renewable Energy* 179: 1–11.

Carr, A.J., de Groot, K., Jansen, M.J. et al. (2015). Tessera: scalable, shade robust module. In: *Proceedings of the 42nd IEEE Photovoltaic Specialist Conference*, 1–5.

Çelik, Ö., Teke, A., and Tan, A. (2018). Overview of micro-inverters as a challenging technology in photovoltaic applications. *Renewable and Sustainable Energy Reviews* 82: 3191–3206.

Das, S.K., Verma, D., Nema, S., and Nema, R.K. (2017). Shading mitigation techniques: state-of-the-art in photovoltaic applications. *Renewable and Sustainable Energy Reviews* 78: 369–390.

Deshkar, S.N., Dhale, S.B., Mukherjee, J.S. et al. (2015). Solar PV array reconfiguration under partial shading conditions for maximum power extraction using genetic algorithm. *Renewable and Sustainable Energy Reviews* 43: 102–110.

DUB, 2022 *Zonnecelonderzoeker Wilfried van Sark wint Science2Business Challenge*, https://dub.uu.nl/nl/artikel/zonnecelonderzoeker-wilfried-van-sark-wint-science2business-challenge (last accessed 18 September 2022)

Gao, X., Deng, F., Wu, G. et al. (2023). Divide and Conquer Q-Learning (DCQL) algorithm based Photovoltaic (PV) array reconfiguration scheme for alleviating the partial shading influence. *Solar Energy* 249: 21–39.

Golroodbari, S.Z.M., de Waal, A.C., and van Sark, W.G.J.H.M. (2018). Improvement of shade resilience in photovoltaic modules using buck converters in a smart module architecture. *Energies* 11: 250.

Golroodbari, S.Z.M., de Waal, A.C., and van Sark, W.G.J.H.M. (2019). Proof of concept for a novel and smart shade resilient photovoltaic module. *IET Renewable Power Generation* 13: 2184–2194.

Hasyim, E.S., Wenham, S., and Green, M. (1986). Shadow tolerance of modules incorporating integral bypass diode solar cells. *Solar Cells* 19: 109–122.

Karakose, M., Baygin, M., and Parlak, K.S. (2014). A new real time reconfiguration approach based on neural network in partial shading for PV arrays. In: *Proceedings of the 3rd International Conference on Renewable Energy Research and Applications (ICRERA2014)*, 633–637.

Karakose, M., Baygin, M., Murat, K. et al. (2016). Fuzzy based reconfiguration method using intelligent partial shadow detection in PV arrays. *International Journal of Computational Intelligence Systems* 9: 202–212.

Karatepe, E. and Hiyama, T. (2010). Simple and high-efficiency photovoltaic system under non-uniform operating conditions. *IET Renewable Power Generation* 4: 354–368.

Krishna, G.S. and Moger, T. (2019). Reconfiguration strategies for reducing partial shading effects in photovoltaic arrays: state of the art. *Solar Energy* 182: 429–452.

La Manna, D., Vigni, V.L., Sanseverino, E.R. et al. (2014). Reconfigurable electrical interconnection strategies for photovoltaic arrays: a review. *Renewable and Sustainable Energy Reviews* 33: 412–426.

Littwin, M., Baumgartner, F., Biba, C. et al. (2020). Performance of new photovoltaic system designs – IEA PVPS Task 13 subtask 1.3. In: *Proceedings of the 37th European Photovoltaic Solar Energy Conference*, 1208–1220.

Littwin, M., Baumgartner, F., Green, M., and van Sark, W. (2021). Performance of new photovoltaic system designs. In: . IEA-PVPS, Report number IEA-PVPS T13-15: 2021, https://iea-pvps.org/wp-content/uploads/2021/03/IEA-PVPS_Task-13_R15-Performance-of-New-PV-system-designs-report.pdfhttps://iea-pvps.org/key-topics/performance-of-new-photovoltaic-system-designs/.

MacAlpine, S.M., Erickson, R.W., and Brandemuehl, M.J. (2013). Characterization of power optimizer potential to increase energy capture in photovoltaic systems operating under nonuniform conditions. *IEEE Transactions on Power Electronics* 28: 2936–2945.

Masson, G. and Kaizuka, I. (2021, 2021). Trends in PV applications 2021, IEA-PVPS, report IEA-PVPS-T1-41:2021. In: . https://iea-pvps.org/trends_reports/trends-in-pv-applications-2021/.

Maxim Integrated, 2022 https://www.maximintegrated.com/content/dam/files/design/technical-documents/white-papers/solar-cell-optimization1.pdf (last accessed 18 September 2022)

Moraitis, P., Kausika, B., and van Sark, W. (2018). Urban environment and solar PV performance: the case of the Netherlands. *Energies* 11: 1333.

Nguyen, D. and Lehman, B. (2008). An adaptive solar photovoltaic array using model-based reconfiguration algorithm. *IEEE Transactions on Industrial Electronics* 55: 2644–2654.

Nihanth, M.S.S., Ram, J.P., Pillai, D.S. et al. (2019). Enhanced power production in PV arrays using a new skyscraper puzzle based one-time reconfiguration procedure under partial shade conditions (PSCs). *Solar Energy* 194: 209–224.

Obane, H., Okajima, K., Oozeki, T., and Ishii, T. (2012). PV system with reconnection to improve output under nonuniform illumination. *IEEE Journal of Photovoltaics* 2: 341–347.

Olalla, C., Clement, D., Rodriguez, M., and Maksimovic, D. (2013). Architectures and control of submodule integrated DC–DC converters for photovoltaic applications. *IEEE Transactions on Power Electronics* 28: 2980–2997.

Pannebakker, B.B., de Waal, A.C., and van Sark, W.G.J.H.M. (2017). Photovoltaics in the shade: one bypass diode per solar cell revisited. *Progress in Photovoltaics: Research and Applications* 25: 836–849.

Pillai, D.S., Ram, J.P., Nihanth, M.S.S., and Rajasekar, N. (2018). A simple, sensorless and fixed reconfiguration scheme for maximum power enhancement in PV systems. *Energy Conversion and Management* 172: 402–417.

Rani, B.I., Ilango, G.S., and Nagamani, C. (2013). Enhanced power generation from PV array under partial shading conditions by shade dispersion using Su Do Ku configuration. *IEEE Transactions on Sustainable Energy* 4: 594–601.

Reich, N.H., Mueller, B., Armbruster, A. et al. (2012). Performance ratio revisited: are PR > 90% realistic? *Progress in Photovoltaics: Research and Applications* 20: 717–726.

Schmidt, H., Rogalla, S., Goeldi, B., and Burger, B. (2010). Module integrated electronics – an overview. In: *Proceedings of the 25th European Photovoltaic Solar Energy Conference*, 3700–3707.

Serna-Garcés, S., Bastidas-Rodríguez, J., and Ramos-Paja, C. (2016). Reconfiguration of urban photovoltaic arrays using commercial devices. *Energies* 9: 2.

Shams El-Dein, M.Z., Kazerani, M., and Salama, M.M.A. (2013). Optimal photovoltaic array reconfiguration to reduce partial shading losses. *IEEE Transactions on Sustainable Energy* 4: 145–153.

Silvestre, S., Boronat, A., and Chouder, A. (2009). Study of bypass diodes configurations on PV modules. *Applied Energy* 86: 1632–1640.

Sinapis, K., Tzikas, C., Litjens, G. et al. (2016). A comprehensive study on partial shading response of c-Si modules and yield modeling of string inverter and module level power electronics. *Solar Energy* 135: 731–741.

Sinapis, K., Rooijakkers, T.T.H., Slooff, L.H. et al. (2017). Towards new module and system concepts for linear shading response. In: *Proceedings of the 44th IEEE Photovoltaic Specialist Conference*, 1081–1085.

Sinapis, K., Pacheco Bubi, R., Slooff, L.,.H. et al. (2018). Outdoor performance characterization of a novel shadow tolerant module. In: *Proceedings of the 35th European Photovoltaic Solar Energy Conference*, 1169–1171.

Sinapis, K., Rooijakkers, T.T.H., Pacheco Bubi, R., and van Sark, W.G.J.H.M. (2021). Effects of solar cell group granularity and modern system architectures on partial shading response of crystalline silicon modules and systems. *Progress in Photovoltaics: Research and Applications* 29: 977–989.

Slooff, L.H., Carr, A.J., de Groot, K. et al. (2017). Shade response of a full size TESSERA module. *Japanese Journal of Applied Physics* 56: 08MD01.

Sunchip, 2022 https://www.sunchip.nl (last accessed 18 September 2022)

Taylor Solar, 2022 https://uploads-ssl.webflow.com/62af59ed13ce6b71eead6655/6311e442c5989195f62f6013_Taylor%20intro%20(EN).pdf (last accessed 18 September 2022)

Tsafarakis, O., Sinapis, K., and Van Sark, W.G.J.H.M. (2019). A time-series data analysis methodology for effective monitoring of partially shaded photovoltaic systems. *Energies* 12: 1722.

Uno, M. and Kukita, A. (2017). Current sensorless equalization strategy for a single-switch voltage equalizer using multistacked buck – boost converters for photovoltaic modules under partial shading. *IEEE Transactions of Industrial Applications* 53: 420–429.

Velasco-Quesada, G., Guinjoan-Gispert, F., Pique-Lopez, R. et al. (2009). Electrical PV array reconfiguration strategy for energy extraction improvement in grid-connected PV systems. *IEEE Transactions on Industrial Electronics* 56: 4319–4331.

Venkateswari, R. and Rajasekar, N. (2020). Power enhancement of pv system via physical array reconfiguration based Lo Shu technique. *Energy Conversion and Management* 215: 112885.

Weaver, J.F. (2022). World has installed 1TW of solar capacity. *PV magazine* (15 March): https://www.pv-magazine.com/2022/03/15/humans-have-installed-1-terawatt-of-solar-capacity/.

Yang, B., Ye, H., Wang, J. et al. (2021). PV arrays reconfiguration for partial shading mitigation: recent advances, challenges and perspectives. *Energy Conversion and Management* 247: 114738.

# 23

# Colored PV Modules

*Lenneke Slooff-Hoek[1] and Angèle Reinders[2,3]*

[1] Department of Solar Energy, TNO Energy and Materials Transition, P.O. Box 15, 1755 ZG, Petten, The Netherlands
[2] Energy Technology & Fluid Dynamics Group, Department of Mechanical Engineering, Eindhoven University of Technology (TU/e), P.O. Box 513, 5600 MB, Eindhoven, The Netherlands
[3] Department of Design Production and Management, Faculty of Engineering Technology, University of Twente, P.O. Box 217, 7500 AE, Enschede, The Netherlands

## 23.1 Introduction

### 23.1.1 What is Color?

It's a well-known secret that ancient Greeks and Romans used to vividly color their white marble statues, however, their color theories were rather immature. Though in the past many scientists as well as artists have marveled colors, most color theories were not well founded in natural sciences until Isaac Newton (1642–1726) (Newton 1998, originally published in 1704) became the first to demonstrate that light alone is responsible for colors. He did this by explaining the rainbow by refracting white light with a prism, which resolved it into its component colors: red, orange, yellow, green, blue and violet. Also, he defined a circular color diagram that became the model for many color systems for artists in the centuries thereafter, see Figure 23.1. Nowadays, we know that these color components represent photons with a certain wavelength, otherwise said photon energy, see Figure 23.2. However, color is not just a property of electromagnetic radiation but a subjective perception of light by a human observer. Namely, light interacts with the human eye, which senses color with the help of cone cells – the photoreceptors – that are active at higher light levels (photopic vision). Three types of cone cells exist in the eye that respond to different wavelength stimuli, with peak wavelengths of 564–580 nm (red color), 534–545 nm (green color), and 420–440 nm (blue). Human color perception is, hence, an additive combination of the stimulation of three types of cone cells, in combination with further processing of these stimuli by the brain. Moreover, the human eyes also have rods besides the cones. Rods are responsible for vision at low light levels (scotopic vision), and they do not mediate color vision. That's why, at night, human visual perception takes place on a grayscale. In the photopic vision mode, any color can be objectively described by a mixed set of values associated with the perception of the three different types of cones in the eyes. These sets of values are called tristimulus values. Given the wide range of possible tristimulus values, in practice, it is not straightforward to identify the direct relationship between illumination light intensity

*Photovoltaic Solar Energy: From Fundamentals to Applications, Volume 2*, First Edition.
Edited by Wilfried van Sark, Bram Hoex, Angèle Reinders, Pierre Verlinden, and Nicholas J. Ekins-Daukes.
© 2024 John Wiley & Sons Ltd. Published 2024 by John Wiley & Sons Ltd.
Companion website: www.wiley.com/go/PVsolarenergy

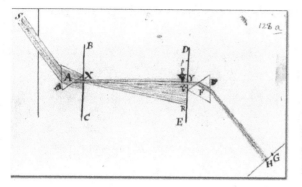

**Figure 23.1** Left: Sketch from Sir Isaac Newton's experiment, 1666–72, on the splitting of light by a first prism (A) in constituent colors and the subsequent bundling of colored beams into white light by the second prism (F) (Anonymous 2022). Right: Color circle by painter Claude Boutet, 1708, which was based on Newton's color theory (Anonymous 2022/https://www.webexhibits.org/colorart/bh.html/last accessed August 30, 2023.).

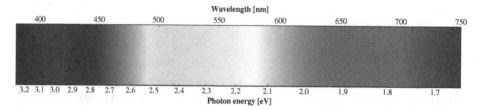

**Figure 23.2** Relation between color of (visible) light, wavelength and its energy in the spectral range of human visual perception, Source: Reinders et al. (2017)/John Wiley & Sons.

upon (or through) an object, the spectrum of reflected (or transmitted) light by the object, and the color identified by the human eye, hence, color impressions in the human brain. Colors are fascinating, and as such, the famous Belgian physicist Marcel Minnaert spent a large part of his life physically explaining color phenomena, especially outdoors, in nature (Minnaert 1954, 1993). There, colors are abundantly available, for instance, in the rainbow, flowers, insects and waves. He explained the different phenomena that cause these colors in relation to human perception.

Colors in solar photovoltaic applications are generally generated by three mechanisms: (1) selective absorption, (2) selective scattering and/or (3) photoluminescence, see Figure 23.3. To start with, the selective absorption color is usually a pigment, which derives its colors from light absorption by electron transitions and, hence, leaves the unabsorbed photons reflected into human eyes. This mechanism of selective absorption also applies to unabsorbed transmitted photons in the case of transparent layers, incl. solar cells. Next, selective scattering is based on (partial) light scattering and reflection on a material's surface, resulting in perception of reflected photons – hence, colors – by the human eye. Thirdly, photoluminescence is light emission from any form of matter after absorption of photons, which cause photoexcitation and subsequent relaxation processes through which other photons are re-radiated (emitted) at specific wavelengths – representing specific colors – different from those of the absorbed photons.

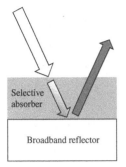 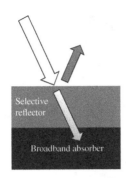 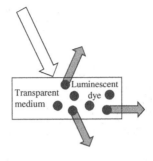

**Figure 23.3** Left: selective absorption. Source: Shang et al. (2020)/American Institue of Physics/CC by 4.0, Middle: selective scattering (Shang et al. 2020) and Right: photoluminescence are some mechanisms to generate color. Please notice that the white arrow represents the incoming light, and the color of the arrows represents the by-material interactions modified color of reflected or transmitted irradiance.

### 23.1.2 CIE Color Space

To avoid subjectivity in color identification, color matching theories aim at numerically specifying a measured color and then later accurately reproducing that measured color (e.g. in print or digital displays). This is different from describing colors as perceived by human observers. The most widely accepted standard in color matching is CIE 1931 (Blankenbach 2016; Westland 2016). It defines a color space map that is described as a range of physically produced colors to the tristimulus values $(X, Y, Z)$. A detailed description of the origin and development of the color-matching functions can be found in Abraham (2017). For a given relative spectral power distribution of the illuminating light $I(k)$, and reflection spectrum $R(k)$, the tristimulus values $(X, Y, Z)$ can be computed by an integral relation across the wavelength range of 380–780 nm that represents the visual range of human perception (Westland 2016). To provide a perceptive color map, the CIE created a two-dimensional chromaticity diagram, Figure 23.4, to show the different color stimuli. Each point $(x, y)$ on the diagram represents a color described by its tristimulus values.

### 23.1.3 Relevance of Color in Photovoltaic Applications

So far, most research on photovoltaic (PV) cells, PV modules and PV systems has focused on increasing their power output. However, for architectural and building applications, five characteristics are required for PV modules: (1) low cost, (2) high efficiency, (3) long life-time, (4) product flexibility and (5) aesthetics view (PVSITES 2016). This aesthetic view includes color, flexibility and transparency (Reinders 2020; Lee and Song 2021) and is still a challenging issue given the fact that the majority of PV products in the present market are still dark-colored. These PV modules are difficult to harmonize with various colored building exterior materials, while a survey on building integrated photovoltaics (BIPV) challenges in the architecture field indicates that the perceived color and decorative features of BIPV modules and systems are considered high-priority factors. For example, a survey among 408 Swiss homeowners revealed that the majority prefer colored PV modules above black ones despite their higher cost (Hille et al. 2018). However, the preferred color for PV

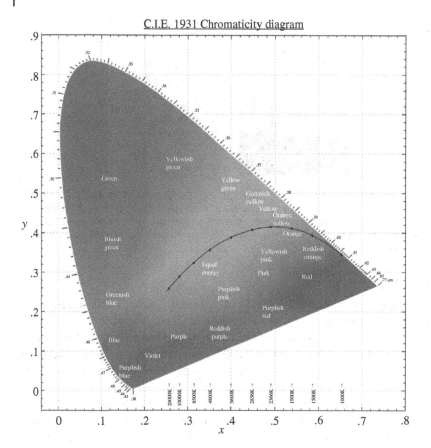

**Figure 23.4** CIE 1931 Color chromaticity diagram. Source: Adapted from Westland (2016).

modules varies by respondent and by type of application. For instance, brown-colored PV modules will be more desired for integration in a terracotta roof, such that the color of the PV system will merge with its surrounding construction materials, while light gray PV modules are more suitable for installation in a concrete façade. On the other hand, instead of using color for camouflage of PV elements in the context of conventional building materials, see Figure 23.5, architects might also like to be able to apply brightly colored PV modules to attract attention to a specific part of the building envelope, Figure 23.6, or simply because their design style usually comprises colorful elements in the final construction. Naturally, as a consequence, if PV modules are aesthetically appealing and can have customized colors, their applications and, hence, the market for PV systems in the built environment will be increased. Therefore, to produce an aesthetic view of PV modules, various types of colored PV modules are more desirable (Moor et al. 2017).

### 23.1.4 Structure of This Chapter

Given the context sketched above, this chapter will address the following topics; to start with, Section 23.2 will cover the physics of colors in PV modules. Next, various options to

**Figure 23.5** Eight showcases the camouflage design of PV modules that mimic existing building materials. Source: Mitrex, 2022.

**Figure 23.6** Explicit colorful designs of PV modules. Left: Orange PV modules in a matte color by Kameleon Solar. Middle: Shiny gold and turquoise PV modules by SWISSINSO. Right: Colorful printed PV modules with a leaf pattern by SolarVisuals.

realize colored silicon PV modules, as the largest market segment for PV modules, will be presented in Section 23.3. Section 23.4 will, in particular, focus on colored graphic designs on PV modules, to be succeeded by Section 23.5 on the performance of these PV modules. This chapter will be completed by conclusions in Section 23.6.

## 23.2 Physics of Colors in PV Modules Explained

### 23.2.1 Photons

In practice, the photon energy, $E$, is a useful variable for color identification. It is perceived as a specific color by the human eye, provided that its wavelength matches with the spectral response of one of the three types of cones in the eye. It is given by the following equation:

$$E = \frac{h\,c}{\lambda} \tag{23.1}$$

where $\lambda$ is the wavelength (in m), $h$ is Planck's constant ($6.626 \times 10^{-34}$ J s), and $c$ is the speed of light in vacuum (exactly 299.792.458 m/s). Since the frequency $v$ (in Hz), wavelength $\lambda$, and speed of light $c$, are related by $\lambda \times v = c$, Eq. (23.1) also can be represented by:

$$E = h \times v \tag{23.2}$$

The energy contained in a photon is rather little and can therefore be better expressed by the unit electron-volt (indicated by eV), according to the following simplified equation:

$$E \text{ (in eV)} = 1.24/\lambda \text{ (in μm)} \tag{23.3}$$

As an example, Figure 23.2 shows the colors of light, visible to the human eye, with corresponding wavelengths (nm) and energy (eV).

### 23.2.2 Reflectance and Transmittance

The wave theory of light is indispensable for the explanation and quantification of transmission, reflection and refraction of irradiance in various media, hence, the representation of optical phenomena that influence color perception.

The speed of light in vacuum, $c$, is a physical constant; however, if light propagates through a medium, which could be, for instance, air, glass, plastic or silicon, its speed is reduced to $v$, the phase velocity of light in media, where $v$ is lower than $c$. The refractive index of a material, $n$, is an indicator for the speed of light in media since it is equal to the ratio of the speed of light in vacuum, $c$, and the phase velocity in the medium, $v$. Logically, in media $n$ is always larger than 1.

The different propagation of light in two different isotropic media results in refraction at the interface of these media according to Snell's law:

$$\frac{\sin \theta_i}{\sin \theta_t} = \frac{n_2}{n_1} \tag{23.4}$$

where $\theta_i$ is the angle of incidence (toward the normal of a surface), $\theta_t$ is the angle of refraction of the outgoing beam, $n_1$ is the refractive index of the medium through which the incident beam of light passes and $n_2$ is the refraction index of the medium through which the refracted beam goes, see Figure 23.7. From Snell's law, it can be derived that when light propagates from a medium with a higher refraction index to a medium with a lower refractive index (hence, $n_1 > n_2$), total internal reflection happens for incidence angles larger than the critical angle, $\theta_{\text{crit}}$, leading to refraction into the medium with the higher refraction index:

$$\theta_{\text{crit}} = arsin \left(\frac{n_2}{n_1}\right) \tag{23.5}$$

The law of refraction states that if light hits an interface between two media with a different refractive index, the angle of reflection, $\theta_r$, is equal to the angle of incidence, $\theta_i$.

Snell's law does not explain which share of the energy contained by the incident irradiance, $I_i$, is subjected to transmission or reflection. However, the law of conservation of energy states that the sum of the transmission coefficient, $T$, and the reflection coefficient, $R$, should be 1: $T + R = 1$ where $T = I_t/I_i$ and $R = I_r/I_i$ for non-magnetic materials.

The values of $T$ and $R$ depend on the polarization of irradiance. If the vector of the electric field happens to be perpendicularly oriented toward the plane that contains both the incident, refracted and reflected irradiance (this is called s-polarization), then the reflection coefficient is indicated by $R_s$. If it happens to be in line with the plane of incidence, so-called

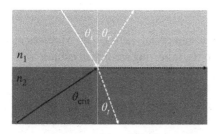

**Figure 23.7** Scheme representing incident, refracted and reflected irradiance at the interface of two media with refraction indices $n_1$ and $n_2$ where $n_2 > n_1$.

p-polarization, we speak about $R_p$. The respective transmission coefficients are given by the law of conservation of energy, namely $T_s + R_s = 1$ and $T_p + R_p = 1$.

$R_s$ and $R_p$ are given by the Fresnel equations for nonmagnetic materials:

$$R_s = \left[\frac{\sin(\theta_t - \theta_i)}{\sin(\theta_t + \theta_i)}\right]^2 = \left[\frac{n_1 \cos(\theta_i) - n_2 \cos(\theta_t)}{n_1 \cos(\theta_i) + n_2 \cos(\theta_t)}\right]^2$$

$$= \left[\frac{n_1 \cos(\theta_i) - n_2 \sqrt{1 - \left(\frac{n_1}{n_2}\sin\theta_i\right)^2}}{n_1 \cos(\theta_i) + n_2 \sqrt{1 - \left(\frac{n_1}{n_2}\sin\theta_i\right)^2}}\right]^2 \tag{23.6}$$

$$R_p = \left[\frac{\tan(\theta_t - \theta_i)}{\tan(\theta_t + \theta_i)}\right]^2 = \left[\frac{n_1 \cos(\theta_t) - n_2 \cos(\theta_i)}{n_1 \cos(\theta_t) + n_2 \cos(\theta_i)}\right]^2$$

$$= \left[\frac{n_1 \sqrt{1 - \left(\frac{n_1}{n_2}\sin\theta_i\right)^2} - n_2 \cos(\theta_i)}{n_1 \sqrt{1 - \left(\frac{n_1}{n_2}\sin\theta_i\right)^2} + n_2 \cos(\theta_i)}\right]^2 \tag{23.7}$$

Equations (23.6) and (23.7) are frequently used to calculate the reflectance and transmittance at the interface between two materials of different refractive indexes at various angles of incidence. This theory lays an important basis for the perception of color in thin film interference and such.

### 23.2.3  Thin Film Interference

In this section, we describe thin-film interference, which is a typical optical process related to colors in surfaces on top of PV modules. This is the underlying mechanism of anti-reflection coatings (Kinoshita et al. 2008), such as $Si_3N_4$, see Figure 23.8, that can be applied to realize, for instance, colored silicon PV cells, see Figure 23.11f.

Thin film interference occurs when a plane wave of light hits a thin film of thickness $d$ and refractive index $n_b$ with the angles of incidence and refraction as $\theta_a$ and $\theta_b$, see Figure 23.8. The irradiance beams reflected at the two surfaces, namely the top and the bottom, of the

**Figure 23.8** Schematic diagram of thin-film interference. Source: Kinoshita et al. (2008)/IOP Science.

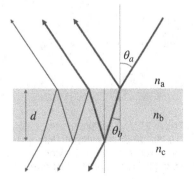

thin film interfere with each other. In general, interference conditions differ depending on whether the thin film is attached to a material having a higher refractive index or not. The former is the case for the anti-reflective coating on top of solar cells, while a typical example of the latter is a soap bubble. The reason for the difference is that the reflection at a surface changes its phase by 180°, when the light is incident from a material with a smaller refractive index to that with a higher one, while it does not in the inverse case. In the anti-reflective coating case, destructive interference occurs, according to:

$$2n_b \, d \, \cos \theta_b = m \, \lambda \tag{23.8}$$

where $\lambda$ is the wavelength giving the maximum reflectivity and m is an integer. Usually, only one-time reflection happens at each surface. However, multiple reflections significantly influence the interference when the material shows higher reflectance at the interface.

### 23.2.4 Multilayer Interference, Plasmonic Coatings

Multilayer interference is qualitatively understood in terms of a pair of thin layers piling periodically. Consider two layers designated as A and B with thicknesses $d_A$ and $d_B$ and refractive indices $n_A$ and $n_B$, respectively, as shown in Figure 23.9.

If we consider a certain pair of AB layers, the phases of the reflected light both at the upper and lower B–A interfaces change by 180°. For now, we assume $n_A > n_B$. Thus, for constructive interference with the angles of refraction in the A and B layers as $\theta_A$ and $\theta_B$, a relation similar to the anti-reflective coating of Eq. (23.8) is applicable as:

$$2(n_A d_A \cos \theta_A + n_B d_B \cos \theta_B) = m\lambda, \tag{23.9}$$

Multilayer interference is only applicable when the difference in the refractive indices of the two layers is small enough. Otherwise, multiple reflections modify the interference condition to a large extent. The quantitative evaluation of the wavelength-dependent reflectivity is rather complex in a general case, but is important for understanding the principle of colors, because most of them originated from multilayer interference or its analog.

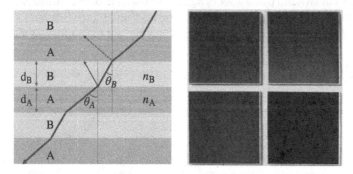

**Figure 23.9** Left: Schematic illustration of multilayer interference. Source: Kinoshita et al. (2008)/IOP Science. Right: Photograph of demonstrator modules with four different multilayer interference coatings showing the colors red, green and blue in black uncoated glass. Source: Bläsi et al. (2017)/Fraunhofer Institute for Solar Energy Systems.

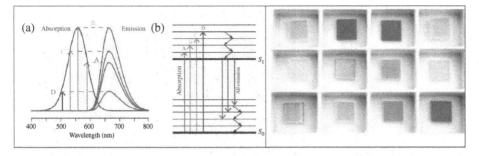

**Figure 23.10** Left: In a photoluminescent process, incoming photons are absorbed and excite electrons, which return to a lower energy state by emitting photons, Source: Schmidt, T. and Ekins-Daukes, E. (2019); (a) absorption and emission curves, and (b) energy levels of excited electrons and emitted photons. Right: Luminescent solar concentrator light guides, which can be applied in photovoltaic devices (Reinders 2020).

### 23.2.5 Photoluminescence

Photoluminescence is a process in which any form of matter, that is to say, an organic molecule, atom or quantum dot, absorbs a photon in the visible region, while exciting one of its electrons to a higher electronic excited state and then radiating a photon as the electron returns to a lower energy state. This radiated photon will be perceived with a certain color according to its wavelength, otherwise said its energy, see Figure 23.10a. Usually, absorption and emission spectra are mirrored. The difference in absorption and emission maxima is known as the Stokes Shift, resulting in different colors of the emitted spectrum, see Figure 23.10b. Photoluminescence is applied in luminescent solar concentrator photovoltaics (Reinders 2020), but also in markers and safety products.

## 23.3 Options to Realize Colored Silicon PV Modules

Kuhn et al. (2020) executed an excellent analysis of various technological design options to realize colored silicon PV modules. This section presents a summary of their findings that resulted in six categories of color options or combinations thereof; see Figure 23.11 for examples.

1. *Colored coatings on surfaces exposed to weather* are described as follows: Ceramic pigments, which are fused into the glass during tempering, which can be applied evenly with a roller coater, by screen printing or in the form of high-resolution digital printing.
2. *Colored bulk materials for front and back covers*, which are described by bulk-colored polymer materials or glass, are sometimes used as the back cover and rarely also as front cover.
3. *Colored coatings on internal surfaces* (not exposed to weather); colors can either be realized with ceramic colors, with spectrally selective interference coatings or other photonic structures.
4. *Colored encapsulants*: Black opaque interlayers are often used behind the cells in glass-glass modules to realize a black background. Pigments can be incorporated into

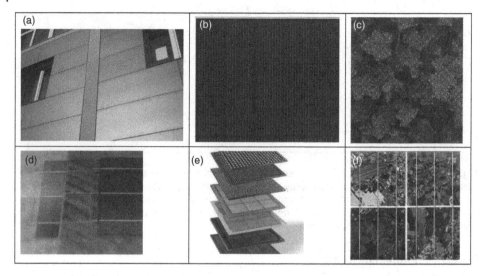

**Figure 23.11** Examples of colored PV module parts, for each of the six identified categories: (a) Colored coatings on surfaces exposed to weather, example from ISSOL, (b) colored bulk materials for front and back covers, example of a Trina PV module with a black backsheet, (c) colored coatings on internal surfaces, example shown by SolarVisuals (d) colored encapsulants by Smart Flex, (e) additional encapsulated colored interlayers, example from Ertex and (f) colored PV cells, example from Kameleon Solar.

colored front encapsulants such as PolyVinylButyral (PVB), ethylene vinyl acetate (EVA), or thermoplastic polyolefin (TPO).

5. *Additional encapsulated colored interlayers*, mainly printed films or colored, semi-transparent, perforated (e.g. woven) textile structures.
6. *Colored PV cells*: colors are realized by additional coatings on standard PV cells or special anti-reflective cell-coatings

These color options coincide with geometrical configurations of front, back and bulk materials as well as solar cells in PV modules, which is illustrated in Figure 23.12. Figure 23.13 shows an example of a PV module in which various coloring options, as presented in Figure 23.12, have been combined.

## 23.4 Colored Front Cover Design of Silicon PV Modules

### 23.4.1 Coloring Techniques

There are several options for coloring the different layers in a PV module, and the inks and pigments that can be used depend on which layer or material the color is applied.

*Coloring of glass* for PV modules can be divided into bulk coloring and surface coloring. In case of bulk coloring, a metallic salt is added to the glass mold, giving the color to the final product. In this way, only homogeneous colored glass sheets can be obtained.

Coloring of PV glass is mostly done by applying a surface coating of ceramic ink using a digital ceramic printing technique (Saretta et al. 2018). The ceramic ink is basically a

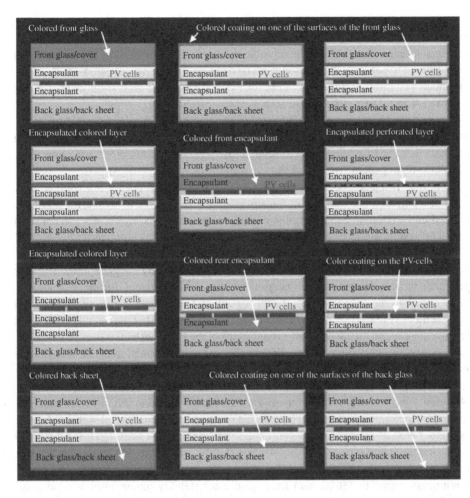

**Figure 23.12** Color options for silicon PV modules. Source: Kuhn et al. (2020)/with permission of Elsevier. The various geometrical design options are shown in the figure. The colored layer or surface is always highlighted in red.

**Figure 23.13** Hexagon pattern of printable colors for PV modules made by Kameleon Solar (Kameleon 2022).

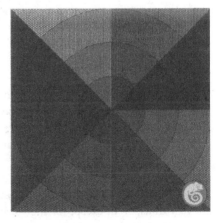

ceramic frit consisting of glass nanoparticles mixed with pigments. After printing, the frit is fused into the glass at temperatures. The technique can be used to print a variety of colors and prints. A point of attention is the reproducibility of the colors. In some cases, this is still rather difficult to achieve. The coloring can be done on the internal or external side of the glass sheet. The internal side is preferred for PV modules as the glass itself will serve as a barrier to UV irradiance, resulting in increased color stability.

Some manufacturers use a technology that generates the color by a multilayer interference coating (Bläsi et al. 2017; Kinoshita et al. 2008; Mertin et al. 2011); see also Section 23.2.3. For this coloring technique, a low-pressure plasma process is often used. The colors are, in general, metallic-like.

Finally, dipcoating can be applied in which the glass slide is dipped into a solution containing the pigment. An example is the sol–gel dip-coating process, where the substrate is dipped into a solution containing the coating materials and is withdrawn at a constant speed. This results in a layer with very homogeneous thickness, as is necessary for interference coatings. This results in homogeneous colors (Habets et al. 2021).

*Colored encapsulants* also offer the benefit of UV protection by the glass layer. They are often made by bulk coloring of the encapsulant by adding a pigment during the manufacturing. But more recently also, digital printing can be used to add, e.g. a colorful picture. The difficulty in the use of colored encapsulants lies in the fact that the resulting color and transparency depend on the lamination conditions of the PV panel lamination process. Furthermore, the lamination process results in softening and deformation of the encapsulant and, thus, the print, thereby blurring the picture. This is less of a problem for monochromatic coloring than for full-color prints.

*Additional encapsulated colored interlayers* are mainly applied in the form of a transparent foil on which the color is added by a digital printing technique (Perret-Aebi et al. 2015; Slooff-Hoek et al. 2017). This enables both monochromatic colors as well as full-color printing. In principle, any type of ink can be used that is compatible with the substrate foil. However, the ink must be able to withstand the lamination conditions of the PV panel, i.e. temperatures up to 150 °C, and have good color stability under outdoor conditions. Water-based inks are not so stable under outdoor conditions, so solvent-based inks are more commonly used, but there are aqueous UV-curable inks that are very stable under outdoor conditions. These inks are dried by UV illumination and can be both organic and inorganic. Organic inks often offer a wider range of available colors but are generally less stable than the inorganic ones.

### 23.4.2 Printing of Patterns

It is important to realize that applying a color to a PV module will always result in a decrease in the efficiency of the module. This is because we can only see color when light is reflected and/or absorbed by an object, and any reflection or absorption of light from a solar module means that less light is entering the solar cell, and thus less light is available for generating electricity.

Hence, a trade-off exists between a colorful PV module and a PV module with highest performance. There are a few ways to tackle this dilemma. One is by developing inks that are very transparent except for the wavelength of the color you want to see. In this way,

**Figure 23.14** Example of a full-color design on a PV system with Kameleon Solar's PV modules.

most of the light will still be able to reach the solar cell, and only a small part is lost through reflection. The drawback of this approach is that it only works for specific colors, e.g. cyan, magenta and yellow, that are used in digital printing. A color that is built up from a combination of these colors will result in a combined wavelength range of the individual colors, and consequently, the efficiency of the solar panel will substantially drop. The more colors will be combined, the larger the performance loss.

Another option is to make use of a multilayer interference coating, as described in Sections 23.2.4 and 23.4.1. Where normal pigments and inks have, besides the reflection, also some absorption that adds to the loss, interference layers result in colors based on superposition of irradiance. This effect is caused by differences in optical properties of the various layers, and the observed color depends both on material properties, such as the refraction index, as well as layer thicknesses. The drawback of this approach is that only homogeneous colors can be obtained and that the creation of ultra-thin multilayer stacks requires dedicated equipment, skills and experience.

A third approach is not to cover the whole surface of the PV module, but to apply a pattern. This can be used for both homogeneous colors as well as full-color prints. With this approach, it is more easy to tune on color or performance, but it has some challenges as well. When printing only on part of the module surface, the perceived color will be the result of the light reflection from the ink but also from the open areas. These open areas are dark blue or black and will result in a more greyish perception of the color. It is thus more difficult to obtain very bright colors as well as to predict the observed color. A few examples by PV module manufacturer Kameleon Solar are shown in Figures 23.13 and 23.14 (Kameleon 2022). Another manufacturer, SolarVisuals, uses an algorithm to generate a dot-like pattern (Slooff-Hoek et al. 2022). This algorithm puts the dots preferably above the metalization of the solar cell to reduce the losses. They offer various patterns and coverages, for which several examples are shown in Figure 23.15.

**Figure 23.15** Left: Full-color designs by SolarVisuals, Right: Application of such as printed PV module in a wall.

## 23.5 Energy Performance of Colored PV Modules

Although in the PV world, it is common to define the energy performance of PV modules as the generated amount of electricity compared to the installed nominal PV power given by the unit kWh/Wp, in BIPV applications, it is becoming more common to express performance as installed power per m$^2$ using the unit Wp/m$^2$. The reason for this approach is that PV modules in BIPV projects will be installed on different surfaces with varying orientations and tilts, which will also result in varying electricity generation. A typical method to determine the energy performance of a (colored) BIPV system would then be:

- Obtain the module power under standard test conditions ($P_{STC}$) per module area ($A_{mod}$) for the modules that will be applied.
- Determine the geographical location of the system.
- With this, determine the nominal energy yield ($E_{nom}$), i.e., the energy yield per installed Wp for a system with optimal orientation and tilt for that location, e.g., see Figure 23.16 for values for the Netherlands.
- Determine the tilt and orientation of the BIPV system.
- With this, determine the loss factor $\eta$ of the system with respect to the optimal orientation from the loss table for the geographical location, as shown in Figure 23.17 for the Netherlands.
- Calculate the generated energy with the following formula:

$$E_y = P_{STC}/A_{mod} \times \eta \times E_{nom}$$

This is a generally accepted approach to estimating the expected energy yield of a BIPV system. The only difference when colored PV is used lies in the fact that the performance (in Wp/m$^2$) for a colored PV module will be different as compared to a conventional non-colored PV panel. When comparing BIPV systems with colored PV modules with different performances it is therefore important to indicate the performance loss compared to a similar non-colored PV module, assuming that besides the PV technology inside the

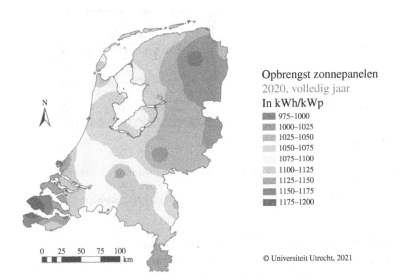

**Figure 23.16** Optimal energy yield for horizontally oriented PV systems in the Netherlands in the year of 2020. Source: Courtesy of N. Nortier and W. van Sark.

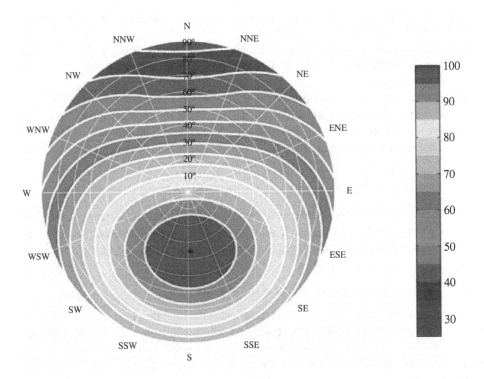

**Figure 23.17** Energy loss yield for PV modules with different orientations (orientation) and tilt angle (roof inclination).

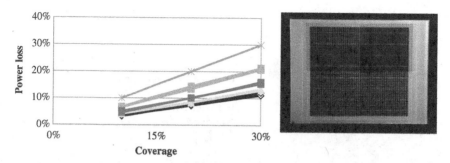

**Figure 23.18** The average power loss of colored prints with a rectangular dot pattern with dot sizes of 1–3 mm as a function of coverage. Source: Slooff-Hoek et al. (2017)/with permission of Elsevier. Also shown in that graph is the 1:1 relation between power loss and coverage (blue line with stars).

module, the performance of colored PV modules strongly depends on the type of applied color or ink/pigment, the ink coverage, the front side structure of the module and the coloring technique used. These aspects will be briefly discussed in the following.

### 23.5.1 Effect of Color or Ink/Pigment

Figure 23.18 shows an example of the differences in loss for different colors for PV test laminates with a rectangular dot pattern.

As can be seen, the cyan ink shows a much higher loss than the black, yellow and white ink. The specific dependency will depend on the ink set that is used and is thus not general for colored PV panels.

As mentioned in the previous paragraph, the perceived color in PV panels with a dot pattern is the result of the mixture of the print color and the bluish/black color of the cells. As a result, a higher print coverage is needed to obtain vivid colors, and these PV modules will generally have lower performance.

### 23.5.2 Effect of Ink Coverage

Figure 23.18 also shows the dependence of the power loss on the ink coverage. The higher the ink coverage, the higher the loss. This is a general trend, but the dependency will be different for different colors, dot patterns and coloring technologies.

### 23.5.3 Effect of Front Side Structure of the Module

Two modules with identical color-printed films inside the module can show a difference in performance depending on the front side material. When a polymer is used instead of glass, the differences in optical constants and, thus, reflection result in a somewhat higher performance for the polymer material. A structured surface of both a glass and a polymer front side results in higher performance as the light is scattered more, and part of that scattered light is able to reach the solar cells, see Figure 23.19. Besides the increase in performance, the observed color changes slightly due to the increased scattering.

**Figure 23.19** The visual effect of flat, highly reflecting glass front side as compared to a structured glass front side has an effect on the energy performance of colored PV modules.

### 23.5.4 Performance of Commercial Panels

Table 23.1 gives an indication of the performance in Wp/m$^2$ for several suppliers (please notice that this is not an extensive overview), as well as the coloring technique they use.

**Table 23.1** Overview of the performance and coloring technique for colored solar modules of several suppliers.

| Facade companies | Number of colors | Coloring Technology | Area coverage | Power (Wp/m$^2$) |
|---|---|---|---|---|
| Beng Solar/Avancis Skala panels | Nine colors; monochrome | Colored glass nanotechnology | Full area | 114–133 |
| Solaxess | Endless number of colors; monochrome | Front foil with nanotechnology | Full area | 110–200 |
| EigenEnergie/ MyEnergySkin | Monochrome and multi color | Colored glass | Full area | 120 W/m$^2$ |
| Kameleon solar | >4000 different colors; monochrome and multicolor | Colored glass | Dot pattern | Up to 150 |
| Mitrex | Endless number of colors: monochrome and multicolor | Colored glass | Full area | 129–179 |
| Pixasolar | 213 RAL K5 classic colors | Colored glass | Dot pattern | 75–170 |
| Solar Visuals | Endless number of colors: monochrome and multicolor | Enclosed foil | Dot pattern | Up to 155 |
| Solarix | 53 standard more possible; mostly monochrome but multicolor possible | Colored glass | Dot pattern | 100–150 |
| Soluxa B.V. | Endless number of colors; monochrome | Colored glass nanotechnology | Full area | |
| Metsolar | 13 standard; others on demand; monochrome | Colored glass | Full area | 110–190 |
| Kromatix from SwissINSO | 10 standard; monochrome | Colored glass nanotechnology | Full area | 149–163 |

## 23.6 Conclusions

With this chapter, we have provided an overview of the basic principles of colors, their visual perception and their applications in PV modules, with a focus on silicon PV modules, which still form the majority of the market.

## Acknowledgments

We would like to thank all PV module developers, researchers, designers and BIPV manufacturers for providing valuable visual materials, such as photos and graphs, through which we could explain several principles and applications of color in PV modules.

## Nomenclature

| Symbol | Variable | Unit |
|---|---|---|
| Electrical Power | $P$ | W |
| Electrical yield loss factor | $\eta$ | – |
| Energy yield per area | $E_y$ | kWh/m$^2$ |
| Power under STC | $P_{STC}$ | Wp |
| Nominal energy yield | $E_{nom}$ | kWh/Wp |

## Abbreviations and acronyms

| Abbreviation | Meaning |
|---|---|
| BIPV | Building Integrated Photovoltaics |

## Author Biographies

**Dr. Lenneke Slooff-Hoek** is the program manager of Solar and Buildings at TNO in The Netherlands. Lenneke studied applied physics at Utrecht University, followed by a Ph. D. in light amplification for telecommunication applications. She started at TNO (formerly known as the Energy Research Center of the Netherlands ECN) in 2001, working on the development of plastic solar cells. During her time at ECN and TNO, she developed her knowledge of solar cells, modules and systems. In the past five years, the focus of her research was directed toward the integration of solar energy systems in the landscape and building environment, with a focus on aesthetic integration. The development of colored modules and their optimization with respect to color perception and performance has been one of the major topics in her research, as well as the development of mathematical models for yield prediction. Lenneke is also an active member of the Dutch national consortium Solar and Buildings.

**Prof. Dr. Angèle Reinders** is a full professor of "Design of Sustainable Energy Systems" at the Department of Mechanical Engineering at Eindhoven University of Technology (TU/e).

In her research, she focuses on the integration of solar energy technologies in buildings and mobility by means of simulation, prototyping and testing. A strong interest exists in designs that enhance user acceptance and reduce environmental impact. Angèle Reinders studied experimental physics at Utrecht University, where she also received her doctoral degree in chemistry. Next, she developed herself in Industrial Design Engineering. Besides TU/e's professorship, she is an Associate Professor at the University of Twente, and she was a full professor in Energy Efficient Design at TU Delft. In the past, she also worked at the Fraunhofer Institute of Solar Energy, World Bank, ENEA in Naples and Center of Urban Energy in Toronto. She is known for her books "Designing with Photovoltaics" (2020), "The Power of Design" and "Photovoltaic Solar Energy From Fundamentals to Applications". In 2010, she co-founded the Journal of Photovoltaics. Also, she initiated many projects and has an extensive publication record.

# References

Abraham, C. (2017). A Beginner's guide to (CIE) Colorimetry. Retrieved from https://medium
.com/hipster-color-science/a-beginners-guide-to-colorimetry-401f1830b65a

Anonymous. (2022). https://www.webexhibits.org/colorart/bh.html. Color Vision and Art

Blankenbach, K. (2016). Color. In: *Handbook of Visual Display Technology*, 3111–3134.

Bläsi, B., Kroyer, T., Höhn, O. et al. (2017). Morpho butterfly inspired coloured BIPV modules.
In: *Paper Presented at the 33rd EU PVSEC, Amsterdam*.

Habets, R., Vroon, Z., Erich, B. et al. (2021). Structural color coatings for high performance
BIPV. In: *Paper Presented at the Crossing Boundaries*.

Hille, S.L., Curtius, H.C., and Wüstenhagen, R. (2018). Red is the new blue – The role of color,
building integration and country-of-origin in homeowners' preferences for residential
photovoltaics. *Energy and Buildings* 162: 21–31:https://doi.org/10.1016/j.enbuild.2017.11
.070.

Kameleon, S. (2022). ColorBlast® by Kameleon Solar. Retrieved from https://kameleonsolar
.com/colorblast/

Kinoshita, S., Yoshioka, S., and Miyazaki, J. (2008). Physics of structural colors. *Reports on
Progress in Physics* 71 (7): 076401. https://doi.org/10.1088/0034-4885/71/7/076401.

Kuhn, T.E., Erban, C., Heinrich, M. et al. (2020). Review of technological design options for
building integrated photovoltaics (BIPV). *Energy and Buildings* 110381.

Lee, H. and Song, H.-J. (2021). Current status and perspective of colored photovoltaic modules.
*WIREs Energy and Environment* 10 (6): e403. https://doi.org/10.1002/wene.403.

Mertin, S., Le Caër, V., Joly, M., and Scartezzini, J.L. (2011). Coloured coatings for glazing of
active solar thermal façades by reactive magnetron sputtering. In: *Paper Presented at the
CISBAT CleanTech for Sustainable Buildings*.

Minnaert, M.G.J. (1954). *The Nature of Light & Colour in the Open Air*. Dover Publications.

Minnaert, M. (1993). *Light and Color in the Outdoors*. New York, NY: Springer.

Moor, T., Egloff, B., Tomovic, T., and Wittkopf, S. (2017). REPEAT – textile design for PV
modules! Design-driven strategies for photovoltaic modules. *The Design Journal* 20 (sup1):
S1879–S1893. https://doi.org/10.1080/14606925.2017.1352706.

Newton, I. (1998, originally published in 1704). *Opticks: Or, a Treatise of the Reflexions, Refractions, Inflexions and Colours of Light. Also Two Treatises of the Species and Magnitude of Curvilinear Figures* (ed. N. Humez). Palo Alto, CA, USA: Octavo.

Perret-Aebi, L.E., Escarré, J., Li, H.Y., and Ballif, C. (2015). When PV modules are becoming real building elements: white solar module, a revolution for BIPV. In: *Paper Presented at the SPIE Optics.*

PVSITES, c. (2016). Formulation of Architectural and Aesthetical Requirements for the BIPV Building Elements to be Demonstrated within the PVSITES Project. Retrieved from https://www.pvsites.eu/

Reinders, A. l. (2020). *Designing with Photovoltaics.* Retrieved from https://search.ebscohost.com/login.aspx?direct=true&scope=site&db=nlebk&db=nlabk&AN=2463679, https://www.taylorfrancis.com/books/e/9781315097923, http://www.vlebooks.com/vleweb/product/openreader?id=none&isbn=9781351578561, https://www.vlebooks.com/vleweb/product/openreader?id=none&isbn=9781351578554 https://doi.org/10.1201/9781315097923

Reinders, A.H.M.E., Verlinden, P.V., Sark, W.v., and Freundlich, A. (2017). *Photovoltaic Solar Energy – From Fundamentals to Applications.* UK: Wiley & Sons.

Saretta, E., Bonomo, P., and Frontini, F. (2018). BIPV meets customizable glass: a dialogue between energy efficiency and aesthetics. In: *Paper presented at the 35th EU PVSEC, Brussels.*

Schmidt, T. and Ekins-Daukes, E. (2019). *Lecture slides of Summer School on Designing with Photovoltaics.* Sydney: UNSW.

Shang, G., Eich, M., and Petrov, A. (2020). Photonic glass based structural color. *APL Photonics* 5 (6):https://doi.org/10.1063/5.0006203.

Slooff-Hoek, L.H., van Roosmalen, J.A.M., Okel, L.A.G. et al. (2017). An architectural approach for improving aesthetics of PV. In: *Paper Presented at the 33rd EU PVSEC, Amsterdam.*

Slooff-Hoek, L.H., Burgers, A.R.T., Carr, A.J. et al. (2022). Model to estimate the performance of full-colored solar panels. In: *Paper presented at the WCPEC-8.*

Westland, S. (2016). The CIE system. In: *Handbook of Visual Display Technology,* 161–169.

# 24

# Building Integrated Photovoltaic Products and Systems

*Francesco Frontini*

*University of Applied Sciences and Arts of Southern Switzerland (SUPSI), Mendrisio, Switzerland*

## 24.1 Introduction

Buildings and the built environment are strategic domains in view of the full decarbonization of our economy, and building skin surfaces represent a huge potential in turning the building stock into a decentralized renewable energy producer. Building integrated photovoltaics (BIPV) are an integral part of construction materials and components, including solar photovoltaic (PV) cells. Regardless of the PV technology used (cSi, thin film or organic PV), as long as the component or system is part of the building construction and holds a building function, we refer to it as BIPV. Current national and international standards (i.e. EN50583: 2016 and IEC 63092:2020 PV in buildings) state that a PV module or system must have (at least) dual functionality as a power generator and a building component to be qualified as building integrated photovoltaics. If the BIPV product is dismounted, it would then have to be replaced by an appropriate construction product.

In recent years, BIPV, from being a niche market, has become a well-established and dynamic sector. Concrete paths are ongoing to bridge the gap between the building and solar industries, architecture and research, as well as design and technology.

## 24.2 Evolution of BIPV

Even if one of the first examples of the integration and innovation of PV in building goes back to 1979–1982 (a solar settlement in Munich designed by the architect Thomas Herzog), it was only in the early 2000s that, driven by incentives and increasing demand, more and more multifunctional solutions, such as PV tiles, or shading elements using standard modules and later curtain walling or ventilated façade cladding systems have started to be developed together with colored and multifunctional systems (Figure 24.1). Until 2009, building-integrated solar systems worldwide accounted for only one percent of the total installed capacity of photovoltaic systems. Although a wide range of solar building products was then available at attractive prices (Frontini et al., 2015; Verberne et al., 2014), the

*Photovoltaic Solar Energy: From Fundamentals to Applications, Volume 2*, First Edition.
Edited by Wilfried van Sark, Bram Hoex, Angèle Reinders, Pierre Verlinden, and Nicholas J. Ekins-Daukes.
© 2024 John Wiley & Sons Ltd. Published 2024 by John Wiley & Sons Ltd.
Companion website: www.wiley.com/go/PVsolarenergy

**Figure 24.1** Evolution of BIPV: 1979–1982: Wohnanlage Richter in Munich (Germany), arch. T. Herzog; 1997: FEAT Building in Lugano (Switzerland), arch. Lo Riso; 2007: Lauper Pilot Installation in Gasel (Switzerland), Solaire Suisse SA; 2008: Kraftwerk B in Bennau (Switzerland), Grab Architekten, 2011: Palazzo Positivo in Chiasso (Switzerland), Tour Baumanagement AG; 2012: Casa Solara in Laax (Switzerland), architect Giovanni Cerfeda; 2014: Historical Residential house in Bern (Switzerland), Wermuth und Partner Architekten; 2015a: CSEM main building in Neuchatel (Switzerland), René Schmid Architekten AG; 2015b: Historical Building in Affoltern (Switzerland), architect Anliker Christian; 2018: Silo Blu in Renens (Switzerland), epure architecture and urbanism SA; 2019: Multi-family house in Affoltern (Switzerland), Viriden & Partner architecture; 2021a: POLIS building in Lugano (Switzerland), Studio Mario Campi SA; 2021b: SOL'CH in Poschiavo (Switzerland), Nadia Vontobel Architekten GmbH. Source: Courtesy of Frontini et al., (2020b)/MDPI Swiss BIPV Competence Centre-SUPSI.

initial investment of BIPV systems was still too expensive at that time to compete with building materials and traditional photovoltaic systems; so, the priority was to maximize annual energy production while reducing costs. Following the development of PV technology, an architectural language based on standard solar panels accompanied the first age of integration in opaque surfaces, mainly building roofs (2010–2015).

However, these conventional PV elements displayed all the limitations of a functional approach, so the drastic change of today's second age involves a shift in technology, complying with the key market-driven demands, such as that of aesthetics, flexibility, building skin performance, durability, and cost-effectiveness.

The development of modern production techniques, increasingly automated and customizable cell stringing, and collaboration with the glass industry have made it possible to create colored or semi-transparent surfaces or multifunctional systems that combine thermal insulation, acoustic protection, and PVs in a single prefabricated component (2015–today) (Frontini et al., 2022).

In more recent years, new colored photovoltaic modules have appeared on the market. Thanks to the special glass processing techniques or the use of special interlayer material (coating or encapsulant), the flexibility of the product design has been improved, providing homogeneous colored surfaces or ad-hoc and custom-designed solutions according to the

**Figure 24.2** Different front and back cover designs to achieve different BIPV appearances. From the left picture: (1) Diffusive front surface with multi-layered coating deposited on the inner glass surface using Kromatix technology; (2) Front glass surface treatment using sandblasting techniques (source: Construct-PV Eu project), (3) Custom-made structured glass from Crea-Glass GmbH assembled with small metallic shiny dots with titanium nitride coating and larger red-yellow color-shifting dots; (4) Digital printing technique with ceramic paints on BIPV modules with textured front glass (source: solarchitecture.ch)

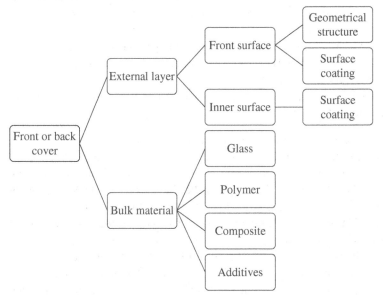

**Figure 24.3** Overview of design parameters for the front and back cover. Source: Adapted from Kuhn et al., (2021).

architect's needs, where the PV cell disappears, thereby making the photovoltaic product a fully customizable building material (see Figure 24.2).

This trend can be referred to as "camouflaged," "customized," or "invisible" PV, where the key material to be designed and engineered is glass or other front materials. Under these banners, there is currently a promising joint effort between PV and the glass industry with the aim of combining high production of solar energy with attractive visual design aesthetics. Moreover, different design options for front and rear cover materials are possible, as presented by Kuhn et al. (2021) and resumed in Figure 24.3.

The available technologies and competitive costs have allowed for the development of different solutions and opportunities for the various applications of PV in the building envelope, as will be presented in the following section, where a classification of BIPV system is presented. Further, the challenges and the ways to evaluate the performance of a BIPV system are presented in Section 24.4.

## 24.3   BIPV as a Part of the Building Skin Elements

BIPV, in contrast to a conventional PV application, is primarily a construction system (belonging to a building or an architecture). Therefore, a classification of BIPV according to this adoption as a building envelope system is fundamental to properly define its guaranteed performance. A possible classification can be done based on the individual functional parts related to the construction work structure that the BIPV module/system will substitute. In conventional constructions, the definition of the main building skin construction systems can be grouped, as reported by Bonomo and Eder (2021) and as shown in Figure 24.4, into:

- Roof: A roof, in traditional building construction with a top distinguishable by the facade, is the building or construction top covering that provides protection and separates the indoor and outdoor environments.
- Facade: A facade, in traditional building construction with parietal walls distinguishable by the roof, is the vertical (or tilted) exterior surface that is the architectural showcase separating the indoor and outdoor environments.
- External integrated devices: Elements and systems of the building skin that are only in contact with the outdoor environment.

For these three groups, several types of applications exist, as presented in Figure 24.4. Each specific application retains particular performances; therefore, a standard classification has been proposed by the standardization bodies to easily provide and identify the requirements of a BIPV module or installation as for the EN50583 or the IEC63092. One of the specific classifications made for glass-based BIPV products is the accessibility of the element from the inside of the building (see Figure 24.5), as it defines the degree of mechanical safety that this element must guarantee.

This variety of products (Corti et al., 2020a; Frontini et al., 2017, 2020b; Petter Jelle et al., 2012) existing today should ease the collaboration among the different actors of the construction design, thereby providing different possibilities to address the diversified

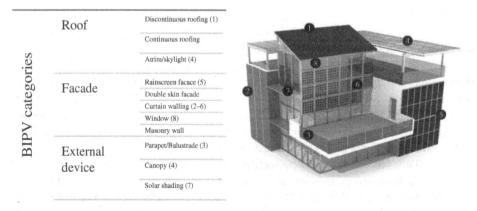

**Figure 24.4**   Technological construction units of BIPV building skins. Source: Adapted from Bonomo and Eder, (2021).

**Figure 24.5** BIPV categorization according to the EN50583 standard, considering glass-based BIPV solution and its accessibility: Cat-A sloped/horizontal non-accessible roof element; Cat-B sloped/horizontal accessible roof element; Cat-C vertical non-accessible façade element; Cat-D vertical accessible façade element; Cat-E externally integrated elements.

requirements from the different stakeholders involved in the value chain: the architects, the client, the engineer, the bank, the installers, etc.

## 24.4 BIPV Performances

During the design and building process, different aspects have to be taken into account to assess the performances of the BIPV system (whether it is a roof, a façade, or an external or accessory element), such as the architectural design or aesthetics, the location constraints [such as the urban density (Lobaccaro et al., 2019; Lobaccaro and Frontini, 2014)], the building typology and function, the thermal properties and module temperature, building installation needs, user behavior, grid connection or energy storage, and passive measures. Some of these performances (such as the building typology and the design) mainly affect the type of integration and, therefore, the aesthetic of the installation; some (such as the type of installation and connection with the building structure and the orientation) will affect the module performances and the capability to produce renewable energy.

### 24.4.1 Design and Aesthetics of BIPV

Architectural integration quality is often perceived as the result of a controlled and coherent integration of the photovoltaic modules simultaneously from all points of view – functional, constructive and formal (aesthetic). For example, when the solar system is integrated into the building envelope (as roof covering, façade cladding, Sun shading, balcony fence, etc.), it must properly satisfy the functions and the associated requirements of the envelope elements it is replacing (constructive/functional qualities) while preserving the global design quality of the building (formal quality). If the design quality is not taken into account (i.e. the system is only constructively/functionally integrated into the building skin without formal control), we can only consider it a building-added system. Thus, the multifunctional use of solar elements taking over one or more envelope functions may require extra effort from building designers; for instance, calling for some modifications in the module's electro-technical or mechanical design in the way it is installed or restraining its use in some parts of the building. On the other hand, it brings the major advantages of global cost reduction and an enhanced architectural quality of the building. Moreover, different Photovoltaic materials can be used to address the different design challenges in terms of color, shape, weight, and transparency.

In addition to the functional compatibility, it is important to ensure that the active multifunctional envelope system meets plenty of other building construction standards, such as:

- transferring the actions to the load-bearing structure through appropriate fixing to also guarantee proper maintenance;
- withstanding different weather conditions and also fire as prescribed by international, national and local regulations;
- resisting wind load and impact stresses to be safe in case of damage;
- avoiding thermal bridges of the fixing such that the global U value of the wall or window should not be negatively affected;
- transferring vapor through the wall in order to avoid condensation layers and allow the wall to dry correctly.

The relevant essential requirements are defined by applicable product standards, building and electrotechnical regulations, as well as by the requirements defined in the single construction works and projects. Quality compliance of BIPV systems can be framed by the existing standards valid for product certification and marking, which defines procedures and methodologies for design, testing, engineering and the construction of each product category and material technology (Bonomo et al., 2022). Moreover, the risks of theft and/or damage related to vandalism should be evaluated, and appropriate measures must be taken, and all the system characteristics affecting building appearance (i.e. system formal characteristics) should be coherent with the overall building design (Probst and Roecker, 2012; Wall et al., 2012).

### 24.4.2 Consideration About the BIPV Module Temperature

One of the main drawbacks of BIPV systems could be the elevated operating temperatures, especially when they are installed close to the building envelope with little or no rear ventilation. In this case, airflow and heat transfer are reduced; thus, the thermal insulation effect on the back of the modules leads to increased module temperatures, affecting the power performance with possible influence on the degradation. As presented in the previous chapters of the book, the module performance is strongly dependent on the cell temperature, as depicted by the following basic equations:

$$P_m = I_m V_m = (FF) I_{sc} V_{oc}$$

In this equation, FF is the fill factor, $I_{sc}$ is the short circuit (SC) current, $V_{oc}$ is open circuit voltage, and the subscript "$m$" refers to the maximum power point in the module $I-V$ curve. Both $V_{oc}$ and FF decrease substantially with temperature, while the short-circuit current increases slightly. Thus, the net effect leads to a linear relation in this form:

$$\eta_c = \eta_{T_{ref}}[1 - \beta_{ref}(T_c - T_{ref}) + \gamma \log_{10} I(t)]$$

in which $\eta_{T_{ref}}$ is the module's electrical efficiency at the reference temperature $T_{ref}$ and at solar radiation of 1000 W/m$^2$. The temperature coefficient, $\beta_{ref}$, and the solar radiation coefficient, $g$, are mainly material properties, having values of about 0.004 K$^{-1}$ and 0.12, respectively, for crystalline silicon modules. The latter, however, is usually taken as zero (Evans, 1981), and the equation reduces to:

$$\eta_c = \eta_{T_{ref}}[1 - \beta_{ref}(T_c - T_{ref})]$$

which represents the traditional linear expression for PV electrical efficiency (Evans and Florschuetz, 1977). Information on how to estimate and calculate $\eta_{T_{ref}}$ and $\beta_{ref}$ (if not available from the manufacturer of the PV module) can be found in the previous chapters of the book.

Under real-world operating conditions, the cell temperature depends on many physical and environmental factors, such as cell technology (i.e. cSi, aSi or CIGS), module construction (i.e. laminated glass, insulated glass unit, etc.), solar irradiance, ambient temperature, wind speed and direction, and mounting configuration (i.e. ventilated installation or non-ventilated installation).

The type of installation and the ventilated or non-ventilated impact of the operative temperature was demonstrated by different experiments (Chatzipanagi et al., 2016; Gok et al., 2020; Özkalay et al., 2022); (Peng et al., 2013; Zhang et al., 2018). Considering roof installation, we could have three types of mounting structures, as depicted in Figure 24.6:

a. open rack non-integrated systems (free-standing PV module on flat or pitched roof);
b. ventilated (integrated system with air gap); or
c. insulated BIPV installation (no air ventilation on the back side of the module).

The temperature differences between these three configuration can vary up to 20–30° at certain period of the year, as presented by some studies (Chatzipanagi et al., 2012; Gok et al., 2020) and in Figure 24.7 and strongly depends apart from the weather conditions from the air gap between the BIPV module and the roof structure or insulation. A minimum air gap of four cm is normally to be considered, even if for a certain type of installation and climate, and a larger air gap must be considered to noticeably cool the module.

Different experiments demonstrate that elevated temperatures not only reduce the instantaneous power output of modules, but may also accelerate the degradation of polymeric and other materials, causing thermo-mechanical stresses because of variations in diurnal temperatures and mismatches in the thermal expansion coefficients of module materials.

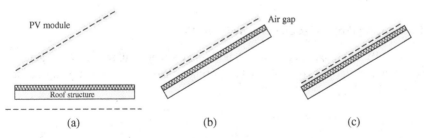

**Figure 24.6** Possible installation typology of roof-mounted PV: (a) Open-rack solution where the module is largely ventilated; not constructive integration is foreseen. (b) Roof integrated Photovoltaic with back ventilation between the BIPV module and the roof structure. (c) Roof Integrated photovoltaic without back ventilation, the BIPV module adheres to the roof structure/insulation.

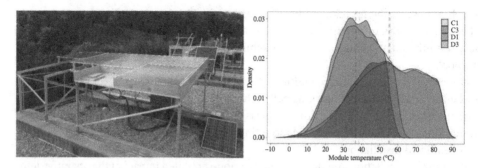

**Figure 24.7** Left picture presents two installation typologies: the two left modules are installed in an open rack configuration, while the two right modules adhere to the roof insulation. Right picture: Density plot for the measured module rear temperatures of four modules mounted in open rack configuration (C1-glass/back sheet, D1-Glass/Glass) and BIPV fully insulated where the module adheres to the roof structure (C3, D3). Source: Gok et al., 2020/IEEE.

### 24.4.3 Consideration About the Orientation and Shadows

Shadows play an important role and have to be considered as a key boundary condition for the proper design of the building and of the BIPV system. We have different types of shadows caused by the neighboring buildings or the urban environment (such as trees or urban furniture), but also the design and shape of the building itself can generate unfavored shadows on the BIPV systems. Shadows can be permanent or a temporary phenomenon, such as from snow, leaves, and soiling. The representations of the most common shadows are presented in Figure 24.8.

In their in situ analysis, the planners must take into account the annual and seasonal variations of the shadows, considering the Sun's altitude on the horizon. A Sun path diagram helps while assessing the surroundings. The horizon should be essentially unobstructed above an elevation of about 10°–15° so that the modules can exploit the solar radiation during most of the daytime during the year. Shadows in the early morning or late evening or on a few winter days are generally accepted, but they should be carefully considered. Different tools are available to properly assess the shadows' paths during the year and on an hourly basis with the aim to properly locate and designing the BIPV system. A list of the available tools and approaches can be found in the report of the IEA PVPS Task 15 (Jakica et al., 2019).

### 24.4.4 Daylighting Properties and Energy Protection

Semi-transparent PV modules integrated into building envelopes allow electricity generation to be combined with solar control and daylighting properties. Three main approaches

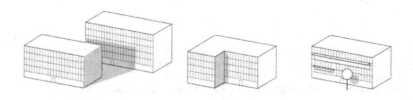

**Figure 24.8** Shadows from surrounding buildings and overshadows from architectural design, building shape and existing obstructions.

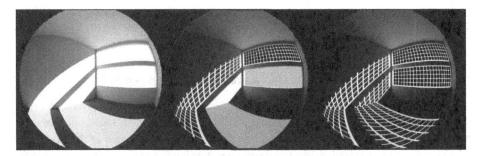

**Figure 24.9** Daylight rendering studies to assess indoor glare for various BIPV window technologies. Source: IEA PVPS T15 Frontini et al., (2020a).

exist that use different technologies, in particular: semi-transparent cSi glass-glass modules, thin film semi-transparent modules (either aSi, CIGS, or CdTe), organic PV, perovskite or luminescent solar concentrators.

When referring to daylight properties for crystalline semi-transparent PV modules made of opaque cells, only the area between the cells must be characterized, while in monolithic thin-film semi-transparent PV modules having a homogeneous appearance, an average representative area should be considered. Weighting between the opaque and transparent/translucent areas should be performed to obtain the final daylighting behavior of the BIPV element as also the indoor glare caused by the transparency of the glass and or the solar cells (Figure 24.9).

The angle-dependent light transmittance of a partially transparent or translucent BIPV module is needed to determine its daylighting properties in a building. Depending on the level of detail expected for the daylighting performance, the angle-dependent light transmittance or the BSDF (bidirectional scattering distribution function) – which includes both the reflectance distribution BRDF and the transmittance distribution BTDF – may be needed for each of the optically differing parts, together with their spatial distribution over the BIPV module surface.

Apart from the daylight and visual properties, while dealing with semi-transparent BIPV elements, solar protection must also be considered. The solar heat gain coefficient (SHGC, also known as g value) is the fraction of incident solar radiation that passes through the element directly ($T_{sol}$) and indirectly after absorption and inward reemission as heat ($q_i$). It is expressed as a percentage or a number between 0 (no solar energy transmission) and 1:

$$SHGC = T_{sol} + q_i$$

In glazing units consisting of one or more glass panes, the SHGC for each optically different area can be determined separately by calculations based on the spectral measurements, as described in two ISO standards (ISO 9050 and ISO 15099) and then can be area-weighted to obtain the SHGC for the complete unit.

The overall SHGC of BIPV depends on the ratio of opaque to transparent areas, the thermal transmittance of the BIPV product and the operating point of the PV module.

Significant variation in the g-value occurs between the open circuit and the maximum power point conditions, as some of the absorbed incident solar energy is removed completely from the glazing system as electricity under closed-circuit conditions. Thus, it is important that the SHGC of a BIPV window product is tested and reported both under open circuit and maximum power point conditions, as reported by (Olivieri et al., 2015, where difference in the SHGC was noticed up to 11% moving from the maximum power point (minimum g-value) to the SC condition (maximum g-value).

### 24.4.5 Design with Photovoltaic: Mechanical Performances and Safety

As photovoltaic becomes a part of the building envelope, the BIPV modules have to fulfill the main essential requirements of building products to be qualified both as part of construction (in compliance with the construction product regulations such as the European CPR 305/2011) and as part of an electrical system according to requirements prescribed by the electrotechnical standards (mainly the IEC 61730 Photovoltaic module safety qualification) to allow for the CE marking and commercialization.

Glass is the principal component in conventional adopted BIPV modules and, therefore, the material that governs the design and construction. It is for this reason that the deployment of BIPV installation is mostly influenced by the rules and approaches conventionally used for glass design and engineering in buildings. Flexible BIPV modules that are normally embedded in synthetic materials are the exception (Bonomo et al., 2021).

Also, new IEC 63092 Photovoltaic in buildings – Part 1: Requirements for building-integrated photovoltaic modules and the EN 50583 Photovoltaics in buildings – Part 1: BIPV modules standard largely refers to the construction standards and requirements related to glass elements or components.

Mechanical resistance and stability are some of the essential requirements set by different Construction Products Regulation authorities (for example, the CPR 305/2011, valid for European countries). There are several standards defining related characteristics, as well as testing and assessment methods, used to verify the fulfillment of these requirements.

The specific loads and forces in the case of photovoltaic systems are typically the following:

- The weight of the photovoltaic system itself and its substructures and fixtures (dead weight)
- Snow loads
- Wind loads
- Seismic loads
- Hard impacts from hail or other external objects falling onto a PV module
- Soft impact caused, e.g. by people falling or leaning onto a PV module
- Thermal expansion and thermal stress
- Relevant single-point or distributed loads, e.g. those from people standing or walking on the photovoltaic modules
- Loads occurring during construction or maintenance work, e.g. while cleaning or replacing PV system components or modules
- Loads from other building parts, where applicable.

BIPV modules must also be characterized by the property of "post-breakage integrity." This is the property that a broken BIPV element must remain safe (e.g. not fall apart or slip out of its frame) under a load for a predefined period of time. This is to ensure that persons cannot be injured by a broken BIPV element falling down onto them and that a

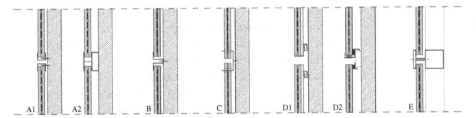

**Figure 24.10** The fixing typologies are essential to guarantee the mechanical performance of a BIPV system and its reliability during time. Different possibilities are presented: A1 linear frame screwed on site, A2 continuous clamping, B individual/point clamping, C point fixing with drilled holes, D1 Adhesive vertical rail with hooks, D2 Adhesive self-weight system with safety mechanical retention, and E continuous post-and-rail façade system.

BIPV module that is used for a barrier function (e.g., as a balcony balustrade) continues to prevent a person from falling through it for a predefined period of time even after breakage.

The requirements relating to mechanical resistance and stability vary extensively depending on the application, whether as a roofing or a façade element or a solar shading device, and also on the building typology (e.g. private or public). Compared to the ground-mounted PV modules, some aspects of the built environment affecting mechanical requirements include higher mounting positions and thus greater wind loads and the greater safety margin needed when building occupants are nearby (e.g. underneath glass-glass modules applied as overhead glazing).

Once the configuration of the modules has been defined (i.e. frameless module or framed module), the proper fixation method has to be identified and designed. When it comes to the glass-based module, a regular form of support for glass construction can be used. Figure 24.10 presents some of the possibilities often used for rainscreen façade or curtain walls. These include linear support systems, point support elements, point fixing drilled through the panes, fixing with adhesive and post-and-rail façade elements to form an insulating unit.

## 24.5 Conclusion and Outlook

BIPV is an important and relevant field for the future. Numerous current developments are continuously improving the boundary conditions for the building industry in which solar building envelopes are becoming the standard. There are many good reasons for integrating photovoltaics into building design, whether to meet requirements or simply realize better buildings. Cost is still a matter of concern due to the complexity of the product; therefore, business models have to be developed, consolidated and used by the stakeholders in order to properly assess and identify the BIPV advantages and competitiveness compared to other technologies (Corti et al., 2020a, 2020b).

Through the solar activation of roof and façade surfaces, PVs as a visible technology will increasingly influence our built environments in terms of design perception. As a result, it is gradually changing from a mere building block for energy generation to an element of building culture, which now needs to be jointly designed for the solar age.

Several technological opportunities based on well-validated technologies, such as crystalline solar cells or novel solutions such as luminescence solar concentrator, exist today, and new ones are under development so that BIPV can become a mainstream technology that can be adopted in the majority of building typologies.

In this chapter, it was not possible to discuss all aspects regarding the integration and use of Photovoltaic as a building component. To be mentioned a further opportunity today, is the role of digitalization. The building industry and the energy sector have begun a fundamental transformation toward a fully digitalized transition, which, in the case of solar building skins, raises the possibility of strong synergies. Digitalization will be an enabler for distributed PV generation in the built environment through the adoption of a more open and collaborative workflow based on data-sharing among process stages, disciplines, and stakeholders. From design to O&M, there are many possibilities for supporting process quality and improving collaboration, coordination and communication while also reducing extra costs, making the global BIPV value chain more competitive.

## List of Acronyms

| Acronym | Description |
| --- | --- |
| aSi | Amorphous silicon |
| BIPV | Building integrated photovoltaic |
| BRDF | Bidirectional reflectance distribution function |
| BSDF | Bidirectional scattering distribution function |
| BTDF | Bidirectional transmittance distribution function |
| cdTe | Cadmium telluride |
| CIGS | Cadmium indium gallium selenide |
| CPR | Construction product regulation |
| cSi | Crystalline silicon |
| O&M | Operation and management |
| OV | Open voltage |
| PV | Photovoltaic |
| SC | Short circuit |
| SHGC | Solar heat gain coefficient |

## Author Biography

**Prof. Dr. Francesco Frontini** has been since 2011 the head of the Building Sector and of the Swiss BIPV Competence Center at the University of Applied Sciences and Arts of Southern Switzerland (SUPSI). In 2019, he co-founded iWin Sagl, an innovative startup developing and producing an innovative multifunctional BIPV window.

After his graduation in Building Engineering and Architecture, he collaborated with Engineer and Architect offices as Project manager. Research activity was always supported by experimental work on the design of actual buildings and solar envelopes. In 2009 he got a PhD cum laude in Building Engineering, where he developed, together with different manufacturers, a new multifunctional BIPV façade for solar control and glare protection.

He worked as a researcher in the Solar façades group at Fraunhofer Institute for Solar Energy Systems (in Germany), one of the largest Research institutes in the World, where he gathered extensive experience in building simulations and in BIPV. He is a member of different international standardization bodies which developed new BIPV standards. Mr. Frontini is co-manager of the IEA PVPS Task 15 dealing with BIPV, and he is the author of different publications in the field of energy in buildings, daylighting, and BIPV systems.

# References

Bonomo, P. and Eder, G. (2021). *Categorization of BIPV Applications*. IEA PVPS.

Bonomo, P., Frontini, F., and Parolini, F. (2021). Safety of laminated BIPV glasses: progresses towards product qualification. In: *Engineered Transparency 2021: Glass in Architecture and Structural Engineering*. Ernst & Sohn.

Bonomo, P., Frontini, F., Parolini, F. et al. (2022). Impact resistance of BIPV systems: proposal of a test procedure to harmonize building and electrical performance assessment. In: *8th World Conference Photovoltaic Energy Conversion, Milano*.

Chatzipanagi, A., Frontini, F., and Dittmann, S. (2012). Investigation of the influence of module working temperatures on the performance of BiPV modules. In: *27th European Photovoltaic Solar Energy Conference and Exhibition*, 4192–4197.

Chatzipanagi, A., Frontini, F., and Virtuani, A. (2016). BIPV-temp: a demonstrative building integrated photovoltaic installation. *Applied Energy* 173: 1–12. https://doi.org/10.1016/j .apenergy.2016.03.097.

Corti, P., Bonomo, P., Frontini, F. et al. (2020a). *Building Integrated Photovoltaics: A Practical Handbook for Solar Buildings' Stakeholders*, 2020the. SUPSI https://solarchitecture.ch/bipv-status-report-2020/.

Corti, P., Capannolo, L., Bonomo, P. et al. (2020b). Comparative analysis of BIPV solutions to define energy and cost-effectiveness in a case study. *Energies* 13: 3827. https://doi.org/10 .3390/en13153827.

Evans, D.L. (1981). Simplified method for predicting photovoltaic array output. *Solar Energy* 27: 555–560. https://doi.org/10.1016/0038-092X(81)90051-7.

Evans, D.L. and Florschuetz, L.W. (1977). Cost studies on terrestrial photovoltaic power systems with sunlight concentration. *Solar Energy* 19: 255–262. https://doi.org/10.1016/ 0038-092X(77)90068-8.

Frontini, F., Wilson, H.R., K. Berger et al., 2020a. *Multifunctional Characterisation of BIPV – Proposed Topics for Future International Standardisation Activities*.

Frontini, F., Chatzipanagi, A., Bonomo, P. et al. (2015). *BIPV Product Overview for Solar Façades and Roofs*. SUPSI https://repository.supsi.ch/11904.

Frontini, F., Zanetti, I., Bonomo, P. et al. (2017). *Building Integrated Photovoltaics: Product Overview for Solar Building Skins*, 2017the. SUPSI.

Frontini, F., Zanetti, I., Bonomo, P., Andreas, H., Studer, D., 2020b. *Solararchitecture - Sun as a Build. Mater.* www.solarchitcture.ch.

Frontini, F., Bonomo, P., Moser, D., and Maturi, L. (2022). Building integrated photovoltaic facades: challenges, opportunities and innovations. In: *Rethinking Building Skins* (ed. E.

Gasparri, A. Brambilla, G. Lobaccaro, et al.), 201–229. Elsevier https://doi.org/10.1016/B978-0-12-822477-9.00012-7.

Gok, A., Ozkalay, E., Friesen, G., and Frontini, F. (2020). The influence of operating temperature on the performance of BIPV modules. *IEEE Journal of Photovoltaics* 10: 1371–1378. https://doi.org/10.1109/JPHOTOV.2020.3001181.

Jakica, N., Yang, R., Eisenlohr, J., Boddaert, S., Bonomo, P., Saretta, E., Frontini, F., Freitas, S., Alamy, P., Leloux, J., 2019. BIPV Design and Performance Modelling: Tools and Methods. Report IEA-PVPS T15-09:2019, ISBN: 978-3-906042-86-2.

Kuhn, T.E., Erban, C., Heinrich, M. et al. (2021). Review of technological design options for building integrated photovoltaics (BIPV). *Energy and Buildings* 231: 110381. https://doi.org/10.1016/j.enbuild.2020.110381.

Lobaccaro, G. and Frontini, F. (2014). Solar energy in urban environment: how urban densification affects existing buildings. *Energy Procedia* 1559–1569. https://doi.org/10.1016/j.egypro.2014.02.176.

Lobaccaro, G., Lisowska, M.M., Saretta, E. et al. (2019). A methodological analysis approach to assess solar energy potential at the neighborhood scale. *Energies* 12: https://doi.org/10.3390/en12183554.

Olivieri, L., Frontini, F., Polo López, C.S. et al. (2015). G-value indoor characterization of semi-transparent photovoltaic elements for building integration: new equipment and methodology. *Energy and Buildings* 101: 84–94. https://doi.org/10.1016/j.enbuild.2015.04.056.

Özkalay, E., Friesen, G., Caccivio, M. et al. (2022). Operating temperatures and diurnal temperature variations of modules installed in open-rack and typical BIPV configurations. *IEEE Journal of Photovoltaics* 12: 133–140. https://doi.org/10.1109/JPHOTOV.2021.3114988.

Peng, J., Lu, L., Yang, H., and Han, J. (2013). Investigation on the annual thermal performance of a photovoltaic wall mounted on a multi-layer façade. *Applied Energy* 112: 646–656. https://doi.org/10.1016/j.apenergy.2012.12.026.

Petter Jelle, B., Breivik, C., and Drolsum Røkenes, H. (2012). Building integrated photovoltaic products: a state-of-the-art review and future research opportunities. *Solar Energy Materials & Solar Cells* 100: 69–96. https://doi.org/10.1016/j.solmat.2011.12.016.

Probst, M.M. and Roecker, C. (2012). Criteria for architectural integration of active solar systems IEA task 41, subtask A. *Energy Procedia* 30: 1195–1204. https://doi.org/10.1016/j.egypro.2012.11.132.

Verberne, G., Bonomo, P., Frontini, F. et al. (2014). BIPV products for Façades and roofs: a market analysis. In: *EUPVSEC 2014*, 3630–3636. https://doi.org/10.4229/EUPVSEC20142014-6DO.7.6.

Wall, M., Probst, M.C.M., Roecker, C. et al. (2012). Achieving solar energy in architecture-IEA SHC Task 41. *Energy Procedia* 30: 1250–1260. https://doi.org/10.1016/j.egypro.2012.11.138.

Zhang, T., Wang, M., and Yang, H. (2018). A review of the energy performance and life-cycle assessment of building-integrated photovoltaic (BIPV) systems. *Energies* 11: 3157. https://doi.org/10.3390/en11113157.

# 25

# PV Module *I–V* Models

*Clifford W. Hansen*

*Photovoltaics and Materials Technologies, Sandia National Laboratories, Albuquerque, NM, USA*

## 25.1 Introduction

Models of photovoltaic (PV) module output translate the irradiance reaching a module's cells to the electrical output from the module, accounting for the temperature of the module's cells. Models can be grouped into several categories, listed in order of decreasing complexity:

- *I–V* **curve models** describe the mathematical relationship between the module's output current *I* and voltage *V*. An *I–V* curve model comprises a set of equations: one equation that gives the *I–V* relationship from a set of parameter values, and additional equations that specify the dependence of each parameter on irradiance and temperature. Examples of *I–V* curve models include the single diode models of De Soto et al. (2006) and in PVsyst (Mermoud and Lejeune 2010).
- **Point value models** use empirical expressions that relate current *I* and voltage *V* to irradiance and temperature at a few specific points on a module's *I–V* curve, typically at maximum power, short circuit, and open circuit. The Sandia Array Performance Model (King et al. 2004) and the Loss Factors Model (Sutterlueti et al. 2011) are examples of point value models.
- **Simple efficiency models** use an empirical expression to predict power at the maximum power point from environmental conditions. These models typically do not separate power into the corresponding current and voltage. Examples are the PVWatts model (Dobos 2014) and the model of Huld et al. (2011).

This chapter focuses on *I–V* curve models. The most used *I–V* curve model, a *single diode* model, is based on representing the PV device as an equivalent circuit comprising a diode and two resistors. We summarize the challenge of estimating parameters for the single-diode equation and for single-diode models and briefly discuss two-diode models, as well as the measurements used to estimate model parameters.

## 25.2 Overview of PV Cell, Module, and Array Physics

A PV cell is fundamentally a semiconductor device that absorbs photons and converts a portion of the absorbed energy to electrical current. For example, in a single-junction conventional solar cell, a p-type semiconductor is placed in contact with an n-type semiconductor to form the junction. Absorption of a photon of a suitable wavelength close to the pn-junction can transfer energy to an electron, exciting the particle to a mobile state in the conduction band. In this state, the electron can move toward the n-type material, leaving behind a corresponding hole, conceptually a carrier of a positive charge, which moves toward the p-type material. The separation of the electron–hole pair comprises electrical current, which is collected through contacts on the outside surfaces of the two semiconductor materials. The current is reduced by recombination of some of the charge carriers within the cell and by resistance to current flow within the cell and at the cell-contact junction. The electrical potential formed at the electrical contacts by the separation of charge carriers provides the PV cell's voltage. Detailed and thorough treatment of the physics of PV devices is found in several sources, e.g. (Gray 2011).

A PV module typically assembles individual cells in series into strings. One or more strings are connected in parallel to form the module's electrical structure. The cell-strings are typically encapsulated for mechanical support, electrical isolation and protection from environmental elements detrimental to the cell materials. The encapsulated cell-strings are sandwiched between back-surface and front-side materials, e.g. a plastic film and glass, respectively, which provide electrical and environmental isolation as well as mechanical rigidity; a metal frame is sometimes added. Electrical connections to the cell-strings are provided through a junction box that is usually mounted outside the module's back-surface.

An *I–V* curve of a PV cell specifies the combinations of current and voltage, which can be produced at the cell's output terminals when illuminated and connected to an external load. Figure 25.1 illustrates a conceptual *I–V* curve and the corresponding power-voltage curve. Key performance points along the *I–V* curve are the short-circuit ($I_{SC}$), open circuit ($V_{OC}$), and maximum power points (MPP). Different combinations of voltage and current can be obtained by changing the resistance of the external load.

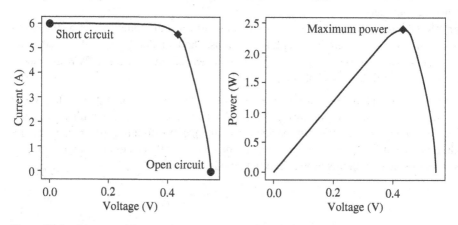

**Figure 25.1** Conceptual *I–V* curve (left) and power-voltage curve (right) for a PV cell.

Assuming all cells behave identically, electrical characteristics for cell-strings, PV modules, and arrays of modules can be determined by scaling the *I–V* curve for a cell. The module's voltage is the sum of the voltage from each series-connected cell, i.e., Vmodule = Ns × Vcell where Ns is the number of cells connected in series, and module current is the sum of the current across the parallel-connected strings, i.e. Imodule = Np × Icell. In reality, some variation is present among a module's cells, and even modules with well-matched cells, cell-to-cell variation reduces module power slightly below the power of the ideal, scaled *I–V* characteristic. This power reduction is termed *module mismatch loss*.

The *I–V* curves for arrays of modules (comprising parallel-connected strings of series-connected modules) are often estimated in a similar manner by scaling the *I–V* curve of a representative module, i.e. Varray = Ms × Vmodule, and Iarray = Mp × Imodule, where Ms and Mp are respectively the number of series-connected modules in a string and the number of parallel-connected strings in the array. Variation among modules and resistances between modules leads to a corresponding power loss termed *array mismatch loss*. Typically, mismatch losses are small compared to other array losses (Lorente et al. 2014; Wurster and Schubert 2014). We henceforth assume the ideal condition that all cells and modules behave identically.

## 25.3 *I–V* Curve Models

An *I–V* curve model comprises an equation that describes the current and voltage combinations in an illuminated condition. The equation is derived by applying Kirchoff's laws to a notional circuit of diodes and resistors that represents current flows in the physical module. The resulting equation is commonly referred to by the number of diodes included in the equivalent circuit. In PV performance modeling, the most used equivalent circuit (Figure 25.2) contains a single diode and two resistors, one in series and one in parallel (shunt) with the diode. Less commonly, two diodes are set in parallel (e.g., Lun et al. 2015), or the circuit in Figure 25.2 is simplified by removing the shunt resistor in parallel with the diodes.

### 25.3.1 The Single-Diode Equation

Figure 25.2 shows a one-diode equivalent circuit for a single-junction PV cell, which includes a source representing the current generated by the PV effect. Terminals represent the cell's contact points at which the current can be delivered to a load. A resistor in series with the load represents voltage drop across internal resistance, e.g. between the cell and

**Figure 25.2** Equivalent circuit for a photovoltaic cell using a single diode.

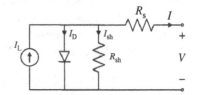

its current-collecting fingers. A diode is represented in parallel with the load to account for current lost due to recombination of charge carriers, and a resistor (the shunt resistor) represents leakage current that flows through paths within the cell rather than through the intended load. Summing current at the indicated point and using Ohm's law for current through the shunt resistor:

$$
\begin{aligned}
I &= I_L - I_D - I_{SH} \\
&= I_L - I_D - \frac{V_D}{R_{SH}}
\end{aligned}
\tag{25.1}
$$

where $V_D$ represents the potential across the diode, which can be expressed in terms of the current and the potential across the cell's external terminals $V$:

$$
V_D = V + IR_S
\tag{25.2}
$$

Using Shockley's law for an ideal diode

$$
I_D = I_O \left( \exp\left( \frac{V_D}{nV_{th}} \right) - 1 \right)
\tag{25.3}
$$

and substituting Eq. (25.2), we obtain the *single-diode equation*

$$
\begin{aligned}
I &= I_L - I_O \left( \exp\left( \frac{V_D}{nV_{th}} \right) - 1 \right) - \frac{V_D}{R_{SH}} \\
&= I_L - I_O \left( \exp\left( \frac{V + IR_S}{nV_{th}} \right) - 1 \right) - \frac{V + IR_S}{R_{SH}}
\end{aligned}
\tag{25.4}
$$

where

- $I_L$ is the photocurrent (A)
- $I_O$ is the dark, or saturation, current (A)
- $R_S$ is the series resistance (ohm)
- $R_{SH}$ is the shunt resistance (ohm)
- $n$ is the diode, or ideality, factor (unitless)

The term $V_{th}$ is the thermal voltage, which is the voltage across the p–n junction of the cell that is produced by the action of temperature on electrons within the semiconductor material. $V_{th}$ is given by

$$
V_{th} = \frac{k}{q}(T_C + 273.15)
\tag{25.5}
$$

where $k$ is the Boltzmann constant ($1.3806 \times 10^{-23}$ J/K), $q$ is the elementary (electron) charge ($1.6022 \times 10^{-19}$ C) and $T_C$ is the cell temperature in °C. At the Standard Test Condition (STC) temperature of 25 °C (see Section 25.5.1) $V_{th}$ is approximately 26 mV.

When $N_S$ identical cells are connected in series and subject to the same irradiance and temperature conditions, each cell conducts an equal current $I$ and produces an equal (external) voltage $V_C$, thus the voltage across the string of cells is $V = N_S V_C$. Consequently, a model for the $I–V$ characteristic of the string of cells is obtained from Eq. (25.4) by inserting $N_S$ and rearranging terms:

$$
I = I_L - I_O \left( \exp\left( \frac{V_C + IR_S}{nV_{th}} \right) - 1 \right) - \frac{V_C + IR_S}{R_{SH}}
$$

$$= I_\mathrm{L} - I_\mathrm{O}\left(\exp\left(\frac{N_\mathrm{S}(V_\mathrm{C} + IR_\mathrm{S})}{N_\mathrm{S}nV_\mathrm{th}}\right) - 1\right) - \frac{N_\mathrm{S}(V_\mathrm{C} + IR_\mathrm{S})}{N_\mathrm{S}R_\mathrm{SH}}$$

$$= I_\mathrm{L} - I_\mathrm{O}\left(\exp\left(\frac{V + I(N_\mathrm{S}R_\mathrm{S})}{N_\mathrm{S}nV_\mathrm{th}}\right) - 1\right) - \frac{V + I(N_\mathrm{S}R_\mathrm{S})}{(N_\mathrm{S}R_\mathrm{SH})} \tag{25.6}$$

Eq. (25.6) is the same as the single-diode equation in Eq. (25.4) except that the per-cell series and shunt resistances, $R_\mathrm{S}$ and $R_\mathrm{SH}$, respectively, and the thermal voltage $V_\mathrm{th}$ are multiplied by the number of cells in series $N_\mathrm{S}$. Henceforth, for convenience, we simply write $R_\mathrm{S}$ and $R_\mathrm{SH}$ in Eq. (25.6) and rely on the context to define whether these resistances are interpreted at the cell or cell-string level. We leave the thermal voltage as defined by Eq. (25.5).

### 25.3.2 Computing Solutions to the Single Diode Equation

The single-diode equation (Eq. 25.4) is implicit in that the argument $I$ appears on both sides of the equation. The equation is also transcendental, in that writing $I = I(V)$ (or $V = V(I)$) involves a transcendental function, Lambert's W function (Jain and Kapoor 2004). Thus, calculation of a solution $(I, V)$ to Eq. (25.4) requires using either a root-finding numerical method operating on Eq. (25.4), or evaluation of expressions involving Lambert's W function (Corless et al. 1996).

Given values for the five constants which appear in Eq. (25.4), i.e., $I_\mathrm{L}$, $I_\mathrm{O}$, $R_\mathrm{S}$, $R_\mathrm{SH}$, and $n$, a value for the thermal voltage $V_\mathrm{th}$ and a voltage $V$, the value $I$ which solves Eq. (25.4) may be obtained by applying fixed-point iteration directly to Eq. (25.4) or by applying Newton's method (or a similar root-finding technique) to solve

$$f(I) = I_\mathrm{L} - I_\mathrm{O}\left(\exp\left(\frac{V + IR_\mathrm{S}}{N_\mathrm{S}nV_\mathrm{th}}\right) - 1\right) - \frac{V + IR_\mathrm{S}}{R_\mathrm{SH}} - I = 0 \tag{25.7}$$

Care must be taken when solving Eq. (25.7) at large voltage as the argument $V + IR_\mathrm{S}/N_\mathrm{S}nV_\mathrm{th}$ may cause numerical overflow when exponentiated.

Analytic expressions $I = I(V)$ and $V = V(I)$ may be obtained by use of the Lambert's W function $W(x)$, which is the solution $y = W(x)$ to $x = ye^y$. To present these expressions, we introduce two functions $\theta(V)$ and $\psi(I)$ for convenience:

$$\theta(V) = \frac{R_\mathrm{S}I_\mathrm{O}}{N_\mathrm{S}nV_\mathrm{th}}\frac{R_\mathrm{SH}}{R_\mathrm{SH} + R_\mathrm{S}}\exp\left(\frac{R_\mathrm{SH}}{R_\mathrm{SH} + R_\mathrm{S}}\frac{R_\mathrm{S}(I_\mathrm{L} + I_\mathrm{O}) + V}{N_\mathrm{S}nV_\mathrm{th}}\right) \tag{25.8}$$

$$\psi(I) = \frac{I_\mathrm{O}R_\mathrm{SH}}{N_\mathrm{S}nV_\mathrm{th}}\exp\left(\frac{(I_\mathrm{L} + I_\mathrm{O} - I)R_\mathrm{SH}}{N_\mathrm{S}nV_\mathrm{th}}\right) \tag{25.9}$$

Solutions for Eq. (25.4) may be written as

$$I = \frac{R_\mathrm{SH}}{R_\mathrm{SH} + R_\mathrm{S}}(I_\mathrm{L} + I_\mathrm{O}) - \frac{V}{R_\mathrm{SH} + R_\mathrm{S}} - \frac{N_\mathrm{S}nV_\mathrm{th}}{R_\mathrm{S}}W(\theta(V)) \tag{25.10}$$

and

$$V = (I_\mathrm{L} + I_\mathrm{O} - I)R_\mathrm{SH} - IR_\mathrm{S} - N_\mathrm{S}nV_\mathrm{th}W(\psi(I)) \tag{25.11}$$

Algorithms for computing values for $W(x)$ are available in many software packages (e.g. `lambertw` in Matlab and in scipy for Python). Risk for numerical overflow is still present at large voltage because the exponentiated arguments for $W(x)$ become exceeding large. These circumstances can be overcome by an appropriate log transformation.

### 25.3.3 The Two-Diode Equation

The two-diode equation (Eq. 25.12) models a single-junction solar cell with an equivalent circuit that has a second diode in parallel with the first diode and the shunt resistor. Current $I_{D1}$ through one diode represents recombination of carriers in the quasi-neutral region and current $I_{D2}$ through the second diode represents recombination in the depletion region. The two-diode equation has five parameters: photocurrent $I_L$ (A), dark (saturation) currents $I_{O1}$ and $I_{O1}$ (A), series resistance $R_S$ ($\Omega$) and shunt resistance $R_{SH}$ ($\Omega$) because the ideality factors are specified as 1 and 2 for the recombination currents in the quasi-neutral and depletion regions, respectively. In a sense, the single-diode equation (Eq. 25.4) approximates the two-diode equation by replacing the two exponential terms with a single term with an unspecified ideality factor [see section 3.4.2 of Gray (2011)].

$$I = I_L - I_{D1} - I_{D2} - \frac{V_D}{R_{SH}}$$

$$= I_L - I_{O1}\left(\exp\left(\frac{V + IR_S}{V_{th}}\right) - 1\right) - I_{O2}\left(\exp\left(\frac{V + IR_S}{2V_{th}}\right) - 1\right) - \frac{V + IR_S}{R_{SH}}$$

$$(25.12)$$

## 25.4 All-Condition Single-Diode I–V Models

By itself, the single-diode equation (Eq. 25.4) is not a complete model for the performance of a PV cell, because it lacks expressions that relate the equation's five parameters (i.e., $I_L$, $I_O$, $R_S$, $R_{SH}$, and $n$) to irradiance and temperature conditions. An all-condition single-diode I–V model comprises the single-diode equation along with auxiliary equations that specify how the single-diode equation's parameters change with irradiance and temperature. These auxiliary equations vary among several popular I–V curve models, which are described below: the De Soto model (De Soto et al. 2006), the CEC model (Dobos 2012) and the PVsyst model (PVsyst 2022). We emphasize again the distinction between the *single-diode equation*, which describes a single I–V curve, and an all-condition *single-diode model* that describes I–V curves at any irradiance and temperature conditions.

### 25.4.1 De Soto Model

De Soto et al. (2006) put forth the following single-diode model of a PV module:

$$I = I_L - I_O\left(\exp\left(\frac{V + IR_S}{N_S n V_{th}}\right) - 1\right) - \frac{V + IR_S}{R_{SH}} \tag{25.13}$$

$$I_L = I_L(E, T_C) = \frac{E}{E_0}\left[I_{L0} + \alpha_{Isc}(T_C - T_0)\right] \tag{25.14}$$

$$I_O = I_O(T_C) = I_{O0}\left[\frac{T_C}{T_0}\right]^3 \exp\left[\frac{1}{k}\left(\frac{E_g(T_0)}{T_0} - \frac{E_g(T_C)}{T_C}\right)\right] \tag{25.15}$$

$$E_g(T_C) = E_{g0}\left(1 - 0.0002677(T_C - T_0)\right) \tag{25.16}$$

$$R_{SH} = R_{SH}(E) = R_{SH0}\frac{E_0}{E} \tag{25.17}$$

$$R_S = R_{S0} \tag{25.18}$$

$$n = n_0 \tag{25.19}$$

The independent variables for the auxiliary equations (Eq. 25.14 through Eq. 25.19) are *effective irradiance E* and *cell temperature* $T_C$. Effective irradiance $E$ (W/m$^2$) is the irradiance that reaches the module's cells and is converted to photocurrent. Effective irradiance differs from broadband plane-of-array irradiance by surface reflections and by the spectral response of the cell's absorber. In Eq. (25.14) through Eq. (25.19) the subscript $\sim_0$ indicates a value at the reference conditions $E_0$ or $T_0$; typically these values are $E_0 = 1000$ W/m$^2$ and $T_0 = 25\,°C = 298$ K. Together, Eq. (25.13) through Eq. (25.19) comprise a complete model that translates from effective irradiance $E$ and cell temperature $T_C$ to current $I$ and voltage $V$.

Eq. (25.14) represents the generally linear response of the photocurrent to effective irradiance and cell temperature. The term $E_g(T_C)$ in Eq. (25.15) and Eq. (25.16) is the band gap, which has units of eV. Consequently, in Eq. (25.15), the Boltzmann constant $k$ must also be specified in units of eV/K, i.e., $k = 8.617$ eV/K. Eq. (25.15) is derived from diode theory (see section 3.4.3 of Gray 2011).

Eq. (25.16) represents the (slight) temperature dependence of the bandgap. Eq. (25.17) is an empirical model of observed changes in shunt resistance $R_{SH}$ with irradiance originating with (De Soto et al. 2006). Eq. (25.18) assumes that series resistance $R_S$ is constant, an assumption that is common to many single-diode models; similarly, Eq. (25.19) represents the assumption that the diode (ideality) factor is constant.

Although the De Soto model is often termed the "five parameter model," for the five parameters appearing in Eq. (25.13), in fact, the model involves a total of seven parameters, i.e. constants representing module-specific properties that should be estimated from measurements of module performance: $I_{L0}$ (A), $\alpha_{Isc}$ (A/°C), $I_{00}$ (A), $E_{g0}$ (eV), $R_{SH0}$ ($\Omega$), $R_{S0}$ ($\Omega$) and $n_0$ (unitless).

### 25.4.2 The CEC Model

The California Energy Commission (CEC) model is a variation on the De Soto model, and it is coded in the System Advisor Model (SAM) (Gilman 2015). The CEC model comprises Eq. (25.13) through Eq. (25.19) but adds an additional parameter, *adjust* (%), which modifies the temperature dependence of short-circuit current:

$$\alpha_{Isc} = \alpha' \left( 1 - \frac{\text{adjust}}{100} \right) \tag{25.20}$$

where $\alpha'$ is the input temperature coefficient (A/K) for short-circuit current. The adjustment is made in order that the predicted temperature coefficient for maximum power, i.e. the derivative of power $IV$ with respect to cell temperature $T_C$ at the maximum power point, can match the observed value for the module (Dobos 2012). In practice, after fitting the CEC model to data, these two quantities may differ by a few percent.

### 25.4.3 PVsyst Model

The popular PVsyst software application implements the following single-diode model (as of PVsyst version 7.2) (PVsyst 2022):

$$I = I_L - I_O \left( \exp\left( \frac{V + IR_S}{N_S n V_{th}} \right) - 1 \right) - \frac{V + IR_S}{R_{SH}} \tag{25.21}$$

$$I_L = I_L(E, T_C) = \frac{E}{E_0}\left[I_{L0} + \alpha_{\mathrm{Isc}}(T_C - T_0)\right] \tag{25.22}$$

$$I_O = I_O(T_C) = I_{00}\left[\frac{T_C}{T_0}\right]^3 \exp\left[\frac{qE_{g0}}{kn}\left(\frac{1}{T_0} - \frac{1}{T_C}\right)\right] \tag{25.23}$$

$$R_{\mathrm{SH}} = R_{\mathrm{SH,base}} + (R_{\mathrm{SH,0}} - R_{\mathrm{SH,base}})\exp\left(-R_{\mathrm{SH\ exp}}\frac{E}{E_0}\right) \tag{25.24}$$

$$R_{\mathrm{SH,base}} = \max\left[\frac{R_{\mathrm{SH,ref}} - R_{\mathrm{SH,0}}\exp(-R_{\mathrm{SH\ exp}})}{1 - \exp(-R_{\mathrm{SH\ exp}})}, 0\right] \tag{25.25}$$

$$R_S = R_{S0} \tag{25.26}$$

$$n = n_0 + \mu_n(T_C - T_0) \tag{25.27}$$

The PVsyst model differs from the De Soto model in the equations for dark (saturation) current $I_O$ (Eq. 25.23), shunt resistance $R_{\mathrm{SH}}$ (Eq. 25.24 and Eq. 25.25) and diode (ideality) factor $n$ (Eq. 25.27). The PVsyst model involves a total of 10 parameters: $I_{L0}$ (A), $\alpha_{\mathrm{Isc}}$ (A/°C), $I_{00}$ (A), $E_{g0}$ (eV), $R_{\mathrm{SH0}}$ ($\Omega$), $R_{\mathrm{SH,\ ref}}$ ($\Omega$), $R_{\mathrm{SH,\ exp}}$ (unitless), $R_{S0}$ ($\Omega$), $n_0$ (unitless) and $\mu_n$ (1/°C). In Eq. (25.24) $E_{g0}$ is the bandgap in eV; the terms $q$ and $k$ are the elementary charge ($1.602 \times 10^{-19}$ C) and the Boltzmann constant ($1.381 \times 10^{-23}$ J/K), respectively. The empirical relationship between shunt resistance $R_{\mathrm{SH}}$ and effective irradiance originates with PVsyst and is compared in Figure 25.3 with the shunt resistance equation from the De Soto model. In practice, the difference in modeled shunt resistance has little effect on simulated energy because maximum power is relatively insensitive to shunt resistance (except at extremely low resistance values), and the two models generally agree at high effective irradiance. PVsyst implements a temperature-dependent diode factor $n$ (Eq. 25.27); one may expect the temperature coefficient $\mu_n$ to be small relative to $n_0$.

## 25.5 Parameter Estimation for *I–V* Models

Parameter estimation for *I–V* models can be separated into two related problems:

1. Estimate values for the five diode equation parameters $I_L$, $I_O$, $R_S$, $R_{\mathrm{SH}}$, and $n$ given data for a single *I–V* curve. We term this "single *I–V* curve fitting"; in literature this process is also termed parameter extraction.

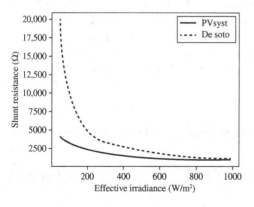

**Figure 25.3** Comparison of equations for shunt resistance in the PVsyst and De Soto models.

2. Estimate the parameters for an all-condition $I$–$V$ model, e.g., for the De Soto model $I_{L0}$, $I_{O0}$, $R_{SH0}$, $R_{S0}$ and $n_0$ given one or more $I$–$V$ curves and information about the PV device, i.e., number of cells in series $N_S$, band gap $E_{g0}$ and short circuit temperature coefficient $\alpha_{Isc}$. We term this "all-condition model fitting."

### 25.5.1  Single *I–V* Curve Fitting

The literature on single $I$–$V$ curve fitting is extensive; articles presenting methods for fitting the single diode equation to data appear regularly. For instance, Ortiz-Conde et al. (2014) cataloged many different methods for parameter extraction for electronic devices with diode-like behavior. Techniques generally include simplifying the single diode equation over portions of the voltage domain, calculus-based global optimization, and, more recently, numerous variations of heuristic optimization approaches. It is useful to consider what sustains the stream of publications. In our view, four factors contribute:

1. The single-diode equation is implicit, and the equation and its analytic solutions (Eq. 25.11 and Eq. 25.12) involve transcendental functions. Consequently, formal analysis of the equation and the behavior of a solution can be challenging.
2. Solution of Eq. (25.4) requires careful attention to numerical precision. The two non-constant terms on the right side of Eq. (25.4) vary greatly over the $I$–$V$ curve's range of voltage. The diode current term is a product of two factors with greatly varying orders of magnitude, for example, for a 72-cell module $I_O$ is typically $\sim 10^{-9}$ A and the exponential factor is typically $\exp(36\,\mathrm{V}/(72 \times 0.025\,\mathrm{V})) = \exp(20)$. The product of these greatly different factors must be a number comparable to the third term (the shunt and series currents), which is typically on the order of 0.01 A. The need for precision complicates numerical routines that select optimal parameters.
3. Optimal parameters depend on the numerical optimization method and on the objective function that is minimized by the method. Most often, the objective function is formulated as a sum of squared differences in current at each voltage point, e.g. (Dobos 2012), although other formulations are published, e.g. (Campanelli and Hamadani 2018). Because of these dependencies, different optimization methods can yield different optimal parameters for the same $I$–$V$ curve data.
4. Verification and validation of the fitting method (numerical optimization and objective function) too often relies only on comparing an $I$–$V$ curve computed with extracted parameters with the input data. This comparison confirms that the computed $I$–$V$ curve resembles the input data but does not verify that the parameters are "correct," because similar $I$–$V$ curves can result from quite different parameter sets, as illustrated in Figure 25.4. A parameter extraction method's accuracy should be verified first by showing that the method can recover known parameters by the following method:
   a. Select several sets of parameters covering a wide range in each value and compute the corresponding $I$–$V$ curves.
   b. Apply the fitting method to each computed $I$–$V$ curve and compare the extracted parameters with the known inputs.

This verification procedure will expose any tendency of the method to bias extracted parameters, for example, by overestimating the $I_O$ parameter and then compensate by

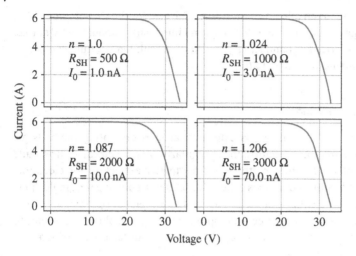

**Figure 25.4** Four similar *I–V* curves resulting from different parameter sets.

underestimating the diode factor *n*. The method's sensitivity to measurement noise can also be evaluated by applying simulated noise to the computed *I–V* curves, extracting parameters from the noisy *I–V* curves, and examining the distribution of the extracted parameters.

### 25.5.2 All-Condition Model Fitting

All-condition *I–V* model fitting estimates parameters for the auxiliary equations that describe how the five single-diode equation parameters change with irradiance and temperature. Thus, a fitting algorithm applies to one specific all-condition *I–V* model, and far fewer algorithms are published than for single *I–V* curve fitting. Often, single *I–V* curve fitting is one of several steps in an all-condition *I–V* model fitting algorithm, and thus, all-condition model fitting inherits the challenges described for single *I–V* curve fitting.

Dobos (2012) provides an algorithm to fit the CEC model, using the *I–V* curve at STC and temperature coefficients for power, short-circuit current and open circuit voltage. The fitting algorithm assumes that the analyst knows the junction band gap at STC ($E_{g0}$). The PVsyst software provides a fitting algorithm for the PVsyst model. Sauer et al. (2015) provide an analysis showing how to optimize parameters fitted using the PVsyst software to improve agreement between measured and modeled efficiency. Hansen (2015) describes a procedure to fit the De Soto model to data, which can be extended to other all-condition models.

A model fitting algorithm should be subject to verification to ensure that the algorithm does not introduce bias in the fitted parameters. Verification can be demonstrated using a procedure similar to that described for single *I–V* curve fitting algorithms:

1. Select known parameters for the all-condition *I–V* model, select a range of irradiance and temperature values, and use the model to compute synthetic *I–V* curves at the selected conditions.
2. Apply the fitting algorithm to the synthetic *I–V* curves to extract estimated model parameters.

3. Compare the estimated model parameters to the known parameters.
4. For each auxiliary equation, compute the diode equation parameter using both the known and estimated model parameters and compare the results.

Besides verification, fitting an all-condition $I$–$V$ model to data from a PV device requires two assumptions:

1. The single-diode equation is an appropriate model for the device's $I$–$V$ characteristics.
2. The auxiliary equations correctly describe how the model parameters change with irradiance and temperature.

Both assumptions should be subject to validation by comparison of the fitted equations to the underlying data. When uncertain about the auxiliary equations, their form may be confirmed (or discovered) from the measured data by the following method:

1. Obtain a set of $I$–$V$ curves for combinations of irradiance and temperature spanning a wide range.
2. Fit each $I$–$V$ curve using a verified $I$–$V$ curve fitting method to obtain values for the five diode equation parameters at each irradiance and temperature combination.
3. Plot each of the five parameters against irradiance and temperature, and fit the plotted data using the candidate auxiliary equation.

Hansen (2015) provides an example of this procedure.

## 25.6 Module Performance Characterization

Module performance characterization comprises a sequence of measurements of a PV module's electrical output at different irradiance levels and cell temperatures. PV modules are characterized for two primarily purposes: module rating, and calibration of module performance models.

Module ratings quantify the module's performance at a set of reference conditions, such as the Standard Test Conditions (STC) or Normal Operating Cell Temperature (NOCT). Performance parameters used for module rating include, for example, module power, open circuit voltage and short circuit current; values are listed on module data sheets to allow for comparison among modules. Rating values are determined by measuring module output as close to the reference conditions as practical and then translating the resulting measurements to the reference conditions. Several translation methods are specified in available standards (e.g. IEC 2021). By contrast, calibration of module performance models can require measurements of $I$–$V$ curves over a wide range of conditions, such as those described in IEC 61853 (IEC 2011b).

Module characterization can be done either indoors using solar simulators or outdoors using two-axis trackers or fixed racking. Indoor testing offers the advantage of controlling some important environmental factors, such as broadband irradiance and temperature, but entails the disadvantage of not representing other factors, such as the varying spectral content of irradiance, because indoor test equipment is typically designed for the primary purpose of testing at STC conditions. Outdoor testing measures module performance in the actual operating environment. However, outdoor testing is subject to local weather conditions, which may offer limited opportunity to observe module performance across the desired range of conditions.

When calibrating a PV performance model, one must pay careful attention to the methods and equipment used to measure environmental conditions to ensure that the measurements are consistent with the terms in the model, and that the model is calibrated to the quantities which will be used to predict performance. In the models described in Section 25.3, the effective irradiance $E$ is that which reaches a module's cells and is converted to photocurrent. Plane-of-array (POA) irradiance measured by a typical pyranometer must be transformed to effective irradiance by adjusting for module surface reflections and the module's spectral response to the irradiance. Models for surface reflections are available, e.g. (Martin and Ruiz 2001). The spectral irradiance adjustment can be determined by comparing the measured broadband POA irradiance with the measured short-circuit current of the module (Hansen et al. 2014). POA irradiance can also be measured with a reference cell, i.e. a PV device with a calibrated translation from measured current or voltage to irradiance. When irradiance is measured with a spectrally matched reference cell, the effects of spectral content and reflections are accounted for in the irradiance measurements.

The temperature $T_C$ that appears in the models described in Section 25.3 is cell temperature, a quantity which requires modification of the module to measure directly. Most frequently, module surface temperature is measured during characterization using a temperature device attached to the back surface. When characterization is done in temperature-controlled, indoor conditions, module and cell temperatures can be assumed to be equal. For outdoor characterization, cell temperature must be estimated from the measured surface temperature, with attention to the potential for variation of surface temperature across the module's area. Alternatively, cell temperature can be estimated from measured open-circuit voltage and short-circuit current (IEC 2011a).

### 25.6.1 Standard Test Conditions (STC)

Standard test conditions (STC) are specified (IEC 2016) to be broadband irradiance $G_{POA}$ of 1000 W/m$^2$ at cell temperature of 25 °C and at the standard solar spectrum. The standard solar spectrum is commonly also termed the air mass 1.5 spectrum. The standard spectrum represents spectral irradiance of natural sunlight on a plane at the earth's surface at sea level inclined to 37° and normal to the Sun, with the atmosphere presenting a specific set of meteorological conditions. The standard spectrum is derived from observations of the extraterrestrial solar spectrum by application of atmospheric transfer modeling (Gueymard 2004). Implementation of the modeling methodology results in two slightly different spectra in standards published by ASTM (ASTM 2012) and IEC (IEC 2008). The differences between these two spectra are small, and for most practical purposes, the spectra can be considered equivalent (Myers 2011); however, care should be taken to record which spectrum is used when test equipment is calibrated.

### 25.6.2 Normal Operating Cell Temperature (NOCT)

Normal operating cell temperature (NOCT) conditions are specified (IEC 2016) as broadband irradiance $G_{POA}$ of 800 W/m$^2$ at an ambient air temperature of 20 °C, wind speed of 1 m/s and air mass 1.5 solar spectrum. Because a cell temperature is not specified by NOCT, this value is either measured or modeled from the ambient air temperature, and its value should be provided along with the module's electrical performance parameters.

### 25.6.3  IEC 61853 Standard

In contrast to the single set of conditions specified by STC and NOCT, the IEC 61853 standard (IEC 2011b) requires measuring module performance at a total of 22 combinations of irradiance and module temperature (the IEC standard does not systematically distinguish between cell temperature and module temperature as measured at the module's back surface.)

## 25.7  Conclusion

We have summarized $I$–$V$ curve models with emphasis on models that represent the PV device with an equivalent circuit comprising a single diode and two resistors. The single diode equation (Eq. 25.4) describes a single $I$–$V$ curve from five parameters that describe characteristics of the equivalent circuit components. A single-diode model comprises a set of equations that relate these five parameters to irradiance and temperature conditions and thus describes $I$–$V$ curves at all conditions. Estimating parameter values, either for the single-diode equation or for a single-diode model, remains a challenge, and progress is likely to be achieved by adopting more consistent and reproducible verification and validation strategies.

## Author Biography

**Dr. Clifford Hansen** is a Distinguished Member of the Technical Staff at Sandia National Laboratories. He leads efforts to develop open-source software for renewable energy systems, including performance modeling and analytics, and taxonomies for data sharing. Clifford received his PhD in mathematics from The George Washington University.

## References

ASTM (2012). *ASTM G173-03, Standard Tables for Reference Solar Spetral Irradiances: Direct Normal and Hemispherical on 37° Tilted Surface*. West Conshohocken, PA: ASTM International.

Campanelli, M.B. and Hamadani, B.H. (2018). Calibration of a single-diode performance model without a short-circuit temperature coefficient. *Energy Science & Engineering* 6 (4): 222–238. https://doi.org/10.1002/ese3.190.

Corless, R.M., Gonnet, G.H., Hare, D.E.G. et al. (1996). On the Lambert W function. In: *Advances in Computational Mathematics*, 329–359.

De Soto, W., Klein, S.A., and Beckman, W.A. (2006). Improvement and validation of a model for photovoltaic array performance. *Solar Energy* 80 (1): 78–88.

Dobos, A. (2012). An improved coefficient calculator for the california energy commission 6 parameter photovoltaic module model. *Journal of Solar Energy Engineering* 134 (2): 021011.

Dobos, A. (2014). *PVWatts Version 5 Manual* (NREL/TP-6A20-62641). Retrieved from Golden, Colorado USA:

Gilman, P. (2015). *SAM Photovoltaic Model Technical Reference* (NREL/TP-6A20-64102). Retrieved from National Renewable Energy Laboratory, Golden, CO, USA.

Gray, J.L. (2011). The physics of the solar cell. In: *Handbook of Photovoltaic Science and Engineering*, 2nde (ed. A. Luque and S. Hegedus). John Wiley and Sons.

Gueymard, C.A. (2004). The sun's total and spectral irradiance for solar energy applications and solar radiation models. *Solar Energy* 76 (4): 423–453. https://doi.org/10.1016/j.solener .2003.08.039.

Hansen, C. (2015). *Parameter Estimation for Single Diode Models of Photovoltaic Modules* (SAND2015-2065). Sandia National Laboratories, Albuquerque, NM. https://www.osti.gov/ servlets/purl/1177157.

Hansen, C.W., Klise, K.A., Stein, J.S. et al. (2014). *Calibration of photovoltaic module performance models using monitored system data*. In: *Paper Presented at the 29th European PV Solar Energy Conference, Amsterdam, The Netherlands*.

Huld, T., Friesen, G., Skoczek, A. et al. (2011). A power-rating model for crystalline silicon PV modules. *Solar Energy Materials and Solar Cells* 95 (12): 3359–3369. https://doi.org/10.1016/j .solmat.2011.07.026.

IEC (2008). *IEC 60904-3 Ed. 2.0: Photovoltaic Devices – Part 3: Measurement Principles for Terrestrial Photovoltaic (PV) Solar Devices with Reference Spectral Irradiance Data*. Geneva, Switzerland: International Electrotechnical Commission.

IEC (2011a). *IEC 60904-5 Ed. 2.0: Photovoltaic Devices – Part 5: Determination of the Equivalent Cell Temperature (ECT) of Photovoltaic (PV) Devices by the Open-Circuit Voltage Method*. Geneva, Switzerland: International Electrotechnical Commission.

IEC (2011b). *IEC 61853-1 Ed. 1.0: Photovoltaic (PV) Module Performance Testing and Energy Rating – Part I: Irradiance and Temperature Performance Measurements and Power Rating*. Geneva, Switzerland: International Electrotechnical Commission.

IEC (2016). *IEC 61215-1 Ed. 1.0: Terrestrial Photovoltaic (PV) Modules – Design Qualification and Type Approval – Part 1: Test Requirements*. Geneva, Switzerland: International Electrotechnical Commission.

IEC (2021). *IEC 60891 Ed. 3.0: Photovoltaic Devices – Procedures for Temperature and Irradiance Corrections to Measured I-V Characteristics*. Geneva, Switzerland: International Electrotechnical Commission.

Jain, A. and Kapoor, A. (2004). Exact analytical solutions of the parameters of real solar cells using Lambert W-function. *Solar Energy Materials and Solar Cells* 81 (2): 269–277. https:// doi.org/10.1016/j.solmat.2003.11.018.

King, D. L., Boyson, E. E., & Kratochvil, J. A. (2004). *Photovoltaic Array Performance Model* (SAND2004-3535). Sandia National Laboratories, Albuquerque, NM, USA: https://www.osti .gov/servlets/purl/919131sca5ep/

Lorente, D.G., Pedrazzi, S., Zini, G. et al. (2014). Mismatch losses in PV power plants. *Solar Energy* 100: 42–49. https://doi.org/10.1016/j.solener.2013.11.026.

Lun, S.X., Wang, S., Yang, G.H., and Guo, T.T. (2015). A new explicit double-diode modeling method based on Lambert W-function for photovoltaic arrays. *Solar Energy* 116: 69–82. https://doi.org/10.1016/j.solener.2015.03.043.

Martin, N. and Ruiz, J.M. (2001). Calculation of the PV modules angular losses under field conditions by means of an analytical model. *Solar Energy* 70: 25–38.

Mermoud, A. and Lejeune, T. (2010). *Performance assessment of a simulation model for pv modules of any available technology*. In: *Paper Presented at the 25th European Photovoltaic Solar Energy Conference, Valencia, Spain*.

Myers, D. (2011). *Review of Consensus Standard Spectra for Flat Plate and Concentrating Photovoltaic Performance* (NREL/TP-5500-51865). National Renewable Energy Laboratory, Golden, CO, USA. https://www.nrel.gov/docs/fy11osti/51865.pdf

Ortiz-Conde, A., García Sánchez, F.J., Muci, J., and Sucre-González, A. (2014). A review of diode and solar cell equivalent circuit model lumped parameter extraction procedures. *Facta Universitatis Series Electronics and Energetics* 27 (1): 57–102. https://doi.org/10.2298/FUEE1401103K.

PVsyst. (2022). *PVsyst Photovoltaic Software*. Retrieved from http://www.pvsyst.com/en/

Sauer, K.J., Roessler, T., and Hansen, C.W. (2015). Modeling the irradiance and temperature dependence of photovoltaic modules in PVsyst. *IEEE Journal of Photovoltaics* 5 (1): 152–158. https://doi.org/10.1109/jphotov.2014.2364133.

Sutterlueti, J., Ransome, S., Kravets, R., and Schreier, L. (2011). *Characterising PV modules under outdoor conditions: what's most important for energy yield*. In: *Paper Presented at the 26th European Photovoltaic Solar Energy Conference and Exhibition, Hamburg, GE*.

Wurster, T.S. and Schubert, M.B. (2014). Mismatch loss in photovoltaic systems. *Solar Energy* 105: 505–511. https://doi.org/10.1016/j.solener.2014.04.014.

**Part Nine**

**PV Systems and Applications**

# 26

# Photovoltaic Performance

*Dirk C. Jordan and Kevin Anderson*

Sandia National Laboratories, 1515 Eubank Blvd SE, Albuquerque, New Mexico 87123, USA

## 26.1 Introduction

The produced energy of photovoltaic (PV) systems is vital for financial success and for its role in the dependable delivery of electricity. The performance of PV systems is measured relative to expected production modeled from the geographical location, size, and technology. It is an extensive field in its own right, covered in the first volume of this book (Reinders et al. 2017). Here we focus only on measured performance and issues that can influence field performance. Many excellent and valuable case studies of field performance exist in the literature; however, they are often limited in number of sites, climate, and technology, making it difficult to assess the state of industry as a whole. With the decreasing cost of data acquisition, storage, and analysis more large-scale PV performance assessments have been published (Leloux et al. 2011; Reich et al. 2012; Esmeijer and van Sark 2013; Lindig et al. 2021). These studies are particularly valuable because they provide a high-level overview of the status of PV fleets in particular locations. This chapter continues that body of work with the analysis and lessons learned from a large number of systems in the United States.

The chapter is organized as follows: first, we disseminate lessons learned from a large-scale investigation of PV systems, then we investigate some common issues that influence performance and finally close with performance loss rates that gradually decrease system performance during many years of field exposure.

## 26.2 Performance

Some of the authors had the opportunity to investigate the performance of approximately 100,000 PV systems across the United States. While the data set was limited to annual performance for five years, comments regarding the performance provided insights into issues that led to underperformance. In Figure 26.1 a cumulative distribution function of the measured to predicted performance ratio is shown for "normal" (top) and known underperforming systems (bottom). The data are parsed and color-coded by system size; residential systems are defined in size between 1–25 kilowatt (kW), commercial from 25–1000 kW, and

*Photovoltaic Solar Energy: From Fundamentals to Applications, Volume 2*, First Edition.
Edited by Wilfried van Sark, Bram Hoex, Angèle Reinders, Pierre Verlinden, and Nicholas J. Ekins-Daukes.
© 2024 John Wiley & Sons Ltd. Published 2024 by John Wiley & Sons Ltd.
Companion website: www.wiley.com/go/PVsolarenergy

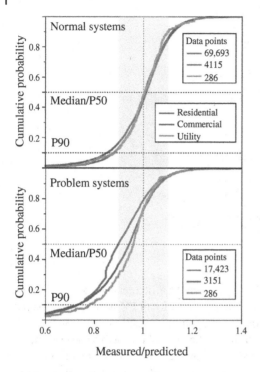

**Figure 26.1** Cumulative distribution function measured to predicted production across the five-year reporting period for normal (top) and problematic (bottom) systems. Different size systems are indicated by colors, residential (1–25 kW) in red, commercial (25 kW–1 MW) in blue, and utility (>1 MW) in green.

utility systems above 1 megawatt (MW). The number of systems for each category is given in the inset. The shaded gray area marks the performance within a 10% region of expected. Expected production values were provided with the data set and were previously shown to be credible (Jordan et al. 2020). Exact mounting configuration was available for systems greater than 5 MW, allowing to generate our own expected production that agreed well with the provided expected values. The apparently normal systems – systems where no known issues were reported – exhibit a performance at the median (horizontal dashed line) as or slightly above expected with a ratio of 1. Most systems, approximately 90%, performed within 10% of the expected production or better indicated by the P90 value (second horizontal dashed line) across the full five-year reporting period. On the other hand, systems with known issues display a marked underperformance (bottom). Much greater is the underperformance for residential systems compared to commercial and utility-scale systems, a finding further examined in the following section.

## 26.3 Failures and Availability

The performance comments were examined and quantified to provide some insights into commonly observed failures and underperformance problems. Color-coded again by system size, and partitioned by separate categories, Figure 26.2 shows the number of occurrences (top) and the estimated lost production (bottom). The lost production can be estimated by comparing the performance during the impacted year relative to the other years. Because the same or different issues may have impacted those same

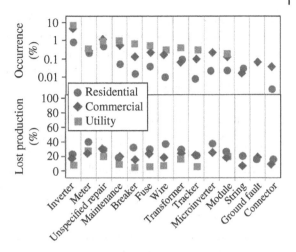

**Figure 26.2** Average occurrence across the five-year reporting period for specific hardware issues (top) and estimated lost production due to the same issues (bottom) partitioned by system size. Different size systems are indicated by colors, residential (1–25 kW) in red, commercial (25 kW–1 MW) in blue, and utility (>1 MW) in green.

systems in the supposedly normal year, the lost production numbers should be viewed as approximate only.

The most common issue inflicting PV systems is inverters, as has been reported before (Golnas 2013). More surprising is the relatively high occurrence of meters indicating potential measurement issues. Unspecified repairs and maintenance are the next two categories occurring about the same, with the exception of residential maintenance events, but lost production is less than repairs. Some of the subsequent categories like breakers, fuses, and wires could be indicative of installation quality or design problems. Problems caused by module failures are relatively low, although the effect of underperforming modules may not have been fully captured. The categories on the far right, ground faults, and connectors – specifically module connectors – may not occur very often, however, they may impose a safety risk and must be taken seriously.

Because inverters were the most common failure category and some other categories suggested installation practices and quality, the inverter category was more closely examined. Figure 26.3 shows the orientation of failed and not failed inverters at residential and commercial buildings. The percentage of inverters that could potentially be shaded by

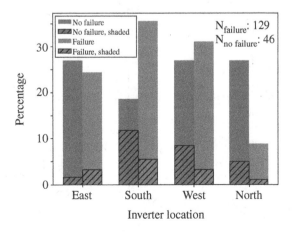

**Figure 26.3** Occurrence of failed and non-failed inverters and the orientation at residential and commercial buildings. The percentage of shaded inverters is indicated by the crosshatch pattern.

vegetation or a nearby building is indicated by the crosshatch pattern. Prolonged exposure to the sun and higher temperatures leads to more failures in the order from the south to the west and east side for systems located in the northern hemisphere. Failures on the north side are markedly reduced. Inverters that did not fail were fairly evenly distributed across the orientations. Certainly, manufacturing quality is important, but the general trend indicates that avoiding full-day sun exposure through best installation practices can have a substantial impact on system reliability.

All hardware failures and issues aggregated can lead to considerable production, as illustrated in Figure 26.4. Unquestionably, different hardware issues can take different lengths to resolve and thus lead to loss of production, however, residential systems are substantially more impacted than commercial and utility systems. At the median, it takes more than a month to resolve hardware problems. In contrast, it takes several weeks and only days for commercial and utility systems, respectively. Larger utility systems are often closely monitored and can even have dedicated maintenance workers such that any hardware problems can be detected and resolved quickly. Supervision and swift detection decrease considerably with decreasing size of the system highlighting an area of future improvement. New monitoring services offered by residential installers may substantially improve this area in the future.

Not only maintenance and operations practices can impact performance and availability, the complement of lost production, of a PV system, but temporary weather events such as snow events can impact overall PV performance. In the PV Fleet initiative program, high-frequency data are collected from primarily large PV installations – average size of

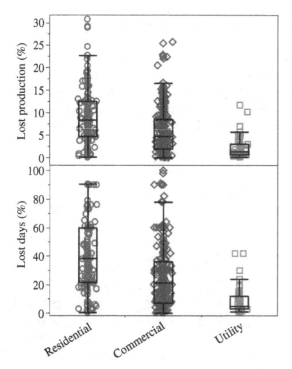

**Figure 26.4** Boxplots lost production days and production aggregated for all hardware issues.

4.1 MW – across the United States totaling more than 7 gigawatts (GW) capacity. The high-fidelity data allows a more thorough examination of availability issues at a fleet scale. Figure 26.5 shows the results of an inverter availability study on a subset of systems based on 15-minute production data instead of maintenance records, covering approximately 250 systems ranging in size from 50 kW to nearly 10 MW (Deline et al. 2021a). In this seven-year dataset, the long-term average monthly energy-weighted availability is roughly 98% with no significant change as the systems age. However, the first few months of the system's lifetime are marked by significantly lower availability. This initial poor performance is probably caused by a combination of initial start-up issues and the fact that larger systems are often commissioned in stages rather than all at once. It is worth emphasizing the fact that Figure 26.5 shows availabilities averaged across many systems, and the monthly availability for any one system might be dramatically different from the average. In fact, the underlying distribution of monthly system-level availability is highly skewed with a median of 100% and a long tail extending down to 0% availability. In contrast, the lost production distribution data in Figure 26.4 were averaged over a longer five-year time period and thus do not extend to 100% loss.

Additionally, for the systems and data represented in Figure 26.5, the monthly performance index was also calculated (Deline et al. 2021b). The performance index is calculated as the ratio of actual generation to expected generation from a simple PVWatts-style power model and National Solar Radiation Database (NSRDB) weather data. The resulting monthly performance index values, after correcting for inverter availability, removing the first few months of data for each system, and removing months with nonzero snowfall are displayed in Figure 26.6. The distribution is centered near 100% and, like Figure 26.1, shows about a 10% chance of performing below 90% of expected. Characterized by a left-skewed underperforming tail, the distribution is fitted well by a Gumbel distribution from the extreme value distribution class that often occurs when multiple mechanisms for underperformance exist. It is interesting to note that a tail remains even after accounting for snow, availability, and start-up issues. Unfortunately, unlike Figure 26.2, the loss

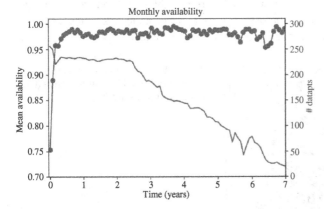

**Figure 26.5** Monthly average energy-weighted inverter availability (blue) across system lifetime, along with the number of values contributing to the average (red). Low initial availability is presumed to be caused by the time it takes to commission new systems and early equipment failures.

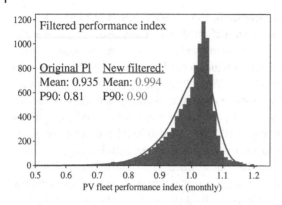

mechanisms driving the remaining tail could not be identified from the time series performance data available in this analysis.

## 26.4 Performance Loss Rates and Degradation

Performance loss rates that gradually reduce the output of a PV system during many years of field exposure are another important part of understanding PV performance. Figure 26.7 compares the performance loss rates from a large number of systems in the United States (blue) with previously published loss rates aggregated from published reports (red) (Jordan et al. 2016). The data from the PV Fleet initiative exhibit a broader distribution than the literature because the literature values were dominated by module performance loss rates. System performance loss rates can approach module performance loss rates but can also be substantially higher, especially if availability issues are not properly considered. Therefore, it is important to differentiate between recoverable and non-recoverable losses (Kiefer et al. 2019).

Balance-of-system (BOS) components such as inverters, cables, and connectors can also gradually degrade and contribute to the overall system performance loss rates. However, quantitatively separating module performance loss rates from other components is difficult.

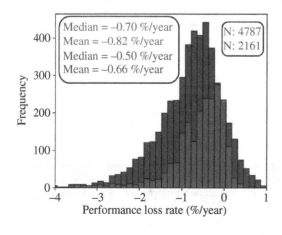

**Figure 26.7** Performance loss rate distribution from PV Fleet initiative (blue) and literature values (red).

**Figure 26.8** Histogram of the difference between AC and DC performance loss rates (a) cumulative distribution function of the difference between AC and DC performance loss rates for specific manufacturers and technologies (b).

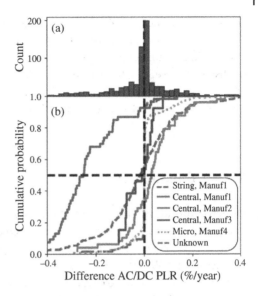

Because of the relatively high inverter failure rates, Figure 26.8 examines the difference in performance loss rates between alternating current (AC) and direct current (DC) performance loss rates. The overall histogram of the difference (blue) is centered on zero, therefore indicating that in general inverters do not appear to contribute to system performance loss rates. Yet, a manufacturer and technology component remains, as the cumulative distribution functions in Figure 26.8b indicate. Evidently, for inverter manufacturer 1 central inverter tends to show higher AC performance loss rates than DC, although the string inverters from the same manufacturer do not exhibit the same trend.

## 26.5 Conclusion

PV systems' performance is critical financially to achieve dependable electricity delivery in sustainable ways. We found in a large study that 90% of systems performed within 10% of expected production. Similarly, a tail of 10% underperforming systems was found for a performing index study using high-frequency data. System failures and underperformance can have many different causes with inverters being the most common. Quantifying and understanding performance and losses will help fulfill the promise of PV to play a major role in dependable and sustainable energy delivery.

## Acknowledgments

The authors would like to thank Chris Deline, the PV Fleet initiative group, and Katherine Jordan. This work was authored in part by Alliance for Sustainable Energy, LLC, the manager and operator of the National Renewable Energy Laboratory for the U.S. Department of Energy (DOE) under Contract No. DE-AC36-08GO28308. Funding was provided by the U.S.

Department of Energy's Office of Energy Efficiency and Renewable Energy (EERE) under Solar Energy Technologies Office (SETO) Agreement Number 30295. The views expressed in the article do not necessarily represent the views of the DOE or the U.S. government. The publisher, by accepting the article for publication, acknowledges that the U.S. government retains a nonexclusive, paid-up, irrevocable, worldwide license to publish or reproduce the published form of this work or allow others to do so, for U.S. government purposes.

## Author Biographies

**Dr. Dirk Jordan** is a distinguished member of research staff at the National Renewable Energy Laboratory, United States. He has been in his current role since 2009. His interests include PV system performance, degradation, quality assurance, and the impact of extreme weather events. Prior to NREL, he worked at Motorola for almost a decade. He received a Ph.D. in physics from Arizona State University in 1999 and a B.S. in Physics from the University of Heidelberg in Germany.

**Kevin Anderson** is a researcher at Sandia National Laboratories focused on photovoltaic (PV) system performance modeling and reliability assessment. His research interests include single-axis tracker modeling and optimization, high-performance computation, and PV degradation rate assessment. He contributes frequently to many open-source software packages for PV applications and co-maintains pvlib python, an open-source software library for photovoltaic performance modeling. He was previously a researcher at the National Renewable Energy Laboratory and a PV performance engineer in industry.

## References

Deline, C., Anderson, K., Jordan, D. et al. (2021a). PV Fleet Performance Data Initiative: Performance Index-Based Analysis. Technical Report, NREL/TP-5K00-78720.

Deline, C., Anderson, K., Jordan, D. et al. (2021b). Performance index assessment for the PV Fleet performance data initiative. *Proceedings of the 48th Photovoltaic Specialist Conference.*

Esmeijer, K.B.D. and van Sark, W.G.J.H.M. (2013). Statistical analysis of PV performance using publically available data in the Netherlands. *Proceedings of 28th European Photovoltaic Solar Energy Conference*, Paris, France.

Golnas, A. (2013). PV system reliability: an operator's perspective. *IEEE Journal of Photovoltaics* 3 (1): 416–421.

Jordan, D.C., Kurtz, S.R., VanSant, K.T., and Newmiller, J. (2016). Compendium of photovoltaic degradation rates. *Progress in Photovoltaics* 24 (7): 978–989.

Jordan, D.C., Marion, B., Deline, C. et al. (2020). PV field reliability status—analysis of 100,000 solar systems. *Progress in Photovoltaics* 28 (8): 739–754.

Kiefer K., Farnung, B., Müller, B. et al. (2019). Degradation in PV power plants: theory and practice. *Proceedings of the 36th European PV Solar Energy Conference*, Marseille, France.

Leloux, J., Narvarte, L., Trebosc, D. (2011). Performance analysis of 10,000 residential PV systems in France and Belgium. *26th European PV Solar Energy Conference*, Hamburg, Germany.

Lindig, S., Ascencio-Vásquez, J., Leloux, J. et al. (2021). Performance analysis and degradation of a large fleet of PV systems. *IEEE Journal of Photovoltaics* 11 (5): 1312–1318.

Reich, N.H., Mueller, B., Armbruster, A. et al. (2012). Performance ratio revisited: is PR>90% realistic? *Progress in Photovoltaics* 20: 717–726.

Reinders, A., Verlinden, P., van Sark, W., and Freundlich, A. (2017). *Photovoltaic Solar Energy: From Fundamentals to Applications*. Wiley.

# 27

# Deployment of Solar Photovoltaic Systems in Distribution Grids in Combination with Batteries

*Elham Shirazi[1], Wilfried van Sark[2], and Angèle Reinders[3,4]*

[1] Faculty of Engineering Technology, Department of Design, Production and Management, University of Twente, P.O. Box 712, 7500 AE Enschede, The Netherlands
[2] Copernicus Institute of Sustainable Development, Utrecht University, Utrecht, The Netherlands
[3] Department of Mechanical Engineering, Energy Technology Group, Eindhoven University of Technology, P.O. Box 513, 5600 MB Eindhoven, The Netherlands
[4] Solliance, Eindhoven, The Netherlands

## 27.1 Introduction

The integration of intermittent solar energy and decentralized energy production in existing electricity grids poses a set of benefits and challenges. Integrating photovoltaic (PV) systems into distribution grids could decrease both transmission and distribution grid losses, reduce cost of energy, lower amounts of $CO_2$ emissions, and postpone investments for a new power plant. Moreover, the expected use of large number of new types of energy products in existing grids, such as photovoltaic systems, electric cars, batteries, and heat pumps will pose big engineering questions in the next decade. So-called smart grids are an important promising solution at transnational, national, and local levels. These are envisaged to support the transition from centralized energy generation to energy systems that contain more distributed energy generation with a high penetration of renewable energy systems and flexible options to manage energy flows via advanced Information and Communications Technology (ICT), see Figure 27.1. This paradigm shift will lead to new requirements toward increased flexibility of distribution grids and the related development of smart energy management algorithms. Chapter 11.1 discusses the basics of grid-connected PV systems, exploring various types and technologies while highlighting key areas of development aimed at enhancing system engineering, performance, protection, and safety (Ball 2017).

In this context, (residential) PV systems are the most promising option to locally generate sustainable and cost-effective energy. In order to connect PV systems to the electrical grid, the generated direct current (DC) power of PV system should be converted to alternating current (AC) power. This is the role of a grid-tied inverter. For an efficient conversion, three distinct functions must be carried out. The primary function involves generating a sinusoidal waveform by modulating the DC input to match the grid frequency and filtering out

*Photovoltaic Solar Energy: From Fundamentals to Applications, Volume 2*, First Edition.
Edited by Wilfried van Sark, Bram Hoex, Angèle Reinders, Pierre Verlinden, and Nicholas J. Ekins-Daukes.
© 2024 John Wiley & Sons Ltd. Published 2024 by John Wiley & Sons Ltd.
Companion website: www.wiley.com/go/PVsolarenergy

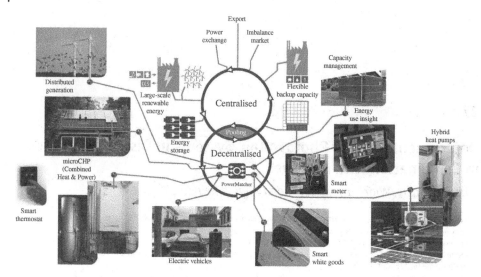

**Figure 27.1** Technical framework of an extensive distributed energy system, called a smart grid, with PV systems, local energy storage, and many other energy components, smart appliances, and electric vehicles, managed by smart ICT solutions, in the Netherlands Source: Courtesy: Powermatching City, DNV GL.

higher-order harmonics. Another crucial function of the device is to implement maximum power point tracking (MPPT) for the DC source, ensuring that the inverter circuit processes the maximum available DC power. The third function involves monitoring both grid and PV sides for suitable interconnection conditions such as anti-islanding and voltage/frequency support. Inverters, power optimizers, and microinverters in grid-tied PV systems are discussed in Section 11.2 (Deline 2017).

One of the major technical issues related to the integration of PV systems into local electricity networks is balancing the mismatch between demand and their intermittent supply of power, see Figure 27.2.

Because daily and seasonal meteorological conditions significantly affect solar energy production, 100% matching of the energy demand with solar power can only be achieved by PV systems in combination with residential storage systems such as batteries, vehicle-to-grid technologies (V2G), or by using community-based storage systems. Energy storage systems such as batteries at consumers' premises, or at other locations in the grid (such as community batteries), can be used to efficiently integrate PV into grid, smooth the power output, increase self-consumption as well as reduce PV curtailment. They can help to spread energy supply over time, and demand and supply can be better matched even on a local, decentralized level. As they are modular, they can be easily scaled up and down. Due to the ability of batteries to react quickly to fluctuations, they can help to support grid stability. In general, batteries may be an essential energy component to increase flexibility so as to maintain a high quality of the power fed into local electricity networks.

In addition to batteries, other solutions for the realization of flexibility also exist, such as demand side management (DSM). DSM refers to initiatives and technologies to improve the

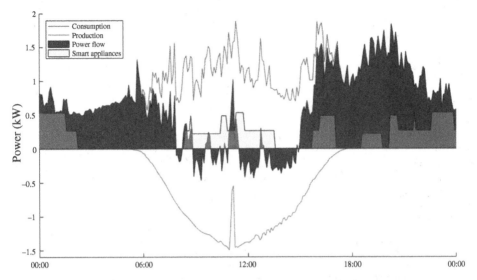

**Figure 27.2** Energy consumption (blue line), energy production by PV modules (red line), and power flow (blue area) for a household with a photovoltaic (PV) system in the Netherlands, 9 September 2012. Negative values indicate power flows into the grid. Yellow areas indicate power consumption by appliances Source: Courtesy: Powermatching City, DNV GL.

energy system by modifying the demand profile. It involves demand modification strategies, through financial incentives or behavioral changes. DSM is associated with increasing system reliability and flexibility, optimizing the injected PV power to the grid, and subsequently reducing grid congestion. These can be achieved through modifying demand to match available resources and, therefore, using as much as possible on-site PV production. Increasing users' awareness and encouraging them to modify their energy demand patterns can also be achieved by applying DSM (Wilkinson et al. 2021). There are various DSM approaches such as load shifting and peak shaving which will be further discussed in Section 27.3.1.

In this chapter, we will explore in more detail the interaction of PV systems and batteries within distribution grids. It will start with challenges in Section 27.2. Several technological developments that support and enhance interaction of PV systems with energy demand, including DSM, batteries, solar charging stations, electric vehicles (EV), and vehicle-integrated PV (VIPV) will be discussed in Section 27.3. The interaction of such technologies with PV and the grid will also be discussed in Section 27.3. Finally, Section 27.4 will conclude this chapter.

## 27.2 Challenges

In conventional power systems, power flows in one direction from a power plant through high-voltage transmission grids to substations, and then through low-voltage distribution grids to consumers. Unlike conventional power plants, the majority of small-scale residential and commercial PV systems are connected to distribution grids, which are operated at

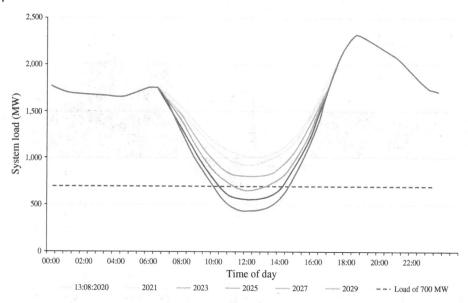

**Figure 27.3** Net load curve of SWIS on September 13, 2020. The curve is deepened as more PV is integrated to the grid. Source: Wilkinson et al. (2021)/with permission of Elsevier.

low or medium voltage levels. Aggregated PV generation in distribution grids can cause an imbalance between peak demand and generation. This imbalance of timing causes the so-called "duck curve" of the net load demand, and as more PV systems are integrated into the grid, it gets more pronounced. The case study of South West Interconnect System (SWIS) in Western Australia (WA) can be seen in Figure 27.3 (Wilkinson et al. 2021). As shown, two steep ramps are happening, one before noon due to abundant PV generation, and the second at the end of the afternoon due to increased demand and reduced PV generation.

High PV power that is injected into the grid during overgeneration periods can lead to an overloading of transformers and distribution grids as well as voltage violations. As illustrated in Figure 27.4, if PV systems are absent, the voltage along the feeder decreases with the increase of distance from the substation (red line). Conversely, in case of grid-connected PV systems, the voltage can potentially increase above normal operational conditions if PV power is fed in at the end of the feeder (blue line) (Coddington et al. 2016).

In addition to that, bi-directional power flows in distribution grids in overgeneration circumstances make it difficult for protection devices to detect short-circuit faults (Shafiullah et al. 2022). This is due to increased fault current levels. Another issue associated with PV systems is the uncertainty involved with prediction of the amount of PV-generated power. This complicates the operation of the power system, as this uncertainty increases the possibility of mismatch between generation and demand, leading to frequency fluctuations as well as voltage instability. Therefore, to connect PV systems to the grid in a safe manner, some standards must be followed such as IEC 62109-2 (IEC 62109-2:2011 2011). In that context, inverters are the interface between PV modules and distribution grids that can help tackle some of these issues. Inverters on the other hand generate harmonics

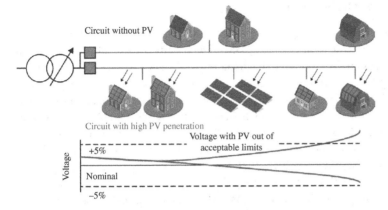

**Figure 27.4** Two circuits illustrating the effect of PV penetration on the voltage over a power line for the case of absence of PV systems (above) and the case of a high penetration of PV systems (below) Source: Coddington et al. (2016)/U.S. Department of Energy/Public Domain.

which can result in power quality issues in the grid. All standards such as IEEE standard 519-1992, require total harmonics distortion (THD) to be less than 5% (Alkahtani et al., 2020).

One efficient approach to mitigating the effect of overgeneration is to increase system flexibility by modifying demand profile and increasing self-consumption which will be discussed in Section 27.3. There are also regulatory and legislation issues together with social acceptance of PV integration into the grid (Litjens et al. 2018), however, this chapter focuses on technical challenges only.

## 27.3 PV, Battery, and Grid Interactions

As has been mentioned previously, integrating PV systems into distribution grids results in many benefits such as reducing grid losses, lowering $CO_2$ emissions, and decreasing energy costs. On the other hand, it will introduce technical issues, for instance, voltage and power fluctuation, frequency deviation, overload, and congestion in distribution grids (Seguin et al. Jan. 2016). In order to mitigate the negative effects of PV integration into distribution grids, the simplest solution would be to curtail PV generation. Curtailment is the deliberate reduction of PV generation below what could have been generated. In that scenario, a great potential is lost, and PV systems cannot be used to their full potential in terms of energy production, economic gain, and environmental benefits. There are various solutions to overcome challenges involved with PV integration such as having an accurate PV forecast, (Litjens et al. 2018) improving self-consumption by integrating either energy storage systems or DSM, designing an efficient protection system for distribution grids, filtering redundant harmonics, installing new transformers, and reinforcing the grid. These solutions are either focused on distribution systems side, on the users' side, or need interactions between these two entities (Raghoebarsing et al. 2022; Borne et al. 2018). Technological development including DSM, batteries, solar charging stations, EV, and VIPV which can help to accelerate adoption of PV will be discussed in the following.

### 27.3.1 Demand-side Management

Overloads and congestions are of the major barriers to effective integration of PV into distribution grids. Demand Side Management (DSM) refers to initiatives and technologies to improve the energy system by modifying the demand profile. DSM can help with the effective integration of PV into distribution grids. One approach to increase the hosting capacity of distribution grids is to flatten the net load curve by increasing self-consumption. Self-consumption is the share of PV generation which is consumed directly on-site (Luthander et al. 2015). The effect of self-consumption is twofold. First, it can increase the financial profit for the owner, especially with decrease in PV subsidies (Raghoebarsing et al. 2022), and second, as mentioned above, it helps in flattening the net load curve by decreasing peak demand from grid and injecting less power to grid.

There exist various approaches for DSM, one is to shift loads from peak times to the times where there is surplus PV generation (Figure 27.5a). This could help to reduce peak demand from the distribution grid as well as inject power into the distribution grid, in other words, it could flatten the net load curve. Another approach is peak shaving (Figure 27.5b), where load is reduced during peak hours, this will improve the load factor, reduce peak demand from distribution grid, and subsequently postpone costly grid expansion programs. The valley filling approach (Figure 27.5c), is where the loads are moved to the hours with minimum load. There are also two other approaches where the load is strategically decreased (Figure 27.5d) or strategically increased (Figure 27.5e) to follow the generation. The last approach would be to have a fully flexible load profile (Figure 27.5f) where increase and decrease in demand are possible at any time. The type of the load should also be considered in DSM strategies, as different load types react differently to different approaches.

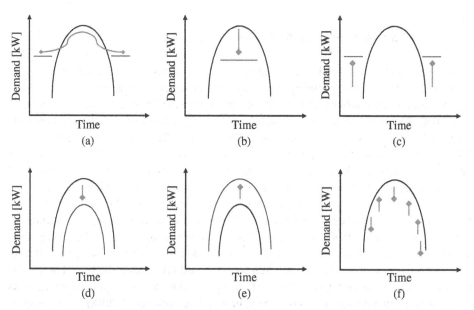

**Figure 27.5** DSM Approaches (a) load shifting, (b) peak shaving, (c) valley filling, (d) strategic decrease, (e) strategic increase, (f) fully flexible demand.

The different types of loads are, for instance, controllable, thermostatically controllable, and fixed loads.

DSM programs can facilitate overcoming many challenges involved with integration of PV systems into distribution grids, such as voltage violation and grid congestion. The effect of DSM will become more apparent if combined with local electricity storage. This will be further explained in the following.

### 27.3.2 PV-battery Systems Integrated in Grids

Co-installing PV and battery systems provides the opportunity to deliver the generated PV electricity from times of high production to times of low/no production. This smoothens out the load profile of the system as the excess PV generation can be stored in a battery and used later during peak hours. Using battery storage together with behavior-modifying mechanisms can increase the hosting capacity of the grid without expanding infrastructures while using PV systems to their full potential.

Various battery technologies are available to be cointegrated with PV systems into distribution grids, among which some electrochemical storage technologies are suitable for residential and commercial electricity storage. The principle of electrochemical energy storage systems is using the difference in cathode electric potentials to convert chemical energy into electrical power. Lithium-ion (Li-ion), lead-acid, and Nickel–cadmium (NiCd) are examples of electrochemical batteries (Luthander et al. 2015), some of which are phased out or no longer used in the context of PV systems, in particular NiCd and lead-acid. While the lead-acid battery is the most mature battery technology, it has a negative environmental impact and is most common in automotive rather than PV system applications. However, in early solar home systems, it was the battery of choice. Li-ion batteries are the most common battery types which are currently used in the PV system context. They have a high energy density and storage efficiency but are more expensive in comparison to other electrochemical technologies (Borne et al. 2018). A comparison of various Li-ion technologies is presented in Table 27.1.

One of the main drawbacks of incorporating batteries is the high cost. Although the costs of batteries have been reduced over past years, and a further reduction to around 100 $/kWh is expected by 2030 (Bajolle et al. 2022), still the major obstacle to integrating batteries is the high cost. In addition to that, there are other factors such as battery capacity, efficiency, cycle life, and lifetime, together with the safety and maturity of the technology which should also be considered. Therefore, optimal sizing of the battery is crucial not only in terms of lifetime, cycle life, and depth of discharge (DoD) but also in terms of upfront investment costs. The optimum size of a grid-tied tied PV-battery system depends on many factors such as load profile, peak demand, and electricity pricing scheme. Many studies have investigated several PV and battery combinations in small and medium-scale PV-battery systems. For example, a study on various PV and battery combinations for a Swedish household found that a 3 kWh battery together with a 5 kWp PV can increase self-consumption by 614 kWh/year while doubling the size of the battery resulted in additional 358 kWh/year (Widén and Munkhammar 2013). Another study considered a range of 5–24 kWh for battery capacity coupled with 2.5 kWp PV, where self-consumption increased by 18% and 33% in case of 5 and 24 kWh batteries respectively

**Table 27.1** Different Li-ion battery technologies.

| Specifications | Li-ion | | |
| --- | --- | --- | --- |
| | Cobalt | Manganese | Phosphate |
| Specific energy density (Wh/kg) | 150–190 | 100–135 | 90–120 |
| Internal resistance (mΩ) | 150–300 7.2 V | 25–75 per cell | 25–50 per cell |
| Life cycle (80% discharge) | 500–1000 | 500–1000 | 1000–2000 |
| Fast-charge time | 2–4 h | 1 h or less | 1 h or less |
| Overcharge tolerance | Low. Cannot tolerate trickle charge | | |
| Self-discharge/month (room temp) | <10% | | |
| Nominal cell voltage (V) | 3.6 V | 3.8 V | 3.3 V |
| Charge cut-off voltage (V/cell) | 4.20 | | 3.60 |
| Discharge cut-off voltage (V/cell, 1 C) | 2.50–3.00 | | 2.80 |
| Peak load current (C-rate) | >3 C | >30 C | >30 C |
| Best result | <1 C | <10 C | <10 C |
| Charge temperature | 0 °C to 45 °C | | |
| Discharge temperature | −20 °C to 60 °C | | |
| Maintenance requirement | Not required | | |
| Safety requirements | Protection circuit mandatory | | |
| In use since | 1991 | 1996 | 1999 |
| Toxicity | Low | | |

Source: Adapted from Diouf (2016).

(Thygesen and Karlsson 2014). A study on a German household investigated the effect of installing a battery on self-sufficiency and self-consumption in a house with a PV installation of 1 kWp/MWh where MWh refers to annual consumption of the household (Weniger et al. 2014). Self-sufficiency and self-consumption were 28% and 30% in the case of no battery, respectively, while they were increased to 56% and 59% in case of battery with capacity of 1 kWh/MWh. Another study on seven Belgian households showed that a battery capacity in the range of 0.4 to 1.5 kWh/MWh is the optimal battery size (Mulder et al. 2010). Reduction in purchased electricity from grid in a household in California, United States, with a PV installation of 4 kWp and various battery sizes was investigated (Huang et al. 2012). For batteries with 0.5, 7, and 10 kWh, the daily reduction in purchased electricity from the grid was 0.79, 7.13, and 9.14 kWh respectively. A typical household in Tokyo, Japan, with a 3.4 kWp of PV installation coupled with a 24 kWh battery, was studied (Osawa et al. 2012). The self-consumption increased from 41% in case of no battery to 79% with a battery. A study on more than 2000 households showed that a household with PV installation of 7 kWp and annual consumption of 20 MWh, requires battery capacities in range of 15 to 24 kWh (Nyholm et al. 2016). Although different studies have yielded different results regarding the extent to which batteries can increase self-consumption, they all confirmed that coupling a battery with PV installation will increase self-consumption and decrease purchased electricity from grid (Litjens et al. 2018).

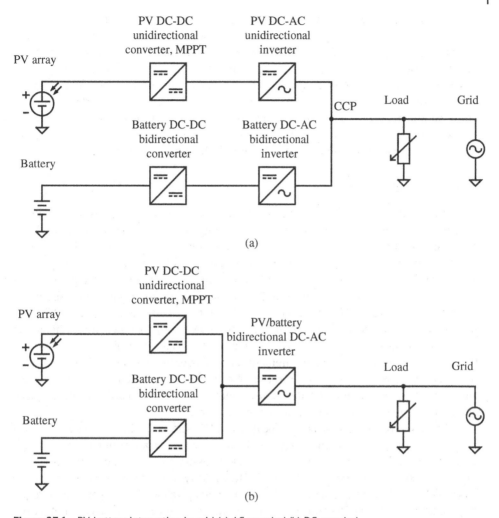

**Figure 27.6** PV-battery integration in grid (a) AC-coupled (b) DC-coupled.

PV battery systems can be coupled to the electrical systems in either the AC or DC mode. In AC architecture (Figure 27.6a), batteries are connected through the inverter to the AC link, while in DC architecture (Figure 27.6b), batteries are connected to the DC link of the inverter (Freitas Gomes et al. 2020).

In the AC-coupled architecture, PV is connected to a unidirectional DC–DC converter equipped with a MPPT. After the DC bus, there is a unidirectional DC–AC inverter which connects PV system to the common coupling point (CCP). The battery system connects to the grid through a bidirectional DC–DC converter and a DC–AC inverter. In the DC-coupled architecture, a PV system is connected to a unidirectional DC–DC converter. Both converters of PV and battery, on the other hand, are connected to a single bidirectional DC–AC inverter.

The architecture could influence efficiency, reliability, system performance, flexibility, and cost of PV battery system (Fu et al. 2018). Considering that the efficiency of a system decreases as the number of power conversion units increases, having an extra inverter in the AC-coupled architecture leads to a lower efficiency and higher cost. However, having separate inverters for the PV and battery will lead to a higher capacity for power delivery, as they can simultaneously supply the load. This will also result in higher flexibility as the power delivered by a PV battery system is not limited to one inverter. Optimum sizing of power conversion devices considering PV battery system size is also important. The cost of this equipment is one of the limiting factors, therefore having an optimal configuration is essential to maximize the financial returns. Battery management systems are of huge importance to manage charging and discharging of the battery in PV battery systems.

### 27.3.3 Control Measures

Among all control measures, active power ($P$) and reactive power ($Q$) control are of a fundamental importance since they are tied to controlling frequency and voltage, respectively. Active and reactive powers could be controlled not only on distribution grid level by grid operators but also locally on the user side. In order to control active power in a distribution grid, an on-load tap changer (OLTC) can be used, which can respond to changes in active power. In order to be able to respond to fast changes in PV production, OLTC should be operated automatically (Michael et al. 2021). To control reactive power, static VAR compensators (SVC) can be used, which inject reactive power into the grid contributing to the voltage control as well (Mahmoud and Hussein 2018). Voltage control can also happen at transformers based on voltage measurements and transformer loading. It should be noted that high PV penetration makes control more complicated, especially in parallel feeders with different penetration levels. Booster transformers can also be used to control voltage along a feeder to prevent extreme voltage drops or rises.

To control active and reactive powers locally, the PV inverter can be used. It controls active power considering the curtailment level and reactive power as a function of local voltage value, as shown in Figure 27.7. Note, the simplest passive curtailment measure is to limit the inverter size to, for example, 50% of PV DC capacity. The reference value for reactive power, $Q_{ref}$ is set based on the actual and reference voltage values via a proportional-integral (PI) controller. Then the PI controller regulates direct and quadrature axis ($d$- and $q$-axis) components of the current (Tamimi et al. 2013).

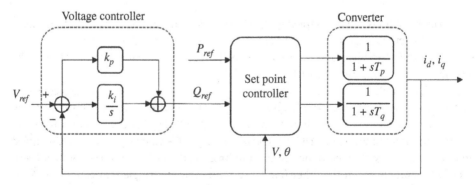

**Figure 27.7** Local active and reactive power control of a PV system. Source: Adapted from Tamimi et al. (2013).

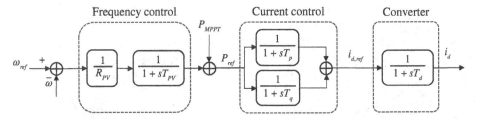

**Figure 27.8** Frequency control of a PV system. Source: Adapted from Ortega and Milano (2018).

As demonstrated in Figure 27.8, the frequency can also be controlled using a droop control. Then the output is added to MPPT reference signal to create the $d$-axis component of the reference current through a PI controller, which will be used as the input signal of the PV converter (Tamimi et al. 2013).

If there is an infrastructure such as supervisory control and data acquisition (SCADA) which enables communication between distribution system operator (DSO) and users, then frequency, voltage, and load control can happen more efficiently. For example, a DSM program with the goal of grid support requires sending specific cost signals which is different from market prices to users in the area with a lot of PV overproduction. Another example is the ability to remotely control users' load and inject active and reactive power to the grid from the PV inverter by the DSO, based on pre-agreed conditions.

Solutions such as reinforcing the distribution grid, reconfiguration of the grid, and operating in mesh or closed loop instead of radial topology can also mitigate the effect of PV integration into distribution grids, but they require large investments.

### 27.3.4 PV-battery Systems Deployment in Transport

The main goal of EVs is to meet transportation demand in an efficient, clean, and reliable way. When an EV is charged from the grid, the grid load increases, especially if charging happens during peak hours. As a potential mean to address this, PV could be integrated into either charging stations or vehicles to meet (part of) the EV electricity demand. This would also lower $CO_2$ emissions. These scenarios are related to PV-battery-grid interactions, and hence serve as a case in this chapter.

#### 27.3.4.1 PV-battery Systems Integrated in Electric Vehicles

VIPV is a concept that refers to the integration of PV into the EV, where PV could provide (1) additional range of 10 to 30 km a day, (2) power to reduce DoD of battery, and (3) part of the demand of auxiliary systems (Lightyear n.d.). The introduction of VIPV plays an important role in taking up electric transportation and creating opportunities for other solar PV applications in transportation sector. VIPV EVs have been introduced by many manufacturers such as Lightyear (Lightyear n.d.), Sono Motors (Solar on every vehicle | Sono Motors n.d.), Toyota (Araki et al. 2021), and Nissan (Nissan n.d.).

The most common technology for PV modules in VIPV is silicon, as the silicon-based module is a good compromise between price and performance. However, lack of flexibility for building a curved module is one limiting factor. Multi-junction solar cells have higher efficiency but are more expensive and have higher mismatching loss (Araki et al. 2021)

**Figure 27.9** PV-powered vehicles (a) Lightyear one, Source: courtesy: Lightyear (Lightyear n.d.), (b) Sion, Source: courtesy: Sono Motors GmbH (Solar on every vehicle | Sono Motors n.d.), (c) Toyota Prius, Source: courtesy: Toyota (Toyota Europe Newsroom n.d.), (d) Nissan e-NV200, Source: courtesy: Nissan (Nissan n.d.).

in comparison to crystalline silicon solar cells. The Lightyear one (Figure 27.9a) has $5\,m^2$ of crystalline silicon PV with $215\,Wp/m^2$ which can provide up to $70\,km/day$. The Sion (Figure 27.9b) can provide up to $34\,km/day$ with $1208\,Wp$ by integrating lightweight crystalline Si PV cells. Toyota and Nissan use III–V multi-junction solar cell with $860\,Wp$ and $1150\,Wp$ PV capacities, respectively (Araki et al. 2021).

Thin film PV cells provide the opportunity to manufacture curved modules at lower price. Amorphous silicon could be one choice, but it has rather low efficiency compared to silicon-based modules. Other technologies such as perovskite cells have both flexibility and high efficiency, however, stability and reliability are of major concern.

The expected benefits of VIPV have been investigated in several case studies. A case study for passenger vehicles in the Netherlands as a country with comparatively low solar irradiance, shows emissions reduction potential of up to $200\,kg\ CO_2$-eq/year as well as 60% reduction in frequency of charging. A case study for light commercial vehicles in Germany shows the total energy yield of $1170\,kWh/year$ for Hamburg, Germany, and $2210\,kWh/year$ for Rome, Italy. The emission factor for integrated PV module is $0.357\ CO_2$-eq/kWh while it is $0.435\ CO_2$-eq/kWh for electric grid (Araki et al. 2021).

### 27.3.4.2 PV-battery Systems Integrated in Charging Stations

Although the integrated PV could provide additional power, it cannot be the only source of power, due to limited PV surface and shadows cast by nearby objects and buildings.

(a)                                                    (b)

**Figure 27.10** Portable and standalone PVCS (a) EV ARC™ 2020 EV charging unit, Source: courtesy: Beam Global (BEAM n.d.), (b) PairTree, Source: courtesy: Paired Power (Paired Power n.d.).

Therefore, at some point, the EV should be connected to an external resource such as grid, charging station, or PV-powered charging station (PVCS) to charge its battery. A considerable reduction in $CO_2$ emission can be achieved through PVCS. An environmental study of PVCS over a course of 10 years in the United States and China shows that scenarios with PV charging and local storage result in $CO_2$ emissions reductions of 60–93% in the United States and 28–93% in China in comparison to a gasoline-fueled vehicle (Sechilariu et al. 2021).

PVCS can operate in grid-connected or standalone mode. Two examples of portable standalone PVCS are EV ARC™ (BEAM n.d.) and PairTree (Paired Power n.d.). EV ARC™ is located in San Diego, United States (Figure 27.10a). It is a portable and a standalone PVCS with a total of 4.3 kWp PV capacity, battery storage options of 24, 32, or 40 kWh, and possible daily range of up to 394 km in San Diego. PairTree follows a similar concept of a portable standalone PVCS, which has 5 kWp of PV capacity and can charge up to 120 km/day (Figure 27.10b).

The other examples of standalone charging stations are from Fastned, in the Netherlands (FASTNED n.d.), and a bus charging station, in Australia (transdev n.d.). The Fastned PVCSs (Figure 27.11a) can have 4–8 kWp of PV capacity with 50 and 175 kW DC charging. The bus charging station (Figure 27.11b) is composed of 250 PV panels generating an average of 438 kWh/day and 135 kWh of storage capacity. Each bus is equipped with 348 kWh battery resulting in daily range of 250–300 km (Sechilariu et al. 2021).

(a)                                                    (b)

**Figure 27.11** Standalone PVCS (a) Fastned, Source: courtesy: Fastned (FASTNED n.d.), and (b) bus charging station, Australia, Source: courtesy: Transdev (transdev n.d.).

(a)                                         (b)

**Figure 27.12**  Grid-connected PVCS (a) Solar canopy, Germany, Source: courtesy: MDT SUN PROTECTION SYSTEMS AG (MDT-tex n.d.), (b) Tesla V3 supercharger, United States, Source: courtesy: Tesla (TESLA n.d.).

Other examples of grid-connected PVCS are the MDT-TEX solar canopy in Germany (MDT-tex n.d.), and the Tesla V3 Supercharger in Nevada, United States (TESLA n.d.)

Each MDT-TEX solar canopy has a square structure of $5.3 \times 5.3$ m (Figure 27.12a). It has a modular design; hence several canopies can be installed next to each other. It is composed of 15 PV panels with in total 4.05 kWp with optional storage capacity of 4.5 kWh. The charging takes place via a 22 kW DC terminal. There are many Tesla V3 superchargers across North America, Europe, and Asia (TESLA n.d.). The charging station (Figure 27.12b) is installed in Nevada consisting of 288 PV panels connected directly to the grid. There are also battery packs with 210 kWh capacity, which are charged through the grid.

Being connected to grid could provide an opportunity for bi-directional power flow, in other words, the EV not only could charge its battery through this connection but also could restore its energy back to the grid when it is needed. This concept is referred to as vehicle-to-grid (V2G) and vehicle-to-home (V2H). Considering that vehicles are parked more than 90% of their lifetime, V2G can be considered as a distributed storage for the grid, helping with smooth integration of solar PV into the grid. To enable V2G there are some requirements, such as having a bi-directional charger and communication interface. The communication between EV owners and aggregators as the intermediators between EV owners and system operators plays a key role in V2G schemes. Specific standards and protocols have been developed to guarantee a reliable communication and interoperability between EV owners and aggregators, including the open charge point protocol (OCPP), ISO 61850, European ISO 15118, IEC 62196, SAE J2836, and CHAdeMO (Schmutzler et al. Dec. 2013; ISO n.d.; Briones et al. 2012). It should be noted that V2G and V2H may reduce EV's battery lifetime, so the financial incentives should be high (Sierra Rodriguez et al. 2020). Then again, V2G could provide flexibility during (local) grid congestion, for which a remuneration scheme would be in place (Brinkel et al. 2022).

Like other applications, having a proper control scheme is essential. The controller will determine the power flow based on predefined objectives such as minimizing charging cost or maximizing self-consumption while maintaining the voltage within a desired range. An example of a control algorithm is shown in Figure 27.13 (Sierra Rodriguez et al. 2020).

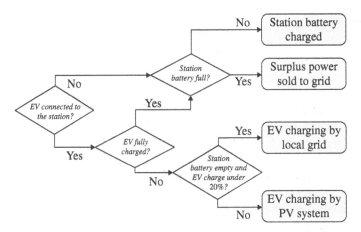

**Figure 27.13** Controlling algorithm of PVCS. Source: Adapted from Sierra Rodriguez et al. (2020).

There are various possible operating modes of a PVCS, considering the available PV power, state of charge (SoC) of EV, and PVCS storage.

There are many factors influencing the synergy of PV-battery-EV including technological development, financial regulations, and user adoption. Technological development can accelerate adoption of PV-battery-EV, while inappropriate financial regulation can be a barrier and impede gaining the potential benefits of this combination. Willingness to pay (WTP) is strongly correlated with socioeconomic variables; however, it can be increased if the technology is offered as bundle of various technologies such as PV-EV-battery (Priessner and Hampl 2020).

## 27.4 Conclusion

Solar PV is a promising technology to meet the growing electricity demand and reduce $CO_2$ emissions, yet the rapid increase in the number of PV systems integrated into the grid impacts the operation and control of the power system. The intermittent nature of PV generation, evolving regulations, and policies together with other technical issues make its integration into the grid at high penetration levels challenging. There should be strategies to balance the goals of facilitating integration of PV and ensuring reliable operation of distribution grids. Therefore, in this chapter, various technical challenges, and some possible approaches to tackle them have been discussed. The following approaches can help to overcome these challenges: accurate PV forecasting methods, enhanced self-consumption by integrating either energy storage systems or DSM, designing an efficient protection system for distribution grid, filtering redundant harmonics, installing new transformers, and reinforcing the grid. Several technological developments such as improved DSM, batteries, EV, PVCS, V2G, and control strategies can result in increasing potential benefits of PV-battery-grid interactions, thus accelerating PV adoption.

## List of Symbols and Acronyms

| Acronym | Meaning |
|---|---|
| AC | Alternating Current |
| CCP | Common Coupling Point |
| DC | Direct Current |
| DOD | Depth of Discharge |
| DSM | Demand Side Management |
| DSO | Distribution System Operator |
| EV | Electric Vehicle |
| ICT | Information and Communications Technology |
| MPPT | Maximum Power Point Tracking |
| OCPP | Open Charge Point Protocol |
| OLTC | On Load Tap Changer |
| PI | Proportional-Integral |
| PV | Photovoltaics |
| PVCS | Photovoltaics Charging Station |
| SCADA | Supervisory Control and Data Acquisition |
| SVC | Static VAR Compensators |
| THD | Total Harmonics Distortion |
| V2G | Vehicle-To-Grid |
| V2H | Vehicle-To-Home |
| VIPV | Vehicle Integrated Photovoltaics |
| WTP | Willingness To Pay |

| Symbol | Description | Unit |
|---|---|---|
| $d$ | Direct axis | – |
| $f$ | Frequency | Hertz |
| $i_{d,ref}$ | Reference value of $d$-axis current | Ampere |
| $i_d$ | $d$-axis current | Ampere |
| $i_q$ | $q$-axis current | Ampere |
| $k_i$ | Integral component of the PI controller | – |
| $k_p$ | Proportional component of the PI controller | – |
| $P_{ref}$ | Reference value of active power | Watt |
| $q$ | Quadrature axis | – |
| $Q_{ref}$ | Reference value of reactive power | Var |
| $s$ | Laplace variable | – |
| $T_d$ | Time constant of transfer function for q-axis current control | – |
| $T_p$ | Time constant of transfer function for active power control | – |
| $T_{PV}$ | Time constant of transfer function for PV frequency control | – |
| $T_q$ | Time constant of transfer function for reactive power control | – |
| $V$ | Voltage | Volt |
| $V_{ref}$ | Reference value of voltage | Volt |
| $\omega$ | Angular frequency | Rad/s |
| $\omega_{ref}$ | Reference value of angular frequency | Rad |
| $\theta$ | Angle of power factor | degree |

## Author Biographies

**Dr. Elham Shirazi (Eli Shirazi)** is an assistant professor at the faculty of Engineering Technology at University of Twente since 2021. She got her PhD in electrical engineering with focus on smart grid control from IUST. Afterwards she joined Electrical Engineering department of KU Leuven as a postdoctoral researcher, she also worked at Energy System group of IMEC. She is a member of advisory board of 4TU. Energy in the Netherlands and also a member of the scientific committee of European and international conferences. She designed the "Energy System Integration" course, which is given as part of the Sustainable Energy Technology program at University of Twente. Her research is multidisciplinary including modeling and control of integrated energy systems, energy management systems, model-driven optimization of energy systems in buildings and districts, grid integration of PV systems and AI supported PV forecast.

**Dr. Angèle Reinders** is a full professor of "Design of Sustainable Energy Systems" at the Department of Mechanical Engineering at Eindhoven University of Technology (TU/e). In her research, she focuses on the integration of solar energy technologies in buildings and mobility by means of simulation, prototyping, and testing. A strong interest exists in designs that enhance user acceptance and reduce environmental impact. Angèle Reinders studied experimental physics at Utrecht University, where she also received her doctoral degree in chemistry. Next, she developed herself in industrial design engineering. Besides TU/e's professorship, since 2023, she is the director of Solliance, based in Eindhoven, and she was a full professor of Energy Efficient Design at TU Delft. In the past, she also worked at Fraunhofer Institute of Solar Energy, World Bank, ENEA in Naples, and Center of Urban Energy in Toronto. She is known for her books "Designing with Photovoltaics" (2020), "The Power of Design," and "Photovoltaic Solar Energy from Fundamentals to Applications." In 2010, she cofounded the Journal of Photovoltaics. Also, she initiated many projects and has an extensive publication record.

## References

Alkahtani, A.A., Alfalahi, S.T.Y., Athamneh, A.A. et al. (2020). Power quality in microgrids including supraharmonics: issues, standards, and mitigations. *IEEE Access* 8: 127104–127122. https://doi.org/10.1109/ACCESS.2020.3008042.

K. Araki, A.J. Carr, F. Chabuel, et al. (2021). *"State-of-the-Art and Expected Benefits of PV-Powered Vehicles* 2021."

Bajolle, H., Lagadic, M., and Louvet, N. (2022). The future of lithium-ion batteries: exploring expert conceptions, market trends, and price scenarios. *Energy Research and Social Science* 93: 102850.

Ball, G.J. (2017). Grid-connected PV systems. In: *Chapter 11.1 in Photovoltaic Solar Energy from Fundamentals to Applications* (ed. A. Reinders, P. Verlinden, W. van Sark, and A. Freundlich), 513–529. Wiley.

BEAM *"Sustainable EV Charging, Lowest TCO and Fastest to Deploy | Beam Global."* Available: https://beamforall.com/product/ev-arc-2020/. [accessed 16 December 2022].

Borne, O., Perez, Y., and Petit, M. (2018). Market integration or bids granularity to enhance flexibility provision by batteries of electric vehicles. *Energy Policy* 119: 140–148.

Brinkel, N., AlSkaif, T., and van Sark, W. (2022). Grid congestion mitigation in the era of shared electric vehicles. *Journal of Energy Storage* 48: 103806.

A. Briones, J. Francfort, P. Heitmann, et al. (2012). *"Vehicle-to-Grid (V2G) Power Flow Regulations and Building Codes Review by the AVTA,"* 2012.

Coddington, M., Miller, M., and Katz, J. (2016). In: Grid-integrated distributed solar: addressing challenges for operations and planning, NREL/FS-6A20-63042. Available: https://www.nrel.gov/docs/fy16osti/63042.pdf).

Deline, C. (2017). Inverters, power optimizers, and microinverters. In: *Chapter 11.2 in Photovoltaic Solar Energy from Fundamentals to Applications* (ed. A. Reinders, P. Verlinden, W. van Sark, and A. Freundlich), 530–538. Wiley.

Diouf, B. (2016). A second life for mobile phone batteries in light emitting diode solar home systems. *Journal of Renewable and Sustainable Energy* 8 (2).

FASTNED *"Fastned – Supersnel laden langs de snelweg en in de stad."* Available: https://fastnedcharging.com/nl/. [accessed 16 December 2022].

Freitas Gomes, I.S., Perez, Y., and Suomalainen, E. (2020). Coupling small batteries and PV generation: a review. *Renewable and Sustainable Energy Reviews* 126: 109835.

R. Fu, T. Remo, and R. Margolis et al. (2018). *"U.S. Utility-Scale Photovoltaics-Plus-Energy Storage System Costs Benchmark,"* 2018.

Huang, S., Xiao, J., Pekny, J.F. et al. (2012). Quantifying system-level benefits from distributed solar and energy storage. *Journal of Energy Engineering* 138 (2): 33–42.

*"IEC 62109-2:2011 Safety of Power Converters for Use in Photovoltaic Power Systems – Part 2: Particular Requirements for Inverters,"* 2011.

*"ISO - ISO 15118-1:2019 – Road Vehicles – Vehicle to Grid Communication Interface – Part 1: General Information and Use-case Definition."* Available: https://www.iso.org/standard/69113.html. [accessed 25 July 2022].

*"Lightyear — Solar Electric Vehicle."* Available: https://lightyear.one/. [accessed 22 July 2022].

Litjens, G.B.M.A., Worrell, E., and van Sark, W.G.J.H.M. (2018). Assessment of forecasting methods on performance of photovoltaic-battery systems. *Applied Energy* 221: 358–373.

Luthander, R., Widén, J., Nilsson, D., and Palm, J. (2015). Photovoltaic self-consumption in buildings: a review. *Applied Energy* 142: 80–94.

Mahmoud, K. and Hussein, M.M. (2018). Combined static VAR compensator and PV-inverter for regulating voltage in distribution systems. In: *2017 19th International Middle-East Power Systems Conference, MEPCON 2017 – Proceedings, vol. 2018-February*, 683–688.

MDT-tex *"Solar Carport with Integrated Charging Station | MDT-tex."* Available: https://www.mdt-tex.com/en/products/solar-canopies/solar-carports. [accessed 16 December 2022].

Michael, P., Thet, A., Tun, P. et al. (2021). Facilitating higher photovoltaic penetration in residential distribution networks using demand side management and active voltage control. *Engineering Reports* 3 (10): e12410.

Mulder, G., de Ridder, F., and Six, D. (2010). Electricity storage for grid-connected household dwellings with PV panels. *Solar Energy* 84 (7): 1284–1293.

*"Nissan e-NV200 Winter Camper brings spark to frosty adventures."* Available: https://global.nissannews.com/en/releases/nissan-e-nv200-winter-camper-concept. [accessed 16 December 2022].

Nyholm, E., Goop, J., Odenberger, M., and Johnsson, F. (2016). Solar photovoltaic-battery systems in Swedish households-Self-consumption and self-sufficiency. *Applied Energy* 183: 148–159.

Ortega, Á. and Milano, F. (2018). Frequency control of distributed energy resources in distribution networks. *IFAC-PapersOnLine* 51 (28): 37–42.

Osawa, M., Yoshimi, K., Yamashita, D. et al. (2012). Increase the rate of utilization of Residential photovoltaic generation by EV charge-discharge control. In: *2012 IEEE Innovative Smart Grid Technologies – Asia, ISGT Asia 2012.*

Paired Power *"About Us | Paired Power | World's first 100% solar powered EV charger."* Available: https://www.pairedpower.com/aboutus. [accessed 16 December 2022].

Priessner, A. and Hampl, N. (2020). Can product bundling increase the joint adoption of electric vehicles, solar panels and battery storage? Explorative evidence from a choice-based conjoint study in Austria. *Ecological Economics* 167: 106381.

Raghoebarsing, A., Farkas, I., Atsu, D. et al. (2022). The status of implementation of photovoltaic systems and its influencing factors in European countries. *Progress in Photovoltaics: Research and Applications* 31 (2): 1–21.

Schmutzler, J., Andersen, C.A., and Wietfeld, C. (Dec. 2013). Evaluation of OCPP and IEC 61850 for smart charging electric vehicles. *World Electric Vehicle Journal* 6 (4): 863–874.

M. Sechilariu, A. Reinders, Y. Krim, et al. (2021). *"PV-Powered Electric Vehicle Charging Stations Preliminary Requirements and Feasibility Conditions,"* 2021.

R. Seguin, J. Woyak, D. Costyk, et al. (2016). *"High-Penetration PV Integration Handbook for Distribution Engineers,"* Jan. 2016.

Shafiullah, M., Ahmed, S.D., and Al-Sulaiman, F.A. (2022). Grid integration challenges and solution strategies for solar PV systems: a review. *IEEE Access* 10: 52233–52257.

Sierra Rodriguez, A., de Santana, T., MacGill, I. et al. (2020). A feasibility study of solar PV-powered electric cars using an interdisciplinary modeling approach for the electricity balance, CO2 emissions, and economic aspects: the cases of The Netherlands, Norway, Brazil, and Australia. *Progress in Photovoltaics: Research and Applications* 28 (6): 517–532.

"Solar on every vehicle | Sono Motors." Available: https://sonomotors.com/. [accessed 22 July 2022].

Tamimi, B., Canizares, C., and Bhattacharya, K. (2013). System stability impact of large-scale and distributed solar photovoltaic generation: the case of Ontario, Canada. *IEEE Transactions on Sustainable Energy* 4 (3): 680–688.

TESLA *"Introducing V3 Supercharging."* Available: https://www.tesla.com/blog/introducing-v3-supercharging. [accessed 16 December 2022].

Thygesen, R. and Karlsson, B. (2014). Simulation and analysis of a solar assisted heat pump system with two different storage types for high levels of PV electricity self-consumption. *Solar Energy* 103: 19–27.

Toyota Europe Newsroom *"All-new Toyota Prius Plug-in Hybrid."* Available: https://newsroom.toyota.eu/all-new-toyota-prius-plug-in-hybrid/. [accessed 16 December 2022].

transdev *"100% Sustainably Powered e-Bus to hit Queensland Roads."* Available: https://www.transdev.com.au/press-release/100-sustainably-powered-e-bus-to-hit-queensland-roads/. [accessed 16 December 2022].

Weniger, J., Tjaden, T., and Quaschning, V. (2014). Sizing of residential PV battery systems. *Energy Procedia* 46: 78–87.

Widén, J. and Munkhammar, J. (2013). Evaluating the benefits of a solar home energy management system: impacts on photovoltaic power production value and grid interaction. In: *ECEEE, Presqu'île de Giens*.

Wilkinson, S., Maticka, M.J., Liu, Y., and John, M. (2021). The duck curve in a drying pond: the impact of rooftop PV on the Western Australian electricity market transition. *Utilities Policy* 71: 101232.

# 28

# Floating PV Systems

*Sara M. Golroodbari[1] and Josefine Selj[2]*

[1] *Copernicus Institute of Sustainable Institute, Utrecht University, Utrecht, The Netherlands*
[2] *Department of Renewable Energy Systems, Institute for Energy Technology (IFE), Kjeller, Norway*

## 28.1  Introduction

Ground-based solar and wind are the dominant candidates for massive renewable deployment and are expected to meet at least 50% of the world's electricity demand by 2040 (Gorjian et al. 2021). However, to enable such a renewable-dominated electricity mix, large areas to harvest wind and solar energy is needed (Dotson et al. 2022; Rosa-Clot and Tina 2018). This is a strong motivation for integrated and hybrid systems that enable multi-purpose utilization of land or water surfaces. Building-integrated PV, agrivoltaics, vehicle integrated PV, and floating PV are all solutions that enable new areas to be turned into viable and value-adding commercial solar installations (Cagle et al. 2020). In successful integration, the solar power plant should be in symbiosis with other agricultural or residential use, rather than in conflict (Golroodbari and van Sark 2020; Spencer et al. 2018). As 71% of the Earth's surface is covered with water, floating photovoltaics (FPV) has immense potential and is a particularly attractive solution for countries with high population density and/or shortage of available areas to expand ground-based solar power installations (Kakoulaki et al. 2022; Holm 2017; Mirbagheri Golroodbari et al. 2018). Although FPV is a relatively new technology, its deployment is accelerating with global installed capacity raising from below 1 MW in 2007 to 1314 MW in 2018, and over 2.4 GW at the end of 2019, and is expected to reach approximately 3 GW by 2022 (Kakoulaki et al. 2022; Lee et al. 2020; World Bank 2019). Globally, FPV is expected to grow by an average of 22% year-over-year through 2024, however, most of the projects are installed in Asia, followed by Europe (Kjeldstad et al. 2022; Cox 2019). The dominant market for FPV has so far been relatively calm, inland waterbodies, but new FPV technologies for use offshore are also under development. Offshore floating photovoltaics (OFPV) will be particularly attractive for small island nations, e.g. Malta (Trapani and Millar 2013), the Maldives (Keiner et al. 2022) or highly populated areas in Asia, such as Singapore (Liu et al. 2018).

OFPV can be deployed in a hybrid system, e.g. with an already installed offshore floating wind system (Golroodbari et al. 2021). A theoretical study for an OFPV system with horizontally placed modules on the North Sea about 50 km off the coast showed an annual yield

*Photovoltaic Solar Energy: From Fundamentals to Applications, Volume 2*, First Edition.
Edited by Wilfried van Sark, Bram Hoex, Angèle Reinders, Pierre Verlinden, and Nicholas J. Ekins-Daukes.
© 2024 John Wiley & Sons Ltd. Published 2024 by John Wiley & Sons Ltd.
Companion website: www.wiley.com/go/PVsolarenergy

enhancement of ~13% compared to a land-based system, which is due to increased offshore irradiance as well as cooling of the modules (Golroodbari and van Sark 2020). While there has been vast progress made in this field in recent years (Rosa-Clot and Tina 2020), economic obstacles may hamper the further deployment of OFPV (Trapani and Santafé 2015).

In this chapter, we will first discuss the mechanical and electrical structure of the FPV systems in Section 28.2. Then, in Section 28.3 the most important aspects of the performance and reliability of FPV systems will be covered. In Section 28.4, FPV system modeling will be reviewed very briefly. Finally, the chapter will be warped up in Section 28.5.

## 28.2 Structure and Design

An installed floating PV system combines both PV plant, and floating technologies. Generally, FPV systems consist of the following components: (i) floating platform, (ii) mooring system, and (iii) electrical components.

### 28.2.1 Floating Platform

The floating structure of an FPV system determines the tilt, the motion of the system in response to waves, and the efficiency of module cooling by wind and water. It significantly impacts the cost, life cycle assessment, and system configuration. Also, it is essential for the reliability of the system and which environments the system can withstand. This is why FPV is often categorized based on the floating structure. There are, however, many different classifications for the platform of FPV systems. By some research, the platforms are categorized as rigid and flexible floaters, but more typically, they are classified into two main categories: pontoon-type, and superficial (Claus and López 2022; Soppe et al. 2022).

#### 28.2.1.1 Pontoon-Type
The main idea behind pontoon-type systems relies on designing a raft or pontoon to establish a stable floating platform for the solar modules to be installed on Claus and López (2022) and Rosa-Clot and Tina (2017). Pontoons-based systems are the most widely used today. The dominant FPV technology manufacturers (Sungrow and Ciel et Terre) use pontoon-based floating structures.

#### 28.2.1.2 Pontoon Concept
Pontoons are however rather expensive and used for small and medium-sized systems. These systems, are normally small to medium-sized plants, with special attention to robustness. The Pontoons like the one depicted in Figure 28.1a, are designed for harsh weather conditions with bigger waves and higher wind speeds. Thus, offshore implementation of these systems is extremely promising for the future of FPV.

#### 28.2.1.3 Rafts Concept
A raft itself might be built out of either plastic or plastic and galvanized steel, the latter option is useful to be assembled to construct a wide platform (Rosa-Clot and Tina 2017; Cazzaniga et al. 2018; Choi et al. 2014; Kim et al. 2012).

(a)            (b)

**Figure 28.1** (a) A pilot system of oceans of energy near the Brouwersdam. Source: (Soppe et al. 2022)/Oceans of Energy and (b) pontoon floater from Ciel & Terre International. Source: (ESMAP and SERIS 2019)/World bank group.

Rafts built of plastic and galvanized steel have been implemented in several projects (Cazzaniga et al. 2018), and these systems are particularly successful due to their durability and flexibility. In 2011, the Colignola plant was created to support three modules and has since been adjusted to implement minor material modifications to add high-density polyethylene (HDPE) and polyvinyl chloride (PVC) components (Golroodbari et al. 2022).

Rafts build exclusively of plastic have further evolved to using rafts exclusively produced using plastics. In 2012, the company Ciel et Terre, a leading company in PV innovation, had created and implemented an entirely HDPE system of mono-panel, modular rafts, illustrated in Figure 28.1b. This system significantly reduces cost through ease of production and transportation, and the Installation costs (Rosa-Clot and Tina 2017). Although the concept has been widely commercialized, the overall PV plant faces issues when undergoing various stresses from wind and waves and the flexibility of retrofitting optimizations is extremely difficult.

#### 28.2.1.4 Truss Concept

In this concept, shown in Figure 28.2a, typically there is a space between the bottom of the deck and the water level, called an air gap. This has three consequences: (i) Prevent waves

(a)            (b)

**Figure 28.2** (a) Truss concept for SolarDuck's floating solar demonstrator installed in IJzendoorn. Source: (Soppe et al. 2022)/SolarDuck and (b) fish farm concept from Norwegian company OceanSun. Source: (OceanSun 2020) Ocean Sun AS.

from hitting the deck. (ii) The wind load in the air gap causes the platform to behave like a kite. (iii) The water cooling effect has less impact due to the air gap in the system. This rigid structure is generally walkable, so there is no need for catwalks.

### 28.2.1.5 Superficial

Some plants were designed to have the PV modules resting on the water surface or even partially submerged. Although submerging the panels could be beneficial in terms of wind and wave load reduction, due to the fact that the water is a light absorber, the total available solar energy would be reduced (Claus and López 2022). Superficial types are classified into rigid and flexible concepts.

### 28.2.1.6 Rigid

A rigid sinkable FPV plant was proposed by Clot et al. (2011)based on this design the photovoltaic (PV) modules can submerge up to 2 m to be able to withstand moderate waves. Considering the marine environment for this structure is essential, as it could be submerged in the water.

### 28.2.1.7 Flexible

The flexible strategy has two approaches, namely, using thin-film flexible modules or using crystalline modules backed with flexible foam (Claus and López 2022; Rosa-Clot and Tina 2017). The latter also called the fish farm concept is shown in Figure 28.2b, which builds on experience from near-shore fish farming. A floating ring equipped with a membrane is moored. The solar panels are mounted on the membrane (Soppe et al. 2022).

## 28.2.2 Mooring System

The FPV system should be adjusted with respect to the water level fluctuations while maintaining its position in a specific direction (Choi et al. 2014; Sahu et al. 2016; Sharma et al. 2015). Figure 28.3 shows a generic overview of various mooring concepts that are being used or proposed for offshore floating platforms (Soppe et al. 2022; Davidson and Ringwood 2017). Based on this overview the mooring system can be categorized as (i) single point mooring, and (ii) spread moorings. Moreover, it depends on if the line is tight or loose it may be called taut or slack/catenary, respectively. Also, if the compliance of the mooring system needs to be increased, then subsurface buoys and clump weights can be attached to the mooring line, which is called the lazy-S configuration. Regarding the characteristics of the water, i.e. water depth, tidal changes, water current, etc., the appropriate type of mooring systems can be selected. Also, new mooring concepts compatible with hybrid floating wind and PV are under development (Soppe et al. 2022).

Basically floats are multiple plastic hollows that guarantee the buoyancy and stability of the FPV system. They are either made from medium density polyethylene (MDPE) or HDPE, which is known for its tensile strength, maintenance-free, and ultraviolet (UV) and corrosion resistance. Glass fiber reinforced plastic (GRP) can also be used for the construction of floating platforms (Sahu et al. 2016; Santafé et al. 2014).

The mooring system for the floating platform can be fixed with nylon wire rope slings, which can be tied to bollards on banks and lashed at each corner (Sahu et al. 2016; Santafé

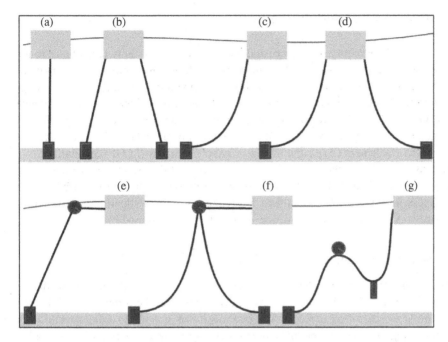

**Figure 28.3** Schematic overview of different mooring concepts: (a) taut; (b) taut spread; (c) catenary; (d) multi-catenary; (e) single anchor leg mooring (SALM); (f) catenary anchor leg mooring (CALM); and (g) Lazy-S. Source:(Davidson and Ringwood 2017)/MDPI/CC BY 4.0.

et al. 2014). An anchoring system holds the floating cover and transmits horizontal forces to the sides of the reservoir.

### 28.2.3 Electrical Components

#### 28.2.3.1 Solar PV Module
PV modules implemented in FPV systems might be flat and rigid modules (Cazzaniga et al. 2012, 2017) or thin-film flexible floating PV (Trapani 2014; Trapani et al. 2013). Almost in all of the FPV systems till now, Si crystalline solar PV modules are used (Sahu et al. 2016). However, for metal corrosion, especially over salty water new designs should be considered.

#### 28.2.3.2 Cables and Connectors
In many FPV systems, cabling is implemented above water (Sahu et al. 2016). However, with an appropriate design, the system could have cables underwater. More considerations are required for cabling in OFPV systems, due to the mechanical loads coming from waves in that circumstances. For electrical devices in an FPV system, it should always be taken into consideration that the water ingress protection (IP) code should be high enough. In Sahu et al. (2016) IP67 is recommended this means the device is both dust-tight and can be immersed in water up to 1 m depth. For IP68 other than being dust-tight, the device can be immersed deeper than 1 m.

## 28.3   Performance and Reliability

As for ground-based PV, the performance and reliability of FPV systems depend on a range of factors, including the quality of all components used, the environment and climate of the site, the handling of components during installation and transport, the system configuration, and the floating structure. The understanding of how different factors influence the performance of the FPV systems is essential to accurately estimating power production. Large uncertainties in predictions of performance directly convert into uncertainties in energy yield assessment (EYA) and levelized cost of electricity (LCOE) calculations.

Certain aspects of the performance and reliability of FPV systems will differ from land-based systems. This includes the shading and soiling loss, the temperature loss, the mismatch loss, and the variation in albedo. For reliability, both the floating structure itself and the stressors it is exposed to, are widely different from land-based PV.

### 28.3.1   Soiling and Shading

For FPV systems, the risk near shading is reduced, and so is generally soiling loss due to dirt and sand. However, soiling loss due to bird droppings can be prominent. Soiling for FPV systems is an active topic of research, to quantify losses, but also to develop technology to prevent soiling or clean the modules more efficiently. However, very little quantitative information about soiling loss rates for FPV systems is available.

### 28.3.2   Mismatch

Mismatch loss in PV systems refers to losses caused by slight differences in the electrical characteristics of interconnected modules. This type of loss also occurs for land-based PV systems, due to differences in string length, nameplate ratings, and shading conditions. However, for FPV, wave-induced movement of the system introduces another origin of mismatch losses. The wave motion can cause modules to be oriented in different directions, hence receiving different irradiance. This results in different maximum power points between the modules and corresponding losses when they are forced to operate at the same current. To separate this effect from the normal mismatch losses, it is often called wave-induced mismatch losses. Wave-induced mismatch losses will depend on the floating structure and configuration of the modules and the sea state. For configurations where all the modules in a string is on the same raft, there will not be differences in orientations between the modules and no wave-induced mismatch will occur. For modules in a pontoon-based system, or with small interconnected rafts, the wave-induced mismatch can be substantial for rough sea states. For calm waters, the mismatch losses are generally small.

### 28.3.3   Operating Temperatures

A frequently stated advantage of FPV over land-based PV is the potentially reduced module temperature due to water cooling, resulting in a higher power conversion efficiency.

FPV has often been reported as having superior performance compared to land-based PV (Golroodbari and van Sark 2020; Sahu et al. 2016; Choi 2014; Lee et al. 2014), but the majority of these claims are either based on comparisons of ground-mounted systems and FPV systems without much monitoring (Choi et al. 2014; Sahu et al. 2016; Lee et al. 2014) or studies that have been undertaken for prototype FPV technologies in direct contact with water, or with floats made from highly conductive materials (Cagle et al. 2020; Golroodbari and van Sark 2020; Ho et al. 2016). Other studies that have found superior performance due to water cooling have utilized water as a cooling agent by either submerging the panels in water, or spraying the panels with water. It is partly this diversity of FPV technologies that have made a general assessment of the cooling effect challenging. In the recently published guidelines from DNV, it is recommended to apply different thermal loss factors for different types of systems and different float technologies. A rough division between FPV technologies where the modules are predominantly cooled by water, and FPV technologies where the modules are predominantly cooled by air is convenient. For modules that are separated from the water by air and/or a floating material with low conductivity, the dominating heat exchange mechanism will be convection. The operating temperature of these modules will, as for land-based systems, predominantly be determined by the mounting structure (which greatly affects the $U$-value), wind, and air temperature. Hence, the effects that reduce the operating temperature of such installations are lower air temperature, and changes to the flow of air underneath the modules. They can be categorized as air-cooled FPV systems. The air-cooled systems can further be subdivided based on the footprint of the floater and how much of the rear side is covered/insulated (Liu et al. 2018).

For FPV technologies where the modules are in direct contact with water, or only separated from the waterbody by a highly heat-conductive material, the dominating heat exchange mechanism will be conduction. As water has significantly higher thermal conductivity than air ($\lambda_{water} = 0.6\,W/m\,K$, $\lambda_{air} = 0.026\,W/m\,K$), the water temperature and water flow will dominate for such architectures. Hence, the operating temperature of the module is therefore likely to depend on the water temperature and water flow in addition to air temperature, wind, and mounting structure. Such FPV technologies can be categorized as water-cooled.

The dominant FPV technologies in today's market, Ciel et Terre and Sungrow, use floats consisting of modular HDPE structures that are linked together and would be categorized as air-cooled FPV technology with a large footprint. Recent publications for this type of floater indicate that the cooling effect is in fact modest (Liu et al. 2018; Mittal et al. 2017; Oliveira-Pinto and Stokkermans 2020) and some studies have even reported a lower module performance on the water compared to land (Kumar et al. 2021). When the reduced operating temperature is observed for air-cooled it is generally due to lower ambient temperatures over water, often reported to be 2–3 °C lower than on land, and higher wind speeds over water bodies. An example of a water-cooled system is the technology developed by Ocean Sun. For this system, located on the West Coast of Norway, it was found that the effect of water cooling gave an increase in yield of 6–10% (Seij et al. 2019). Shown in Figure 28.4, the string of modules that were in direct contact with water was relatively cooler compared to the air-cooled string that was lifted from the canvas.

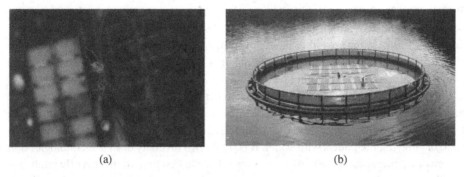

(a)                                (b)

**Figure 28.4** (a) Infrared (IR) image from the installation at Skaftå. The difference in temperature between water-cooled (right) and air-cooled (left) is clearly visible and (b) ocean Sun's pilot located at the west coast of Norway. Source: (Seij et al. 2019).

### 28.3.4 Dynamic Albedo

If one wants to compare, the water albedo is relatively lower with respect to the different land coverage. However, a study about dynamic albedo may change this perspective.

One important factor in dynamic albedo variation is the formation of white caps on the sea surface which are more clearly effective during months with larger solar zenith angles (SZAs). Formation of the whitecap depends on the wind speed threshold, which changes the roughness of the water surface, once the wind speed exceeds this value the albedo starts increasing, while for values below this threshold, the albedo gets smaller. The energy yield of a floating PV system for the two different albedo scenarios has been compared by implementing data of the year 2016 at Hoek van Holland (southwest of the Netherlands) (Golroodbari and van Sark 2022). The calculated PV system performance is larger with dynamic albedo compared with a constant value of $\alpha_{OSA} = 0.06$ (American National Snow and Ice Data (NSID), 2023). The performance has a larger value in all months throughout the year, with an annual average of about 1.03%, without a clear seasonal effect. It is stated that the increased performance is due to the increased Global Tilt Irradiance (GTI). Summing the months, the annual GTI would be 14.8 kWh/m$^2$ higher if the dynamic albedo is considered. This difference will lead to an increase of 13.4 kWh/kWp, which is equal to a 1.03% difference in performance in the offshore FPV system modeled in Golroodbari and van Sark (2020).

### 28.3.5 Reliability

With respect to reliability, FPV systems are very different from land-based systems. The floating structure and mooring are critical and must be suitable for the sea conditions at the site. The stressors experienced by the FPV system will depend greatly on the sea state, and both mechanical loads (wind, wave, mounting), humidity, salinity, and the presence of liquid water can give very different environmental conditions. Currently, no standards exist for modules used in FPV applications. Research on performance loss rates and reliability of FPV systems is still very limited, as a solid analysis of reliability requires a time series of several years. Thus, there is a need for long-term performance and degradation studies of floating PV systems with different technologies to identify these aspects.

### 28.3.6    OFPV in Worldwide Perspective

It is important to understand exactly which sites globally can expect to have the most pronounced advantage for the use of OFPV, and these would then become the sites at which further exploration becomes feasible first. Therefore, 20 port sites across the world, as depicted in Figure 28.5 are chosen by a group of researchers to compare this metric. An hourly temporal resolution for 10 years (between 2008 and 2018) of data based on satellite imagery has been obtained via the NASA POWER service (Sparks 2018). Then, a generic mathematical model of heat transfer is implemented to make the analysis possible.

The analysis of temperature differences between offshore and inland sites for the different locations showed that in almost all of the cases, the module temperatures at offshore sites were lower compared to the land-based sites. The maximum value for the average cell temperature difference is 0.32°C which belongs to Port Coquitlam, shown by number 10 in Figure 28.5, located in Canada. The minimum value is −14°C belonging to Puerto Colombia, as shown by number 16 in Figure 28.5 located in Colombia. The positive value for the Port Coquitlam site could be a result of the effect of the Alaska Current (Walsh et al. 2018). However, the large negative temperature difference at Puerto Colombia is due to a strong outbreak of cold air from the north called "northers" which not only brings the sea surface and ambient temperature on the offshore side down but also may cause gales and very strong waves toward the coastal areas (Devis-Morales et al. 2017; Ortiz-Royero et al. 2013). In Puerto Colombia, specific structural design and extra consideration are needed due to potentially catastrophic weather.

The cell temperature only with wind cooling effect (initial temperature), and final equilibrium operating cell temperature are shown in Figure 28.6a. As expected, for all of the locations the average of the final cell temperature is lower than its initial value. Figure 28.6b shows the average irradiation during the period of this study for all the sites. In 70% of the locations, the average value for irradiation at the offshore site is higher than at the land-based site. The maximum and minimum irradiation level difference between

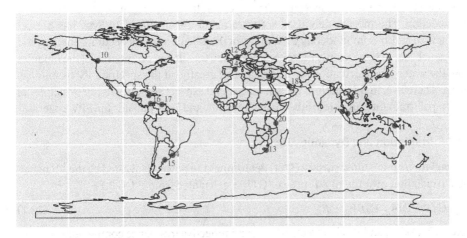

**Figure 28.5**  The locations of the 20 sites for this study.

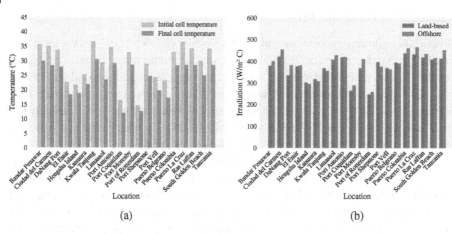

**Figure 28.6** Average (a) initial and final cell temperature of the offshore side and (b) irradiation levels, for all locations for both offshore and inland sides.

the land-based and offshore systems are found for the DaNang Port, 13.94%, and Port Shepstone, −5.42%, shown by number 3 and 13 in Figure 28.5, respectively.

Based on this research, the best and worst-case results for the energy yield advantage belong to DaNang port, and Port Shepstone, respectively. For DaNang port, the irradiation level difference is much higher between land-based and OFPV systems, compared to the Port Shepstone. Moreover, the dry-bulb temperature of DaNang port for the land-based system is relatively higher compared to Port Shepstone. This means that the land-based system in Port Shepstone performs better. Considering this research, the water cooling effect is much more tangible for the environments where minimum temperature is higher (Golroodbari et al. 2023).

## 28.4 FPV System Modeling

There is currently no widely accepted commercial software that considers all aspects of FPV modeling. The most widely used software in the PV business, PVSyst, has not yet implemented features for modeling FPV. Many of the calculations for accurate energy yield assessment become significantly more complex when the tilt and orientation of the modules follow the water. The calculations will also depend greatly on the specific FPV technology utilized. Therefore, in this section, we discuss general aspects and underlying principles of modeling of irradiance, dynamic albedo, dynamic tilt, and heat transfer for FPV systems.

### 28.4.1 Irradiation and Dynamic Albedo

The irradiation on the tilted surface (GTI) determines the power generated by the PV panel. Global irradiation over the tilted surface (GTI) is calculated from Eq. (28.1):

$$GTI = DIR_\phi + DIF_\phi + R_\phi \tag{28.1}$$

where $DIR_\phi$, $DIF_\phi$, and $R_\phi$ are direct, diffuse, and reflective irradiance components, respectively, and $\phi = \{\theta, \gamma\}$, where $\theta$ is surface tilt angle and $\gamma$ is azimuth angle.

**Direct tilted irradiance**, $\text{DIR}_\phi$, is calculated as

$$\text{DIR}_\phi = B_h \times r_b \tag{28.2}$$

where $B_h$ is direct normal irradiance (DNI), and $r_b$ is the direct irradiance conversion factor which is a function of surface tilt, solar zenith, and azimuth angles (Klucher 1979; Gulin et al. 2013).

**Diffuse irradiance**, $\text{DIF}_\phi$, based on anisotropic model called Klucher is estimated as

$$\text{DIF}_\phi = D_h \times R_{\text{dif}}, \quad R_{\text{dif}} \geq 0 \tag{28.3}$$

where $D_h$ is diffuse horizontal irradiance (DHI), and $R_{\text{dif}}$ is diffuse irradiance conversion factor and a function of surface tilt, solar zenith, and azimuth angles (Klucher 1979; Gulin et al. 2013).

**Reflected irradiance**, $R_\phi$, could be calculated via the classical approach modeling assuming that reflected rays are diffuse and coefficients of reflection of the direct and diffuse rays are identical (Gulin et al. 2013). Therefore, we have

$$R_\phi = \rho G_h R_h \tag{28.4}$$

where $\rho$ is the foreground and $R_h = 0.5(1 - \cos\beta)$ is the transposition factor for ground reflection.

In the oceans, the fraction of solar radiation penetrating the subsurface is controlled by the ocean surface albedo (OSA, or $\alpha_{\text{OSA}}$) (Séférian et al. 2018). Despite its importance, OSA is a parameter that receives insufficient attention from both an observational and modeling point of view, and in most studies, it is assumed to be a constant ($\alpha_{\text{OSA}} \simeq 0.06$) for the open ocean surface (Payne 1972). It is also reported that the SZA is the most prominent driving parameter for OSA, for instance in Huang et al. (2019) several parameterizations are compared. Following Séférian et al. (2018), ocean surface albedo $\alpha_{\text{OSA}}$ can be defined as shown in Eq. (28.5).

$$\alpha_{\text{OSA}} = \text{Fr}_{\text{DIR}} \times \alpha_{\text{O,DIR}} + \text{Fr}_{\text{DIF}} \times \alpha_{\text{O,DIF}} \tag{28.5}$$

where $\text{Fr}_i, \alpha_{\text{O},i}\ i \in \{\text{DIR}, \text{DIF}\}$ refer to the fraction of incident radiation and surface albedo, respectively, for direct and diffuse irradiation components. In Golroodbari and van Sark (2022), a detailed study of this topic is presented.

The fraction of whitecaps ($f_{\text{WC}}$) can be generated from the disturbance coming from the breaking of waves due to the wind generating the foam at the sea surface, which can change the albedo considerably (Séférian et al. 2018; Frouin et al. 1976). Equation (28.6) is implemented to formulate the $f_{\text{WC}}$ as a function of wind speed ($v$ [m/s]) at a height of 10 m above the sea surface.

$$f_{\text{WC}}(v) = 3.97 \times 10^{-2} \times v^{1.59} \quad | \quad v \in [2 \quad 20]\,\text{m/s} \tag{28.6}$$

Considering the components of total OSA, i.e. $\alpha_{\text{O,DIR}}$ and $\alpha_{\text{O,DIF}}$, one can calculate the total OSA using the ratio of direct and diffuse irradiation, Eq. (28.5). Like $\alpha_{\text{O,DIR}}$ and $\alpha_{\text{O,DIF}}$ these are wavelength dependent and are shown in Eq. (28.7) (Séférian et al. 2018).

$$\begin{aligned}
\alpha_{\text{O,DIR}}(\lambda, \zeta, v) &= \alpha_{\text{O,DIR}} \times (1 - f_{\text{WC}}(v)) + f_{\text{WC}}(v)\alpha_{Rf_{\text{WC}}}(\lambda) \\
\alpha_{\text{O,DIF}}(\lambda, \zeta, v) &= \alpha_{\text{O,DIF}} \times (1 - f_{\text{WC}}(v)) + f_{\text{WC}}(v)\alpha_{Rf_{\text{WC}}}(\lambda)
\end{aligned} \tag{28.7}$$

Regarding Eq. (28.7) the effect of whitecaps on the direct and diffuse albedo is formulated with $(1 - f_{WC})$ as coefficient which means that when $f_{WC}$ is increasing the effect of whitecaps on albedo is becoming dominant.

### 28.4.2 Dynamic Tilt

Waves impact both the yield and the performance of the FPV system by altering the tilt angles of the modules. To quantify this effect, we need to map the dynamic sea surface condition to the tilt angle of the floaters. A wave can be described using frequency ($f$), wavelength ($\lambda$), time period ($T$), amplitude ($a$), and height ($H$), which is double the amplitude.

A generic mathematical model considering the wave spectrum is discussed in Golroodbari and van Sark (2020). In this method, energy density and power density parameters for irregular waves or real waves are calculated by considering the wave characteristics at a fully developed open sea. As the case study was at the North Sea, the so-called Joint North Sea Wave Project (JONSWAP) spectrum was implemented in that study (Hasselmann et al. 1976; Yu et al. 2017). This spectrum, $S(\omega)$, is a fetch-limited wind wave spectrum, which was developed for the North Sea by the offshore industry and is used extensively.

Looking at the 2D motion for the floater with dimensions shown in Figure 28.7, once the wave crest touches the floater, a perpendicular force makes a torque as in Eq. (28.8):

$$\tau = \vec{F} \times \vec{r_1}, \quad |r_1| = b \tag{28.8}$$

Hence, the variation for $\theta$ can be described as follows:

i) When the crest of the wave touches the floater as shown in Figure 28.8a torque is calculated as mentioned above, which causes the floater to rotate clockwise at an angle $\theta$.
ii) The wave moves forward and touches the middle of the floater, this situation can be translated using linear algebra to the following equation,

$$\tau = \vec{F} \times (\vec{r_1} + \vec{r_2}), \quad |r_1| = |r_2| = b/2 \tag{28.9}$$

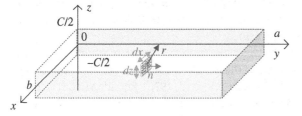

**Figure 28.7** Coordinates and the pontoon dimensions.

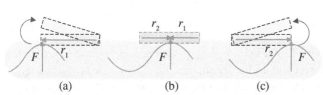

(a)          (b)          (c)

**Figure 28.8** Force and radius vector in different positions when wave crest touches for the first time (a) the left side, (b) middle, and (c) right side, of the floater.

due to the symmetrical characteristics of the floater, the vectors $\vec{r_1}$ and $\vec{r_2}$ are in opposite directions and have equal magnitude so the total torque is equal to zero, which means that the normal vector for the floater at this scenario is $\pi/2$ and the floater makes an angle of $\theta = 0$. In this time window, only the floater altitude could be changed.

iii) This scenario is similar to scenario (a), but the $\vec{r_2}$ is in the opposite direction which makes the floater to rotate anticlockwise at angle $\theta$. In comparison with scenario (a), the absolute value of the tilt angle is the same, but the orientation of the floater is different.

### 28.4.3 Heat Transfer

Air-cooled FPV systems can be fairly well represented in the most commonly used PV simulation software by adapting the $U$-value. Water-cooled FPV systems on the other hand cannot be accurately modeled in the existing commercial software. However, even for air-cooled FPV systems, the most appropriate $U$-values will be technology-dependent and no standard values exist as it does for land-based PV. The current best practice procedure is to determine the thermal loss factors for FPV systems through the application of technology-specific measurements (Det Norske Veritas 2021).

A combination of modeling and experimental data has resulted in a set of widely used $U$-values for different mounting modes for land-based PV systems. PVsyst proposes a $U$-value of 29 W/m² K for freestanding systems, a $U$-value of 15 W/m² K for fully insulated systems, and a value of 20 W/m² K for intermediate cases.

Variations in temperature, both over an individual module, and over the rows of modules, are expected. Figure 28.9 shows simulated temperature variations over a single module and a row of modules in a Ciel & Terre FPV system located on a small dam in South Africa. The temperature variations over an individual module are substantial. The temperature distribution will depend significantly on the wind direction.

A 1D heat transfer model may be implemented to calculate the operating solar cell temperature considering the ambient and sea surface temperatures, for the system consisting of a PV module, floater, and water. The heat conductivity of the floater material which is in direct contact with solar panels plays an important role in this calculation. Thus, for a material with a high heat conductivity, e.g. steel, the initial condition could be assumed as

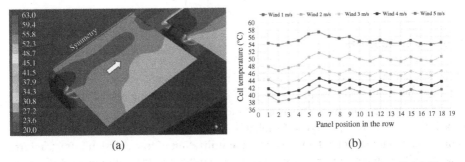

(a)          (b)

**Figure 28.9** (a) The temperature distribution over a single module mounted on a Ciel & Terre floater in South Africa. For symmetry reasons, and ease of calculations, only half the module is included. (b) The temperature distribution over several rows of modules in the same system. Source: (Lindholm et al. 2022)/Elsevier/CC BY 4.0.

the temperature of the floater is equal to the water surface temperature. However, for other materials, e.g. polymers or plastics, it is essential to develop another heat transfer model between the floater and water surface to calculate the floater temperature at time $t = 0$ for the analysis.

As the solar panels during the day will have higher temperatures compared to the floater and sea surface temperature we consider the heat flux flow to go from the solar module toward the water. The equation of the 1D heat transfer analysis is defined as

$$Q_{PV} = Q_{PO} \tag{28.10}$$

where $Q_i$, $i \in$ PV, PO is the heat flow rate (W/m²), and PV and PO denotes the PV module and the pontoon, respectively. The heat flow rates are shown in Eq. (28.10)

$$Q_i = U_i \times A_i \times \Delta T_i \tag{28.11}$$

where $U_i$ is thermal transmittance, $A_i$ is module/pontoon area, and $\Delta T_i$ is the temperature difference.

The $U$-value for the solar module could be estimated using an experimental method. One example of that is discussed in Sánchez-Palencia et al. (2019) where the $U$-value for the module is a function of solar irradiation. Moreover, calculating the $\Delta T_{PV}$ is based on Eq. (28.12),

$$\Delta T_{PV} = T_{PV,i} - T_{PV,eq} \tag{28.12}$$

where $T_{PV,i}$ is the initial cell temperature and $T_{PV,eq}$ is the equilibrium cell temperature, which is also the operating cell temperature (Golroodbari et al. 2022).

Based on these discussions, we need to consider the initial solar cell temperature. To this end, one can either use the measured data from a setup or implement estimation techniques. For the latter option, the effect of the ambient temperature, irradiation, relative humidity, wind speed, etc. should be taken into consideration. In Schwingschackl et al. (2013) several methods for cell temperature estimation are discussed, however, in this study the wind effect played one and only important role. The equilibrium temperature is required to be calculated considering the thermodynamics of the module/floater/water system. In Golroodbari and van Sark (2020), the equilibrium cell temperature is introduced based on considering relative humidity, irradiation, and also wind speed.

## 28.5 Conclusions

In this chapter, the FPV system was studied, which is a promising new technology developed to complement traditional land-based PV systems by making new areas available for renewable energy production. This is critical to achieving the installation rate and scale to reach the global climate targets. Installation of FPV can reduce conflict with residential or commercial use of land and improve the utilization of transmission infrastructure. For some FPV systems, performance is similar to land-based PV, although slightly lower ambient temperatures, higher wind speeds, and water cooling effects can result in relatively lower operating temperatures. Offshore FPV is also a promising solution for maintaining energy security for small islands or nations with comparatively large coastal areas.

The cost of FPV is currently higher than the land-based PV, but as an emerging technology, cost reductions based on economics of scale and continuous improvements are expected. Finally, it is worth considering that FPV is a rapidly growing renewable energy alternative, aiding the energy transition from fossil fuels.

## Author Biographies

**Dr. Sara M. Golroodbari** is an assistant professor at the Copernicus Institute of Sustainable Development, Utrecht University. She did her PhD in solar energy integration at the same institute. Her research focus is on new solar energy integration technologies such as floating PV, Agrivoltaic, BiPV, etc.

**Dr. Josefine Selj** is the principal scientist at the Solar Power Systems department at Institute for Energy Technology. She holds a PhD in electronics and has been working with research on PV and PV systems for the past 15 years. She has had research stays at NREL and Stellenbosch University and 7 years as associate professor II at the University of Oslo. Currently, she is heading the Floating PV research group at IFE and is on the board of the Institute.

## References

American National Snow and Ice Data (NSID) (2023). https://nsidc.org (accessed 2 November 2023).

Cagle, A.E., Armstrong, A., Exley, G. et al. (2020). The land sparing, water surface use efficiency, and water surface transformation of floating photovoltaic solar energy installations. *Sustainability* 12 (19): 8154.

Cazzaniga, R., Rosa-Clot, M., Rosa-Clot, P., and Tina, G.M. (2012). Floating tracking cooling concentrating (FTCC) systems. *2012 38th IEEE Photovoltaic Specialists Conference*, 000514–000519, June 2012.

Cazzaniga, R., Cicu, M., Rosa-Clot, M. et al. (2017). Compressed air energy storage integrated with floating photovoltaic plant. *Journal of Energy Storage* 13: 48–57.

Cazzaniga, R., Cicu, M., Rosa-Clot, M. et al. (2018). Floating photovoltaic plants: performance analysis and design solutions. *Renewable and Sustainable Energy Reviews* 81: 1730–1741.

Choi, Y.-K. (2014). A study on power generation analysis of floating PV system considering environmental impact. *International Journal of Software Engineering and its Applications* 8: 75–84.

Choi, Y.-K., Lee, N.-H., Lee, A.-K., and Kim, K.-J. (2014). A study on major design elements of tracking-type floating photovoltaic systems. *International Journal of Smart Grid and Clean Energy* 3 (1): 70–74.

Claus, R. and López, M. (2022). Key issues in the design of floating photovoltaic structures for the marine environment. *Renewable and Sustainable Energy Reviews* 164: 112502.

Clot, M.R., Clot, P.R., and Carrara, S. (2011). Apparatus and method for generating electricity using photovoltaic panels. US Patent App. 13/061,365, July 14 2011.

Cox, M. (2019). *Floating Solar Landscape*. Wood Mackenzie Power and Renewables. https://www.greentechmedia.com/articles/read/the-state-of-floating-solar-bigger-projects-and-climbing-capacity (accessed 27 December 2023).

Davidson, J. and Ringwood, J.V. (2017). Mathematical modelling of mooring systems for wave energy converters—a review. *Energies* 10 (5): 666.

Det Norske Veritas (2021). Design, Development and Operation of Floating Solar Photovoltaic Systems. *Technical report*. https://www.dnv.com/energy/standards ... (accessed 2 November 2023).

Devis-Morales, A., Montoya-Sánchez, R.A., Bernal, G., and Osorio, A.F. (2017). Assessment of extreme wind and waves in the Colombian Caribbean sea for offshore applications. *Applied Ocean Research* 69: 10–26.

Dotson, D.L., Eddy, J., and Swan, P. (2022). Climate action and growing electricity demand: meeting both challenges in the 21st century with space-based solar power delivered by space elevator. *Acta Astronautica* 198: 761–766.

Energy Sector Management Assistance Program and Solar Energy Research Institute of Singapore (2019). *Where Sun Meets Water: Floating Solar Handbook for Practitioners*. World Bank.

Frouin, R., Schwindling, M., and Deschamps, P.-Y. (1976). Spectral reflectance of sea foam in the visible and near-infrared: in situ measurements and remote sensing implications. *Journal of Geophysical Research: Oceans* 101 (C6): 14361–14371.

Golroodbari, S.Z. and van Sark, W. (2020). Simulation of performance differences between offshore and land-based photovoltaic systems. *Progress in Photovoltaics: Research and Applications* 28 (9): 873–886.

Golroodbari, S. and van Sark, W. (2022). On the effect of dynamic albedo on performance modelling of offshore floating photovoltaic systems. *Solar Energy Advances* 2: 100016.

Golroodbari, S.Z.M., Vaartjes, D.F., Meit, J.B.L. et al. (2021). Pooling the cable: a techno-economic feasibility study of integrating offshore floating photovoltaic solar technology within an offshore wind park. *Solar Energy* 219: 65–74.

Golroodbari, S., Fthenakis, V., and van Sark, W.G.J.H.M. (2022). 1.32 - Floating photovoltaic systems. In: *Comprehensive Renewable Energy*, 2e (ed. T.M. Letcher), 677–702. Oxford:Elsevier.

Golroodbari, S.Z., Ayyad, A.W.A., and van Sark, W. (2023). Offshore floating photovoltaics system assessment in worldwide perspective. *Progress in Photovoltaics: Research and Applications* 31 (11): 1061–1077.

Gorjian, S., Sharon, H., Ebadi, H. et al. (2021). Recent technical advancements, economics and environmental impacts of floating photovoltaic solar energy conversion systems. *Journal of Cleaner Production* 278: 124285.

Gulin, M., Vašak, M., and Baotic, M. (2013). Estimation of the global solar irradiance on tilted surfaces. *17th International Conference on Electrical Drives and Power Electronics (EDPE 2013)*, Volume 6, 347–353.

Hasselmann, K., Ross, D.B., Müller, P., and Sell, W. (1976). A parametric wave prediction model. *Journal of Physical Oceanography* 6: 200–228.

Ho, C.J., Chou, W.-L., and Lai, C.-M. (2016). Thermal and electrical performances of a water-surface floating pv integrated with double water-saturated MEPCM layers. *Applied Thermal Engineering* 94: 122–132.

Holm, A. (2017). *Floating Solar Photovoltaics Gaining Ground*. National Renewable Energy Laboratory.

Huang, C.J., Qiao, F., Chen, S. et al. (2019). Observation and parameterization of broadband sea surface albedo. *Journal of Geophysical Research: Oceans* 124: 4480–4491.

Kakoulaki, G., Gonzalez-Sanchez, R., Amillo, A.G. et al. (2022). Benefits of Pairing Floating Solar Photovoltaic with Hydropower Reservoirs in Europe (National and Regional Level). Available at SSRN 4074448.

Keiner, D., Salcedo-Puerto, O., Immonen, E. et al. (2022). Powering an island energy system by offshore floating technologies towards 100% renewables: a case for the Maldives. *Applied Energy* 308: 118360.

Kim, H.I., Lee, N.H., Lee, E.C. et al. (2012). Groundwork Research for Commercialization of Floated Photovoltaic System. *K-water research report*.

Kjeldstad, T., Nysted, V.S., Kumar, M. et al. (2022). The performance and amphibious operation potential of a new floating photovoltaic technology. *Solar Energy* 239: 242–251.

Klucher, T.M. (1979). Evaluation of models to predict insolation on tilted surfaces. *Solar Energy* 23 (2): 111–114.

Kumar, M., Niyaz, H.M., and Gupta, R. (2021). Challenges and opportunities towards the development of floating photovoltaic systems. *Solar Energy Materials and Solar Cells* 233: 111408.

Lee, Y.-G., Joo, H.-J., and Yoon, S.-J. (2014). Design and installation of floating type photovoltaic energy generation system using FRP members. *Solar Energy* 108: 13–27.

Lee, N., Grunwald, U., Rosenlieb, E. et al. (2020). Hybrid floating solar photovoltaics-hydropower systems: benefits and global assessment of technical potential. *Renewable Energy* 162: 1415–1427.

Lindholm, D., Selj, J., Kjeldstad, T. et al. (2022). CFD modelling to derive u-values for floating PV technologies with large water footprint. *Solar Energy* 238: 238–247.

Liu, H., Krishna, V., Leung, J.L. et al. (2018). Field experience and performance analysis of floating PV technologies in the tropics. *Progress in Photovoltaics* 26: 957–967.

Liu, H., Krishna, V., Leung, J.L. et al. (2018). Field experience and performance analysis of floating PV technologies in the tropics. *Progress in Photovoltaics: Research and Applications* 26 (12): 957–967.

Mirbagheri Golroodbari, S.Z., De Waal, A.C., and Van Sark, W.G. (2018). Improvement of shade resilience in photovoltaic modules using buck converters in a smart module architecture. *Energies* 11 (1): 250.

Mittal, D., Saxena, B.K., and Rao, K.S.V. (2017). Floating solar photovoltaic systems and their feasibility at Kota in Rajasthan. *International Conference on Circuits Power and Computing Technologies*, 1–6.

OceanSun (2020). Ocean Sun: towards a clean energy future.

Oliveira-Pinto, S. and Stokkermans, J. (2020). Assessment of the potential of different floating solar technologies –overview and analysis of different case studies. *Energy Conversion and Management* 211: 112747.

Ortiz-Royero, J.C., Otero, L.J., Restrepo, J.C. et al. (2013). Cold fronts in the Colombian Caribbean sea and their relationship to extreme wave events. *Natural Hazards and Earth System Sciences* 13 (11): 2797–2804.

Payne, R.E. (1972). Albedo of the sea surface. *Journal of the Atmospheric Sciences* 29 (5): 959–970.

Rosa-Clot, M. and Tina, G.M. (2017). *Submerged and Floating Photovoltaic Systems: Modelling, Design and Case Studies.* Academic Press.

Rosa-Clot, M. and Tina, G.M. (2018). Front matter. In: *Submerged and Floating Photovoltaic Systems* (ed. M. Rosa-Clot and G.M. Tina), i–iii. Academic Press.

Rosa-Clot, M. and Tina, G.M. (ed.) (2020). *Floating PV Plants.* Academic Press.

Sahu, A., Yadav, N., and Sudhakar, K. (2016). Floating photovoltaic power plant: a review. *Renewable and Sustainable Energy Reviews* 66: 815–824.

Sánchez-Palencia, P., Martín-Chivelet, N., and Chenlo, F. (2019). Modeling temperature and thermal transmittance of building integrated photovoltaic modules. *Solar Energy* 184: 153–161.

Santafé, M.R., Soler, J.B.T., Romero, F.J.S. et al. (2014). Theoretical and experimental analysis of a floating photovoltaic cover for water irrigation reservoirs. *Energy* 67: 246–255.

Schwingschackl, C., Petita, M., Wagner, J.E. et al. (2013). Wind effect on PV module temperature: analysis of different techniques for an accurate estimation. *Energy Procedia* 40: 77–86.

Séférian, R., Baek, S., Boucher, O. et al. (2018). An interactive ocean surface albedo scheme (OSAv1. 0): formulation and evaluation in ARPEGE-Climat (V6. 1) and LMDZ (V5A). *Geoscience Model Development* 11: 321–338.

Seij, J.H., Lereng, I.H., De Paoli, P. et al. (2019). The performance of a floating PV plant at the west coast of Norway. *Proceedings of 36th European Photovoltaic Solar Energy Conference (EUPVSEC)*, 1763–1767, Marseille, France.

Sharma, P., Muni, B., and Sen, D. (2015). Design parameters of 10 KW floating solar power plant. *Proceedings of the International Advanced Research Journal in Science, Engineering and Technology (IARJSET), National Conference on Renewable Energy and Environment (NCREE-2015)*, Volume 2, Ghaziabad, India.

Soppe, W., Kingma, A., Hoogeland, M. et al. (2022). Challenges and Potential for Offshore Solar. *Technical report*. TKI Wind op ZEE.

Sparks, A.H. (2018). nasapower: a NASA POWER global meteorology, surface solar energy and climatology data client for R. *The Journal of Open Source Software* 3 (30): 1035.

Spencer, R.S., Macknick, J., Aznar, A. et al. (2018). Floating photovoltaic systems: assessing the technical potential of photovoltaic systems on man-made water bodies in the continental United States. *Environmental Science & Technology* 53 (3): 1680–1689.

Trapani, K. (2014). Flexible floating thin film photovoltaic (PV) array concept for marine and lacustrine environments. PhD thesis. Laurentian University of Sudbury.

Trapani, K. and Millar, D.L. (2013). Proposing offshore photovoltaic (PV) technology to the energy mix of the Maltese islands. *Energy Conversion and Management* 67: 18–26.

Trapani, K. and Santafé, M.R. (2015). A review of floating photovoltaic installations: 2007–2013. *Progress in Photovoltaics: Research and Applications* 23: 524–532.

Trapani, K., Millar, D.L., and Smith, H.C.M. (2013). Novel offshore application of photovoltaics in comparison to conventional marine renewable energy technologies. *Renewable Energy* 50: 879–888.

Walsh, J.E., Thoman, R.L., Bhatt, U.S. et al. (2018). The high latitude marine heat wave of 2016 and its impacts on Alaska. *Bulletin of the American Meteorological Society* 99 (1): S39–S43.

World Bank (2019). Where Sun Meets Water: Floating Solar Handbook for Practitioners. *Technical report*. The World Bank.

Yu, Y., Pei, H., and Xu, C. (2017). Parameter identification of JONSWAP spectrum acquired by airborne LIDAR. *Journal of Ocean University of China* 16 (6): 998–1002.

# 29

# Agrivoltaics

*Christian Braun and Matthew Berwind*

*Fraunhofer Institute for Solar Energy Systems, Division Photovoltaics, Freiburg, Germany*

## 29.1 History and Status

### 29.1.1 Introduction

With the growing need for renewable energy sources in the transition to a low-carbon energy grid, competition with other land uses as well as the impact of solar energy projects on local ecosystems is increasing steadily in societal relevance. In this arena, the possibility of using that land for multiple purposes has gained traction in the area of integrated photovoltaics, which in its essence attempts to use every bit of suitable surface area for the distributed generation of green energy. The dual use of land for agricultural purposes and solar electricity generation is commonly denoted as agrivoltaics (AV) or agrisolar, but we will refer to this practice as agrivoltaics in this chapter for detailed definitions (DIN SPEC 91434 n.d.). In the first part of the chapter, the historical development from idea to realized projects in different parts of the world will be outlined.

### 29.1.2 Goetzberger/Zastrow

The first known publication (Potatoes under the collector -"Kartoffeln unter dem Kollektor" (Goetzberger and Zastrow 1981)) was published by the founder of the Fraunhofer Institute of Solar Energy Systems (ISE) in Freiburg, Adolf Goetzberger, and his colleague Armin Zastrow as well as an international proceedings publication (Goetzberger and Zastrow 1981) in the institute's founding year 1981.

The following year, the first international paper from the same authors appeared (Goetzberger and Zastrow 1982) ("On the coexistence of solar energy conversion and plant cultivation"). Then, nothing more was heard of APV until the next German publication titled Energy Farming (Goetzberger 2006) appeared in 2006.

### 29.1.3 First Practical Experiences in Germany

In Bavaria, the electrical engineer Guggenmoos privately experimented since 2005 with plants under solar modules that were mounted on an uplifted static construction (Figure 29.1).

*Photovoltaic Solar Energy: From Fundamentals to Applications, Volume 2*, First Edition.
Edited by Wilfried van Sark, Bram Hoex, Angèle Reinders, Pierre Verlinden, and Nicholas J. Ekins-Daukes.
Companion website: www.wiley.com/go/PVsolarenergy

**Figure 29.1** APV Guggenmoos, Bavaria.

After the first pioneering papers listed above, the later attempts in 2009 and 2010 by ISE to realize APV research projects in Germany were rejected on the grounds of not being innovative enough.

### 29.1.4 Solar Sharing, Japan

Akira Nagashima formulated thoughts about "sharing of sunlight between agriculture and power generation," called it "Solar Sharing" and filed a patent in 2004 which was granted in 2005 (Nagashima 2015). In May 2010, he bought an old farmhouse in Ichihara and completed the first installation with friends' help to show the feasibility of the concept. In June 2011, he started a website called "Solar sharing recommendations" and the concept became more widely known through an article in the Tokyo Shimbun newspaper.

Then, the voluntary "Solar Sharing Promotion Group" published a first book and through contact with former Prime Minister Naoto Kan, Solar Sharing found its way into the "Guidelines to implement Solar Sharing in high-quality farmland (grade A)" by the Ministry of Agriculture, Forestry and Fisheries of Japan.

Under this and other programs, the solar sharing has had widespread success in a large number of small systems. It has been estimated that in 2019, more than 1992 APV farms were installed (Toledo and Scognamiglio 2021), leading to installed capacity of 2.8 GWp in January 2020.

### 29.1.5 France

Christian Dupraz and colleagues developed the Land Equivalent Ratio (LER Mead and Willey 1980; see also Chapter 3) to evaluate the efficiency of AV and AF (Agroforestry) systems (Dupraz et al. 2011). The LER is a measure for comparing the relative yield (relative to the yield in monosystems, i.e. PV systems for power, not for dual land use) of the components. Dupraz in 2011 found LERs in AVs (Dupraz et al. 2011) to be 1.3–1.6.

Furthermore, they compared the efficiency of energy from biomass (liquid biofuels) to energy from the photovoltaic process where the latter was found to be much more efficient (Marrou et al. 2013).

**Figure 29.2** APV plant in Montpellier.

Marrou and colleagues were interested in the effect of partial shading on crop growth (Marrou et al. 2013) as well as on the actual evapotranspiration (AET) from the plants (Marrou et al. 2013). For this, they analyzed lettuce growth under a test system from the University of Montpellier, installed in 2010. They found out that the efficiency of water use can be optimized by carefully selecting crop species and varieties with a rapid soil covering.

The Montpellier APV system has been constructed in designing areas with three different shade densities to be able to directly compare the plant results. From the results of the Montpellier tests, an AV field was built in the east Pyrenean region. This plant with 2.2 MWp PV covers 4.5 ha of vineyards.

Then, the company Akuo Energy started so-called Agrinergie plants on the island of La Reunion with lemongrass plantations between module rows (Figure 29.2). Akuo is developing new concepts for PV-covered greenhouses and taking into account landscape preservation in designing their APV plants (https://www.akuoenergy.com/en/solar/agrivoltaics).

### 29.1.6 Italy

In Italy, the first "Conto Energia"-Law in 2005 introduced a FIT scheme for PV (Sgroi et al. 2014). The following regulatory framework prohibited PV installations on agricultural land. This led to an increase in mounting PV panels on greenhouses in order to avoid this restriction. Starting in 2012, the so-called "agrovoltaico"-systems have been installed in 2012 in Castelvetro Piacentino (1.3 MWp) and Monticelli d'Ongina (3.2 MWp), which were the first ones in Italy with a solar tracking system covering agricultural land (Toledo and Scognamiglio 2021).

### 29.1.7 USA

In 2011, the University of Massachusetts-Amherst started experimenting with elevated solar panels to allow crops to grow underneath and on a commercial level the Convergence Energy Solar Farm in Denavan, Wisconsin started growing sunflowers under elevated PV panels (NREL 2013) (Here, APV is called co-location). On the other hand, the use of brownfields (revegetation, e.g. the SunEdison plant at the NREL site in Colorado), cost saving in the installation of standard PV plants with methods like accommodating existing topography instead of land grading, lessening of soil disturbance, and other environmental factors led to a closer look at co-location (NREL 2019). The University of Arizona researches the growth of vegetables in an AV plant (https://news.arizona.edu/newsletter/041817-ua-researchers-plant-seeds-make-renewable-energy-more-efficient). The state of Massachusetts was the first in the United States to introduce policy exclusive to AVs.

### 29.1.8 China

The Chinese PV industry has rapidly developed as a supplier of PV modules in the European and US markets. This industry ran into major oversupply problems in 2011–2012 when those markets absorbed much fewer products (based on very different causes) from the Chinese production. In 2012, the PV industry was put into the group of emerging strategic industries, and in the years 2013 and 2014, the PV industry saw the implementation of a series of support policies (Xue 2017). As an example, in October 2014, a PV-supporting scheme for poor rural areas and agricultural greenhouses was issued. Since then, AV plants have emerged in the form of PV combined with ecorestauration, crop farming, fishery, and livestock farming. It is estimated that several gigawatts of AV power have been installed, examples are the 30MW installation in covering rice plantations or the 1 GW plant covering goji berry bushes in Binhe new district, Ningxia province.

## 29.2 Technology and Classifications

### 29.2.1 APV System Types

The title of the first scientific publication on AVs "Potatoes under the collector" indicates that the original idea of a dual land use referred to a placement of PV modules above plants to harvest electricity and to cultivate food crops on the ground below (DIN SPEC 91434 n.d.). This leads to the idea that the overhead horizontal arrangement could be considered a "classical" design, but this is by far not the only possible layout that promotes dual land use. Other notable system types include inter-row agriculture with ground-mounted modules or even vertical module arrangements. Another example includes greenhouses outfitted with solar modules, which also face many of the same light limitation issues that classical AV systems are confronted by. These systems can roughly be broken into two categories – interspace and overhead AVs, with example illustrations found in Figure 29.3. With all AV systems, agricultural light availability plays a critical role. Systems can and arguably should be designed with the light requirements of specific plant species taken

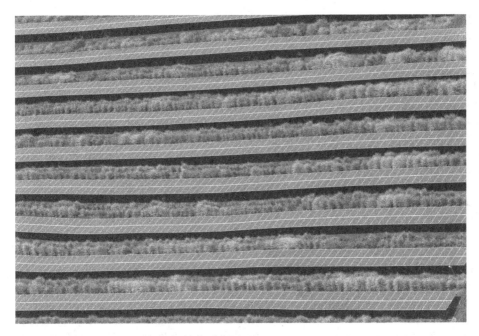

**Figure 29.3** APV plant in Pierrefonds, La Reunion.

into consideration. Without adequate light availability and thus agricultural yields, AVs risk public perception catastrophes. Negative perception events can be avoided, for example, by simulating the impact of AV systems on the availability of direct solar irradiation at crop level with ray-tracing, whereby diffuse light contributions can also be considered using different sky models (Perez et al. 1993) and allowing for considerable reflections in the analysis (Kang et al. 2019).

### 29.2.2 Overhead Agrivoltaics

Overhead PV refers to the cultivation of crops below the PV module rows with tracked or fixed PV modules. The clearance of the AV modules typically depends on the height of the plants being harvested underneath and the size of agricultural machinery used. Typical vertical clearances range between 2 and 6 m. The land use efficiency of overhead PV systems can be higher than that of other PV systems. While the higher cost of overhead PV systems for the support structure remains a challenge in countries with no or insufficient AV support schemes, there is a range of advantages of overhead PV compared to interspace PV systems. The higher vertical clearance of the systems allows the installation of inverters and other electrical components at heights such that potentially no fencing or further security measures are required for insurance purposes. Further benefits result from sheltering effects of PV modules. Additionally, rainwater harvest and irrigation management approaches are likely to be integrated more efficiently compared to other PV systems. Figure 29.3 shows illustrative diagrams of some common systems, including an overhead design.

### 29.2.2.1 Arable Land Use

Open-air farming with plowed, large-area crops can be one of the most technically challenging AV system designs, particularly when considering the techno-economic requirements for feasibility. One bottleneck here is the size of the agricultural machinery used in working the land, irrigation, and harvesting. Agricultural machinery can require clearance heights even above 4 m, with large area harvesters needing vast horizontal breadths (~20 m) as well. This reality mandates robust static constructions that can sustain high wind, snow loads, or other extreme weather events. Achieving such structures with steel truss designs is possible but limits the economic performance of such installations by increasing investment capital. Furthermore, higher steel amounts also have larger $CO_2$ footprints. Static optimization can aid in material use reductions, but other approaches include steel tensioning cables to decrease the amount of steel in construction. Examples of an AV tensile structure system in arable farming are the "Agrovoltaico" pilots (Amaducci et al. 2018).

The aforementioned mechanical loading due to wind or other weather events might also necessitate foundations more robust than what is typically found in ground-mounted PV installations. Ramming is common practice for ground-mounted PV due to the speed and reliability of the construction process, but higher shear and torsional forces on the foundation require increased stability. This might lead project planners to consider concrete foundations, but this is detrimental to agricultural use of the land beyond the photovoltaic system's lifetime. For longevity and agricultural concerns, fully reversible foundations are preferred.

### 29.2.2.2 Greenhouse Photovoltaics

The integration of solar modules in greenhouse systems presents some challenges quite similar to other overhead PV systems, but in general, the role of mounting structures plays a less critical role since large agricultural machinery need not pass under or between any structural elements. That being said, design for light availability is arguably more complicated due to the full roofing enclosure found in many plastic-based greenhouse designs. Until now, the opaque crystalline silicon PV modules were frequently used in greenhouse applications. Their well-known properties and low cost made these modules attractive (Cossu et al. 2020). However, the lack of light transparency of the opaque modules creates challenges for the plant growth and yield because of the reduction in photosynthetic active radiation (PAR) caused by the shading of modules and sunlight heterogeneity inside the greenhouse. Even though the shade tolerance of the plants varies from one plant to another, the cover ratio of the roof by opaque PV panels should not exceed 20–30% in order to not harm crop production remarkably (Emmott et al. 2015).

Furthermore, there are different approaches and studies to improve the light availability and uniformity inside the opaque PV greenhouse such as using different installation patterns of the modules on the roof, rearranging the cell densities and alignment within the module in order to increase the transparency as well as the light uniformity inside the greenhouse (Yano et al. 2010; Li et al. 2018). In fact, the artificial shading is already applied in hot and tropical regions to protect crops from excessive sunlight during the hot months of the year. By doing so, it was reported that the yield and quality of crops enhanced when compared with the ones that were not shaded (Ahemd et al. 2016). Thus, the shade resultant from PV panels could be taken advantage of by an appropriate assessment and implementation. This will not only protect the crops but also reduce the cooling loads in addition to energy generation (Roslan et al. 2018).

Nevertheless, in practice, the lack of the agricultural considerations combined with the willingness to take advantage of governmental subsidies for renewable energy generation has led to PV over-coverage on the roof. This fundamental problem of a combined agricultural and solar power system is not unique to the PV greenhouse brand of AVs and highlights the need for PV greenhouse-specific studies (Cossu et al. 2020).

### 29.2.3  Interspace Agrivoltaics

Interspace AVs refer to the cultivation of crops between PV module rows with tracked or fixed modules. In most cases, the mounting structure is very similar to ground-mounted systems, which can be attractive to existing companies well-established in grid-scale power plants. Interspace cropping systems typically differ from overhead PV AV approaches in that they have lower ground clearances. For this reason, agricultural usage of the land is limited to the free strips between PV rows, with a notable exception being in use cases with grazing range animals.

A special type of interspace cropping is presented by vertical interspace arrangements as seen in Figure 29.3. Here, bifacial PV modules are used on an east–west orientation. This results in power peaks in the morning and evening hours of a day, which can be useful for grid peak smoothing issues typically seen with south-facing PV (Kopecek and Libal 2021). Due to the vertical arrangement of modules, a smaller footprint is associated with vertical AVs compared to other low-ground clearance, interspace AV settings.

### 29.2.4  Module Technologies

The module technologies employed in AV systems are as diverse as the PV sector in general. Due to market factors driving the costs of crystalline silicon modules to highly competitive levels, the majority (approximately 95%) of PV as well as AV installations rely on this technology (Fraunhofer 2020). Some demonstration projects are also being conducted with spatially distributed cell arrangements Figure 29.4 as well as other emerging module technologies that address light availability at the crop level through partial transparencies. Due to the unique geometric conditions of many AV installations, whereby the rear module sides typically would receive increased diffuse irradiation compared to roof-mounted or ground-mounted PV with relatively low ground clearances, bifacial modules can be particularly interesting and present unique economic advantages such as reduced LCOE and BOS costs. This suggests that bifacial modules in AVs might even outpace the general market share of bifacial technologies, which is expected to surpass 35% by 2038 (ITRPV 2018). The most highly commercialized bifacial modules of today revolve around PERC and PERT technologies, with bifacial factors ranging from 70% to 95% dependent on the specific technology (Podlowski 2019; Liang et al. 2019).

### 29.2.5  Emerging Technological Approaches

Novel solar cell and/or module concepts designed specifically around the photosynthetic needs of the agricultural crops in AV systems present potential solutions to fundamental

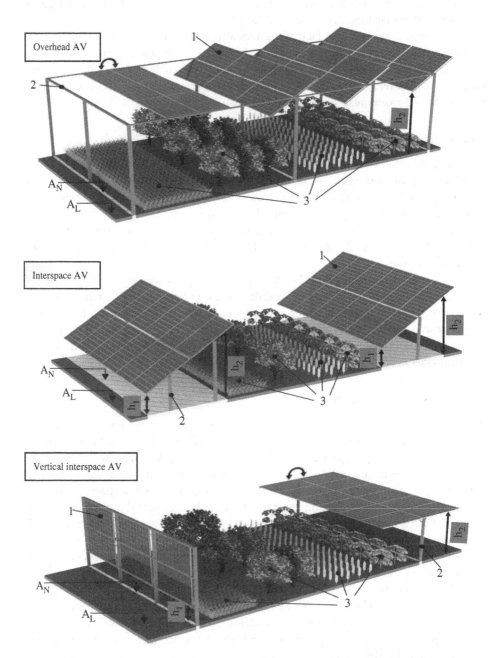

**Figure 29.4** Illustrative examples of overhead and interspace agrivoltaic designs. $A_L$ denotes areas useable for agricultural purposes. $A_n$ denotes non-usable areas. $h_1$ and $h_2$ show clearance heights below and above 2.1 m. 1, 2, and 3 illustrate example PV modules, mounting structures, and crops, respectively.

light concern issues. In general, these lab-ready but not yet fully commercialized technologies attempt to achieve improved light distribution and availability through transparency, be it resultant of system control or intrinsic photonic absorption material properties. These engineered solutions aim to balance light availability for agricultural and photovoltaic purposes so that both subsystems function at acceptable or even ideal levels.

### 29.2.5.1 Wavelength Selective Solar Cells

One subset of emerging AV solar cells is composed of materials that absorb light in wavelengths that are principally less relevant to photosynthesis, that is, photovoltaic absorption is primarily in wavelengths offset from plant chlorophyll or carotenoid (Meitzner et al. 2021; Shi et al. 2019; Peretz et al. 2019). Some candidate solar cell types of this class are organic photovoltaics, that profit from their constituent polymer's optical property tunability, and luminescent solar concentrators (Loik et al. 2017; Corrado et al. 2016), which utilize a luminescent material and waveguides to split different light spectra for different purposes.

### 29.2.5.2 Dynamic Module Technologies/Solar Trackers

Another way of achieving fine-tuned control of available light at plant level is through the use of tracked or otherwise actuated systems. The most readily available examples of this are single- and dual-axis tracking systems, whereby modules typically track the path of the Sun to increase or maximize photovoltaic yield, but the tracking logic behind these systems can also be employed to *increase* agriculturally available light during seasons or stages critical to plant growth (Sun'Agri 2021; Agostini et al. 2021). That might be during the bloom of fruit trees, for example, at the time adequate light levels strongly determine fruiting harvest. These dynamic systems can be integrated with smart monitoring to maintain adequate agricultural conditions within a large meteorological parameter space and are a particularly active area of research.

Another approach implements the actuation of opaque absorbing and transparent module layers to allow light transmittance under specific conditions. These technologies may or may not include concentrating optics to increase PV yield (Figures 29.5 and 29.6) (Insolight 2021; Roccaforte 2021).

**Figure 29.5** An overhead agrivoltaic system installed in Germany showing spatially-segmented cells above rows of apple trees. In the project, APV-Obstbau, eight different apple cultivars are being investigated for their tolerance of the heterogeneous shading conditions. Source: Image copyright of the Fraunhofer Institute for Solar Energy Systems (ISE).

**Figure 29.6** Tracked overhead agrivoltaic system installed in Straßkirchen, Germany. Such systems, this one funded in the European Union project HyPErFarm, reduce system investment costs by replacing steel beams with tensile cables.

## 29.3 Benefits from Dual Land Use

In Section 29.1.5, we introduced a measure to compare the land use efficiency from so-called mono-production (only agricultural or only electrical production) to multiuse (here **dual** use for electricity and agricultural production). Therefore, the LER has been introduced (Mead and Willey 1980) and here we are applying this measure to the case of the APV plant in Heggelbach (Figure 29.7).

The LER is defined as the relative land area that is required for mono-production to achieve the yield of dual land use. LER <1 would indicate that combined crop and electrical production is less than in a mono (separate) production. If we use $a$ for the agricultural and $e$ for the electrical production and deduct 8.3% of the land that has been used for the mounting structure in Heggelbach, we get the following definition:

$$\text{LER [\%]} = \frac{\text{Yield}_a(\text{dual}) - 8.3\%}{\text{Yield}_a(\text{mono})} + \frac{\text{Yield}_e(\text{dual})}{\text{Yield}_e(\text{mono})}$$

Now, the results of the agricultural and electrical production (which is reduced for APV compared to GM-PV) for 2017 and 2018 show a wide variation in the agricultural production. Remarkably, for APV, more than 100% agricultural productivity compared to the mono-use can be reached depending on the crop. For 2018, this resulted in an increased yield of 3%, 11%, and 12% for winter wheat, potato, and celeriac.

Land use ratios in the range of 160% to almost 190% are an extraordinarily high productivity level and show what is possible for the dual use of land with one of the first thoroughly researched APV plants in Germany.

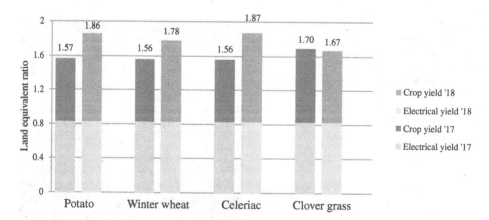

**Figure 29.7** LER of four crops in two consecutive years in Heggelbach.

## 29.4 Conclusion

### 29.4.1 Research Prospects

Deployment of photovoltaic systems in the agricultural sector is globally expanding. As economies of scale drive the costs of decentralized photovoltaic systems down, the economic viability of employing AV systems improves. This can be particularly interesting in developmental areas with inadequate power supply systems, but general plant protection synergies render AVs broadly appealing. Such AV systems can be installed in a horizontal overhead manner such as in open-field applications on arable land or in orchards. They can also be installed in ways similar to typical ground-mounted photovoltaics with interrow spacing increased for agricultural usage, even being erected in a vertical manner to reduce lost agricultural space. Greenhouse integration of AV concepts also promises robust advances in agriculture. While the dual-generation concept of AV systems is generally attractive in areas with land scarcity (Myers et al. 2014; Droulia and Charalampopoulos 2021; Kogo et al. 2021), the complex light distribution beneath or between such systems presents challenges and opportunities for widespread international deployment. As climate change has clear and strong impacts on the agricultural realm, the protective effects seen in AV systems can synergistically protect crop species from excessive temperatures and extreme weather events, leading to improvements in food security.

However, these positive synergetic benefits are not uniformly expected across all plant species. The effects of heterogeneous light conditions, such as the shading found beneath overhead horizontal AV systems, result in changes in agricultural plant yield and quality, dependent on the species' shade tolerance. A current focus in AVs research lies on the phenological indicators that can be used to predict changes in harvest. A better understanding of these effects is necessary for more detailed and accurate system potential analyses as well as yearly yield forecasts, which are necessary to attract investors or design systems to fulfill the yield loss regulations that some countries have passed for AVs.

Apart from enhanced plant knowledge, further research could address the combination of energy storage, organic PV foil, employment of electrical agricultural machines, rainwater harvest, AF, and solar water treatment and distribution. Another vision is "swarm farming" with smaller, automated, solar-powered agricultural machines working under AV systems. Machines for autonomous weeding or the elimination of pests using lasers already exist today – without using chemicals or contaminating soils or groundwater. The mounting structure and availability of electricity of AV systems facilitate the integration of such smart farming elements.

### 29.4.2 Outlook

In the upcoming decades, PV will become one of the main technologies to meet the global renewable energy demand. Climate change and increasing water scarcity demand for mitigation and adaptation measures in agriculture as a sector are both contributing largely to greenhouse gas emissions and suffering from increased temperatures and extreme weather events. With respect to land use competition between PV systems and agriculture, AVs enable an expansion of PV capacity while conserving farmland as a resource for food production. A dual use of farmland considerably increases land use efficiency. Additionally, PV modules can protect soils and crops that are exposed to increasing and more frequent severe weather events such as heat, heavy rain, or drought. AVs can also provide climate-friendly energy to cover the demand of agricultural operations and diversify income of farmers by providing additional revenues through the power generated which, in turn, can strengthen and vitalize rural economies. To tackle the challenges of climate change, it is essential to take the effective steps towards our energy and climate policy goals and to foster resilience in food production. To do so, a wide range of tools must be employed. Considering the opportunities of AVs for climate mitigation and adaptation, there is little doubt that the technology is among the most promising of these tools.

## List of Symbols

| APV | Agri-photovoltaics |
|---|---|
| AVS | Agrivoltaic systems |
| BMY | Biomass yield |
| c-Si | Crystalline silicon |
| GHI | Global horizontal irradiation |
| $G_{hor}$ | Global horizontal irradiation |
| GM-PV | Ground-mounted photovoltaics |
| h | Vertical clearance |
| ha | Hectare |
| kW | Kilowatt |
| kWh | Kilowatt hours |
| LCOE | Levelized cost of electricity |
| MW | Megawatt |
| MWh | Megawatt hours |
| PR | Performance ratio |
| PV | Photovoltaics |
| RT-PV | Rooftop PV |
| Si | Silicon |
| STC | Standard Test Condition |

## Author Biographies

**Dipl. Phys. Christian Braun (M)**, senior project engineer for the Fraunhofer Institute for Solar Energy Systems, develops hard- and soft-ware projects in solar monitoring including statewide solar measurement networks. He has around 150 PV power plants under long-term technical supervision and develops new monitoring concepts and sensor technologies for APV-& FPV-plants and mobile irradiance measuring campaigns.

His prior experience of working eight years for a CPV company has led him to start a research facility in the United States and gain extensive onsite measurement experience including in South Africa and the Middle East.

Christian brings a physics background from the University of Karlsruhe and enjoys developing groundbreaking solar projects with an enthusiastic team of people bringing with them a huge range of knowledge.

**Dr.-Ing. Matthew Berwind (M)**, subject manager for photovoltaic simulations at the Fraunhofer Institute for Solar Energy Systems, focuses on developing modeling methodologies to adapt to the ever-changing solar energy sector. Through numerous public and industry projects, he has implemented simulative approaches to better quantify the dual-yield value proposition presented in AV systems, implementing such techniques in the ISE's yield analysis software Zenit™, which is used for approximately 300–400 industry yield simulations per year. With a further focus on solar trackers and specifically optimizing their control algorithms in bifacial systems, he has guided the construction of and experimentation with six unique pilot systems for AV testing.

Matthew attained a doctorate in engineering from the Albert Ludwig University of Freiburg in microsystems engineering while working in experimental mechanics and metamaterials.

# References

Agostini, A., Colauzzi, M., and Amaducci, S. (2021). Innovative agrivoltaic systems to produce sustainable energy: An economic and environmental assessment. *Applied Energy* 281: 116102.

Ahemd, H.A., Al-Faraj, A.A., and Abdel-Ghany, A.M. (2016). Shading greenhouses to improve the microclimate, energy and water saving in hot regions: A review. *Scientia Horticulturae* 201: 36–45.

Amaducci, S., Yin, X., and Colauzzi, M. (2018). Agrivoltaic systems to optimise land use for electric energy production. *Applied Energy* 220: 545–561.

Corrado, C., Leow, S.W., Osborn, M. et al. (2016). Power generation study of luminescent solar concentrator greenhouse. *Journal of Renewable and Sustainable Energy* 8 (4): 043502.

Cossu, M., Yano, A., and Solinas, S. et al. (2020). Agricultural sustainability estimation of the European photovoltaic greenhouses. *European Journal of Agronomy* 118: 126074.

DIN. Agri-photovoltaic systems (2021). Requirements for primary agricultural use: English translation of DIN SPEC 91434: 2021–05; 2021.

Droulia, F. and Charalampopoulos, I. (2021). Future climate change impacts on European viticulture: A review on recent scientific advances. *Atmosphere* 12 (4): 495.

Dupraz, C., Marrou, H., Talbot, G., Dufour, L., Nogier, A., and Ferard, Y. (2011). Combining solar photovoltaic panels and food crops for optimising land use: Towards new agrivoltaic schemes. *Renewable Energy* 36 (10): 2725–2732.

Dupraz, C., Talbot, G., Marrou, H. et al. (2011). To mix or not to mix: evidences for the unexpected high productivity of new complex agrivoltaic and agroforestry systems. In: *5th World Congress of Conservation Agriculture incorporating 3rd Farming Systems Design Conference*, September 2011, Brisbane, Australia, 202–203.

Emmott, C.J.M., Röhr, J.A., Campoy-Quiles, M. et al. (2015). Organic photovoltaic greenhouses: a unique application for semi-transparent PV? *Energy & Environmental Science* 8 (4): 1317–1328.

Fraunhofer ISE. (2020). *Photovoltaics Report.*

Peretz, M. et al. (2019). Testing organic photovoltaic modules for application as greenhouse cover or shading element. *Biosystems Engineering* 184: 24–36.

Goetzberger, A. (2006). Energy farming. *Solarzeitalter.*

Goetzberger, A. and Zastrow, A. (1981). Kartoffeln unter dem Kollektor. *Sonnenenergie* 3.

Goetzberger, A. and Zastrow, A. (1981). On the simultaneous use of land for solar energy conversion and agriculture. *Proceedings of the International Solar Energy Society Congress.*

Goetzberger, A. and Zastrow, A. (1982). On the coexistence of solar-energy conversion and plant cultivation. *International Journal of Solar Energy* 1: 5569. https://doi.org/10.1080/01425918208909875.

Insolight (2021). *Technical Features.* https://insolight.ch/technology/ (accessed 28 February 2024).

ITRPV (2018). *International Technology Roadmap for Photovoltaic (ITRPV) – 2017 Results.*

Kang, J., Jang, J., Reise, C., and Lee, K. (2019) Practical Comparison Between view factor method and raytracing method for bifacial PV system yield prediction. In: *Proceedings of the 36th European Photovoltaic Solar Energy Conference and Exhibition,* (9-13 September 2019), Marseille, France.

Kogo, B.K., Kumar, L., and Koech, R. (2021). Climate change and variability in Kenya: a review of impacts on agriculture and food security. *Environment, Development and Sustainability* 23: 23–43.

Kopecek, R., and Libal, J. (2021). Bifacial photovoltaics 2021: status, opportunities and challenges. *Energies* 14 (8): 2076.

Li, G., Xuan, Q., Pei, G., Su, Y., and Ji, J. (2018). Effect of non-uniform illumination and temperature distribution on concentrating solar cell-a review. *Energy* 144: 1119–1136.

Liang, T.S., Pravettoni, M., Deline, C. et al. (2019). A review of crystalline silicon bifacial photovoltaic performance characterisation and simulation. *Energy & Environmental Science* 12 (1): 116–148.

Loik, M.E. et al. (2017). Wavelength-selective solar photovoltaic systems: powering greenhouses for plant growth at the food-energy-water nexus. *Earth's Future* 5 (10): 1044–1053.

Macknick, J., Beatty, B., and Hill, G. (2013). *Overview of opportunities for co-location of solar energy technologies and vegetation* (No. NREL/TP-6A20-60240). National Renewable Energy Lab. (NREL), Golden, CO (United States).

Marrou, H., Wery, J., Dufour, L., and Dupraz, C. (2013). Productivity and radiation use efficiency of lettuces grown in the partial shade of photovoltaic panels. *European Journal of Agronomy* 44: 54–66. https://doi.org/10.1016/j.eja.2012.08.003.

Marrou, H., Dufour, L., and Wery, J. (2013). How does a shelter of solar panels influence water flows in a soil–crop system? *European Journal of Agronomy* 50: 38–51. https://doi.org/10.1016/j.eja.2013.05.004.

Mead, R. and Willey, R.W. (1980). The concept of a 'land equivalent ratio' and advantages in yields from intercropping. *Experimental Agriculture* 16: 217–228. https://doi.org/10.1017/S0014479700010978.

Meitzner, R., Schubert, U.S., and Hoppe, H. (2021). Agrivoltaics—the perfect fit for the future of organic photovoltaics. *Advanced Energy Materials* 11 (1): 1–7.

Myers, S.S., Zanobetti, A., Kloog, I., et al. (2014). Increasing $CO_2$ threatens human nutrition. *Nature* 510 (7503): 139–142.

Nagashima, A. (2015). *Solar Sharing*. Tokyo: Access Int.

NREL (2019). *Co-location of Agriculture and Solar*. https://www.nrel.gov/docs/fy19osti/73696
.pdf (accessed 28 February 2024).

Perez, R., Seals, R., and Michalsky, J. (1993). All-weather model for sky luminance
distribution–preliminary configuration and validation. *Solar Energy* 50 (3): 235–245.

Podlowski, L. (2019). *Bifacial PV Technology: Ready for Mass Deployment*. PI Berlin.

Roccaforte, G. (2021). *Agrivoltaics 2021, AIP Conference Proceedings*.

Roslan, N., Ya'acob, M.E., Radzi, M.A.M. et al. (2018). Dye Sensitized Solar Cell (DSSC)
greenhouse shading: New insights for solar radiation manipulation. *Renewable and
Sustainable Energy Reviews* 92: 171–186.

Sgroi, F., Tudisca, S., Trapani, A. et al. (2014). Efficacy and efficiency of Italian energy policy:
the case of PV systems in greenhouse farms. *Energies* 7: 3985–4001. https://doi.org/10.3390/
en7063985.

Shi, H., Xia, R., Zhang, G. et al. (2019). Spectral engineering of semitransparent polymer solar
cells for greenhouse applications. *Advanced Energy Materials* 9 (5): 1–8.

Sun'Agri (2021). *Agrivoltaics Viticulture*. https://sunagri.fr/en/project/nidoleres-estate/
(accessed 28 February 2024).

Toledo, C. and Scognamiglio, A. (2021). Agrivoltaic systems design and assessment: a critical
review, and a descriptive model towards a sustainable landscape vision (three-dimensional
agrivoltaic patterns). *Sustainability* 13: 6871. https://doi.org/10.3390/su13126871.

Tonking New Energy (2018). *Changshan PV station*. http://tonkingtech.com/english/news_
show.aspx?newsCateid=117&cateid=117&NewsId=137. Accessed 9 February 2018.

Xue, J. (2017). Photovoltaic agriculture – new opportunity for photovoltaic applications in
China. *Renewable and Sustainable Energy Reviews* 73: 1–9. https://doi.org/10.1016/j.rser
.2017.01.098.

Yano, A., Kadowaki, M., Furue, A. et al. (2010). Shading and electrical features of a
photovoltaic array mounted inside the roof of an east-west oriented greenhouse. *Biosystems
Engineering* 106 (4): 367–377.

# 30

# Vehicle Integrated Photovoltaics

*Neel Patel[1,2], Karsten Bittkau[1], Kaining Ding[1], Wilfried van Sark[3], and Angèle Reinders[2]*

[1] *Forschungszentrum Jülich, IEK 5, Photovoltaics, Jülich, Germany*
[2] *Eindhoven University of Technology, Energy technology, Eindhoven, The Netherlands*
[3] *Copernicus Institute of Sustainable Development, Utrecht University, Utrecht, The Netherlands*

## 30.1 Introduction

Photovoltaics (PV) has become a mainstream source of energy generation (IEA 2020), bringing a dramatic price reduction and efficiency gains due to economies of scale (ITRPV 2022). This has led to a rethinking of PV in niche applications that were not viable. Vehicle-integrated PV (VIPV) is one such application. Its prototype dates back to the 1950s when a vintage model 1912 Baker electric car was equipped with a PV panel of around 10,640 cells for energy generation (Automostory 2022). VIPV never materialized since the PV technology of the time was costly and inefficient at converting irradiance to usable power. Also, the automotive industry was dominated by internal combustion engine vehicles, making the integration of PV into the vehicle redundant. Since electric vehicles (EVs) are becoming mainstream (IEA 2022) and PV has become cheaper and more efficient, VIPV has gained some traction. Although our majority focus for this chapter is VIPV for road transport vehicles, VIPV applications also exist for boats and airplanes.

For most parts of the world, the road transport sector faces two significant hurdles: greenhouse gas (GHG) emissions and energy security. As seen in Figure 30.1, the road transport sector contributes considerably to the total world (GHG) emissions (Ritchie et al. 2020). Geopolitical tensions can lead to dwindling fuel supply and higher fuel prices, affecting energy security. Battery-EVs are seen as a solution to both of these problems. However, an EV's emissions are as clean as a nation's electricity grid, ranging up to 0.275 kg $CO_2$-equivalent per kWh of electricity in the European Union (EEA 2022). Since the majority of the nation's electricity grids still rely on fossil fuels, which also raises the question of energy security, and transitioning them to sustainable energy sources will take time, VIPV can be viewed as a part of a solution. Energy generated by VIPV will lead to a lowering of $CO_2$ emissions and can be utilized directly for driving.

*Photovoltaic Solar Energy: From Fundamentals to Applications, Volume 2*, First Edition.
Edited by Wilfried van Sark, Bram Hoex, Angèle Reinders, Pierre Verlinden, and Nicholas J. Ekins-Daukes.
© 2024 John Wiley & Sons Ltd. Published 2024 by John Wiley & Sons Ltd.
Companion website: www.wiley.com/go/PVsolarenergy

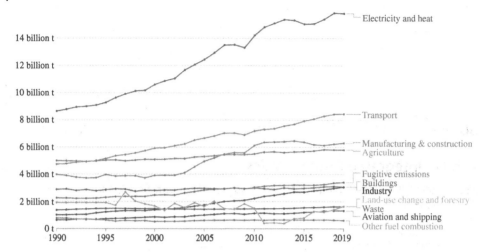

**Figure 30.1** Global annual GHG emissions for different sectors in $CO_2$-equivalents (in tons).

## 30.2 State-of-the-Art: Industry

PV for the automotive sector can be broadly divided into VIPV and vehicle-added PV (VAPV). VIPV is a part of the structural design that is aesthetically more pleasing and generates electricity, whereas VAPV's sole purpose is electricity generation, not aesthetic design. A similar segmentation exists for building integrated PV (BIPV) and building added PV (BAPV), which can be used as an analogy further to explain the more nuanced differences between VIPV and VAPV and can be found in Ritzen et al. (2017). Based on these broad definitions of VIPV and VAPV, the products announced for the automotive sector can be further divided into original equipment manufacturer (OEM) products (which are purpose-built for a specific vehicle) and after-market kits or services. The OEM products are passenger cars, while after-market kits or services are available for caravans and light and heavy commercial vehicles.

Established automotive manufacturers like Toyota and Hyundai have already announced VIPV options for their vehicles. Toyota Prius has an option of a 180 $W_P$ PV panel, which can charge the traction battery while parking and the auxiliary battery while driving (Lambert 2016). Hyundai released the IONIQ 5 with a solar roof option (McCann 2021). Start-up companies like Sono Motors, Lightyear, and Aptera are also entering the market with announcements of purpose-built VIPV products. Sono Motors has announced the Sion passenger car, as seen in Figure 30.2, with a 1200 $W_P$ VIPV system of 456 half-cut cells capable of directly charging the traction battery of 54 kWh capacity (Sono Motors 2022). Sono Motors also offers after-market PV products for commercial vehicles like vans, trucks, and buses. Lightyear has announced Lightyear 0, as seen in Figure 30.2, with a 1005 $W_P$ VIPV system with interdigitated back contact (IBC) cells and a battery capacity of 60 kWh (Lightyear 2022). Aptera has announced a highly aerodynamic car with a 700 $W_P$ VIPV charging system (Aptera 2022).

Apart from OEMs, after-market service providers tailor a VAPV solution on a case-by-case basis or provide plug-and-play VAPV products. IM efficiency (IM Efficiency 2022) offers

**Figure 30.2** Sono Sion (left, Credit: Sono Motors GmbH) and Lightyear 0 (right, Credit: Lightyear).

a business case for the VAPV installation and hardware and installation service for truck trailers. Solbian (Solbian Solar 2022) provides components, from PV modules and charge controllers to adhesives needed to add PV to different vehicles. There are also niche VIPV solution providers like Gochermann-solar (Gochermann Solar Technology 2022) and more, which custom design modules per the necessity for a specific vehicle, such as a solar race car or a yacht.

Most VIPV/VAPV products mentioned mainly use silicon PV technology, which provides a better performance and price trade-off than other contemporary PV technologies. The products and service providers discussed here are just a part of a growing list. More details about onboard vehicle PV products and prototypes can be found in Commault et al. (2021). There is also an ongoing collaboration effort called Alliance for Solar Mobility (ASOM) (ASOM 2022) among various companies and research institutes to discuss the status of onboard vehicle PV, which provides a good overview of the industry's progress.

## 30.3 State of the Art: Academia

The upward trend in the VIPV field can also be found in academia, with the number of journal and conference articles increasing in recent years. Research is ongoing on different aspects of VIPV, like module technology, control electronics design, performance and energy yield modeling, measurements and standardization, and analysis of potential benefits of VIPV, to name just a few.

### 30.3.1 Module Technology

PV modules for vehicles have unique requirements compared to steady-state applications: they should be highly efficient, bendable (in the case of passenger cars), relatively lightweight (especially for commercial vehicles), aesthetically matching and pleasing, tolerant of fast dynamic shading during vehicle motion (requiring special maximum power point tracking strategies) and partial shading during vehicle parking, and highly durable to withstand the stresses and vibrations during vehicle motion to name a few. Silicon-based PV technologies have been dominating so far in the space of VIPV. Most of the announced VIPV/VAPV products use some variation of silicon technologies like passivated emitter and rear cell (PERC), IBC, and silicon heterojunction (SHJ) with efficiencies in the range of 17–22%. Research has been ongoing to address the requirements mentioned earlier. Hayakawa et al. (2021) studied the effect of bypass diodes (BPDs) on 3D curved modules. They conclude that the effectiveness of BPDs is affected by module curvature radius and the diffuse component of the irradiance. For a smaller curvature radius and diffuse irradiance component, a higher number of BPDs results in a higher power. However, if the curvature radius and diffuse irradiance component increase, BPDs have a negligible effect. They argue that the design of the curved VIPV modules should be optimized using BPDs and the series and parallel connection of the modules' cells. Tayagaki et al. (2021) also highlight this sentiment, arguing that the design process should rely on the curved surface and solar irradiation profile. Yamaguchi et al. (2021) and Yamaguchi et al. (2021) have analytically studied the low light performance and temperature coefficients of several PV cell technologies and concluded that the III–V-based InGaP/GaAs/InGaAs 3-junction cell has the best performance. The PV coloring method described in Blasi et al. (2021) has been used to make VIPV module prototypes, as seen in Coloured PV, Fraunhofer ISE (2022). Yamada (2020) has shown a method to shape-tailor PV modules for VIPV applications using triangle-cut Si and CIGS cells. Kim et al. (2021) study different materials and approaches to make lightweight PV modules for VIPV applications. Kutter (2020) has proposed an integrated PV module technology and manufacturing process for commercial vehicles. Their technology has a glass fiber-reinforced plastic structure that can be integrated directly into commercial vehicles box bodies.

### 30.3.2 Control Electronics

Control electronics consist mainly of the maximum power point tracker (MPPT), which tracks the module's power output under ambient conditions. Wetzel et al. (2021) have shown that the irradiance falling on the VIPV modules is highly variable during the driving state. They report that the changes in irradiance typically have a frequency of 1 Hz but can sometimes reach frequencies as high as 100 Hz. This directly impacts the selection of the switching frequencies of the MPPT circuit. Kano et al. (2021) have shown a circuit design of a fast MPPT and battery charge controller for a solar race vehicle with an efficiency of 98.5%. However, it is still an ongoing research topic to determine the best switching frequency of VIPV MPPT, fast or slow, which could satisfy the tracking needs while driving and parking.

### 30.3.3 Measurement Campaigns

Available measurements from traditional stationery PV systems can only sometimes be used to estimate VIPV potential as the vehicle position keeps changing, needing new measurements. Measurement parameters of interest for VIPV are mainly the vehicle position, irradiance on all sides, temperature, and wind speed. A description of such a measurement system can be found in Ota et al. (2019). Analysis from solar irradiance measurement campaigns can be found in Eitner et al. (2017) and (Ota et al. (2021). Peibst et al. (2021) published an energy flow analysis of a demonstrator vehicle outfitted with PV modules and electronic components necessary to inject VIPV energy yield into the vehicle traction battery. Sovetkin et al. (2022) published the first known open-source VIPV irradiance measurement dataset in the form of a data challenge. PV2Go (PV2Go 2022) is a citizen measurement campaign in which measurement kits are provided to private citizens for measurements, and data is collected and stored centrally. Lade-PV (Lade-PV 2022) is another demonstrator project where a heavy commercial vehicle is outfitted with a VIPV system and measurement setup. Results from these campaigns have shown that the available irradiance for the VIPV system fluctuates significantly from the stationery systems, at least for the vehicle motion phase.

### 30.3.4 Energy Yield

Energy yield modeling of VIPV systems consists of multiple calculation steps, where irradiance modeling has the highest complexity, given continuous change in location, fast dynamic shading from changing terrain, and curved nature of the plane of the array. Various approaches have been proposed to calculate the irradiance values on a curved or flat module. Araki et al. (2020) and Mont et al. (2020) have proposed methods to estimate the irradiance distribution on a curved module without considering the shading impact from the surrounding terrain. Centeno Brito et al. (2021) have used the Area Solar Radiation model in ArcGIS (ArcGIS 2022) and the digital surface model obtained using LiDAR to calculate the VIPV irradiation potential of Lisbon, Portugal. Sovetkin et al. (2021) have made an open-source modeling framework for irradiance calculation using LiDAR data. Kühnel et al. (2017) have proposed a complete system simulation scheme for the VIPV system without considering the shading impact on power production. VIPV temperature modeling efforts are also presented by Wheeler et al. (2019) and Hayakawa et al. (2022).

### 30.3.5 Testing and Standardization

In terms of testing and standardization procedures, VIPV has to pass a double test where it must satisfy not only the standard PV industry regulations but also the stringent automotive regulations before becoming a truly mass-market technology. Several efforts are being put into standardizing the VIPV technology. Markert et al. (2021) proposed a safety qualification program for VIPV modules, combining the testing procedures from regular PV testing and the automotive sector. Araki et al. (2018) published a list of recommendations for performing irradiation measurements for VIPV modules, environmental and qualification testing, and power and energy prediction. A collaboration among researchers from various institutes was formed as Task 17 of the IEA-PVPS (IEA-PVPS Task 17 PV & Transport 2022), providing an overview and status of the progress (Araki et al. 2021).

### 30.3.6 Social Acceptance and Sustainability

For any new technology to be successful, high public adoption is necessary. Surveys of a representative demographic population analyze public sentiment to determine their willingness to pay (WTP) for a particular technology. Ghasri et al. (2021) conducted such a social acceptance survey in Australia for a VIPV option for EVs. Their analyses indicate that the WTP for a VIPV option is 18.13$/(km/day) of added range and 1383.98$ for a VIPV system with an aesthetically matching color option. The cost of such a system is high but still relatively low compared to the vehicle cost. Since the combination of PV and EVs is promoted as a sustainable technology, it is essential to determine its environmental impact in various scenarios. Quantifying whether charging with VIPV would have lower global warming potential than grid charging scenario analyses for specific use cases is crucial. Kanz et al. (2020) did such an analysis for a light commercial EV with a 930 Wp PV system in Germany for an operational period of eight years. Considering a shading loss of 30%, it was estimated that the emissions from an onboard PV system would be 0.357 kg $CO_2$-equivalent for a kWh of generated electricity compared to 0.435 kg $CO_2$-equivalent for a kWh of electricity from the German electric grid.

## 30.4 Potential Impact

The electric passenger vehicles projected to be sold in 2030 according to the Stated Policies Scenario (STEPS) and Announced Pledges Scenario (APS) of IEA (IEA 2022), as seen in Figure 30.3, give us a sense of how big of a market there could be for VIPV. Assuming a module power of 200 W/m² and roof sizes of 20, 25, 10, and 2 m² for buses, trucks, vans, and cars, respectively, the combined VIPV demand in 2030 could be as high as 18 $GW_p$ for STEPS and 26 $GW_p$ for APS assuming all of these vehicles come equipped with PV systems. However, it should be noted that these numbers show a maximum potential for VIPV demand and are in no case a forecast of it.

Although VIPV demand will be much smaller than overall PV demand, it will likely be a high-value market.

Various research groups and automotive companies have estimated range extension numbers for VIPV. Heinrich (2020) calculate a solar range of around 1900–3400 km or 13–23% of the average annual driving distance of 15,000 km in Freiburg, Germany, using a 583 $W_p$ VIPV system. Analysis by Carr (2020) estimates a solar range of 3300 km in Amsterdam, 3000 km in Oslo, and 5200 km in Melbourne using an 800 $W_p$ VIPV system on a Tesla Model 3 vehicle. Centeno Brito et al. (2021) calculated the irradiation potential for roads and parking spaces in Lisbon and reported that for a driving pattern with 30 km/day of travel with an EV with a consumption of 13 kWh/100 km and a one kWp VIPV system, a 33% reduction in energy demand could be expected. Sono Sion (Sono Motors 2022), with its 1.2 kWp VIPV system, is projected to add an average of 112 km of the range weekly, which amounts

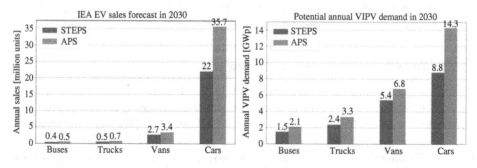

**Figure 30.3** IEA EV sales forecast and corresponding VIPV demand.

to around 44% annual mileage coverage in Munich, Germany. Lightyear 0 (Lightyear 2022) claims to add around 11,000 km of annual range in southern Spain. Aptera (Aptera 2022) reports about 48 km of daily solar range in a location like California. The added range numbers reported here from various sources are just theoretical maximum estimates. The actual range extension numbers are expected to be lower, considering the number of dynamics associated with VIPV energy yield and geographical location (or parking in a parking garage). Nevertheless, VIPV exemplifies a symbiotic relationship between the PV and automotive sectors.

Another advantage of PV-powered vehicles is their characteristic of resilience in natural disasters. A PV-powered EV is a portable power-generating source. This sets them up to be a good utility during a disaster event when the electric grid stops working. Also, with the vehicle-to-grid (V2G) capabilities of modern EVs, a VIPV power source could help power a household (vehicle-to-X, V2X) during such events in conjunction with a probable rooftop PV system.

Since VIPV is in its infancy, more accurate analyses of the potential benefits and pitfalls of the technology are needed. These analyses could help make informed investments in VIPV technology. Making technology prototypes to analyze various scenarios is always an option, but being cost and time-intensive, the development of simulation models could provide an effective solution. Since VIPV differs from traditional steady-state PV applications in several ways, new validated simulation tools to estimate the energy yield of the VIPV system are needed to execute such analysis. The state of VIPV simulation tools is highlighted in the next section.

## 30.5 Energy Yield Modeling

VIPV energy yield modeling can be done by bottom-up and top-down approaches. The bottom-up approach represents a simulation of an individual vehicle to estimate the VIPV energy yield. The steps to do so are shown in Figure 30.4. In a top-down approach, rather than modeling individual vehicles, the VIPV yield potential of a region is estimated for roads and parking places. Centeno Brito et al. (2021) shows an example of this approach. This section further explains the bottom-up approach.

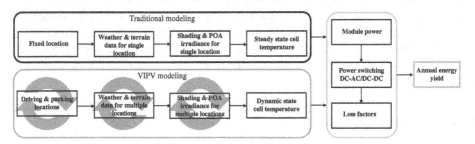

**Figure 30.4** Modeling flow diagram for traditional and VIPV systems. The green arrows in the VIPV modeling section indicate the repetition of respective steps for each location the vehicle has been to. The calculations of irradiance and cell temperature differ significantly for both systems, but from thereon, the calculations have many similarities.

VIPV has two distinct operating states, dynamic and quasi-static, where dynamic is when the vehicle is in motion, and quasi-static is the vehicle parking for which a fixed location cannot always be guaranteed. These two operating states are quite different from the traditional PV system, where once the site is selected, it will stay in place for the entirety of its operational life. This fundamental difference leads to higher complexity in simulating VIPV system energy yield and necessitates a precise location of the vehicle and the elevation of the surroundings to estimate the amount of shade that will be cast on the PV modules throughout the day in a similar manner as with BIPV simulations. This can be achieved by tapping into the vast amount of geospatial data available in the public domain. Figure 30.4 depicts the critical differences in the modeling steps required for traditional PV and VIPV systems. The significant complexity is in calculating the irradiance encountered by the vehicle, as shown in the VIPV modeling block, where the calculations have to be done for multiple locations. Once accurate irradiance and cell temperature are known, the calculations become straightforward. Since the VIPV system will always be coupled with a DC battery system, the inverter modeling is irrelevant.

### 30.5.1 Input Data

During the vehicle motion, the amount of irradiance encountered by the vehicle changes rapidly due to fast dynamic shades created by the changing surrounding environment, vehicle orientation, clouds, and sun position. This becomes better when the vehicle is parked and the surrounding environment becomes static, but the rate of change will still be affected by clouds and the changing Sun's position. To model such a dynamic system, information such as vehicle position, characteristics of the surrounding terrain, and weather are needed. The information about a vehicle's position during motion and parking can be obtained from map services, OpenStreetMap (OpenStreetMap 2022) being the open-source option and Google (Google Maps API 2022) or Bing (Bing Maps API 2022) maps being a few closed-source options. Characteristics of the surrounding terrain can be inferred from geographic elevation datasets like region-specific open-source LiDAR datasets, many of which are available (LiDAR Data NRW 2022) or global datasets like ASTER (ASTER Global DEM 2022) or ground-level pictures from Google Street View (Google Street View Static API 2022). Open-source weather data is available from services like CAMS (CAMS Radiation Service 2022) or NSRDB (NSRDB 2022), which provide almost global coverage. The simulation timestep can be as small as one second or even smaller if desired. However, the trade-off between simulation timestep to accuracy for a more extended timeframe simulation, for example, one year, has yet to be discovered. The simulation approach can be further simplified if the vehicle parking time is significantly higher than the motion time for an annual scenario.

### 30.5.2 Effective Irradiance

Calculating effective irradiance on a VIPV module with shading losses incorporated requires large-scale geospatial data, as mentioned earlier, vehicle orientation and relative sun position. Rapidly handling large quantities of data requires a parallel computing framework developed in the open-source library PV-GRIP (PV-GRIP 2022). PV-GRIP gathers and processes this data to calculate irradiance using either SSDP (SSDP 2022) or r.sun (GRASS r.Sun 2022) models. SSDP can project the sky dome on a VIPV module surface, given its tilt, orientation, height, and the surrounding topography in the form of a point cloud or a regular grid. The r.sun model also calculates the surface irradiance, considering the local "topography's shadowing effect. However, it does not account for the height of the VIPV system.

Another approach presented during the PV in Motion Conference in 2021 is using Google Street View pictures (Google Street View Static API 2022). The ground-level images are binarized using an open-source library like OpenCV (OpenCV 2022), where 0 indicates a shadow from objects in the image and 1 indicates an open sky. This binarization helps calculate the sky view factor from the "vehicle's position, making it possible to calculate the irradiance value after considering the partial shading effect.

The mentioned approaches calculate the irradiance at fixed points on the surface of the flat module. To account for partial shading and its impact on the module power output, it would be necessary to calculate irradiance values for multiple points on the module. This helps to estimate the amount of module surface shaded and what cells will be bypassed given the specification of the BPDs in the module. This method of calculating multipoint irradiance is computationally demanding and could be replaced by estimating the power loss due to shading mismatch using an empirically derived loss factor.

Determination of irradiance on a curved module depends on the angular distribution of the irradiance and requires azimuth and tilt angle of individual cells. A simplification of this approach is the determination of a curve correction factor as described in Ota et al. (2022), which is a ratio of solar irradiance on a curved to a flat surface. Once empirically defined for a particular module, the curve correction factor can be utilized to estimate the irradiance on a curved module by calculating the irradiance on a flat surface. This is a simplification to save computation time.

### 30.5.3 Cell Temperature

Traditional PV systems are generally free-standing and allow for convective airflow over and under the modules, and the modules themselves have a low thermal mass. This can be modeled using a steady state model, many of which exist. The VIPV module is mounted on a vehicle, which can be considered a greenhouse, especially in the case of cars. The temperature of a car cabin can easily reach as high as 76 °C depending on location and time of year (Grundstein et al. 2009). It is unknown at this point how much this will affect the PV module mounted on the car, as the thermal mass of the combined system will become much higher than the traditional PV system. Vehicle temperatures rise when parked under the sun for a longer time and start cooling rapidly when they begin moving due to high headwinds. This heating and cooling cycle of the vehicle and, hence, the VIPV module involves a thermal time constant for heating up and cooling down. Hayakawa et al. (2022) experimentally determined the convective cooling rate of a VIPV module to be −16.5 °C/min. It is not concretely known whether such a thermal system can be modeled by the steady-state PV temperature models or will require the development of a new dynamic model.

### 30.5.4  Module Power

The equivalent circuit diode models like the Desoto or PVsyst are used to estimate the power of PV modules as described in Stein (2017). The only prerequisite is that information about the irradiance, cell temperature, and module specifications are known. For flat plate modules, specifications are easy to characterize since testing standards exist. Curved modules, however, provide some additional challenges in this regard. Specifications are harder to determine as characterizing standards are being determined by a working group of IEC (Araki et al. 2019).

### 30.5.5  Loss Factors

Apart from the power output mismatch due to the varying irradiance profile on a VIPV module, as mentioned earlier, there is also a power mismatch due to cell level inconsistencies that come from manufacturing or variable soiling levels on individual cells. Connecting a group of cells at a similar tilt and azimuth to an individual MPPT could be a probable solution to both mismatches. Stein (2017) mentioned that the magnitude of losses occurring due to the latter power mismatch is deducted as a system derate factor estimated from the module design outside the simulation. DC resistive losses are calculated from the module system design outside the simulation. Module degradation losses due to light induced degradation (LID) should be the same as the traditional PV systems. Still, losses due to aging effects induced by high vibrations during vehicle operation are unknown.

EV traction systems have a DC electrical system that operates at 400 V or higher. Since it might not always be possible to match the VIPV module voltage to the battery voltage, considering the vehicle surface area limitations or interconnection strategies to reduce shading impact, injecting the VIPV energy yield into the traction system will require a voltage boost for the VIPV power using a DC–DC converter. The voltage-boosting efficiency depends on the input and output power and the "converter's operating temperature. The efficiency of such a converter can be assumed to be 90% and is verified by the VIPV demonstrator system mentioned in Peibst et al. (2021). Depending on the design of the EV battery, there will be a charging/discharging efficiency of around 80–90%, which should be considered for estimating the final VIPV system efficiency.

### 30.5.6  Uncertainties

The highest uncertainty in VIPV modeling comes from the input data sources. Weather data sources available today have a temporal and spatial resolution that does not equate well with the dynamics of a VIPV system, which occur at an average frequency of 1 Hz or even higher. While weather data sources are very accurate on clear sky days, their accuracy suffers on cloudy days. Location data from map services sometimes place vehicles in unlikely places, leading to erroneous vehicle position calculations. Elevation datasets and Google Street View pictures used for shading calculations are not continuously updated to represent the present situation of the terrain. Despite these uncertainties, the simulation model results provide a compelling insight into VIPV systems.

## 30.6   Outlook

With rising climate change and the shift toward transport electrification, VIPV has become a possible contender to ameliorate the burden of added electricity needs. Rising efficiencies and decreasing costs of PV will no doubt help in this regard. Module technologies that can be curved, lightweight, tolerant to fast shades, and aesthetically pleasing are possible. From a control electronics perspective, it is clear that to accommodate the entire operation cycle of VIPV, special MPPT (fast or slow) is needed to extract the most power. Energy yield models for VIPV with varying levels of accuracy have been proposed. Various measurement campaigns have been conducted, and standardized tests have been proposed. Potential impacts of VIPV technology in terms of demand, range extension, sustainability, and social acceptance have emerged.

Although much progress has been made, much work remains for VIPV to become a mass-market technology. With all these existing and emerging PV technologies, methods need to be developed for selecting the best-suited one for a specific VIPV application. From a module design optimization perspective, efforts must be made to provide guidelines for the best module design considering BPDs and cell interconnections to extract the maximum power. Open-source and well-validated energy yield models are needed for transparency in analyzing the VIPV system. More extended measurements must be conducted and open-sourced, highlighting accurate irradiation and temperature profiles. Comprehensive standardization testing should be agreed upon to guarantee the functioning of a VIPV system for the entirety of a vehicle's lifetime. Finally, the most crucial aspect for any technology to become successful is the cost and sustainability aspect, which need to be estimated for various envisioned use cases of the technology. This will be achieved by a framework that can simulate these VIPV use cases with verifiable accuracy.

With all the progress made and the probable improvement targets already identified, the future looks promising for VIPV-powered electric mobility.

## List of Acronyms

| Acronym | Description |
| --- | --- |
| AC | Alternating current |
| API | Application programming interface |
| APS | Announced pledges scenario |
| ASOM | Alliance for solar mobility |
| BAPV | Building added photovoltaic |
| BIPV | Building integrated photovoltaic |
| BPD | Bypass diodes |
| CAMS | Copernicus atmosphere monitoring service |
| CIGS | Copper indium gallium selenide |
| DC | Direct current |
| DEM | Digital elevation model |

| Acronym | Description |
| --- | --- |
| EV | Electric vehicle |
| GRASS | Geographic resources analysis support system |
| IBC | Interdigitated back contact |
| IEA | International energy agency |
| IEA-PVPS | International energy agency photovoltaic power systems |
| IEC | International electrotechnical commission |
| ITRPV | International technology roadmap for photovoltaic |
| LID | Light-induced degradation |
| LiDAR | Light detection and ranging |
| MPPT | Maximum power point tracking |
| NSRDB | National solar radiation database |
| OEM | Original equipment manufacturer |
| PERC | Passivated emitter and rear cell |
| PV | Photovoltaic(s) |
| PV-GRIP | Photovoltaic – geographic raster image processor |
| SHJ | Silicon heterojunction |
| SSDP | Simple sky dome projector |
| STEPS | Stated policies scenario |
| V2G | Vehicle-to-grid |
| V2X | Vehicle-to-anything |
| VAPV | Vehicle added photovoltaic |
| VIPV | Vehicle integrated photovoltaic |
| WTP | Willingness to pay |

## Author Biography

**Neel Patel** received his bachelor of engineering (BE) degree in electrical engineering from the Gujarat Technological University in India and his M.Sc. degree in renewable energy engineering from the University of Freiburg in Germany in 2015 and 2019, respectively. He is working toward a doctorate focusing on modeling and analyzing onboard vehicle PV systems at the Eindhoven University of Technology in the Netherlands and the Forschungszentrum Jülich in Germany.

## References

Aptera. (2022). https://aptera.us/vehicle/

Araki, K., Ji, L., Kelly, G., and Yamaguchi, M. (2018). To do list for research and development and international standardization to achieve the goal of running a majority of electric vehicles on solar energy. *Coatings* 8 (7): 251. https://doi.org/10.3390/coatings8070251.

Araki, K., Liu, Z., Limpinsel, M. et al. (2019). Modeling and standardization researches and discussions of the car-roof PV through international web meetings. In: *2019 IEEE 46th Photovoltaic Specialists Conference (PVSC)*, 2722–2729. https://doi.org/10.1109/PVSC40753.2019.8980555.

Araki, K., Ota, Y., and Yamaguchi, M. (2020). Measurement and modeling of 3D solar irradiance for vehicle-integrated photovoltaic. *Applied Sciences* 10 (3): 872. https://doi.org/10.3390/app10030872.

Araki, K., Karr, A.J., Chabuel, F. et al. (2021). *State-of-the-Art and Expected Benefits of PV-Powered Vehicles*. International Energy Agency – PhotoVoltaic Power Systems.

ArcGIS. (2022). https://pro.arcgis.com/en/pro-app/latest/tool-reference/spatial-analyst/area-solar-radiation.htm

ASOM. (2022). https://asom.solar/

ASTER Global DEM. (2022). https://cmr.earthdata.nasa.gov/search/concepts/C1711961296-LPCLOUD.html

Automostory (2022). *First Solar Car*. Automostory https://www.automostory.com/first-solar-car.htm.

Bing Maps API. (2022). https://docs.microsoft.com/en-us/bingmaps/rest-services/routes/?toc=https%3A%2F%2Fdocs.microsoft.com%2Fen-us%2Fbingmaps%2Frest-services%2Ftoc.json&bc=https%3A%2F%2Fdocs.microsoft.com%2Fen-us%2FBingMaps%2Fbreadcrumb%2Ftoc.json

Blasi, B., Kroyer, T., Kuhn, T., and Hohn, O. (2021). The MorphoColor concept for colored photovoltaic modules. *IEEE Journal of Photovoltaics* 11 (5): 1305–1311. https://doi.org/10.1109/JPHOTOV.2021.3090158.

CAMS Radiation Service. (2022). https://www.soda-pro.com/web-services/radiation/cams-radiation-service

Carr, A.J. (2020). *Vehicle Integrated Photovoltaics – Evaluation of the Energy Yield Potential Through Monitoring and Modelling*. 5. doi: 10.4229/EUPVSEC20202020-6DO.11.3.

Centeno Brito, M., Santos, T., Moura, F. et al. (2021). Urban solar potential for vehicle integrated photovoltaics. *Transportation Research Part D: Transport and Environment* 94: 102810. https://doi.org/10.1016/j.trd.2021.102810.

Coloured PV, Fraunhofer ISE. (2022). https://www.ise.fraunhofer.de/en/business-areas/photovoltaics/photovoltaic-modules-and-power-plants/integrated-photovoltaics/vehicle-integrated-photovoltaics/vehicle-integrated-cars.html

Commault, B., Duigou, T., Maneval, V. et al. (2021). Overview and perspectives for vehicle-integrated photovoltaics. *Applied Sciences* 11 (24): 11598. https://doi.org/10.3390/app112411598.

EEA. (2022). https://www.eea.europa.eu/data-and-maps/daviz/co2-emission-intensity-12#tab-googlechartid_chart_11

Eitner, U., Ebert, M., Zech, T., Schmid, C., Watts, A., & Heinrich, M. (2017). *Solar Potential on Commercial Trucks: Results of an Irradiance Measurement Campaign on 6 Trucks in Europe and USA*. 4. https://doi.org/10.4229/EUPVSEC20172017-6DO.12.4

Ghasri, M., Ardeshiri, A., Ekins-Daukes, N.J., and Rashidi, T. (2021). Willingness to pay for photovoltaic solar cells equipped electric vehicles. *Transportation Research Part C: Emerging Technologies* 133: 103433. https://doi.org/10.1016/j.trc.2021.103433.

Gochermann Solar Technology. (2022). https://www.gochermann.com/

Google Maps API. (2022). https://developers.google.com/maps/documentation/directions

Google Street View Static API. (2022). https://developers.google.com/maps/documentation/streetview/overview

GRASS r.sun. (2022). https://grass.osgeo.org/grass82/manuals/r.sun.html

Grundstein, A., Meentemeyer, V., and Dowd, J. (2009). Maximum vehicle cabin temperatures under different meteorological conditions. *International Journal of Biometeorology* 53 (3): 255–261. https://doi.org/10.1007/s00484-009-0211-x.

Hayakawa, Y., Baba, M., and Yamada, N. (2021). Effect of bypass diode on power generation of three-dimensional curved Si photovoltaic module. *IEEE Journal of Photovoltaics* 12 (1): 388–396. https://doi.org/10.1109/JPHOTOV.2021.3119254.

Hayakawa, Y., Sato, D., and Yamada, N. (2022). Measurement of the convective heat transfer coefficient and temperature of vehicle-integrated photovoltaic modules. *Energies* 15 (13): 4818. https://doi.org/10.3390/en15134818.

Heinrich, M. (2020). *Potential and Challenges of Vehicle Integrated Photovoltaics for Passenger Cars*. 6. https://doi.org/10.4229/EUPVSEC20202020-6DO.11.1

IEA (2020). *Renewables 2020*. IEA https://www.iea.org/reports/renewables-2020.

IEA (2022). *Global EV Outlook 2022*. IEA https://www.iea.org/reports/global-ev-outlook-2022.

*IEA-PVPS Task 17 PV & Transport*. (2022). https://iea-pvps.org/research-tasks/pv-for-transport/

IM Efficiency. (2022). https://imefficiency.com/index

ITRPV. (2022). https://www.vdma.org/international-technology-roadmap-photovoltaic

Kano, F., Kasai, Y., Kimura, H. et al. (2021). Buck-boost-type MPPT circuit suitable for vehicle-mounted photovoltaic power generation. *IEEJ Transactions on Electrical and Electronic Engineering* 16 (9): 1229–1238. https://doi.org/10.1002/tee.23421.

Kanz, O., Reinders, A., May, J., and Ding, K. (2020). Environmental impacts of integrated photovoltaic modules in light utility electric vehicles. *Energies* 13 (19): 5120. https://doi.org/10.3390/en13195120.

Kim, S., Holz, M., Park, S. et al. (2021). Future options for lightweight photovoltaic modules in electrical passenger cars. *Sustainability* 13 (5): 2532. https://doi.org/10.3390/su13052532.

Kühnel, M., Hanke, B., Geißendörfer, S. et al. (2017). Energy forecast for mobile photovoltaic systems with focus on trucks for cooling applications: energy forecast for mobile photovoltaic systems. *Progress in Photovoltaics: Research and Applications* 25 (7): 525–532. https://doi.org/10.1002/pip.2886.

Kutter, C. (2020). Integrated lightweight, glass-free pv module technology for box bodies of commercial trucks. In: *37th EUPVSEC*. https://doi.org/10.4229/EUPVSEC20202020-6DO.11.6.

Lade-PV. (2022). https://www.ise.fraunhofer.de/en/research-projects/lade-pv.html

Lambert, F. (2016). *Toyota Brings Back the Solar Panel on the Plug-In Prius Prime – But Now It Powers the Car*. Electrek https://electrek.co/2016/06/20/toyota-prius-plug-prime-solar-panel/.

LiDAR Data NRW. (2022). https://www.bezreg-koeln.nrw.de/brk_internet/geobasis/ hoehenmodelle/3d-messdaten/index.html

Lightyear. (2022). https://lightyear.one/lightyear-0

Markert, J., Kutter, C., Newman, B. et al. (2021). Proposal for a safety qualification program for vehicle-integrated PV modules. *Sustainability* 13 (23): 13341. https://doi.org/10.3390/ su132313341.

McCann, J. (2021). *Hyundai's New Electric Car Has a Solar Panel Roof and Can Charge Other EVs.* Techradar https://www.techradar.com/news/hyundais-new-electric-car-has-a-solar-panel-roof-and-the-ability-to-charge-other-evs.

Mont, S. N., Heinrich, M., Pfreundt, A., Kutter, C., Tummalieh, A., & Neuhaus, H. (2020). *Energy Yield Modelling of 2D and 3D Curved Photovoltaic Modules.* 6. https://www.eupvsec-planner.com/presentations/c50125/energy_yield_simulation_of_3d_curved_vipv_modules .htm

NSRDB. (2022). https://nsrdb.nrel.gov/

OpenCV. (2022). https://github.com/opencv/opencv-python

OpenStreetMap. (2022). https://www.openstreetmap.org

Ota, Y., Masuda, T., Araki, K., and Yamaguchi, M. (2019). A mobile multipyranometer array for the assessment of solar irradiance incident on a photovoltaic-powered vehicle. *Solar Energy* 184: 84–90. https://doi.org/10.1016/j.solener.2019.03.084.

Ota, Y., Araki, K., Nagaoka, A., and Nishioka, K. (2021). Evaluating the output of a car-mounted photovoltaic module under driving conditions. *IEEE Journal of Photovoltaics* 11 (5): 1299–1304. https://doi.org/10.1109/JPHOTOV.2021.3087748.

Ota, Y., Araki, K., Nagaoka, A., and Nishioka, K. (2022). Curve correction of vehicle-integrated photovoltaics using statistics on commercial car bodies. *Progress in Photovoltaics: Research and Applications* 30 (2): 152–163. https://doi.org/10.1002/pip.3473.

Peibst, R., Fischer, H., Brunner, M. et al. (2021). Demonstration of feeding vehicle-integrated photovoltaic-converted energy into the high-voltage on-board network of practical light commercial vehicles for range extension. *Solar RRL* 6 (5): 2100516. https://doi.org/10.1002/ solr.202100516.

PV-GRIP. (2022). https://github.com/IEK-5/PV-GRIP.

PV2Go. (2022). https://pv2go.org/.

Ritchie, H., Roser, M., & Rosado, P. (2020). *$CO_2$ and Greenhouse Gas Emissions.* https:// ourworldindata.org/emissions-by-sector

Ritzen, M., Vroon, Z., and Geurts, C. (2017). Building integrated photovoltaics. In: *Photovoltaic Solar Energy* (ed. A. Reinders, P. Verlinden, W. van Sark, and A. Freundlich), 579–589. John Wiley & Sons, Ltd. https://doi.org/10.1002/9781118927496.ch51.

Solbian Solar. (2022). https://solbian.solar/en/

Sono Motors. (2022). https://sonomotors.com/

Sovetkin, E., Patel, N., Gerber, A., & Pieters, B. (2021). *Open-Source Geospatial Data Service with an Application in Irradiance Modelling for VIPV.* 9. https://www.eupvsec-planner.com/ presentations/c50097/open-source_geospatial_data_service_with_an_application_in_ irradiance_modelling_for_vipv.htm

Sovetkin, E., Noll, J., Patel, N. et al. (2022). Vehicle-integrated photovoltaics irradiation modeling using aerial-based LIDAR data and validation with trip measurements. *Solar RRL* 2200593. https://doi.org/10.1002/solr.202200593.

SSDP. (2022). https://github.com/IEK-5/ssdp

Stein, J.S. (2017). Energy prediction and system modeling. In: *Photovoltaic Solar Energy* (ed. A. Reinders, P. Verlinden, W. van Sark, and A. Freundlich), 564–578. John Wiley & Sons, Ltd. https://doi.org/10.1002/9781118927496.ch50.

Tayagaki, T., Haruya, S., Ayumi, S. et al. (2021). Comparative Study of Power Generation in Curved Photovoltaic Modules of Series- and Parallel-Connected Solar Cells. *IEEE Journal of Photovoltaics* 11 (3): 708–14. https://doi.org/10.1109/JPHOTOV.2021.3060399.

Wetzel, G., Salomon, L., Krügener, J. et al. (2021). High time resolution measurement of solar irradiance onto driving car body for vehicle integrated photovoltaics. *Progress in Photovoltaics: Research and Applications* 3526. https://doi.org/10.1002/pip.3526.

Wheeler, A., Leveille, M., Anton, I. et al. (2019). Determining the operating temperature of solar panels on vehicles. In: *2019 IEEE 46th Photovoltaic Specialists Conference (PVSC)*, 3592–3597. https://doi.org/10.1109/PVSC40753.2019.9311292.

Yamada, N. (2020). Vehicle-integrated 3D-PV module with III-V and Si solar cells. In: *2020 47th IEEE Photovoltaic Specialists Conference (PVSC)*, 2324–2326. https://doi.org/10.1109/PVSC45281.2020.9300929.

Yamaguchi, M., Masuda, T., Araki, K. et al. (2021). Analysis of temperature coefficients and their effect on efficiency of solar cell modules for photovoltaics-powered vehicles. *Journal of Physics D: Applied Physics* 54 (50): 504002. https://doi.org/10.1088/1361-6463/ac1ef8.

Yamaguchi, M., Ozaki, R., Nakamura, K. et al. (2021). Development of high-efficiency solar cell modules for photovoltaic-powered vehicles. *Solar RRL* 2100429. https://doi.org/10.1002/solr.202100429.

# 31

# Yield Assessments for PV Bankability

*David Moser*

Institute for Renewable Energy, Eurac Research, Bolzano, Italy

## 31.1   Introduction

The energy transition relies on the capacity of the photovoltaic (PV) sector to mobilize investments to unprecedented levels. In fact, the annual added capacity will have to increase from around 150/200 GW/year to an annual TW market, where this milestone should be reached by 2030 (Vartiainen et al. 2020). It is in this scenario that it is important for investors to perceive PV projects as low risk, and to this extent, professional risk assessments must be established.

The H2020 project Solar Bankability (H2020 Solarbankability 2017) defined the risk assessment as "an active quality management process where all stakeholders in the approval process of a PV project attempt to identify potential legal, technical and economic risks through the entire project lifecycle. These risks need to be quantitatively and qualitatively assessed, managed and controlled. Despite a wide overlap in this quality management process, the focus and the assessment criteria will vary depending on whether the stakeholder represents an investor, a bank, an insurance company or a regulatory body" (Tjengdrawira et al. 2017).

The different stakeholders have to develop their own individual risk management strategies along the lifecycle of a PV project using the four-step process of risk identification, risk assessment, risk management and risk controlling. Best-practice guidelines and concrete tools to better manage technical risks throughout the PV project lifetime are emerging in various projects. The ultimate responsibility for project risks remains with the owner and operator of the PV plant. With the help of a professional risk management plan, they can significantly reduce and transfer the initial risks associated with a PV project to other stakeholders.

With increasing experience gained from more and more projects, a more mature and professional PV industry, and established professional and standardized processes, these related residual risks can be low. This paper focuses on how to reduce risk during the design phase by looking at the state of the art in yield assessments (YAs) and long-term yield predictions (LTYPs) and the implications to PV bankability. It presents an updated methodology on key measures as part of the technical risk management during the design

*Photovoltaic Solar Energy: From Fundamentals to Applications, Volume 2*, First Edition.
Edited by Wilfried van Sark, Bram Hoex, Angèle Reinders, Pierre Verlinden, and Nicholas J. Ekins-Daukes.
© 2024 John Wiley & Sons Ltd. Published 2024 by John Wiley & Sons Ltd.
Companion website: www.wiley.com/go/PVsolarenergy

phase of a PV project that helps to reduce technical risks and increase the probability of economic successful investment.

## 31.2 Introduction to Yield Assessment

One important step for the calculation of the bankability of a project is to assess the production of a given PV system in a given location. In a nutshell, the process requires knowledge of the site and on the chosen technology, where from weather and climate data (long time series of irradiance, temperature, wind speed, etc.) and from technology data (nominal power, efficiency curves, temperature coefficients, etc.) the assessor can through various modeling steps convert irradiance into electricity production.

YAs and LTYPs are a prerequisite for business decisions on long-term investments into PV power plants. Together with cost and financial data (CAPEX, OPEX, and discount rate), the output of a YA and LTYP (covering the utilization rate, performance loss rate (PLR) and lifetime) provides financial investors with the parameters needed for the calculation of the levelized cost of electricity (LCOE) and to assess the cash flow model of an investment with relative internal rate of return (IRR) and net present value (NPV).

There are various stakeholders involved during a PV project quality lifecycle in the assessment of yield, from the basic design (site identification) to the preliminary design (site survey and solar resource assessment) to the investment decision before the execution design (yield-assessment based on solar resources). In all these phases, the preparation of a YA and LTYP report typically relies on numerical modeling and prediction of the expected energy yield based on experience with previous PV power plants, laboratory measurements and knowledge gained in the PV community over the past years and decades.

In the IEA PVPS TASK 13 report "Uncertainties in PV System Yield Predictions and Assessments" (Reise 2018), the authors presented a comprehensive investigation of the uncertainties related to this task. The report collected some insights into the field of uncertainties of several technical aspects of PV system yield prediction and assessment, investigating several of the modeling steps for gains and losses in a PV system: the solar resource – including long-term trends, PV module properties, system output and performance. In a subsequent report, "PV Yield Assessment and LCOE" (Moser 2021), the authors also highlighted that the main challenge in YA and LTYP relates to the trustworthiness of site-specific information. In a global market, it is, in fact, not uncommon to assess the yield of a PV plant to be located in areas that are not familiar to the yield assessor, and local knowledge is thus of extreme importance. In Moser (2021), the authors also built upon the previous work by running benchmarking exercises in order to be able to quantify the differences between existing approaches.

In the presence of long-term trends, the question for solar resource assessments is no longer "what is the 'true' climatological value?", but "what is the best predictor for the project lifetime?". A suitable estimator should be a recent time period that is long enough to filter the influence of single years with high anomalies, but which is short enough to minimize the influence of past trends. For more insights about high-quality solar resource assessments and, in particular, on the development of enhanced analysis of long-term inter-annual variability and trends in the solar resource, we invite the reader to look at

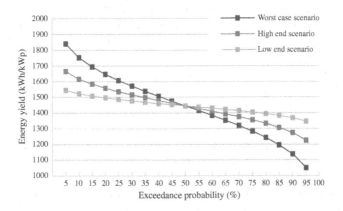

**Figure 31.1** Exemplary figure to show the impact of uncertainty in a yield assessment.

the output of the IEA PVPS Task 16 "Solar resource for high penetration and large scale application" (Solar Resource for High Penetration and Large Scale Applications 2022).

With respect to uncertainties, the different influencing effects (irradiance level, operating temperature, transposition model, DC electricity model etc.) are typically represented by one individual factor per effect. The influences are assumed to be independent. Furthermore, these factors are often used in integrated form, e.g. over one year. As previously mentioned (Reise 2018), a framework for the calculation of uncertainty for a complete LTYP is presented. as an effort to standardize the procedure for uncertainty calculation of predicted energy yields of PV systems in order to properly estimate financial investment risk.

Jahn et al. (2016) also reported different uncertainty scenarios presenting mitigation measures on how to reduce the uncertainty in a YA and thus increase confidence in the exceedance probability output defined as P50/P90 values. Figure 31.1 provides an exemplary figure to show the impact of uncertainty on a YA with the same P50 value. The impact of the P90 value can be as high as 20% in terms of yield reduction due to different quality of information available to the yield assessor.

## 31.3 The Key Elements of Yield Assessments

In this chapter, the aim is to provide a short introduction to the key elements on which a YA is based. Not all elements are included, for more exhaustive info, please refer to e.g. Moser et al. (2016) and Reise (2018). It is important to highlight that yield assessors should provide details on how each step of the modeling chain was considered together with assumptions and uncertainties. Failing to do so will hinder the possibility for investors to compare YAs provided by different assessors.

### 31.3.1 Long-Term Insolation and Trends

Different sources of long-term insolation data are available worldwide: measured data from meteorological institutions, interpolated values (e.g. Meteonorm (METEOTEST

Genossenschaft 2014) and estimates from satellite-derived images (e.g. SoDa HC-3, SolarGIS, SatelLight, PVGIS, and NASA). These databases not only use irradiation data obtained by different methods but also often cover different periods. As shown by Richter et al. (2015) and Muller et al. (2015), given the long-term variations of irradiance, the time period used to estimate the irradiation for a typical year often has an important influence that has to be accounted for. Müller and Kiefer (2019) show an average increase of in-plane irradiance (global tilted irradiance [GTI]) of 1.1%/year or about 11%/decade over the period 2008 to 2018 for 44 rooftop installations in Germany.

Significant differences can be observed when comparing the databases between each other or against reference meteorological observations. Consequently, the insolation uncertainty depends to a large extent on the source of the data and the reference period used. In Moser et al. (2016) examples are given to illustrate this by taking Belgium as a case study. The comparison of the mean values provided by these databases against the 10-year centered moving average of the meteorological station highlights the effect of the time period used to estimate the irradiation for a typical year. In Moser (2021), similar analyses have been carried out for countries in Northern Europe showing different trends depending on the analyzed period and the specific location.

Research has thus revealed that the irradiation in several places across Europe showed a dimming period followed by a significant brightening trend starting from around 1990 (Wild 2009; Müller et al. 2014; Wild et al. 2005). In one case, a study of long-term global horizontal irradiation (GHI) measurement records from eight stations in Germany found a brightening trend of +3.3% per decade, starting from around 1984 (Müller et al. 2014). The Solar Bankability project team performed a similar analysis using the long-term GHI measurement records from 32 meteorological stations of the Royal Meteorological Institute of The Netherlands (KNMI) covering the period from 1958 to 2015. The trends were found by calculating the slope for each possible combination of 10 years (moving window). Since around 1990, a clear brightening trend has been observed with a slope of +2.63% per decade.

Often, different databases are combined to reduce the uncertainty in solar resource estimation. If ground measurements are available for a short period (e.g. one year), this data can be combined with long-term satellite estimations using the measure-correlate-predict (MCP) methodology, described in, e.g. Thuman et al. (2012) or Gueymard and Wilcox (2009). The MCP method is a widely established and recognized methodology for wind resource assessments, and its application is gaining ground for solar resource assessment. The purpose of the MCP methodology is to combine the data of a short period of record but with site-specific seasonal and diurnal characteristics with a data set having a long period of record with not necessarily site-specific characteristics. Assuming a strong correlation, the strengths of both data sets are captured, and the uncertainty in the long-term estimate can be reduced.

In Tjengdrawira and Richter (2016), the authors validated the application of the MCP methodology on 32 meteorological stations in the Netherlands. The study concluded that the MCP methodology can yield high accuracies with uncertainties below 2% (bias) if the common reference period used is at least one year. However, if the bias of the satellite is not constant over the year, the application of the MCP methodology based on periods shorter than one year can have considerably lower accuracy. This can be improved by using more advanced site adaptation methods as proposed, e.g. by Polo et al. (2016).

### 31.3.2 Annual Insolation Variability

The annual insolation variability or "year-to-year variability" is defined as the ratio of the standard deviation ($\sigma$) to the average GHI over a long-term period (typically more than 10 years) (Scharmer and Greif 2000). On average, the standard deviation of the yearly sums of GHI is mostly in the range of 4% to 7%, as shown, for example, by Richter et al. (2015) and Suri et al. (2007). Interestingly, the variability over other continents can be much lower than in Europe. In Moser (2021) the case of Africa is represented where still some important variations can occur; for example, 2017 was not a very good year for some countries in the sub-Saharan region like Zambia and Mozambique, where the irradiation over 2017 was approx. 6% lower than the P50. On the contrary, 2016 was a much better year in terms of solar resources for all sub-Saharan countries. For risk assessments, the annual insolation variability may become the main source of uncertainty when analyzing the risk associated with the cash flow during a single year (Vartiainen et al. 2015). However, when calculating the lifetime accumulated income, this uncertainty has a relatively small effect since the years with less irradiation are generally compensated for by other years with more irradiation.

### 31.3.3 Temperature

Different models are available for estimating the cell temperature of a PV module. From simple models that neglect both thermal dynamics and wind speed up to advanced models that take into account both dynamics and wind speed effects are available (Maturi et al. 2014; Schwingshackl et al. 2013). Validation results of different temperature models reported in Richter et al. (2015) show that the uncertainty in the cell temperature estimation using an advanced model that takes dynamics and wind speed into account can be as low as 1°C, though most traditional models show uncertainties of 2°C and higher. Typical RMSE of the k factor relating ΔT and irradiance G is around 2%. As reported in King et al. (2002), the effect of the operating temperature on annual energy production was found to be dependent on both module technology and site environmental conditions, with an influence on the annual energy production of −2% to −10%.

### 31.3.4 Parameters Used in Power Calculation

The power calculation in PV modeling software not only depends on the software's algorithms, but it also requires that components (modules, inverters) have been correctly parametrized. The uncertainties on (sub)components in the PV power modeling chain are often relatively low through the implementation of peer-reviewed methods. However, modeling risks can occur through errors in module or inverter files, which can negatively affect the yield, and with it, the financial viability of PV plants. Therefore, it becomes more and more the common industry standard that PV modules or inverters are subjected to additional characterization by independent laboratories upon instruction by investors to ensure that the power plant model is bankable. Special care must be taken during this phase in terms of number of modules to be tested and the selection procedure in order to obtain a reliable mean value of electrical parameters to be used in power calculation. The creation of digital twins and electrical hierarchy can further help in reducing the error in the power plant model (Horvath 2022).

### 31.3.5 Availability

PV power plant effective availability can be split into two components: a portion that can be influenced by the PV asset owner through its O&M contractor (such as ensuring that all power-producing equipment is "available" to generate power through appropriate procedures) and the portion which cannot be controlled, such as availability of the grid to accept power, or grid operator signals to curtail power. O&M contractors can guarantee availability values of 99% or higher (Jahn et al. 2020), with exceptions for PV plants in remote regions. Hunt et al. (2015) reported an average effective availability of 99.5% for 27 plants for a 5-year period. Tjengdrawira and Richter (2016) reported mean yearly unavailability of the analyzed portfolio (41 rooftop PV plants) of around 2% ± 2%. A more detailed analysis has been carried out, extracting the mean monthly availability of a portfolio of 533 PV plants installed mostly in Belgium, Germany, and Switzerland in Moser (2021). The monitored data show the highest mean unavailability of 6.85% in February, followed by 4.28% in March and 3.48% in January, which coincide with harsher weather conditions (snow, winds, etc.) during the European winter. For the rest of the year, the mean monthly unavailability ranges from 1.16% to 1.81%. Furthermore, it has been seen that unavailability has decreased over time, possibly based on improved O&M protocols.

The initial performance and availability of PV plants can be affected by a ramp-up to full generation period, which could be as high as 6 months, as shown by McLeod et al. (2019). Similarly, curtailment or other grid operator signals (on top of forced outages) can reduce the availability of the grid for established PV plants. Uncertainties in availability values thus can have a long tail: while typical values are 99% availability ±1% (assuming a suitable O&M contract and contractor), PV plants in areas with constrained grids can have availabilities below 90% for weeks (e.g. during spring/autumn shoulder periods with high PV production and low demand, or due to grid strength issues).

### 31.3.6 Degradation and Performance Loss Rates for LTYP

In the IEA PVPS Task 13 report "Assessment of Performance Loss Rate of PV Power Systems" (French et al. 2021), the authors focused their efforts on the determination of reliable PLR calculation approaches. PLR is defined as the sum of losses on the system and module level and includes both reversible and irreversible effects. The term degradation instead is often used to verify an irreversible loss on module level stemming from appearing degradation modes. In general, PLR calculations follow a strict order of steps, which includes input data acquisition, data filtering, performance metric selection and the application of statistical methods to obtain the final PLR, which is given in %/year. Lindig et al. (2021) suggest that there is no outperforming combination, rather one should

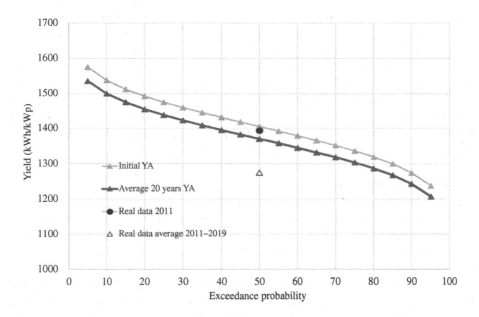

**Figure 31.2** Comparison between the initial yield assessment, the average 20-year yield assessment and operational data after 10 years of operation.

apply several combinations to the same dataset to find an average value for PLR, which might approximate best the "true" value (which remains unknown). However, certain filter/metric/statistical method combinations should be avoided as they produce PLR values with high deviations from the "true" value.

Moser (2021) showed the impact of wrong assumptions in the PLR parameter for a real PV plant in operation since 2010. The yield assessor included only PV module with −0.25%/year degradation rate, while the system shows a PLR of −0.9%/year. The implication is that although the initial YA was in line with the operational data, the average 20-year YA is already around 10% higher than the average production from the first 10 years. This will result in an overoptimistic LTYP (Figure 31.2).

### 31.3.7 Yield Assessments of Novel PV System Typologies

The aforementioned impacts are typical of any PV system installation type. Special considerations will have to be added to other PV system's typology, for e.g. floating PV (FPV), vehicle integrated PV (VIPV), building integrated PV (BIPV), and agri-PV, etc. Oliveira-Pinto and Stokkermans (2020) published for example, a publication on the use of PVsyst with modifications to account for specific effects typically found in FPV, i.e. albedo, water cooling, and soiling losses.

#### 31.3.7.1 Irradiance Variability

While a typical PV system is fixed, in VIPV, vehicle location changes with each trip, so annual yield simulation cannot be performed with single location data. Depending on annual drive profile of a vehicle, input data for multiple locations is needed. Furthermore, VIPV modeling has two aspects: steady state for parking conditions and dynamic state for driving conditions. VIPV modeling requires thus large-scale geospatial data from various sources, which must include spatial and temporal assignment. Araki et al. (2020) modeled solar irradiance by a random distribution of shading objects and car orientation with the correction of the curved surface of the PV modules. The measurement of the solar irradiance onto the car roof and car body was done using five pyranometers in five local axes on the car for one year. The measured dynamic solar irradiance onto the car body and car roof was used for validation of the solar irradiance model in the car.

### 31.3.7.2 Temperature Effects

In BIPV, for example, the temperature model is of particular importance as the PV modules might not benefit from retro ventilation and wind cooling effects, see for example Maturi et al. (2014). In FPV, enhanced performance of FPV due to water cooling is widely claimed (Lawrence et al. 2018), but this effect is so far poorly quantified and documented in the scientific literature. Kjeldstad et al. (2021) assessed the effect of water cooling for a system consisting of a floating membrane with horizontally mounted PV modules, allowing for thermal contact between the modules and the water. By applying a thermal model that incorporates the effect of heat transport from the module to the water to estimate the module temperature, they found that the water-cooled string had on average, 5–6% higher yield compared to an air-cooled string. Oliveira-Pinto and Stokkermans (2020) proposes to adjust the PVsyst ("PVsyst Website," 2019) heat loss factor referred to as U-value to account for the cooling effect. Liu et al. (2018) proposed three different U-values depending on the typology of FPV structure: free-standing, small footprint, or large footprint.

In VIPV, wind speed will have a major role as cooling effect for the operating temperature of PV modules. Kronthaler et al. (2014) studied the impact of wind cooling for different scenarios, estimating an increase in energy yield of up to 8%.

## 31.4 The Impact of Yield Assessments on Bankability

### 31.4.1 Impact of Decisions on Yield Assessments

Mitigation measures must be identified along the PV value chain and assigned to various technical risks. Typical mitigation measures during the design phase are linked to the component selection (e.g. standardized products, products with known track record), O&M-friendly design (e.g. accessibility of the site, state-of-the art design of the monitoring system), LCOE optimized design (e.g. tracker vs. fixed tilt, central versus string inverter, quality check of solar resource data). Mitigation during the transportation and installations are linked to the supply chain management (e.g. well-organized logistics, quality assurance during transportation), quality assurance (e.g. predefined acceptance procedures), grid connection (e.g. knowledge of grid code). These mitigation measures positively affect the uncertainty of the overall energy yield, increase the initial energy yield and reduce the cost of O&M during the operational phase (e.g. faster replacement of components, lower cost of site maintenance, and lower occurrence and severity of defect).

As we have seen in previous chapters, the planning of a PV plant requires the assignment of a set of input parameters to predict its final yearly energy production and its long-term performance. This is usually done using specific software. Each input has a given uncertainty, depending on the availability and quality of the information that the planner has. PV projects relying on low information of the input parameters will have the highest uncertainties of the output values and, therefore, will be less attractive for investments.

By applying mitigation measures, Jahn et al. (2018) and Moser (2021) quantified the uncertainty reduction from a base case of 8.7% down to 4.55%. In Moser (2021), seven highly specialized yield assessors were asked to provide a YA and LTYP for two locations. The results show remarkable differences in terms of both energy yield and long-term performance. The differences were analyzed, and the causes were mainly due to the choice of the irradiance database for location in the alpine area affected by high interannual variability and complex far shading environment, while for a desert location, the key issues were to be found along the electricity modeling steps, from initial degradation to the impact of soiling. Further to this, the assumptions on the degradation were all underestimating the measured PLRs coming from real measurements over an 8+ years monitoring period.

### 31.4.2 Uncertainty Scenarios

When planning a PV system, the available information can be different from case to case, also affecting the value of the associated uncertainty. For example, when considering the solar resource parameter, which accounts for most quote of the energy yield uncertainty as shown, e.g. in Tjengdrawira and Richter (2016), the planner could have:

- Different time series ranges, i.e. different insolation variability: e.g. 1, 5, 20 etc., years of available data
- Different insolation resources: e.g. measured, satellite-retrieved, or a combination of long-term satellite data and short-term series of measured data (Woyte et al. 2016; Gueymard and Wilcox 2009; Polo et al. 2016)
- Datasets of irradiance on the plane of the array or the need to use plane-of-array transposition models from global and diffuse horizontal irradiance.

Further information is needed to define the value and the uncertainty of, among the most important, the ambient temperature variability, the temperature coefficients or temperature effects, the PLR, the soiling losses, the shading losses, the spectral mismatch gain/losses, the module nominal power and the inverter efficiency.

Jahn et al. (2016) and Moser (2021) show that YAs with low uncertainties typically refer to the use of long-time series of either ground measurements or satellite estimates of insolation. The temporal range of the available insolation data seems, therefore, to be the most important factor affecting the uncertainty of the yield estimation. For a location in continental climate, there is, for example, a 5% reduction in the overall uncertainty when using 20 years instead of 5 years of GTI data. The best case corresponds to the use of 20 years of measured values of GTI, showing also that a lower uncertainty is assured when (i) reliable ground measurements are used in place of satellite estimates and (ii) time series of plane-of-array irradiance is available without the need to apply transposition models. Unfortunately, the availability of trustworthy ground-measured data (especially at the right tilt angle) is not common especially for long periods of time. Using a combination of a long time series of satellite data with a short series of measured data is recommended rather than just using satellite data. In the case a PV plant is to be installed in a location with high insolation variability, the uncertainty of the yield estimation is also negatively affected, with at least an additional 3% to be added to the final yield uncertainty. Besides the insolation variability and the solar resource quantification uncertainty, the uncertainties related to shading and soiling effects and the use of transposition models also play a role in the overall uncertainty of the final yield.

### 31.4.3  The Link with Economic KPIs

Various definitions of LCOE can be found in the literature. In general terms, it is defined as the average generation cost, i.e. including all the costs involved in supplying PV electricity at the point of connection to the grid. The PV LCOE includes all the costs and profit margins of the whole value chain, including manufacturing, installation, project development, O&M, inverter replacement, etc. PV LCOE also includes the cost of financing but excludes the profit margin of electricity sales and thus represents the generation cost, not the electricity sales price, which can vary depending on the market situation. In the calculation of the LCOE, the typical definition considers the degradation at the PV module level as a result of irreversible degradation modes. However, performance losses evolve over time not only for PV modules but also for the balance of system components. The value of initial yield used in the LCOE calculation is typically defined as the P50 yield and comes from the YA and should include already an estimation of typical system unavailability, soiling, shading and other effects contributing toward performance reduction. In the LCOE calculation, it is important to make sure that not only module degradation is included, but at the same time, double counting of effects must be avoided (e.g. unavailability or soiling included in both the initial yield and PLRs taken from the literature). Particular attention must be given to how the PLR is included in the LCOE equation, as both formulations exist with linear degradation and with exponential decay.

Vartiainen et al. (2020) showed that the parameter with the largest impact on the LCOE is the cost of capital. Risk reduction in PV projects will lead to easier access to capital and thus to lower weighted average cost of capital (WACC) or to projects with similar WACC. Technical parameters will thus become more important for the comparison of PV projects; the variation of technical parameters becomes essential in the benchmarking of LCOE. It is in this context that the competitiveness of PV in the future will be (i) closely linked to the uncertainty and variability of the input parameter determining the initial yield value and (ii) cost-effectiveness and improvements of parameters related to the downstream sector of the PV value chain will have an impact. The assumptions are considered during the EPC phase but with an impact on the operational phase and O&M in terms of required lifetime, PLR, performance ratio, availability, and operational costs. Quality along the whole value chain will thus become essential in order to match the expectations generated in the YA. Parameters like lifetime, degradation or PLR, performance ratio, and operational expenditures can have an overall impact on the LCOE, which can be as high as 100% (doubling of LCOE).

### 31.4.4 The Impact on LCOE and NPV

The output of YAs can have a large impact on the LCOE and the cash flow of business models. Moser (2021) showed the impact when the p50 or p90 is used for a specific PV project using data coming from different yield assessors. As expected, the impact of different assumptions in terms of PLR is limited compared to the impact of the use of different insolation databases.

## 31.5 Conclusions

In this chapter, the uncertainty framework in YA and the implication in real implementations were presented together with the assessment of the impact that uncertainties can have on lifetime yield predictions, on the LCOE and on the cash flow. One of the most relevant questions is related to the reliability of YAs, especially in emerging fields such as VIPV, FPV, and agri-PV. Typically, investors require one YA. In some cases, more YAs might be requested if the results are unclear. The various YAs can be averaged to assign a purchase value to a given project. In any case, the question remains unanswered: why do different assessors obtain different answers? Is one YA more reliable than others? Investors know that past performance is no guarantee for future results. This maxim also applies to long-term YAs and the LCOE that can be determined from these, also within the context of a changing climate. YA is an essential step in a PV project, as it helps to determine whether a system will be funded or not. However, the YA is not only about the software used, it is mainly about the user. YAs may not be as reliable as expected, and literature demonstrated how highly skilled specialists might not arrive at the same result, having been provided the same detailed inputs. The difference stems from personal experience and assumptions by the modeler.

The direct flow-on consequence from this is that LCOE values will also exhibit a variance on top of the additional modeling assumptions that can be employed for LCOE calculations. Determining P50 and P90 values for LCOE results and highlighting the assumptions/modeling chain will be important. From an industry perspective, it would be beneficial if more "live" postmortem analyses (i.e. comparison of the LTYP and measured data, at, e.g. every five years of system life) would be made and published. These can then be used as crucial feedback and inputs for YA modelers, financiers, and insurers. To conclude, in this chapter, important key elements to understand if one YA is more reliable than another were critically discussed, and which input and output data must be provided by the assessor to reach this conclusion.

## List of Acronyms

| | |
|---|---|
| IRR | Internal rate of return |
| LCOE | Levelized cost of electricity |
| LTYP | Long-term yield prediction |
| MCP | Measure-correlate-predict |

NPV         Net present value
POA         Plane of array
YA          Yield assessment

## Nomenclature and Definitions

*Yield assessment*   Assessment of the expected energy yield (in kWh/kWp) of a defined PV
system at a specified location. It can be calculated for the 1st year of
operation (initial yield assessment), for a specific year or as an average
over the lifetime of a PV project.

*Irradiance*   Irradiance is an instantaneous measurement of solar power over some
area. The unit of irradiance is watts per square meter (W/m²).

*Insolation*   Insolation is a measurement of the cumulative energy measured over
some area for a defined period of time (e.g. annual, monthly, and daily).
The common unit of insolation is kilowatt hours per square meter
(kWh/m²).

*Uncertainty*   Defined as the contribution of each Modeling step toward the overall
uncertainty of a yield assessment. The overall uncertainty is typically
calculated as the root mean square of the sum of the errors. Considering
the standard deviation ($\sigma$), the lower threshold for the interval ($E \pm \sigma$) is
in correspondence with P84.1.

*Long-term*   Yield Assessment calculated over the lifetime of a PV project including
*yield prediction*   considerations about the degradation of performance and the evolution
of uncertainty due to the reduction of site-dependent variability
(irradiation, temperature, etc.).

*PX*   Exceedance probability is defined as the X % probability that a certain
value will be exceeded.
A P90 value corresponds to a number that has a 90% probability of being
exceeded.

*LCOE:*   The levelized cost of electricity is the cost of generating 1 kWh considering the initial investment (Capital Expenditures), the discounted
operational expenditures and the discounted utilization rate over the
lifetime of the power plant. Its unit is usually euro per kWh (€/kWh)

*NPV*   The net present value is the discounted difference between the present
value of cash inflows and the present value of cash outflows over a
period of time, and it is used to analyze the profitability of an investment
or project over its financial lifetime, usually given in euros (€)

## Author Biography

**David Moser** (PhD in physics). He is responsible for the research group "Photovoltaic
Energy Systems" at the Institute for Renewable Energy, EURAC, Bolzano, Italy. Coordinator
of several EU H2020/Horizone Europe projects (Solar Bankability, TRUST-PV, etc) dealing

with the economic impact of technical risks in PV projects. Coordinator for EURAC of the activities within the IEA PVPS Task 13 "Performance and Reliability of PV modules and systems." He has experience in the performance and reliability of PV modules and installations, with a special focus on building integrated solutions and storage. His expertise varies from modeling and simulations to indoor/outdoor PV systems characterization. He is a vice-chair of the European Technology and Innovation Platform Photovoltaics (ETIP-PV).

# References

Araki, K., Ota, Y., and Yamaguchi, M. (2020). Measurement and modeling of 3D solar irradiance for vehicle-integrated photovoltaic. *Applied Sciences* 10 (3): 872. https://doi.org/10 .3390/app10030872.

French, R.H., Bruckman, L.S., Moser, D. et al. (2021). Assessment of Performance Loss Rate of PV Power Systems [IEA-PVPS T13-22: 2021] IEA PVPS Task 13.

Gueymard, C.A. and Wilcox, S.M. (2009). Assessment of spatial and temporal variability in the solar resource: from radiometric measurements and predictions from models using ground-based or satellite data. *Solar Energy* 85: http://www.sciencedirect.com/science/ article/pii/S0038092X11000855.

H2020 Solarbankability (2017). http://www.sciencedirect.com/science/article/pii/ S0038092X11000855 http://solarbankability.org/home.html (accessed 28 February 2024).

Horvath, I. (2022). Automated PV Digital Twin Based Yield Simulation Framework.

Hunt, K., Blekicki, A., and Callery, R. (2015). Availability of utility-scale photovoltaic power plants. In: *2015 IEEE 42nd Photovoltaic Specialist Conference (PVSC)*, 1–3. https://doi.org/10 .1109/PVSC.2015.7355976.

Jahn, U., Herz, M., Ndrio, E. et al. (2016). Minimizing Technical Risks in Photovoltaic Projects. http://www.solarbankability.org/results.html (accessed 28 February 2024).

Jahn, U., Herz, M., Moser, D. et al. (2018). Managing technical risks in PV investments: how to quantify the impact of risk mitigation measures for different PV project phases? *Progress in Photovoltaics: Research and Applications* 26 (8): 597–607. https://doi.org/10.1002/pip.2970.

Jahn, U., Micheli, L., Tjengdrawira, C. et al. (2020). *Guidelines for Operation and Maintenance in Different Climates* (IEA-PVPS T13-25).

King, D.L., Boyson, W.E., and Kratochvil, J. (2002). Analysis of factors influencing the annual energy production of photovoltaic systems. In: *Photovoltaic Specialists Conference, 2002. Conference Record of the 29th IEEE*, 1356–1361. http://ieeexplore.ieee.org/xpls/abs_all.jsp? arnumber=1190861 (accessed 28 February 2024).

Kjeldstad, T., Lindholm, D., Marstein, E., and Selj, J. (2021). Cooling of floating photovoltaics and the importance of water temperature. *Solar Energy* 218: 544–551. https://doi.org/10 .1016/j.solener.2021.03.022.

Kronthaler, L., Alberti, L., and Moser, D. (2014). The effect of head-wind cooling on the electrical and thermal behaviour of vehicle-integrated photovoltaic (VIPV) systems. In: *29th European Photovoltaic Solar Energy Conference and Exhibition*, 3541–3544. https://doi.org/10 .4229/EUPVSEC20142014-6BO.9.6.

Lawrence, C., Kamuyu, W., Lim, J.R. et al. (2018). Prediction model of photovoltaic module temperature for power performance of floating PVs. *Energies* 11 (2): 447. https://doi.org/10.3390/en11020447.

Lindig, S., Moser, D., Curran, A.J. et al. (2021). International collaboration framework for the calculation of performance loss rates: data quality, benchmarks, and trends (towards a uniform methodology). *Progress in Photovoltaics: Research and Applications* 29 (6): 573–602. https://doi.org/10.1002/pip.3397.

Liu, H., Krishna, V., Lun Leung, J. et al. (2018). Field experience and performance analysis of floating PV technologies in the tropics. *Progress in Photovoltaics: Research and Applications* 26 (12): 957–967. https://doi.org/10.1002/pip.3039.

Maturi, L., Belluardo, G., Moser, D., and Del Buono, M. (2014). BiPV system performance and efficiency drops: overview on PV module temperature conditions of different module types. *Energy Procedia* 48: 1311–1319. https://doi.org/10.1016/j.egypro.2014.02.148.

McLeod, L., Scheltus, D., Miller, M. et al. (2019). Lessons from large-scale solar in Australia. In: *36th European Photovoltaic Solar Energy Conference and Exhibition*, 1906–1913. https://doi.org/10.4229/EUPVSEC20192019-7DO.7.2 (accessed 28 February 2024).

METEOTEST Genossenschaft (2014). *METEONORM v7.12 (V7.12) [Windows]*. METEOTEST Genossenschaft. http://meteonorm.com/ (accessed 28 February 2024).

Moser, D. (2021). *Uncertainties in Yield Assessments and PV LCOE*. IEA PVPS Task 13. https://iea-pvps.org/key-topics/uncertainties-yield-assessments/ (accessed 28 February 2024).

Moser, D., Del Buono, M., Jahn, U. et al. (2016). *Identification of Technical Risks in the PV Value Chain and Quantification of the Economic Impact*. https://doi.org/10.4229/EUPVSEC20162016-5DP.1.1

Müller, B., and Kiefer, K. (2019). Long-term trends of in-plane-irradiance, energy yield and performance for PV systems. *29th International Photovoltaic Science and Engineering Conference*, Xi'an (China) (7 November 2019). https://doi.org/10.13140/RG.2.2.16442.54721

Müller, B., Wild, M., Driesse, A., and Behrens, K. (2014). Rethinking solar resource assessments in the context of global dimming and brightening. *Solar Energy* 99: 272–282.

Muller, B., Heydenreich, W., Reich, N. et al. (2015). Investment risks of utility-scale PV: opportunities and limitations of risk mitigation strategies to reduce uncertainties of energy yield predictions. In: *2015 IEEE 42nd Photovoltaic Specialist Conference (PVSC)*, 1–5. https://doi.org/10.1109/PVSC.2015.7355942.

Oliveira-Pinto, S. and Stokkermans, J. (2020). Assessment of the potential of different floating solar technologies – overview and analysis of different case studies. *Energy Conversion and Management* 211: 112747. https://doi.org/10.1016/j.enconman.2020.112747.

Polo, J., Wilbert, S., Ruiz-Arias, J.A. et al. (2016). Preliminary survey on site-adaptation techniques for satellite-derived and reanalysis solar radiation datasets. *Solar Energy* 132: 25–37. https://doi.org/10.1016/j.solener.2016.03.001.

PVsyst Website (2019). *PVSyst*. Retrieved March 20, 2019. from https://www.pvsyst.com/.

Reise, C. (2018). *iea-pvps.org – Uncertainties in PV System Yield Predictions and Assessments*. http://www.iea-pvps.org/index.php?id=477 (accessed 28 February 2024).

Richter, M.E., Schmidt, T., Kalisch, J. et al. (2015). Uncertainties in PV modelling and monitoring. In: *31st European Photovoltaic Solar Energy Conference and Exhibition*, 1683–1691. https://doi.org/10.4229/EUPVSEC20152015-5BO.12.5.

Scharmer, K. and Greif, J. (2000). *The European Solar Radiation Atlas: Database and Exploitation Software*, vol. 2. Ecole des Mines de Paris.

Schwingshackl, C., Petitta, M., Wagner, J.E. et al. (2013). Wind effect on PV module temperature: analysis of different techniques for an accurate estimation. *Energy Procedia* 40: 77–86. https://doi.org/10.1016/j.egypro.2013.08.010.

Solar Resource for High Penetration and Large Scale Applications. (2022). https://iea-pvps.org/research-tasks/solar-resource-for-high-penetration-and-large-scale-applications/ (accessed 28 February 2024).

Suri, M., Huld, T., Dunlop, E. et al. (2007). Uncertainties in solar electricity yield prediction from fluctuation of solar radiation. In: *22nd European Photovoltaic Solar Energy Conference*. http://hal.archives-ouvertes.fr/hal-00468620/ (accessed 28 February 2024).

Thuman, C., Schnitzer, M., and Johnson, P. (2012). Quantifying the accuracy of the use of measure-correlate-predict methodology for long-term solar resource estimates. Proceedings of the World Renewable Energy Forum (WREF) 2012, American Solar Energy Society, Denver, CO, USA. 2012.

Tjengdrawira, C., and Richter, M. (2016). Review and Gap Analyses of Technical Assumptions in PV Electricity Cost – Report on Current Practices in How Technical Assumptions are Accounted in PV Investment Cost Calculation (Solar Bankability WP3 Deliverable D3.1; EC-Funded Solar Bankability Project Grant Agreement #649997).

Tjengdrawira, C., Moser, D., Jahn, U. et al. (2017). *PV Investment Technical Risk Management: Best Practice Guidelines for Risk Identification, Assessment and Mitigation* [D5.8]. www.solarbankability.org (accessed 28 February 2024).

Vartiainen, E., Masson, G., Breyer, C. et al. (2015). PV LCOE in Europe 2014-30. In: *European PV Technology Platform*. http://scholar.google.com/scholar?cluster=6631314094761898237&hl=en&oi=scholarr (accessed 28 February 2024).

Vartiainen, E., Masson, G., Breyer, C. et al. (2020). Impact of weighted average cost of capital, capital expenditure, and other parameters on future utility-scale PV levelised cost of electricity. *Progress in Photovoltaics: Research and Applications* 28 (6): 439–453. https://doi.org/10.1002/pip.3189.

Wild, M. (2009). Global dimming and brightening: a review. *Journal of Geophysical Research* 114: https://doi.org/10.1029/2008JD011470.

Wild, M., Gilgen, H., Roesch, A. et al. (2005). From dimming to brightening: decadal changes in solar radiation at earth's surface. *Science* 308 (5723): 847–850. https://doi.org/10.1126/science.1103215.

Woyte, A., De Brabandere, K., Sarr, B., and Richter, M. (2016). The quality of satellite-based irradiation data for operations and asset management. In: *32nd European Photovoltaic Solar Energy Conference and Exhibition*, 1470–1474.

# Part Ten

# Sustainability

# 32

# Terawatt Sustainability

*Masafumi Yamaguchi*

*Toyota Technological Institute, Nagoya, Japan*

## 32.1 Introduction

Solar Photovoltaics (PVs) are expected to play a massive role in the world in order to solve climate change. Figure 32.1 shows modeled cumulative installed capacity for Shell Sky scenario (Shell 2018) and Breyer model (Breyer et al. 2017, Haegel et al. 2019). Cost-efficient climate change mitigation requires installing a total of 20–60 TW PVs until 2050 and 70–170 TW until 2100. Although solar PV is thought to be ready to become one of the main energy sources (Victoria et al. 2021), efficient usage of PV is very important in order to realize terawatt scale PVs (Victoria et al. 2021; Choudhary and Srivastava 2019). In addition, considerations for broader impacts of supply chain, manufacturing, deployment on the environment and on societies with various levels and models of PV uptake and penetration toward terawatt scale PVs (Goldschmidt et al. 2021).

In Section 32.2, a brief overview of sustainability of PVs by summarizing references is presented from the technical viewpoints. In Section 32.3, discussion about sustainability of material resources for PVs is shown. In Section 32.4, effectiveness of concentrator photovoltaic (CPV) is presented in order to solve material limitations of various solar cell modules.

## 32.2 Brief Overview of Sustainability of Photovoltaics

The European Strategic Research & Innovation Agenda (SRIA) for PVs (Topic et al. 2021) is thought to be very useful for realizing terawatt PVs. Table 32.1 shows 5 challenges by the SRIA. In this section, technologically important issues for TW PV are overviewed by referring to some papers.

Several works (Victoria et al. 2021; Choudhary and Srivastava 2019; Goldschmidt et al. 2021; Topic et al. 2021; Wagner et al. 2021; Gervals et al. 2021; Jones-Albertus 2022) have pointed out that an increase in installation of PV systems, as shown in Figure 32.1, increases problems. Table 32.2 shows the summary (Wagner et al. 2021; Gervals et al. 2021) of problems of PVs, resource demand estimated, and technology solutions for terawatt scale PVs.

*Photovoltaic Solar Energy: From Fundamentals to Applications, Volume 2*, First Edition.
Edited by Wilfried van Sark, Bram Hoex, Angèle Reinders, Pierre Verlinden, and Nicholas J. Ekins-Daukes.
© 2024 John Wiley & Sons Ltd. Published 2024 by John Wiley & Sons Ltd.
Companion website: www.wiley.com/go/PVsolarenergy

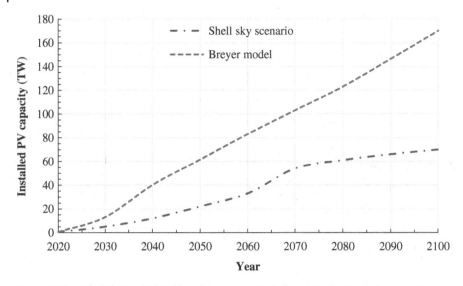

**Figure 32.1** Modeled cumulative installed photovoltaics capacity for Shell sky scenario and Breyer model.

The primary energy demand for production and installation of a PV system is estimated (Wagner et al. 2021; Gervals et al. 2021) at 40 EJ/year (that is about 7% of the global primary energy consumption 585 EJ of 2017) in the case of 5 TW PV production in 2010. Decrease in energy consumption of module production such as development of high-efficiency solar cells and modules, low-temperature processing of materials, solar cells, and modules, reduction in wafer thickness and material intensity, and so forth, is very important in order to realize sustainability of PVs. As a result of usage of such technological solutions, decrease in primary energy demand for PV production and installation from 20 GJ/kW in 2010 to 4.1–7.7 GJ/kW in 2050 has been estimated (Wagner et al. 2021).

Cumulative green gas emission in the case of 5 TW PV production is estimated (Wagner et al. 2021; Gervals et al. 2021) at 12 GtCO$_2$-eq. that is approximately 4% of the CO$_2$ budget 290 Gt to contain the anthropogenic global warming below 1.5 °C. A further removal of CO$_2$ from the atmosphere is necessary to lead to larger budget being available for PVs. Increasing share of renewables in the energy mix is essential to reducing fossil-fuel-based energy consumption. As a result of advancement of renewable energies, annual greenhouse gas emission has been estimated (Wagner et al. 2021) to drop from 0.5 GtCO$_2$-eq./year in 2030 to 0 in 2060.

Total module area is estimated (Wagner et al. 2021) to increase from 4400 km$^2$ in 2020 to 360,000 km$^2$ (252,000–466,000 km$^2$) in 2100. The module area values for 2100 correspond roughly to the land area of the United Kingdom or Sweden, respectively. In addition to development of high-efficiency solar cells and modules, importance of effective applications such as agri-PVs combining agriculture land and PV, vehicle-integrated PV placing on car roofs and so forth, and floating PV placing water bodies as well as building and industry has been pointed out by the SRIA (Topic et al. 2021; Victoria et al. 2021; Goldschmidt et al. 2021; Becca Jones-Albertus 2022), DOE, has presented that buildings, industry, and

**Table 32.1** 5 challenges by the European Strategic Research & Innovation Agenda (SRIA) for PV (Topic et al. 2021).

| Challenge | Items | Objectives |
| --- | --- | --- |
| 1 | Performance enhancement and cost reduction [through advanced PV technologies, manufacturing, and applications] | 1. PV modules with higher efficiencies and lower costs<br>  * Si PV modules<br>  * Perovskite PV modules<br>  * Thin-film (non-perovskite) PV modules<br>  * tandem PV modules<br>2. System design for lower LCoE of various applications<br>  * Balance of Systems (BoS) and energy yield improvement<br>3. Digitalization of PVs (potential of performance improvement and cost reduction by information management)<br>  * Digitalization of PV manufacturing<br>  * Digitalization of PV systems |
| 2 | Lifetime, reliability, and sustainability enhancements [through advanced PV technologies, manufacturing, and applications] | 1. Sustainable and circular solar PV<br>  * Reduce: Low environmental impact materials, products, and processes<br>  * Reuse: Design systems and O&M for reuse<br>  * Recycle<br>  * Technologies for sustainable manufacturing<br>  * Eco-labeling and energy-labeling<br>2. Establishing a lifetime relevant quality assurance along the whole value chain at component and system level<br>  * Quality assurance to increase lifetime and reliability<br>  * Increased field performance and reliability<br>  * Bankability, warranty, and contractual terms<br>  * Interoperability and communication standardization |

*(Continued)*

**Table 32.1** (Continued)

| Challenge | Items | Objectives |
|---|---|---|
| 3 | New applications through integration of PVs [for diversified and double-use deployment and enhanced value] | Physical integration of PV into other applications, infrastructure, products, and environments/landscapes<br>* PV in buildings<br>* Vehicle Integrated PV<br>* Agrivoltaics and landscape integration<br>* Floating PV<br>* Infrastructure Integrated PV<br>* "Low-power" energy harvesting PV |
| 4 | Smart energy system integration of PVs [for large-scale deployment and high penetration] | Energy system integration<br>* More intelligence in distributed control<br>* Improved efficiencies by integration of PV systems in DC networks<br>* Hybrid systems including demand flexibility (PV+Wind+Hydro with embedded storage+batteries +green hydrogen/fuel cells, or gas turbines)<br>* Aggregated energy and virtual power plants (VPPs)<br>* Interoperability in communication and operation of RES smart grids |
| 5 | Socioeconomic aspects of high contribution of PV [to the clean energy transition] | 1. Higher awareness of benefits that solar PV bridge (potential for promotion of PV and creation of energy policy)<br>  * Wide societal involvement and participation in solar PV deployment<br>  * Developing a PV hotbed for urban implementation<br>2. Economic and sustainability benefits |

**Table 32.2** Summary (Wagner et al. 2021; Gervals et al. 2021) of problems of PVs, resource demand estimated, and technology solutions for terawatt scale PVs.

| Problems | | Resource demand estimated | | | Technology solutions |
|---|---|---|---|---|---|
| | | Current (180 GW production) | 2050 (1 TW production) | 2100 (5 TW production) | |
| Energy consumption (EJ/year) | | 2 | 10 | 23.6 | * High-efficiency modules<br>* Low-temperature processing of materials, cells, and modules |
| Cumulative $CO_2$ emission (Gt$CO_2$-eq) | | 1.5 | 10 | 12 | * High-efficiency modules<br>* Low-temperature processing of materials, cells, and modules<br>* Increasing share of renewables in the energy mix |
| Module area (km$^2$) | | 4400 (in 2020) | 90,000 | 360,000 | * High-efficiency modules<br>* Effective applications such as Agrivoltaics, VIPV, Floating PV |
| Material consumption | Glass (Mt) | 20 | 50 (single glass)<br>100 (double glass) | 122 (single glass)<br>330 (double glass) | * Thinner glass<br>* Glass foil<br>* Polymer foil<br>* High-efficiency modules<br>* Recycling of modules |
| | Ag (t) | 1959 (in 2018) | 12,000 | 18,000 | * Further reduction in Ag<br>* Usage of Cu as alternative<br>* High-efficiency modules<br>* Recycling of modules |

transportation will share 55%, 25%, and 25% as end uses in cumulative PV capacity of 1.6 TW (45% of electricity demand) in the United States in 2050.

Currently, almost all commercial PV modules use at least one glass sheet. Glass consumption for PV has been estimated (Wagner et al. 2021) to exceed annual glass production of 84 Mt in 2074 from about 20 Mt in 2020. As technology solutions, thinner glass, glass foil, polymer foil, and high-efficiency solar cell modules are effective for reduction in the usage of glass consumption. Recycling of modules is very effective.

Ag is widely used as front metallization of Si solar cells. Annual Ag consumption of PV industry was 1959 t in 2018 (Wagner et al. 2021). That corresponds to 7.9% of the total annual Ag mine production of 28,037 t (Wagner et al. 2021). If the Ag consumption per cell remains constant, the Ag demand of the PV industry will exceed today's global Ag production in the year 2051. As technology solution, further reduction in Ag, usage of Cu as alternative contacts, and high-efficiency modules are effective for reduction in usage of Ag consumption. Recycling of modules is very effective.

Detail discussions about material resources are presented in the Section 32.3. As common technology solution to solve several problems for sustainability of PVs, development of high-efficiency solar cell modules is one of key issues for solving such problems. Figure 32.2 shows scenario (Wagner et al. 2021) for improving solar cell module efficiency. Multi-junction solar cells (Yamaguchi et al. 2021), Si tandem solar cells (Yamaguchi et al. 2018), and concentrator solar cells (Yamaguchi and Luque 1999) are very attractive for realizing higher solar cells and modules. Although the maximum efficiency limit of single-junction solar cells is 29.4%. (Richter et al. 2013), 39.5% (France et al. 2022) under 1-sun with III-V 3-junction solar cell, 35.9% (Essig et al. 2017) and 36.1% (Schygulla et al. 2024) under 1-sun with III-V/Si 3-junction tandem solar cell, 33.9% (Green et al. 2024) under 1-sun with perovskite/Si 2-junction tandem solar cell, and 47.1% (Geisz et al. 2020) under 143-suns with 6-junction solar cell have been demonstrated.

Figure 32.3 shows calculated efficiency potential of multi-junction solar cells under 1-sun operation as a function of number of junctions (Yamaguchi et al. 2021). The 2-junction

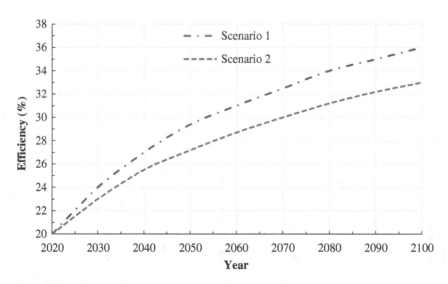

**Figure 32.2** Scenario for improving solar cell module efficiency.

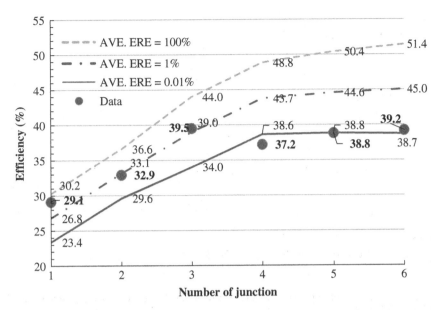

**Figure 32.3** Calculated efficiencies of III–V compound multi-junction solar cells under 1-sun condition as a function of the number of junctions and average external radiative efficiency (ERE) in comparison with efficiency data (best laboratory efficiencies Green et al. 2024). Source: (Reproduced with permission from Yamaguchi et al. (2021). Updated).

and 3-junction solar cells have higher potential efficiencies of more than 36% and 44%, respectively.

Recycling is also very important for maintaining sustainability of PVs. The WEEE (World Electrical and Electronic Equipment) directive specifies targets for PV of 85% recovery and 80% preparation for reuse and recycling (Gervals et al. 2021). These targets are currently achieved by recycling Al, glass, and Cu (Gervals et al. 2021). High-value recycling for Si-based modules is still in the experimental phase but already shows positive results (Heath et al. 2020) with the recovery of Ag with 74–94% for 99% purity, and Si with 90–97% for metallurgical Si. Importance of the development of long lifetime 50-year modules (Heath et al. 2020), compared to today's industry standard 25- or 30-year lifetimes, by understanding and addressing the factors that cause wear out and failure in various types of PV modules. Development of long lifetime modules will reduce the materials consumption.

Importance of continuous improvements in efficiency, cost, and reliability of solar cell modules, effective utilization, and yield improvements of materials and storage, recycling, intelligent integration of infrastructures, circular economy, and sustainable investment and growth have been pointed out by Haegel (2020). In order to upgrade grids, stability with inverter-based resources, transmission, and storage is very important. In addition, demand management, new infrastructure and intelligent integration, policy, and research and development are also very important for TW challenge (Haegel 2020).

## 32.3 Discussion About Sustainability of Material Resources

PVs will play a key role in a future net zero greenhouse gas emission energy systems. According to the International Technology Roadmap for PVs (ITRPV 15th Edition)

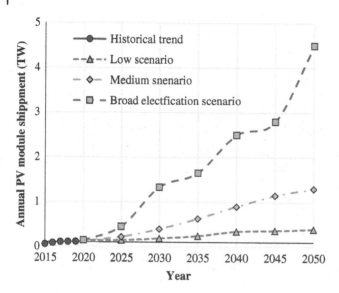

**Figure 32.4** ITRPV scenario for annual PV module shipment (International Technology Roadmap for Photovoltaics (ITRV) 2024).

(International Technology Roadmap for Photovoltaics (ITRV) 2024), shipments of PV modules with 0.4 TW (medium scenario) and 1.3 TW (broad electrification scenario) in 2030 and 1.3 TW (medium scenario) and 4.5 TW (broad electrification scenario) in 2050 are estimated as shown in Figure 32.4. However, there are limitations of solar cell module production capacity that are dependent on annual supply of materials for solar cells and modules. The CPV (Yamaguchi and Luque 1999; Swanson 2000; Fraas 2015; Wiesenfarth et al. 2018) has great potential for higher efficiency and lower cost compared to conventional crystalline Si PV and thin-film PV. In this chapter, sustainability of material resources for various solar cells and effectiveness of CPV to overcome limitations of solar cell module production capacity are presented.

Material availability and maximum production capacity of Si solar cells have been discussed by Zhang et al. (2021). Assuming an upper material consumption limit as 20% of global supply, manufacturing capacities (Zhang et al. 2021) of passivated emitter and rear cell (PERC) with Ag consumption of 16.3 mg/W, tunnel oxide passivated contact (TOP-Con) with Ag consumption of 26.2 mg/W, and silicon heterojunction (SHJ) solar cells with Ag consumption of 35.4 mg/W are shown to be 380, 230, and 170 GW, respectively, in the Ag-limited case as shown in Figure 32.5. In the case of In consumption-limited, 35 and 170 GW are estimated for SHJ with In consumption of 10.7 mg/W, and Si tandem solar cells with In consumption of 3.7 mg/W. By reduction in Ag and In consumption and development of alternative earth-abundant materials such as Cu instead of Ag, Al-doped zinc oxide (AZnO) instead of ITO (indium-tin oxide), and so forth, multi-TW scale manufacturing of Si solar cells is shown to be possible. Finally, global supply of mother elements is thought to be a limitation factor for manufacturing capacities of various solar cells. In this study, manufacturing capacities of various solar cells are analyzed by using global supply.

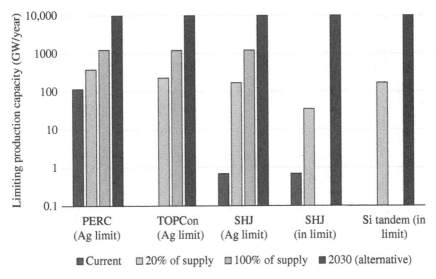

**Figure 32.5** Current shipment volume and limiting production capacities of Si solar cells (Victoria et al. 2021) and updated values

**Table 32.3** Global supply of various materials

| Material | Annual supply (t) | Year (References) |
|---|---|---|
| Ag | 2900 | 2019 (Zhang et al. 2021) |
| Al | 5.87E+07 | 2017 (JOGMEC 2018) |
| As | 59000 | 2016 (Rech et al. 2021) |
| Cd | 25000 | 2020 (Rech et al. 2021) |
| Cu | 2.04E+07 | 2018 (JOGMEC 2018) |
| Ga | 315 | 2017 (JOGMEC 2018) |
| Ge | 134 | 2017 (JOGMEC 2018) |
| In | 2100 | 2019 (Zhang et al. 2021) |
| S | 7.8E+07 | 2020 (Rech et al. 2021) |
| Se | 3300 | 2020 (Rech et al. 2021) |
| Si | 7.43E+06 | 2017 (JOGMEC 2018) |
| Sn | 3.71E+05 | 2017 (JOGMEC 2018) |
| Te | 440 | 2020 (Rech et al. 2021) |
| Zn | 1.2E+07 | 2020 (Rech et al. 2021) |

Table 32.3 shows global supply of elements for various solar cells (Zhang et al. 2021; JOG-MEC 2018; Rech and Schatz 2021). Ag, Ga, Ge, In, Se, and Te are rare materials. Maximum manufacturing capacities of various solar cells were estimated by assuming that the PV industry can sustainably use 20% and 100% of global material supply, and efficiencies and

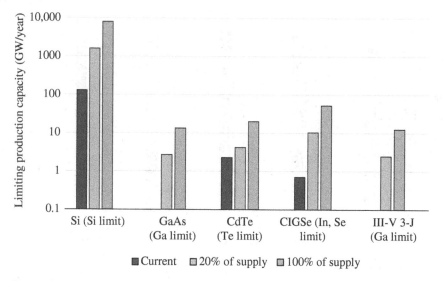

**Figure 32.6** Current shipment volume and limiting production capacities of various solar cells.

layer thicknesses with 25% and 100 μm for Si, 25% and 2 μm for GaAs, 20% and 2 μm for CdTe, 20% and 2 μm for CIGSe, and 35% and 2 μm for III–V 3-junction (3-J) solar cells.

Figure 32.6 shows current shipment volume and limiting production capacities of various solar cells. Although Si solar cell module shipment volume was about 130 GW in 2020, maximum manufacturing capacities for Si solar cells are estimated to be 1.6 and 8 TW in the cases of 20% and 100% consumptions of global supply. However, maximum manufacturing capacities of the other solar cells are estimated to be 2.7 GW (20% consumption of global supply) and 13.3 GW (100% consumption of global supply) for GaAs, 4.2 GW (20% consumption) and 21 GW (100% consumption) for CdTe, 10.4 GW (20% consumption) and 52.1 GW (100% consumption) for CIGSe, 2.5 GW (20% consumption) and 12.5 GW (100% consumption) for III–V 3-J solar cells as shown in Figure 32.6, and are quite low compared to maximum manufacturing capacities of Si solar cells. Alternative approaches are necessary in order to realize multi-TW scale manufacturing for the other solar cells.

## 32.4 Effectiveness of CPV to Overcome Limitation of Solar Cell Module Production Capacity

The CPV has great potential for higher efficiency and reduction potential of solar cell materials (Yamaguchi and Luque 1999). Figure 32.7 shows comparison of calculated and experimental efficiencies of Si concentrator solar cell reported by Amonix Slade and Garboushain (2005)), InGaP/GaAs/InGaAs 3-junction concentrator solar cell reported by Sharp (Sasaki et al. 2013; Yamaguchi et al. 2016), GaAs concentrator solar cell reported by FhG-ISE (Schilling et al. 2018), and CIGS concentrator solar cell reported by NREL (Ward et al. 2014) as a function of concentration ratio (Yamaguchi and Araki 2019). As shown in Figure 32.7, high concentrating operations of various solar cells with 50-suns

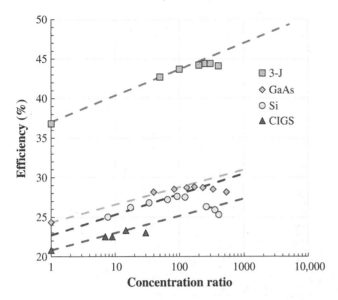

**Figure 32.7** Comparison of calculated and experimental efficiencies of Si concentrator solar cell reported by Amonix, InGaP/GaAs/InGaAs 3-junction concentrator solar cell reported by Sharp, GaAs concentrator solar cell reported by FhG-ISE, and CIGSe concentrator solar cell reported by NREL as a function of concentration ratio.

for CIGS, 100-suns for Si, 200-suns for GaAs, and 300-suns for III–V 3-J solar cells are possible. The CPV is thought to be very useful for reduction in consumption of solar cell materials.

By applying CPV, higher power-generating systems are thought to be produced. By multiplying maximum manufacturing capacities of various solar cells and considering efficiency improvements under concentrator operation shown in Figure 32.7, their possible production capacities were estimated by considering efficiency improvements and by multiplying concentration factors such as 50-times for CIGS, 100-times for Si, 200-times for GaAs, and 300-times for III–V 3-J solar cells under concentration operation. Although CPV utilizes direct sunlight, the ratio of direct solar irradiance compared to global solar irradiance in the case of Japan is distributed from 56.4% to 86.6% (Araki and Yamaguchi 2003), solar capacities were analyzed in order to compare with the ITRPV scenario for annual PV module shipment (International Technology Roadmap for Photovoltaics (ITRV) 2024) as shown in Figure 32.4 in this study. Figure 32.8 shows limiting production capacities of various solar cells by concentrator operation. Even in the case of 20% consumption of global supply, larger electricity generations with 192 TW for Si (Si limit), 650 GW for GaAs (Ga limit), 260 GW for CdTe (Te limit), 650 GW for CIGSe (In limit), and 960 GW for III–V 3-J (Ga limit) solar cells are expected. Especially, concentrating operation of various solar cells are thought to be very attractive in order to realize ITRPV's broad electrification scenario as shown in Figure 32.4, and to create zero carbon emission society. In addition, reuse and recycling of materials composing solar cells, modules, and PV systems are also very important for future large-scale deployment of PVs.

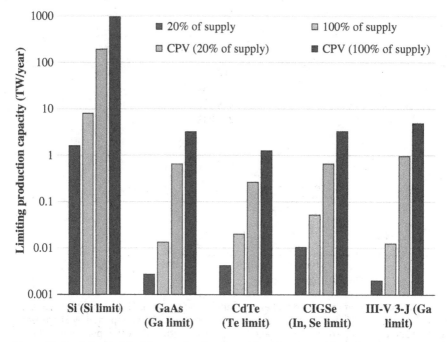

**Figure 32.8** Estimated limiting production capacities of various solar cells.

## 32.5 Conclusion

PVs will play a key role in a future net zero greenhouse gas emission energy systems. In this chapter, sustainability of PVs was overviewed. For realizing terawatt sustainability of PVs, continuous improvements in efficiency, cost, and reliability of solar cell modules, effective utilization, yield improvements of materials and storage, recycling, intelligent integration of infrastructures, circular economy, and sustainable investment and growth are very important. In order to upgrade grids, stability with inverter-based resources, transmission, and storage is very important. In addition, demand management, new infrastructure, intelligent integration, policy, and research and development are also very important for TW challenge. Regarding the limitation of solar cell module production capacity due to global material supply, effectiveness of the CPV was discussed. The CPV has great potential for higher efficiency, lower cost, and potential to overcome sustainability of material resources compared to conventional crystalline Si PV and thin-film PV. This chapter presented impact of material consumption on annual production capacities of various solar cells by using CPV. Merits of CPV compared to conventional PV (1-sun use) were estimated to be 190/1.6 TW for Si (Si limit), 650/2.7 GW for GaAs (Ga limit), 250/4.2 GW for CdTe (Te limit), 650/10.4 GW for CIGSe (In, Se limits), and 950/2 GW for III–V 3-junction solar cells (Ga limit). In addition, reuse and recycling of materials composing solar cells, modules, and PV systems are also very important for future large-scale deployment of PVs.

## Acknowledgments

The author expresses thanks to the NEDO for supporting studies, and to Dr. Y. Zhang and Prof. B. Hallam, UNSW, Prof. K. Araki, Prof. Y. Otha, and Prof. K. Nishioka, Univ. Miyazaki and Ms. I. Kaizuka, RTS for their fruitful discussion.

## Author Biography

**Dr. Masafumi Yamaguchi** is Professor Emeritus and Invited Research Fellow at the Toyota Technological Institute, and the Chairman of the PV R&D Review Committee of the New Energy and Industrial Technology Development Organization (NEDO). He has contributed to the development of high-efficiency multi-junction, Si tandem, concentrator and space solar cells and vehicle integrated photovoltaics. He has received numerous awards such as the Becquerel Prize from the European Commission in 2004, the William Cherry Award from the IEEE in 2008, the PVSEC Award in 2011 and the WCPEC Award in 2014.

## References

Araki, K. and Yamaguchi, M. (2003). Sunshine environment and spectrum analysis for concentrator PV systems in Japan. *Solar Energy Materials & Solar Cells* 75, 715. https://doi.org/10.1016/S0927-0248(02)00139-3.

Breyer, C., Bogdanov, D., Gulagi, A. et al. (2017). On the role of solar photovoltaics in global energy transition s cenarios. *Progress in Photovoltaics* 25: 727. https://doi.org/10.1002/pip.2885.

Choudhary, P. and Srivastava, B.K. (2019). Sustainability perspectives- a review for solar photovoltaic trends and growth opportunities. *Journal of Cleaner Production* 227: 589. https://doi.org/10.1016/j.jclepro.2019.04.107.

Essig, S., Allebe, C., Remo, T. et al. (2017). Raising the one-sun conversion efficiency of III-V/Si solar cells to 32.8% for two junctions and 35.9% for three junctions. *Nature Energy* 2: 17144. https://doi.org/10.1038/nenergy.2017.144.

Fraas, L.M. (2015). Photovoltaics: comparison of PV cell types. In: *Handbook of Clean Energy Systems*. Wiley https://doi.org/10.1002/9781118991978.hces063.

France, R.M., Geisz, J.F., Song, T. et al. (2022). Triple-junction solar cells with 39.5% terrestrial and 34.2% space efficiency enabled by thick quantum well superlattices. *Joule* 6: 1121. https://doi.org/10.1016/j.joule.2022.04.024.

Geisz, J.F., France, R.M., Schulte, K.L. et al. (2020). Six-junction III-V solar cells with 47.1% conversion efficiency under 143 Suns concentration. *Nature Energy* 5: 326. https://doi.org/10.1038/s41560-020-0598-5.

Gervals, E., Heroeg, S., Nold, S., and Welk, K.-A. (2021). Sustainability strategies for PV: framework, status and needs. *EPJ Photovoltaics* 12: 5. https://doi.org/10.1051/epjpv/2021005.

Goldschmidt, J.C., Wagner, L., Pietzcker, R., and Friedrich, L. (2021). Technological learning for resource efficient terawatt scale photovoltaics. *Energy & Environmental Science* 14: 5147. https://doi.org/10.1039/d1ee02497c.

Green, M.A., Dunlop, E.D., Yoshita, M. et al. (2024). Solar cell efficiency tables (Version 63). *Progress in Photovoltaics* 32: 3.

Haegel, N. (2020). Toward 100% renewable energy: trajectories and challenges to terawatt scale electricity generation with PV. In: *Presented at the 47th IEEE Specialists Conference, Virtual, June.*

Haegel, N.M., Atwater, H., Barnes, T. et al. (2019). Terawatt-scale photovoltaics: Transform global energy. *Science* 364: 836. https://doi.org/10.1126/science.aaw1845.

Heath, G.A., Silverman, T.J., Kempe, M. et al. (2020). Research and Development Priorities for Silicon Photovoltaic Module Recycling to Support a Circular Economy. *Nature Energy* 5: 502. https://doi.org/10.1038/s41560-020-0645-2.

International Technology Roadmap for Photovoltaics (ITRV) (2024). itrpv.vdma.org (accessed 26 March 2024).

JOGMEC (2018). mric.jogmec.go.jp/wp-content/uploads/2019/03/material_flow.

Jones-Albertus, B. (2022). Technologies for a solar-powered future. In: *Presented at the 49th IEEE Photovoltaic Specialists Conference, June 5-10, Philadelphia, USA.*

Rech, C. and Schatz, M. World Mining Data 2021, *Minerals Production*, Volume 36 (Vienna, 2021). ISBN 978-3-9010-74-50-9.

Richter, A., Hermilr, M., and Glunz, S.W. (2013). Reassessment of the Limiting Efficiency for Crystalline Silicon Solar Cells. *IEEE Journal of Photovoltaics* 3: 1184. https://doi.org/10.1109/JPHOTOV.2013.2270351.

Sasaki, K., Agui, T., Nakaido, K. et al. (2013). Development of InGaP/GaAs/InGaAs inverted triple junction concentrator solar cells. *AIP Conference Proceedings*, vol. 1556, 22. AIP Publishing http://proceedings.aip.org.

Schilling, C.L., Hohn, O., Micha, D.N. et al. (2018). Combining Photon Recycling and Concentrated Illumination in a GaAs Heterojunction Solar Cell. *IEEE Journal of Photovoltaics* 8: 348. https://doi.org/10.1109/JPHOTOV.2017.2777104.

Schygulla, P., Müller, R., Lackner, D. et al. (2024). Wafer-bonded two-terminal III-V//Si triple-junction solar cell with power conversion efficiency of 36.1% at AM1.5g. *Progress in Photovoltaics*. https://doi.org/10.1002/pip3769.

Shell (2018). *Sky Scenario*. www.shell.com/skyscenario.

Slade, A. and Garboushain, V. (2005). 27.6% Efficient Silicon Concentrator Solar Cells for Mass Production. *Technical Digest of the 15th International Photovoltaic Science and Engineering Conference, Shanghai, China*, 701. https://www.researchgate.net/publication/267779112.

Swanson, R.M. (2000). The promise of concentrators. *Progress in Photovoltaics* 8, 93. https://doi.org/10.1002/(SICI)1099-159X (200001/02)8:1<93::AID-PIP303>3.0.CO;2-S.

Tockhorn, P., Sutter, J., Cruz, A. et al. (2022). Nano-optical designs enhance monolithic perovskite/silicon tandem solar cells toward 29.8% efficiency. *Research Square* https://doi .org/10.21203/rs.3.rs-1439562/v1.

Topic, M., Drozdowski, R., and Sinke, W. (2021). European strategic research & innovation agenda (SRIA) for photovoltaics – fit for 55% and climate neutrality. In: *Presented at the 38th*

*European Photovoltaic Conference and Exhibition, Sep., Virtual.* https://media.etip-pv.eu/filer_public/90/53/90536774-1311-4c54-aea7-05da84a2c1fb/sria_presentation_eu_pvsec_2021.pdf.

Victoria, M., Haegel, N., Peters, I.M. et al. (2021). Solar Photovoltaics is Ready to Power a Sustainable Future. *Joule* 5: 1041. https://doi.org/10.1016/j.joule.2021.03.005.

Wagner, L., Friedrich, L., Hinsch, A. et al. (2021). The $CO_2$ and resource footprint of tomorrow's PV industry. In: *Presented at the 38th European Photovoltaic Conference and Exhibition, Sep. Virtual.*

Ward, J.S., Egaas, B., Noufi, R. et al. (2014). Cu(In,Ga)Se2 solar cells measured under low flux optical concentration. *Proceedings of the 40th IEEE Photovoltaic Specialists Conference (IEEE, New York)*, 2934. https://doi.org/10.1109/PVSC.2014.6925546.

Wiesenfarth, M., Anton, I., and Bett, A.W. (2018). Challenges in the design of concentrator photovoltaic (CPV) modules to achieve highest efficiencies. *Applied Physics Reviews* 5: 041601. https://doi.org/10.1063/1.5046752.

Yamaguchi, M. and Araki, K. (2019). Concentrated solar cells. In: *Encyclopedia of Sustainability Science and Technology* (ed. R.A. Meyers). Springer Nature https://doi.org/10.1007/978-1-4939-2493-6_1062-1.

Yamaguchi, M. and Luque, A. (1999). High efficiency and high concentration in photovoltaics. *IEEE Transactions on Electron Devices* 46: 2139. https://doi.org/10.1109/16.792009.

Yamaguchi, M., Takamoto, T., Araki, K., and Kojima, N. (2016). Recent results for concentrator photovoltaics in Japan. *Japanese Journal of Applied Physics* 55: 04EA05. https://doi.org/10.7567/JJAP.55.04EA05.

Yamaguchi, M., Lee, K.-H., Araki, K., and Kojima, N. (2018). A review of recent progress in heterogeneous silicon tandem solar cells. *Journal of Physics D: Applied Physics* 51: 133002. https://doi.org/10.1088/1361-6463/aaaf08.

Yamaguchi, M., Dimroth, F., Geisz, J.F., and Ekins-Daukes, N.J. (2021). Multi-junction solar cells paving the way for super high-efficiency. *Journal of Applied Physics* 129: 240901. https://doi.org/10.1063/5.0048653.

Zhang, Y., Kim, M., Wang, I. et al. (2021). *Energy & Environmental Science* 14: 5587. https://doi.org/10.1039/dlee01814k.

# 33

# Life Cycle Assessment of Photovoltaics

*Vasilis Fthenakis and Enrica Leccisi*

*Earth and Environmental Engineering Department, Center of Life Cycle Analysis, Columbia University, New York, NY, USA*

## 33.1 Introduction

Life Cycle Assessment (LCA) is a comprehensive framework for quantifying the environmental impacts caused by material and energy flows in each of the stages of the "life cycle" of a product or an activity. It can describe all the life stages, from "cradle to grave," thus from raw materials extraction to end of life. The cycle typically starts from the mining of materials from the ground and continues with the processing and purification of the materials, manufacturing of the compounds and chemicals used in processing, manufacturing of the product, transport, if applicable, installation, use, maintenance, and eventual decommissioning and disposal and/or recycling. To the extent that materials are reused or recycled at the end of their first life into new products, then the framework is extended to "cradle to cradle." The most common metrics used in comparative life-cycle evaluations of the environmental impacts of energy systems are energy payback time (EPBT), energy return on energy investment (EROI), greenhouse gas (GHG) emissions, toxic emissions, land use, and water use.

## 33.2 Methodology

The methodology employed in photovoltaics (PV) LCA studies should conform to the ISO standards 14040 and 14044 (ISO 2006a,b) and should adhere to the IEA PVPS Guidelines (Fthenakis et al. 2011). The ISO standards prescribe four steps for conducting an LCA: (a) goal and scope definition; (b) life cycle inventory (LCI); (c) life cycle impact assessment (LCIA); and (d) interpretation. The first step calls for a clear definition of the research objective and the boundaries of the system used for the analysis. The inventory step quantifies the flows of materials, energy and emissions in each stage of the life cycle of a PV system. In LCIA, the impact on the environment of energy consumption, resource consumption, pollutant emissions, and GHG emissions are quantified, and their cumulative impact on the environment is estimated. The fourth step of the LCA, as defined by the ISO, is an interpretation of the results; this is extremely important as different assumptions can result in very

*Photovoltaic Solar Energy: From Fundamentals to Applications, Volume 2*, First Edition.
Edited by Wilfried van Sark, Bram Hoex, Angèle Reinders, Pierre Verlinden, and Nicholas J. Ekins-Daukes.
© 2024 John Wiley & Sons Ltd. Published 2024 by John Wiley & Sons Ltd.
Companion website: www.wiley.com/go/PVsolarenergy

divergent results. To this end, the International Energy Agency (IEA) Task 12 (Fthenakis et al. 2011) adds specificity regarding what assumptions are valid in stationary PV LCAs and the interpretation of the results.

## 33.3 Cumulative Energy Demand (CED) During the Life of a PV System

The life-cycle stages of PV's involve (1) the production of raw materials, (2) their processing and purification, (3) the manufacture of solar cells, modules, and balance of system (BOS) components, (4) the installation and operation of the systems, and, (5) their decommissioning, disposal, or recycling (Figure 33.1). Typically, separate LCAs are undertaken for the modules and the BOS components (inverters, transformers, mounting, supports, and wiring) as the module's technologies evolve more rapidly and entail more options than the BOS structures.

The life-cycle cumulative energy demand (CED) (Frischknecht et al. 2007) of a PV system is the sum of the (renewable and nonrenewable) primary energy harvested from the geo-biosphere in order to supply the direct energy (e.g. fuels, electricity) and material (e.g. Si, metals, and glass) inputs used in all its life-cycle stages (excluding the solar energy directly harvested by the system during its operation). Thus,

$$\text{CED [MJ}_{\text{PE-eq}}] = E_{\text{mat}} + E_{\text{manuf}} + E_{\text{trans}} + E_{\text{inst}} + E_{\text{EOL}} \tag{33.1}$$

where,

$E_{\text{mat}}$ [MJ$_{\text{PE-eq}}$]: Primary energy demand to produce materials comprising PV system
$E_{\text{manuf}}$ [MJ$_{\text{PE-eq}}$]: Primary energy demand to manufacture PV system
$E_{\text{trans}}$ [MJ$_{\text{PE-eq}}$]: Primary energy demand to transport materials used during the life cycle
$E_{\text{inst}}$ [MJ$_{\text{PE-eq}}$]: Primary energy demand to install the system
$E_{\text{EOL}}$ [MJ$_{\text{PE-eq}}$]: Primary energy demand for end-of-life management

The life-cycle nonrenewable CED (NR-CED) is a similar metric in which only the nonrenewable primary energy harvested is accounted for; details are given elsewhere (Fthenakis et al. 2011).

### 33.3.1 Energy Payback Time (EPBT)

Energy pay-back time (EPBT) is defined as the period required for a renewable energy system to generate the same amount of energy (*in terms of equivalent primary energy*) that was used to produce (and manage at end-of-life) the system itself (Fthenakis et al. 2011).

$$\text{EPBT [years]} = \text{CED}/((E_{\text{agen}}/\eta_{\text{G}}) - E_{\text{O\&M}}) \tag{33.2}$$

where

$E_{\text{mat}}, E_{\text{manuf}}, E_{\text{trans}}, E_{\text{inst,}}$ and $E_{\text{EOL}}$ are defined as above; additionally,
$E_{\text{agen}}$ [MJ$_{\text{el}}$/yr]: Annual electricity generation
$E_{\text{O\&M}}$ [MJ$_{\text{PE-eq}}$]: Annual primary energy demand for operation and maintenance

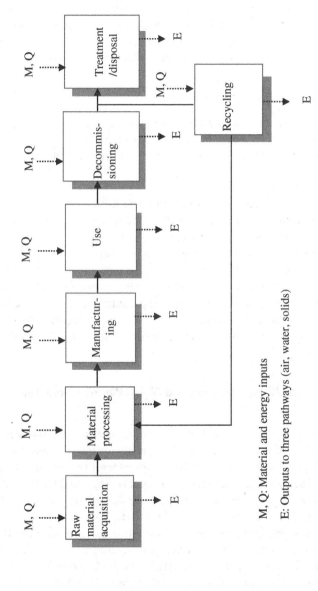

**Figure 33.1** Schematic of energy, materials, and waste inputs/output in each of PV's life-cycle stages.

M, Q: Material and energy inputs

E: Outputs to three pathways (air, water, solids)

$\eta_G$ [MJ$_{el}$/MJ$_{PE-eq}$]: Grid efficiency, i.e. the average life-cycle primary energy to electricity conversion efficiency at the demand side

The annual electricity generation ($E_{agen}$) is converted into its equivalent primary energy, based on the efficiency of electricity conversion at the demand side, using the grid mix where the PV plant is being installed. Thus, calculating the primary-energy equivalent of the annual electricity generation ($E_{agen}/\eta_G$) requires knowing the life-cycle energy conversion efficiency ($\eta_G$) of the country-specific energy-mixture used to generate electricity and produce materials. The average $\eta_G$ for the United States of America and Western Europe are respectively approximately 0.30 and 0.31 (Dones 2003; Franklin Associates 1998).

### 33.3.2 Energy Return on Investment (EROI)

The CED of a PV system may be regarded as the energy investment that is required in order to be able to obtain an energy return in the form of PV electricity.

The overall energy return on (Energy) investment may thus be calculated as:

$$EROI_{PE-eq} [MJ_{PE-eq}/MJ_{PE-eq}] = T/EPBT \qquad (33.3)$$

where $T$ is the period of the system operation; both $T$ and EPBT are expressed in years.

$EROI_{PE-eq}$ and EPBT provide complementary information. $EROI_{PE-eq}$ looks at the overall energy performance of the PV system over its entire lifetime, whereas EPBT measures the point in time ($t$) after which the system will provide a net energy return.

Further discussion of the EROI methodology can be found in an IEA PVPS Task 12 report (Raugei et al. 2015). It is noted that some research groups are disseminating invalid assertions regarding PV having a low EROI by using outdated data, worst conditions of deployment and assigning financial burdens to the CED of PV. A critical overview of an ongoing debate on this issue can be found in Carbajales-Dale et al. (2015) and Fthenakis et al. (2022).

### 33.3.3 Greenhouse Gas (GHG) Emissions and Global Warming Potential (GWP)

The overall global warming potential (GWP) due to the emission of a number of GHGs along the various stages of the PV life cycle is typically estimated using an integrated time-horizon of 100 years (GWP$_{100}$), whereby the following $CO_2$-equivalent factors are used: 1 kg $CH_4$ = 28 kg $CO_2$-eq, 1 kg $N_2O$ = 296 kg $CO_2$-eq, and 1 kg chlorofluorocarbons = 4600–10,600 kg $CO_2$-eq. Electricity and fuel use during the production of the PV materials and modules are the main sources of the GHG emissions for PV cycles, and specifically, the technologies and processes for generating the upstream electricity play an important role in determining the total GWP of PVs, since the higher the mixture of fossil fuels is in the grid, the higher are the GHG (and toxic) emissions. For example, during 1999–2002, the GHG emission factor of the average US electricity grid was 676 g $CO_2$-eq/kWh, that of Germany was 539 g $CO_2$-eq/kWh, and that of China was 839 g $CO_2$-eq/kWh (US Department of Energy, Energy Information Administration 2007).

In this chapter, we present the most up-to-date estimates of EPBT, GHG emissions, and heavy metal emissions from the life cycles of the currently commercial PV technologies (e.g. monocrystalline Si (sc-Si), multicrystalline Si (mc-Si), CdTe, and CIGS). Estimates of other environmental impact indicators (e.g. ecotoxicity, human toxicity, acidification, ozone depletion, and eutrophication potentials) can be found in Leccisi et al. (2016) and Fthenakis and Leccisi (2021a,b). Detailed LCIs can be found in an IEA PVPS Task 12 report (Frischknecht et al. 2020). BOS data refer to a ground-mounted installation (Leccisi et al. 2016).

## 33.4 Results

### 33.4.1 Energy Payback Time and Energy Return on Investment

The energy payback time of PV's has been reduced by almost two orders of magnitude over the last three decades, as material use, energy use and efficiencies have been constantly improving (Fthenakis 2012a).

Currently, for sc-Si PV systems, it takes from approximately 0.5 to 1.3 years (depending on the irradiation at the location of deployment) to return an amount of electricity that is equivalent to the primary energy invested. For CdTe PV systems, the EPBT ranges from 0.4 to 0.9 years, as shown in Figure 33.2 (Leccisi et al. 2016; Fthenakis and Leccisi 2021a,b).

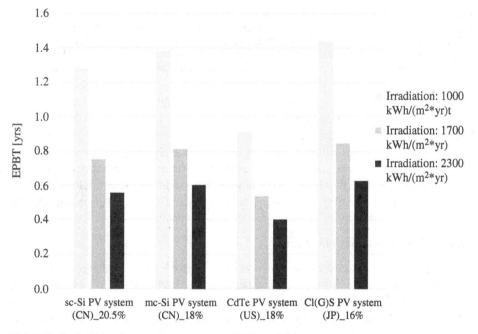

**Figure 33.2** Energy payback times (years) of single- and multicrystalline silicon PV systems produced in China, CdTe PV produced in the United States, and CIGS PV systems produced in Japan, assuming three irradiation levels for the location where the systems operate: 1000 kWh/(m² yr); 1700 kWh/(m² yr); 2300 kWh/(m² yr). These results correspond to ground-mounted latitude-tilt BOS, performance ratio of 0.85, $\eta_{grid}$ = 0.30 (Leccisi et al. 2016; Fthenakis and Leccisi 2021a,b).

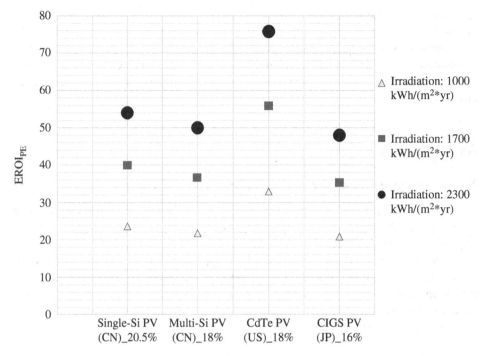

**Figure 33.3** Energy return on investment (in terms of primary energy) of single- and multicrystalline silicon PV systems produced in China, CdTe PV produced in the United States, and CIGS PV systems produced in Japan, under three irradiation levels: 1000 kWh/(m² yr); medium symbols: 1700 kWh/(m² yr); and large symbols: 2300 kWh/(m² yr). Performance ratio: 0.85. Operation lifetime: 30 years. $\eta_{grid} = 0.30$ (Leccisi et al. 2016; Fthenakis and Leccisi 2021a,b).

The corresponding $EROI_{PE\text{-}eq}$ ratio, assuming a 30-year lifetime and 20.5% efficiency for sc-Si PV systems, ranges from approximately 21 up to 52; for CdTe, from 31 to 76 (see Figure 33.3).

EPBT decreases (and EROI increases) as the solar irradiation levels increases; for example, in Phoenix, Arizona, US Southwest (latitude optimal irradiation of 2300 kWh/(m² yr), the EPBTs of crystalline silicon- and cadmium telluride-PV in fixed tilt ground mount utility installations, respectively, are 0.5 and 0.4 years respectively.

### 33.4.2 Greenhouse Gas Emissions

The GHG emissions of Si PV systems are ~23 g $CO_2$-eq/kWh range for a ground-mounted application under Southern European insolation of 1700 kWh/(m² yr) and a performance ratio[1] (PR) of 0.85 (Figure 33.4).

The GHG emissions of CdTe PV systems total 13 g $CO_2$-eq/kWh for irradiation levels of 1700 kWh/m²/yr (Leccisi et al. 2016). The manufacturing of the module and upstream mining/smelting/purification operations comprise most of the energy and greenhouse burdens.

1 Ratio between the DC rated and actual AC electricity output.

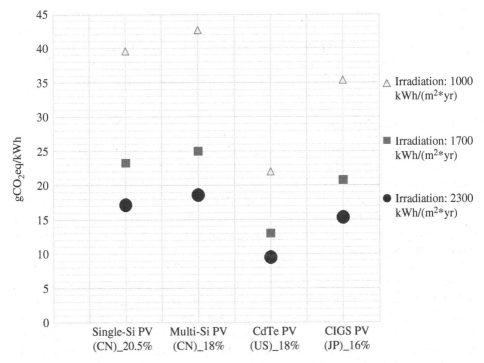

**Figure 33.4** Life-cycle greenhouse gas (GHG) emissions from single- and multicrystalline silicon PV systems produced in China, CdTe PV produced in the United States, and CIGS PV systems produced in Japan, assuming three irradiation levels. Small symbols: 1000 kWh/(m² yr); medium symbols: 1700 kWh/(m² yr); and large symbols: 2300 kWh/(m² yr). These results correspond to ground-mounted latitude-tilt BOS, performance ratio of 0.85, and a lifetime of 30 years (Leccisi et al. 2016); Source: Adapted from Fthenakis and Leccisi (2021a,b).

All the recently published LCA studies found CdTe PV has the lowest EPBT and GHG emissions among the currently commercial PV technologies as it uses less energy in its material processing and module manufacturing (Leccisi et al. 2016; Bhandari et al. 2015; Raugei and Fthenakis 2010). This is explained by the lower thickness of the high-purity semiconductor layer (i.e. 2–3 μm for CdTe versus 150–200 μm for c-Si) and by the fact that CdTe module processing has fewer steps and is much faster than c-Si cell and module processing.

Comparing the GHG emissions from the lifecycle of PV with those of conventional fuel-burning power plants, results reveal the environmental advantage of using PV technologies. The majority of GHG emissions come from the operational stage for the coal-, natural gas-, and oil-fuel cycles, while the material and device production accounts for nearly all the emissions for the PV cycles. With over 50% contributions, the GHG emissions from the electricity demand in the lifecycle of PV are the most impactful input. Therefore, the LCA results strongly depend on the available electricity mix. The GHG emissions from the nuclear-fuel cycle are mainly related to fuel production, i.e. mining, milling, fabrication, conversion, and the enrichment of uranium fuel. The details of the US nuclear fuel cycle are described elsewhere (Fthenakis and Kim 2007).

Other comparisons between the life cycles of PVs and conventional power generation cover land use (Fthenakis and Kim 2009) and water use (Fthenakis and Kim 2010). Accounting for the land occupation in coal mining, it is shown that the life cycles of PV in the US-SW occupy about the same or less land and orders of magnitude less water on an electricity-produced (GWh) basis than the average coal-based power generation.

### 33.4.3 Toxic Gas Emissions

The emissions of toxic gases (e.g. $SO_2$, $NO_x$) and heavy metals (e.g. As, Cd, Hg, Cr, Ni, Pb) during the life cycle of a PV system are largely proportional to the amount of fossil fuel consumed during its various phases, in particular, processing and manufacturing PV materials. Figure 33.5 shows estimates of $SO_2$ and $NO_x$ emissions. Heavy metals may be emitted directly from material processing and PV manufacturing and indirectly from generating the energy used at both stages. For the most part, they originate as trace metals in the coal-based power in current electricity grids.

Direct emissions of cadmium in the life-cycle of CdTe PV have been assessed in detail (Fthenakis 2004). They total 0.016 g per GWh of PV-produced energy under irradiation of 1700 kWh/(m$^2$ yr); this includes emissions during fires on roof-top residential systems, quantified in experiments at Brookhaven National Laboratory that simulated roof fires (Fthenakis et al. 2005). These experiments were designed to replicate average conditions, and the estimated emissions were calculated by accounting for U.S. fire statistics pointing to 1/10,000 houses catching fire over the course of a year in the United States, where most houses have wood frames, by assuming that all fires involve the roof. The indirect Cd emissions from electricity usage during the life-cycle of CdTe PV modules (i.e. 0.24 g/GWh) are an order of magnitude greater than the direct ones (routine and accidental) (i.e. 0.016 g/GWh) (ISO 2006a; US Department of Energy, Energy Information

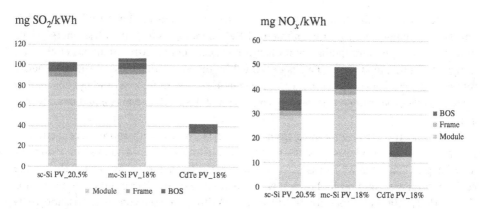

**Figure 33.5** Life-cycle emissions of $SO_2$ and $NO_x$ from silicon and CdTe PV modules, wherein BOS is the BOS that is, module supports, cabling, and power conditioning. The estimates are based on rooftop-mounted installation, Southern European insolation of 1700 kWh/m$^2$/yr[1], and a lifetime of 30 years. These estimates are based on "Fthenakis et al., Emissions from photovoltaic life cycles, Environ. Sci. Technol., 42(6), 2168–2174, 2008". Source: Adapted from Fthenakis et al. (2008), updated with current module efficiencies.

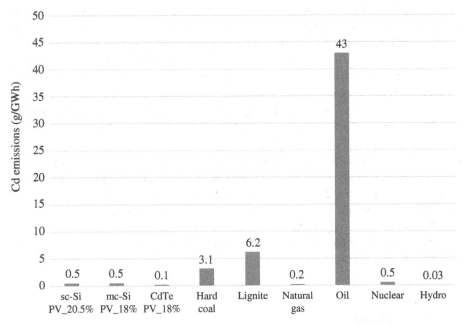

**Figure 33.6** Life-cycle atmospheric Cd emissions for PV systems from electricity and fuel consumption, normalized for a Southern Europe average insolation of 1700 kWh/m²/yr[1], a performance ratio of 0.8, and a lifetime of 30 years. A ground-mounted balance of system is assumed for all PV systems. Adapted from Fthenakis et al. 2008, updated with current module efficiencies.

Administration 2007).[2] Cadmium emissions from the electricity demand for each module were assigned, assuming that the life-cycle electricity for the silicon and CdTe-PV modules was supplied by the Union for the Coordination of Transmission of Energy's (European) grid. The complete life-cycle atmospheric Cd emissions, estimated by adding those from the usage of electricity and fuel in manufacturing and producing materials for various PV modules and BOS, were compared with the emissions from other electricity-generating technologies (Figure 33.6) (Fthenakis et al. 2008). Undoubtedly, displacing the others with Cd PV markedly lowers the amount of Cd released into the air. Thus, every GWh of electricity generated by CdTe PV modules can prevent around 5 g of Cd air emissions if they are used instead of, or as a supplement to, the UCTE electricity grid. In addition, the direct emissions of Cd during the life-cycle of CdTe PV are ten times lower than

---

2 Indirect emissions of heavy metals result mainly from the trace elements in coal and oil. According to the US Electric Power Research Institute's (EPRI's) data, under the best/optimized operational- and maintenance-conditions, burning coal for electricity releases into the air between 2 and 7 g of Cd/GWh. In addition, 140 g/GWh of Cd inevitably collects as fine dust in boilers, baghouses, and electrostatic precipitators (ESPs). Furthermore, a typical US coal-powered plant emits per GWh about 1000 tons of $CO_2$, 8 tons of $SO_2$, 3 tons of $NO_x$, and 0.4 tons of particulates. The emissions of Cd from heavy-oil burning power plants are 12–14 times higher than those from coal plants, even though heavy oil contains much less Cd than coal (~0.1 ppm); this is because these plants do not have particulate-control equipment.

the indirect ones due to the use of electricity and fuel in the same life-cycle, and about 30 times less than those indirect emissions from crystalline PVs. The same applies to total (direct and indirect) emissions of other heavy metals (e.g. As, Cr, Pb, Hg, Ni); CdTe PV has the lowest CED and, consequently, the fewest heavy-metal emissions. Regardless of the particular PV technology, these emissions are extremely small compared to the emissions from the fossil-fuel-based plants that PV will replace. Furthermore, the external environmental costs of PVs are negligible in comparison to the external costs of fossil fuel life cycles (Fthenakis and Alsema 2006; Sener and Fthenakis 2014; Fthenakis 2015).

Ongoing LCA studies include evaluating replacing building materials with PV in building-integrated photovoltaics (BIPV) (Perez et al. 2012 and recycling of PV modules at the end of their useful life (Fthenakis 2000, 2012b; Fthenakis and Wang 2006; Corcelli et al. 2018). Emerging PV technologies, such as single-junction and tandem perovskite solar cells, could further support the reduction of life-cycle environmental impacts (Leccisi et al. 2022; Leccisi and Fthenakis 2019, 2020, 2021; Fthenakis and Leccisi 2021a,b; Billen et al. 2019; Khalifa et al. 2023).

## 33.5 Conclusion

This chapter gives an overview of the life-cycle environmental performance of PV technologies. EPBT is a basic metric of this performance; it measures the time it takes for a PV system to generate as much electricity as could be generated by the current electric mix when using the same amount of primary energy that is used for the production of the PV system. The lower the EPBT, the lower will also be the emissions to the environment, because emissions mainly occur from using fossil-fuel-based energy in producing materials, solar cells, modules, and systems. These emissions differ in different countries, depending on that country's mixture in the electricity grid and the varying methods of material/fuel processing. Under average U.S.- and Southern European conditions (e.g. $1700\,kWh/(m^2\,yr)$), the EPBT of crystalline Si-, and CdTe-ground mounted PV systems, were estimated respectively to be 0.7 and 0.5 years correspondingly. Under U.S. SW irradiation (e.g. Phoenix, AZ, $2300\,kWh/(m^2\,yr)$ at a fixed latitude tilt), the EPBT of ground-mount installations is estimated to be 0.5 years and 0.5 years, respectively, for crystalline silicon- and cadmium telluride-PV.

The environmental impacts of the life-cycle of PVs, as assessed by the common metrics of GHG emissions, toxic emissions and heavy metal emissions, are very small in comparison to those of the power generation technologies they replace.

## Acknowledgment

This material is based upon work supported by the U.S. Department of Energy's Office of Energy Efficiency and Renewable Energy (EERE) under the Solar Energy Technologies Office Award Number DE-EE0009830.

# Disclaimer

This report was prepared as an account of work sponsored by an agency of the United States Government. Neither the United States Government nor any agency thereof, nor any of their employees, makes any warranty, express or implied, or assumes any legal liability or responsibility for the accuracy, completeness, or usefulness of any information, apparatus, product, or process disclosed, or represents that its use would not infringe privately owned rights. Reference herein to any specific commercial product, process, or service by trade name, trademark, manufacturer, or otherwise does not necessarily constitute or imply its endorsement, recommendation, or favoring by the United States Government or any agency thereof. The views and opinions of authors expressed herein do not necessarily state or reflect those of the United States Government or any agency thereof.

# List of Symbols

| Symbol | Description | Unit |
| --- | --- | --- |

# List of Acronyms

| Acronym | Meaning |
| --- | --- |
| GWP | Global Warming Potential |
| CED | Cumulative Energy Demand |
| IEA | International Energy Agency |
| LCIA | Life Cycle Impact Assessment |
| LCI | Life Cycle Inventory |
| LCA | Life Cycle Assessment |
| EROI | Energy Return on Energy Investment |
| EPBT | Energy Payback Time |

# Author Biographies

**Vasilis Fthenakis** has been working for more than 30 years on PV environmental assessments, leading the PV Environmental Research Center at Brookhaven National Laboratory and, as of 2006, the Center of Life Cycle Analysis at Columbia University. Fthenakis has published more than 300 articles on PV sustainability issues, and those were cited more than 15,000 times. He is the recipient of several honors and awards, including the

2018 IEEE William Cherry Award "for his pioneering research at the interface of energy and the environment that catalyzed photovoltaic technology advancement and deployment world-wide" and the 2022 Karl Boer Solar Energy Medal of Merit for distinguished contributions to the quest for sustainable energy.

**Enrica Leccisi** completed her PhD on LCA of future scenarios of large-scale penetration of PV in electric grids. She has worked with the PV industry, carrying out feasibility and resource assessments and as a Postdoctoral Research Scientist at the Center for Life Cycle Analysis with Fthenakis. She authored and co-authored ~30 scientific papers on PV LCA in peer-reviewed international Journals and Book of Proceedings.

# References

Bhandari, K., Collier, J., Ellingson, R., and Apul, D. (2015). EPBT and EROI of PV: a systematic review and meta-analysis. *Renewable and Sustainable Energy Reviews* 47: 133–141.

Billen, P., Leccisi, E., Dastidar, S. et al. (2019). Comparative evaluation of lead emissions and toxicity potential in the life cycle of lead halide perovskite photovoltaics. *Energy* 166: 1089–1096.

Carbajales-Dale, M., Raugei, M., Barnhart, C.J., and Fthenakis, V. (2015). Energy return on investment (EROI) of solar PV: an attempt at reconciliation. *Proceedings of the IEEE* 103 (7): 995–999.

Corcelli, F., Ripa, M., Leccisi, E. et al. (2018). Sustainable urban electricity supply chain—indicators of material recovery and energy savings from crystalline silicon photovoltaic panels end-of-life. *Ecological Indicators* 94: 37–51.

Dones, R. (2003). *Sachbilanzen von Energiesystemen*. Final report ecoinvent 2000, vol. 6. Swiss Centre for LCI, PSI.

Electric Power Research Institute (EPRI) (2002). PISCES Data Base for US Power Plants and US Coal.

Franklin Associates (1998). USA LCI Database Documentation. Prairie Village, Kansas.

Frischknecht, R., Jungbluth, N., Althaus, H.-J. et al. (2007). *Implementation of Life Cycle Impact Assessment Methods*. Ecoinvent Report No. 3, v2.0. Dübendorf, CH: Swiss Centre for Life Cycle Inventories.

Frischknecht, R., Stolz, P., Krebs, L. et al. (2020). Life Cycle Inventories and Life Cycle Assessment of Photovoltaic Systems. International Energy Agency (IEA) PVPS Task 12, Report T12-19:2020.

Fthenakis, V.M. (2000). End-of-life management and recycling of PV modules. *Energy Policy* 28 (14): 1051–1058.

Fthenakis, V.M. (2004). Life cycle impact analysis of cadmium in CdTe PV production. *Renewable and Sustainable Energy Reviews* 8 (4): 303–334.

Fthenakis, V. (2012a). PV energy ROI tracks efficiency gains. *Solar Today, American Solar Energy Society (ASES)* 26 (4): 24–26. http://www.solartoday-digital.org/solartoday/201206#pg24.

Fthenakis, V.M. (2012b). Sustainability metrics for extending thin-film photovoltaics to terawatt levels. *MRS Bulletin* 37 (4): 425–430.

Fthenakis, V. (2015). Considering the total cost of electricity from sunlight and the alternatives. *Proceedings of the IEEE* 103 (3): 283–286.

Fthenakis, V. and Alsema, E. (2006). Photovoltaics energy payback times, greenhouse gas emissions and external costs: 2004—early 2005 Status. *Progress in Photovoltaics: Research and Applications* 14: 275–280.

Fthenakis, V.M. and Kim, H.C. (2007). Greenhouse-gas emissions from solar electric- and nuclear power: a life-cycle study. *Energy Policy* 35: 2549–2557.

Fthenakis, V. and Kim, H.C. (2009). Land use and electricity generation: a life-cycle analysis. *Renewable & Sustainable Energy Reviews* 13: 1465–1474.

Fthenakis, V. and Kim, H.C. (2010). Life cycle uses of water in U.S. electricity generation. *Renewable & Sustainable Energy Reviews* 14: 2039–2048.

Fthenakis, V. and Leccisi, E. (2021a). Updated sustainability status of crystalline silicon-based photovoltaic systems: life-cycle energy and environmental impact reduction trends. *Progress in Photovoltaics: Research and Applications* 29 (10): 1068–1077.

Fthenakis, V. and Leccisi, E. (2021b). Life-cycle analysis of tandem PV perovskite-modules and systems. In: *2021 IEEE 48th Photovoltaic Specialists Conference (PVSC)*, 1478–1485. IEEE.

Fthenakis, V.M. and Wang, W. (2006). Extraction and separation of Cd and Te from cadmium telluride photovoltaic manufacturing scrap. *Progress in Photovoltaics: Research and Applications* 14: 363–371.

Fthenakis, V.M., Fuhrmann, M., Heiser, J. et al. (2005). Emissions and encapsulation of cadmium in CdTe PV modules during fires. *Progress in Photovoltaics: Research and Applications* 13: 713–723.

Fthenakis, V., Kim, H.C., and Alsema, E. (2008). Emissions from photovoltaic life cycles. *Environmental Science & Technology* 42 (6): 2168–2174. https://doi.org/10.1021/es071763q.

Fthenakis, V., Frischknecht, R., Raugei, M. et al. (2011). Methodology Guidelines on Life Cycle Assessment of Photovoltaic Electricity, 2. IEA PVPS Task 12, International Energy Agency Photovoltaic Power Systems Programme. Report IEA-PVPS T12-03:2011. http://www.iea-pvps.org (accessed 21 February 2024).

Fthenakis, V., Raugei, M., Breyer, C. et al. (2022). Comment on Seibert, MK; Rees, WE Through the eye of a needle: an eco-heterodox perspective on the renewable energy transition. Energies 2021, 14, 4508. *Energies* 15 (3): 971.

ISO (2006a). *14040—Environmental Management. Life Cycle Assessment. Principles and Framework*. International Organization for Standardization.

ISO (2006b). *14044—Environmental Management. Life Cycle Assessment. Requirements and Guidelines*. International Organization for Standardization.

Khalifa, S.A., Spatari, S., Fafarman, A.T. et al. (2023). Fate and exposure assessment of Pb leachate from hypothetical breakage events of perovskite photovoltaic modules. *Environmental Science & Technology* 57 (13): 5190–5202.

Leccisi, E. and Fthenakis, V. (2019). Critical review of perovskite photovoltaic life cycle environmental impact studies. In: *2019 IEEE 46th Photovoltaic Specialists Conference (PVSC)*, vol. 2, 1–6. IEEE.

Leccisi, E. and Fthenakis, V. (2020). Life-cycle environmental impacts of single-junction and tandem perovskite PVs: a critical review and future perspectives. *Progress in Energy* 2 (3): 32002.

Leccisi, E. and Fthenakis, V. (2021). Life cycle energy demand and carbon emissions of scalable single-junction and tandem perovskite PV. *Progress in Photovoltaics: Research and Applications* 29 (10): 1078–1092.

Leccisi, E., Raugei, M., and Fthenakis, V. (2016). The energy and environmental performance of ground-mounted photovoltaic systems – a timely update. *Energies* 9 (8): 622. https://doi.org/10.3390/en9080622.

Leccisi, E., Lorenz, A., and Fthenakis, V. (2022). Life-cycle analysis of crystalline-Si "Direct Wafer" and tandem perovskite PV modules and systems. *IEEE Journal of Photovoltaics* 13 (1): 16–21.

Perez, M., Fthenakis, V., Kim, H.C., and Pereira, A. (2012). Façade-integrated photovoltaics: a lifecycle and performance assessment case study. *Progress in Photovoltaics: Research and Applications* 20 (8): 975–990.

Raugei, M. and Fthenakis, V. (2010). Cadmium flows and emissions from CdTe PV: future expectations. *Energy Policy* 38 (9): 5223–5228.

Raugei, M., Frischknecht, R., Olson, C. et al. (2015). Methodological Guidelines on Net Energy Analysis of Photovoltaic Electricity. IEA-PVPS Task 12, Report T12-XX:2015. http://www.iea-pvps.org (accessed 27 February 2024).

Sener, C. and Fthenakis, V. (2014). Energy policy and financing options to achieve solar energy grid penetration targets: accounting for external costs. *Renewable and Sustainable Energy Reviews* 32: 854–868.

US Department of Energy, Energy Information Administration (2007). Form EIA-1605 (2007), OMB No. 1905-0194 Voluntary Reporting of Greenhouse Gases; Appendix F. Electricity Emission Factors.

# 34

# PV Recycling – Status and Perspectives

*Frank Lenzmann*

*TNO, Energy Transition Studies, 1043 NT Amsterdam, The Netherlands*

## 34.1 Introduction

The global deployment of solar photovoltaics (PV) continues to achieve new records year by year, and installation milestones are reached earlier in reality than anticipated in many scenarios. In 2022, a major milestone was reached, namely 1 TWp of cumulative globally installed PV capacity, producing well over 1000 TWh of electricity representing approximately 3.5% of global electricity production of ~28,000 TWh.[1] One of the implications of this encouraging growth is that there is an increasingly urgent need to implement viable recycling schemes for these installations in order to avoid a waste problem in the future and, instead, achieve a high level of sustainability throughout the entire value chain, including the end-of-life (EoL) of PV modules, further also called EoL PV modules. Due to multiple reasons, PV recycling is rather challenging, in particular when the goal is to reclaim high-purity materials for circular or other high-value applications. One reason is unfavorable economics due to currently still low PV waste volumes and a critical balance between the cost and revenues of recycling processes. Another reason is technological issues caused by transparent polymeric encapsulants applied in the production process. These intimately glue together the inside components (solar cells and internal electric wiring) with the front glass and polymeric back sheet and thus prevent an easy disassembly and clean separation of materials at the EoL. Current recycling schemes are, therefore, in essence, basic downcycling schemes where only some materials are reclaimed mainly for low-value applications. R&D in the area of advanced, high-value PV recycling has only emerged in recent years. This is due to the above-mentioned low PV waste volumes to date, which result in a relatively low sense of urgency. Also, until recently, PV research was very much focused on cost reduction and efficiency improvement only instead of recycling, making it a rather new, timely topic.

Section 34.2 of this chapter first provides information about the enormous dimension of the emerging amount of PV waste in the coming decades. Then, Section 34.3 describes compositional and economic characteristics of EoL PV modules as a basis for a discussion of the

---

1 https://www.statista.com/statistics/270281/electricity-generation-worldwide/

*Photovoltaic Solar Energy: From Fundamentals to Applications, Volume 2*, First Edition.
Edited by Wilfried van Sark, Bram Hoex, Angèle Reinders, Pierre Verlinden, and Nicholas J. Ekins-Daukes.
© 2024 John Wiley & Sons Ltd. Published 2024 by John Wiley & Sons Ltd.
Companion website: www.wiley.com/go/PVsolarenergy

challenges and need of PV recycling in Section 34.4. Finally, Section 34.5 provides a description of the contemporary downcycling practice and an overview of future high-value PV recycling technologies currently under development (Section 34.6). Two examples of corresponding industrial initiatives are also included in this chapter. The scope of this chapter is limited to current PV module technologies for crystalline silicon solar cells, which represent a market share of more than 90%.

## 34.2 Evolution of PV Waste in the Coming Decades

Today, the amount of EoL PV modules is merely in the order of a few 10,000 (metric) tons per year (Komoto et al. 2022), which is a negligible number as compared with, for example, the annual global amount of electronic waste which is currently in the order of 50 million tons per year. However, this small amount of PV waste is just a short, transient phenomenon and results from the fact that annual global PV deployment at a GWp scale started only around 2005, when the total installed PV capacity amounted to only 5 GWp. Most of the systems installed since then are simply still operational as the service life of PV systems easily reaches around 20–30 years.

But, looking into the near future, a dramatic increase in EoL PV modules can be readily predicted. Many of the modules installed in the 10-year time period of 2005–2015, i.e. 220 GW in total and hence on average 22 GW per year, will retire by 2030–2040. With an assumed PV module weight and power of resp. 20 kg and 300 Wp, 22 GWp corresponds to approximately 1.5 million tons of PV modules. This is thus the annual amount of EoL PV modules to be expected by 2030–2040, which is two orders of magnitude more than today's.

At the time of writing this chapter (2022), it is expected that by the early 2030's the annual PV additions will reach the TW scale (PV magazine, Goldschmidt). These growth projections translate into a further increase of the expected amount of EoL PV modules by an extra two orders of magnitude, resulting in 100 million tons per year by 2055. So, within the coming decades, the amount of PV waste (or rather EoL PV) can be quite safely predicted to increase by as much as four orders of magnitude (see Figure 34.1) as compared to

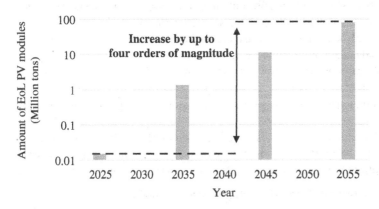

**Figure 34.1** Expected increase of global PV waste in the coming decades.

2022. This prediction is based on current growth numbers for PV system installations and an average service lifetime of 25 years.

It should be noted that Figure 34.1 represents only one possible scenario. Service lifetimes of PV modules in reality can differ from the assumed 25 years, i.e. they can be either longer or shorter. One of the aspects that may prolong service lifetime is repair and reuse. While this aspect will not be covered by this chapter, there are significant R&D as well as commercial initiatives in this area, and some estimates are that up to half of EoL PV modules might be suitable for repair and reuse. On this basis, the picture shown in Figure 34.1 may overestimate the amount of PV waste by a factor of about two on the time scale shown. But eventually, PV panels with an extended service lifetime will reach the waste phase. So, the enormous scale of PV waste in the coming decades is a given, but the timeline is somewhat uncertain.

## 34.3  Compositional and Economic Characteristics of EoL PV Modules

Prior to the discussion of the challenges of PV recycling in the next section, it is helpful to look at the structure of a PV module and its main components (Figure 34.2 and Table 34.1). These consist of the following key materials (in order of decreasing mass fraction): glass, aluminum, polymers, silicon, copper, and silver. Based on this structure, Figures 34.3–34.5

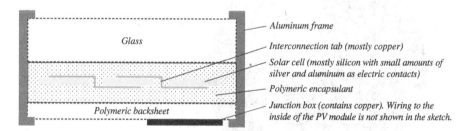

**Figure 34.2**  Structure of a PV module and its key components.

**Table 34.1**  List of key materials and their mass in an average EoL PV module.

| Material | Mass (kg) |
| --- | --- |
| Glass (front pane) | 14 |
| Aluminum (frame) | 3.6 |
| Polymers (encapsulant and backsheet) | 1.3 |
| Silicon (solar cell wafers) | 0.7 |
| Copper (internal and external wiring) | 0.2 |
| Silver (internal electric contacts) | 0.01 |

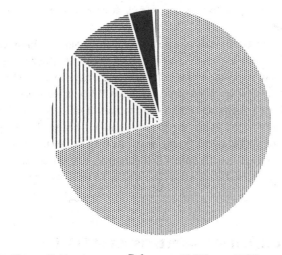

⠿ Glass  ‖ Aluminum  ≡ Polymers  ■ Silicon  ▨ Silver  ⫴ Copper

**Figure 34.3** Materials breakdown by mass of an average EoL PV module (total is around 20 kg).

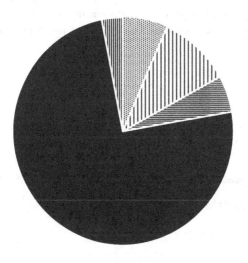

⠿ Glass  ‖ Aluminum  ≡ Polymers  ■ Silicon  ▨ Silver  ⫴ Copper

**Figure 34.4** Materials breakdown by embodied energy of an average EoL PV module (total is around 2000 MJ).

show the corresponding materials composition by mass, embodied energy and economic value.

Considering the module components by mass (Figure 34.3), it becomes immediately clear that is dominated first and foremost by the front glass pane (70%), followed by the aluminum frame and the polymers (backsheet and encapsulants). The silicon from the solar

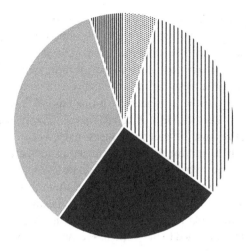

:: Glass  ıı Aluminum  ≡ Polymers  ■ Silicon  ≡ Silver  ıı Copper

**Figure 34.5** Materials breakdown by economic value of an average EoL PV module (total is around 10 €).

cells is only a minor contributor to the mass of the module, and the silver doesn't even show up in the pie chart representation due to its very small mass fraction.

In contrast to that, the breakdown by embodied energy (Figure 34.4) is strongly dominated by the silicon despite its low mass fraction. This is due to energy-intensive process steps involved in the wafer production, in particular the purification from metallurgical to solar grade silicon as well as the ingot growth.

Considering the breakdown by economic value (Figure 34.5), it is also two of the low mass fraction compounds, silver and silicon-from the solar cells-which show large contributions. Together, they represent more than half of the potential economic value. This assumes that the recycled materials are recoverable at a high purity level (meaning solar grade for silicon). The high mass fraction compound aluminum from the frame is the third large contributor to the economic value.

## 34.4 Challenges Versus Need for PV Recycling

As pointed out in Section 34.2, the global amount of PV waste to date is still limited to only a few tens of thousands of tons per year. This is about the scale of a single electronic waste recycling plant. Assuming that economically viable PV waste facilities may be characterized by a similar scale, it is clear that dedicated PV recycling is economically still challenging today due to the limited quantity of input materials. The capacity usage of corresponding recycling plants would simply risk being too low. However, the previous chapter also showed that PV waste will increase dramatically in the near future, so risks of insufficient capacity usage will soon not be an issue anymore.

A different challenge is that the economic value of EoL PV modules is comparatively low,[2] around 8–14 €/module (Tao et al. 2020; Späth et al. 2022). At the same time, the weight is relatively high (~20 kg/module) and hence also costs for transport/logistics. These two factors (value and weight) can be aggregated into a value-to-weight ratio, which is in the order of 400–700 €/ton for PV waste and compares rather unfavorably with more than 1000 €/ton for electronic waste, for example. A further complication is that most of the economic value of EoL PV modules is contained in the silver and silicon components of the solar cells (see Figure 34.5). These are, in turn, the most difficult components to extract.

PV modules are designed and built to survive weather extremes in temperature and humidity where the encapsulant materials, typically ethyl vinyl acetate (EVA), holding the module together also completely encase the solar cells and protect them from moisture ingress and corrosion. The cells become firmly embedded into the transparent polymeric encapsulant during the lamination process of module fabrication. Via the EVA, which acts as a glue, the solar cells are also strongly attached to the front glass as well as the polymeric back sheet (see Figure 34.2). It is these very same protection mechanisms that make the disassembly of the PV module and separation of the solar cells extremely difficult. It is for this reason that contemporary PV recycling schemes, which are described in Section 34.5, abstain from the recovery of the solar cells altogether. In these schemes, the solar cells end up in residual waste streams instead, which means that a lot of the economic value and precious resources are being lost (Komoto et al. 2022; Deng et al. 2019).

With these significant challenges in mind, we can question ourselves, why (advanced) PV recycling would be so urgently needed?

The relevant driver in this context is sustainability, which is also an important factor in the current broad societal support for PV.

Regarding sustainability, it was shown in Figure 34.4 that silicon wafers are the dominant factor in the embodied energy of EoL PV modules accounting for up to 75%.[3] The development of advanced recycling methods able to reclaim silicon from PV waste can help to avoid some of the energy-intensive steps in the production of silicon wafers from virgin quartz ore. Wafers made from recycled silicon will have a significantly lower embodied energy level and associated carbon footprint. The expected rapid increase of annual PV production to the TW scale in the coming 10–15 years implies that PV production will weigh heavily on global electricity demand in the coming decades (Goldschmidt et al. 2021, thereby making the reduction of embodied energy in silicon wafers and PV modules below current values an important goal.

Even larger constraints are anticipated for the availability of sufficient quantities of silver if solutions are not implemented in time. Already in 2018, PV production (then on the order of 100 GW) required almost 10% of the annual silver mine production (Goldschmidt et al. 2021). Other recent publications foresee that the PV sector's demand for silver could exceed production by more than 200% in 2030 (Urbina 2023). So, TW scale PV production could result in serious silver availability constraints. Recycling silver from EoL PV modules is

---

2 The economic value of an EoL PV varies with price fluctuations of materials like silver, silicon, aluminum, etc. and is therefore by no means fixed or certain.
3 This value relates only to the materials components of the PV module only, i.e. does not include energy consumption for the processing of silicon wafers into solar cells and for PV module production.

certainly one of the needed solutions to resolve this constraint (combined with a further decrease in silver consumption per PV module).

Finally, another important factor related to sustainability is to avoid risks of environmental pollution by emission of problematic substances into soil, water or air in inadequate EoL schemes. These risks are first and foremost associated with the following two components in PV modules: electrical contacts on solar cells, which contain lead (Pb) and tin (Sn) and the standard multilayer backsheets, which contain fluorinated polymers, typically polyvinyl fluoride (PVF). Emissions to the environment can result, e.g. from landfilling (leading to leaching of Pb and Sn to soil and/or water) and from thermal treatments without adequate scrubbers, leading to HF emissions to the air. Advanced PV recycling schemes need to make sure that such environmental pollution risks are avoided. R&D in this area is still in its infancy, and related publications are scarce. Therefore, it can't be covered in further detail in this chapter, but it is expected to gain increasing momentum in the coming years.

## 34.5  Current Downcycling Technologies

According to recent review papers (Lunardi et al. 2018; Deng et al. 2019), only about 10% of the global PV module waste is actually recycled at all today, while 90% end up in landfills. The main reason is the lack of PV-specific waste regulations since, in most parts of the world, EoL PV is typically handled under regulatory frameworks for general waste treatment and disposal. An exception to this is Europe, which has established a PV-specific waste regulation through the Waste Electrical and Electronic Equipment (WEEE) directive 2012/19/EU. Under this directive, the collection, transport and (basic) recycling of EoL PV modules has been regulated since 2014.

Current technologies for recycling PV modules can be described as simple downcycling approaches based on crushing, shredding, milling, sieving and sorting. Typically around 80% of the PV module mass is recovered. This high percentage is somewhat misleading, though, because the key materials, silver and silicon, which occur in the PV module only in small mass fractions but which are at the same time most relevant in terms of economic value and environmental importance, are actually not recovered. Instead, the recovered materials comprise only the bulk components, mainly glass, aluminum and copper (see Figure 34.6). The recovery of aluminum is achieved readily by manual or

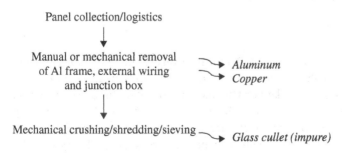

**Figure 34.6** Scheme showing process steps and recovered materials for contemporary downcycling technology.

mechanical detachment of the module frame. Similarly, copper can be recovered from readily detachable electrical wiring and junction boxes. The glass front pane, which represents by itself 70% of the module mass, is recovered as a cullet after mechanical crushing/shredding/sieving of the (often) de-framed and de-wired modules. The cullet is not pure glass, though; it contains remaining fragments of solar cells, encapsulant and back sheet. Due to these impurities, it is not suitable for higher-value applications. Instead, it is typically used in lower-value insulation materials like foam glass and glass wool, grinding material or sub-bases for roads. All these processes are carried out in already existing recycling facilities for other waste streams, such as laminated glass, electronics or metals (Wambach et al. 2017). An example in Europe where EoL PV module downcycling takes place is the ecycling facility of the company BNE Trading & Recycling B.V. in Belgium.

## 34.6 Future High-Value PV Recycling Technology

There are an increasing number of R&D initiatives that aim at transforming the contemporary low-value downcycling practice into an improved high-value one. This section provides a generic overview of principles and methods but does not describe details of specific processes. For such details, the reader is referred to more extensive recent review papers and original references therein (Tao et al. 2020; Deng et al. 2019; Deng et al. 2022; Farrell et al. 2020; Isherwood 2022). The goals of high-value PV recycling are as follows:

- Additional recovery of silver and silicon next to glass, aluminum and copper
- High purity & high-value recovery of materials (re-/upcycling)
- Avoidance of environmental pollution by emissions of Pb, Sn and HF to soil, water or air

These goals are driven by the discussed sustainability motives as well as economic ones. The latter results from the fact that silver and high purity (= solar-grade) silicon represent together more than half of the recoverable value of an EoL PV panel, which makes their recovery potentially attractive. The economic challenge is to shift the cost/revenue ratio of processes developed to this end to the revenue side.

To achieve the goals of high-value PV recycling, more PV-specific disassembly and recovery technologies are required, which implies extra and more complex processing steps (see Figure 34.7). These can be subdivided into two main categories, namely delamination and recovery of materials from solar cell residues.

### 34.6.1 Delamination

The purpose of delamination – which occurs after the removal of aluminum frame, junction box and wiring – is to separate the solar cells from the remaining PV module components, i.e. from the glass, back sheet, encapsulant and interconnection tabs. This is a precondition to enable the additional recovery of silver and silicon in the subsequent solar cell recycling step. A wide range of delamination technologies are currently under development, including mechanical, thermal and chemical methods, which will be shortly addressed in the following paragraphs.

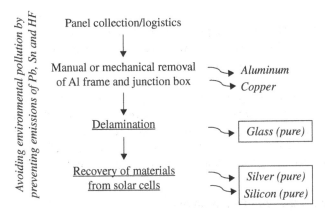

**Figure 34.7** Scheme showing process steps and recovered materials of future high-value recycling.

### 34.6.1.1 Mechanical Methods

Mechanical methods are widely established in already existing recycling industries for other wastes, such as electronics, metals or laminated glass. Compared with other methods, they are characterized by relatively high throughput and moderate cost. There is a large variety of mechanical methods. They range from the already mentioned approach of crushing/shredding/sieving/sorting to methods like scraping away layers from the PV module using waterjets (Deng et al. 2022) or slicing it open using a hot-knife (Latunussa et al. 2016), contactless fragmentation methods using electrohydraulic principles (Akimoto et al. 2018) etc. For most of these processes, only a few details have been published since they are typically under proprietary development. For example, it sometimes remains unclear whether additional chemical or thermal process steps are applied after the mechanical delamination process in order to remove, e.g. remaining encapsulant residues from the often fragmented solar cells and glass. Some mechanical methods, for example, the hot-knife method, allow recovery of the front glass pane intact. This could be an advantage in case there should be a future market demand for intact, recycled PV glass panes, which remains to be seen.

### 34.6.1.2 Thermal Methods

Thermal methods refer to incineration and pyrolysis. The fundamental difference between the two is the presence or absence of oxygen as well as the typical temperatures applied. Incineration, a method widely applied in municipal waste incineration, is carried out at temperatures of at least 850 °C in the presence of oxygen and thus leads to the complete combustion of any organics, leaving inorganics behind in bottom ash. Pyrolysis, on the other hand, is carried out in the absence of oxygen, which means that organics are not combusted, but thermally decomposed instead. Temperatures are typically significantly lower, on the order of 400–500 °C. So, pyrolysis is the milder method, and various literature sources describe it as a promising method to cleanly and gently separate Si cells from the PV module and thus to enable routes for the recovery of silver and (solar grade) silicon (Deng et al. 2022; Tao et al. 2020; Farrell et al. 2020).

### 34.6.1.3 Chemical Methods

Chemical methods for the removal of the EVA encapsulant by dissolution and/or swelling are also under investigation. A wide range of organic and inorganic solvents have been studied, e.g. nitric acid, toluene, cyclohexane, ethanol/KOH and other solvents (Deng et al. 2019; Deng et al. 2022. The application of microwaves or ultrasonication has been found to significantly accelerate the often slow chemical delamination processes, which typically last several days in the absence of such acceleration mechanisms. Processing times of one to two hours have been achieved in this way (Pang et al. 2021; Kim and Lee 2012). Disadvantages of chemical methods are the generation of chemical waste and high cost. It remains to be seen whether competitiveness with mechanical and/or thermal methods can be established.

### 34.6.2 Recovery of Materials from Solar Cell Residues

The emphasis in the current literature is mostly on silver and silicon. Practically all explored technologies are variations of wet chemical methods, such as leaching/etching in combination with filtration, precipitation, electrolysis or metal replacement. These technologies are well-established in contemporary extractive industries, such as hydrometallurgy. Mineral acids such as $HNO_3$, $HCl$, $H_2SO_4$, and $HF$ and bases such as $NaOH$ and $KOH$ are typically employed. For the most effective processes described in the literature, recovery rates of well above 90% of silicon and silver have been reported (Latunussa et al. 2016).

A noteworthy variant of chemical methods is the methane-sulfonic acid (MSA) technology for the recovery of Ag, which aims at reducing chemical waste. The promise of MSA is that it can be reused for multiple cycles, i.e. it acts as a "chemical shuttle" which can extract and release Ag repeatedly, thereby reducing chemical consumption and waste (Yang et al. 2017).

## 34.7  First Industrial Initiatives for High-Value PV Recycling

Recently pioneering industrial initiatives for high-value PV recycling, often on the basis of collaborations with research institutes and other partners, are underway. There are a number of examples in Europe and across the world. However, a common characteristic of all these initiatives is that very few details have been published about, e.g. specific recycling technologies developed by them or plant scales, let alone yields or economic viability. Two examples, Rosi Solar in France and Reiling in Germany, are briefly sketched here below:

The recycling process by Rosi Solar has been described in recent news coverage by PV Magazine to be based on pyrolysis for delamination followed by chemical processing allowing to recover high purity glass, silicon, silver and other metals next to aluminum from the frames. The scale of the plant is designed to reach a capacity of 3000 tons per year.

Press releases about the recycling process by Reiling refer to established mechanical delamination and sorting processes followed by wet chemical etching for the recovery of high-purity silicon (presumably next to silver, other metals, glass and aluminum). A remarkable achievement of this initiative is the production of new passivated emitter rear contact (PERC) solar cells from silicon ingots grown entirely from recycled silicon, with an efficiency of 19.7%, i.e. not much below current premium PERC solar cells (22.2%). This is an indicator of the promising quality of the recycled silicon and, therefore, its potential for circular reuse in the PV production industry.

## 34.8 Conclusions

As the global amount of PV module waste from decommissioned PV systems is expected to increase dramatically within the coming years and decades, PV recycling is coming more and more into the spotlight. Current downcycling practices in facilities not specifically dedicated to PV module recycling only recover the aluminum frame, external copper wiring and junction box as well as impure glass cullet, the latter being suitable merely for low-value applications. The recovery of the economically most valuable and environmentally critical materials, in particular silver and silicon, is not feasible in these facilities. In this situation, an increasing amount of R&D, as well as pioneering industrial initiatives, are getting off the ground with the goal of bringing the current practice to a higher level. To this end, they develop dedicated PV recycling processes for the recovery of high-purity materials from PV module waste. Most importantly, delamination techniques allow the separation of solar cells (fragmented or entire) from the module as well as subsequent wet chemical techniques for the recovery of individual materials, e.g. silver and silicon, at a maximum purity level from the solar cells. The cost/revenue balance of PV recycling is challenging, and supporting regulation may be required to achieve the transformation of the current downcycling practice into a high-value one with the goal of enabling circular reuse of materials in the future.

## Author Biography

**Dr. Frank Lenzmann** works at TNO, the Dutch research center for applied science. His research in the area of PV has focused on sustainability and PV recycling in the recent past. He has been a delegate in the IEA's expert group on PV sustainability (IEA PVPS Task 12) until 2020.

## References

Akimoto, Y., Iizuka, A., and Shibata, E. (2018). High-voltage pulse crushing and physical separation of polycrystalline silicon photovoltaic panels. *Minerals Engineering* 125: 1–9.

Deng, R., Chang, N.L., Ouyang, Z., and Chong, C.M. (2019). Technological learning for resource efficient terawatt scale photovoltaics. *Renewable and Sustainable Energy Reviews* 109: 532–550.

Deng, R., Zhuo, Y., and Shen, Y. (2022). Recent progress in silicon photovoltaic module recycling processes. *Resources, Conservation and Recycling* 187: 106612.

Farrell, C.C., Osman, A.I., Doherty, R. et al. (2020). Technical challenges and opportunities in realizing a circular economy for waste photovoltaic modules. *Renewable and Sustainable Energy Reviews* 128: 109991.

Goldschmidt, J.C., Wagner, L., Pietzcker, R., and Friedrich, L. (2021). A techno-economic review of silicon photovoltaic module recycling. *Energy & Environmental Science* 14: 5147–5160.

Isherwood, P.J.M. (2022). Reshaping the module: the path to comprehensive photovoltaic panel recycling. *Sustainability* 14: 1676.

Kim, Y. and Lee, J. (2012). Dissolution of ethylene vinyl acetate in crystalline silicon PV modules using ultrasonic irradiation and organic solvent. *Solar Energy Materials & Solar Cells* 98: 317–322.

Komoto, K., Held, M., Agraffeil, C. et al. (2022). Status of PV module recycling in selected IEA PVPS Task 12 countries. *Report IEA-PVPS T12–24:2022*.

Latunussa, C.E., Ardente, F., Blengini, G.A., and Mancini, C. (2016). Life Cycle Assessment of an innovative recycling process for crystalline silicon photovoltaic panels. *Solar Energy Materials & Solar Cells* 156: 101–111.

Lunardi, M.M., Alvarez-Gaitan, J.P., Bilbao, J.I., and Corkish, R. (2018). A review of recycling processes for photovoltaic modules. Chapters. In: *Solar Panels and Photovoltaic Materials* (ed. B. Zaidi). IntechOpen.

Pang, S., Yan, Y., and Wang, Z. et al. (2021). Enhanced separation of different layers in photovoltaic panel by microwave field. *Solar Energy Materials & Solar Cells* 230: 111213.

Späth, M., Wieclawska, S., Sommeling, P., and Lenzmann, F. (2022). Balancing costs and revenues for recycling end-of-life PV panels in The Netherlands. *TNO report 2022 R10860*.

Tao, M., Fthenakis, V., and Ebin, B. et al. (2020). Major challenges and opportunities in silicon solar module recycling. *Progress in Photovoltaics: Research and Applications* 28: 1077–1088.

Urbina, A. (2023). Sustainability of photovoltaic technologies in future net-zero emissions scenarios. *Progress in Photovoltaics: Research and Applications* 31: 1255–1269.

Wambach, K., Heath, G., and Libby, C. (2017). Life cycle inventory of current photovoltaic module recycling processes in Europe. *Report IEA-PVPS T12:12-2017*.

Yang, E.H., Lee, J.K., Lee, J.S. et al. (2017). Environmentally friendly recovery of Ag from end-of-life c-Si solar cell using organic acid and its electrochemical purification. *Hydrometallurgy* 167: 129–133.

# 35

# Environmental Impacts of Integrated Photovoltaic Modules in Light Utility Vehicles

*Olga Kanz[1], Karsten Bittkau[1], Kaining Ding[1], Johanna May[2], and Angèle Reinders[3]*

[1] *IEK-5 Photovoltaik, Forschungszentrum Jülich GmbH, Jülich, Germany*
[2] *Institute for Electrical Power Engineering (IET) and Cologne Institute for Renewable Energy (CIRE), Cologne University of Applied Sciences, Cologne, Germany*
[3] *Energy Technology Group, Eindhoven University of Technology, Eindhoven, The Netherlands*

## 35.1 Introduction

Electric mobility is expected to play a crucial role in the decarbonization of the transport sector. However, battery electric vehicles (BEVs) are only preferable to petrol and diesel vehicles if charged by low-emission renewable energy technologies such as photovoltaic (PV) systems (Helmers 2017). This could happen by charging stations powered by regular PV systems or modules directly built into a car's body parts, also called "Vehicle Integrated Photovoltaics" (VIPV). Especially for trucks and light utility electric vehicles (LUV), VIPV can be an attractive feature due to vehicles' significantly large and flat roof surfaces. With a gross vehicle weight of no more than 3.5 metric tons, LUVs are optimized to be tough-built and are often used for short-distance delivery operations. VIPV on trucks and LUVs can potentially yield sufficient amounts of solar power and lower grid consumption for battery charging, thus reducing the operation costs.

Prior studies have indicated that VIPV will result in lower global warming potential (GWP) and contribute to sustainable mobility. A newly published report by IEA analyzed the expected emission reduction of PV-powered passenger vehicles, light commercial vehicles, and trucks in Germany, Netherlands, Spain, and Japan (IEA 2021). The report confirmed the benefits of VIPV to ordinary BEV, especially using the optimized PV charging strategy. The reduction of $CO_2$ emissions was calculated by subtracting the emissions of onboard PV from the reduction of grid power used for charging multiplied by the emission factor of grid electricity. New Energy and Industrial Technology Development Organization (NEDO) used a similar approach for estimating the emissions of PV-Powered Vehicles in Japan, demonstrating the benefits of VIPV compared to the grid (NEDO 2018). A detailed life-cycle assessment (LCA) based on a GaBi model of VIPV confirmed the outcome of earlier studies for the LUV case in Germany (Kanz et al. 2020).

*Photovoltaic Solar Energy: From Fundamentals to Applications, Volume 2*, First Edition.
Edited by Wilfried van Sark, Bram Hoex, Angèle Reinders, Pierre Verlinden, and Nicholas J. Ekins-Daukes.
© 2024 John Wiley & Sons Ltd. Published 2024 by John Wiley & Sons Ltd.
Companion website: www.wiley.com/go/PVsolarenergy

**Figure 35.1** StreetScooter Work L © Superbass/ Wikimedia Commons/CC BY-SA 4.0.

As an example of VIPV LCA on trucks, the case of a German VIPV is presented in this chapter (see Figure 35.1). Each part of the LCA based on a case study of the StreetScooter is described in detail and evaluated based on existing literature (Kanz et al. 2020). The StreetScooter study was completed in the framework of a project called STREET, funded by the German Ministry for Economic Affairs and Energy and a German logistics company of Deutsche Post DHL Group StreetScooter, which is currently working on the integration of PV on the electric LUVs.

The outcome of this LCA is not mandatory, equal to other LCA studies of VIPV systems. Nevertheless, the chapter can be beneficial for the reader to recognize the overall impact of VIPV on global warming. Moreover, a general understanding of the basics of the environmental impacts will allow the reader to respond to the increasing market pressure driven by investors and stakeholders, consumers, the media, and politics.

## 35.2 Methodology

### 35.2.1 Life Cycle Assessment Method

LCA is a valuable tool to quantify environmental performance, considering a holistic perspective. LCA is generally understood as a compilation and evaluation of a product or system's inputs, outputs, and potential environmental impacts throughout its life cycle. LCA studies consist of four main phases, covered through ISO standards DIN 14040 and DIN 14044; see Figure 35.2 (DIN ISO 14044 2006; DIN ISO 14040 2006). The environmental impact assessments are often completed at the mid-point level, where greenhouse gas emissions ($kgCO_2eq$) are used as an indicator of climate change contribution. GWP during the life cycle stages is estimated as an equivalent of $CO_2$ containing all the significant emissions $CO_2$ (GWP = 1), $CH_4$ (GWP = 25), $N_2O$ (GWP = 298) and chlorofluorocarbons (GWP = 4750 – 14,400). The calculations can be performed using LCA software GaBi with Ecoinvent v2.2+ as a background database (Sphera Solutions 2022). Sphera's Life Cycle Assessment (GaBi) databases are created on robust data from primary sources, including official data from industry associations.

The LCA methodology for VIPV can be divided into four sections. In the first section, Functional Unit, the system boundaries and input parameters for the operation must be clarified. The second section is dedicated to compiling an inventory of energy and material inputs and outputs over the life cycle of the VIPV. Based on the inventory, a GaBi model can be developed. In the third part, the simulation of VIPV's contribution to the charging battery is finalized. An energy flow model is constructed to simulate the reduction of grid power

**Figure 35.2** LCA Framework (DIN ISO 14040 2006; DIN ISO 14044 2006).

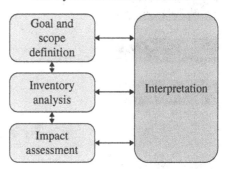

Life Cycle Assessment Framework

demand. The emissions of grid charging are calculated by multiplying the power used for charging by the grid electricity emission factor. In the last part, potential GWP related to identified inputs and releases is evaluated and compared to grid charging.

### 35.2.2 Functional Unit, Goal, and Scope

For the use case of an electric LUV Work L of StreetScooter, the functional unit is 1 kWh of electricity supplied by the PV system to the battery of the LUV. Compared to the functional unit of 1 km driven, the emissions of 1 kWh can be calculated more precisely. Furthermore, the functional unit of 1 kWh allows a direct comparison of the car's charging by PV modules to grid charging. Thus, the emissions of VIPV and grid-charged LUV can be evaluated by referring to the same functional unit. Since the location impacts the grid's emissions, the operation case is set in Cologne, Germany. Within this project's scope, the PV system configuration is based on the first generation of VIPV panels for the STREET project with heterojunction silicon PV modules manufactured in China. The system of VIPV electricity includes raw material extraction, wafer, crystalline silicon-based heterojunction solar cell and module manufacture, mounting structures manufacture, inverters manufacture, system installation, and operation. The analyzed VIPV configuration includes three panels and three control units, including the cables mounted on the vehicle roof. The overall roof capacity of the VIPV system is 930 Wp. In later configurations, the maximal roof capacity was changed to five panels, which is not analyzed in this LCA (Peibst 2021).

### 35.2.3 Input Parameters of the Inventory

Similarly to every LCA, the input data of each step in the inventory have to be chosen separately since the units are usually produced in different locations and under unique conditions (e.g. electricity and heat consumption). For the STREET project, a production process of a typical commercial crystalline silicon solar cell was assumed based on the existing datasets (Frischknecht et al. 2015). The background data of the PV control unit (PVCU) manufacturing the vehicle integration process was adjusted based on internal communication in the project (Peibst 2021). The electricity consumption on all process levels was modeled following specific electricity mixes corresponding to China (CN) or Germany (DE), respectively, based on GaBi (Sphera Solutions 2022).

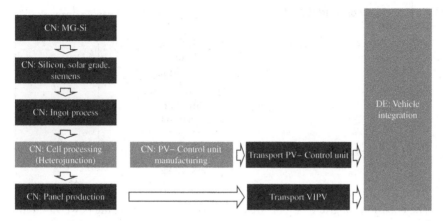

**Figure 35.3** VIPV system value chain: process flow diagram. CN – China, DE – Germany.

Figure 35.3 demonstrates single units of the supply chain of the VIPV before operation starts.

All input parameters of the manufacturing and vehicle integration process are described in Table 35.1. The exact location of the manufacturing plants is undocumented. However, it can be anticipated that the location of this plant is somewhere within China. Modeling of the transportation of the VIPV system was based on the standard distances as suggested in the Report for PV LCA (Frischknecht et al. 2015). Thus, metal parts were reported with 200 km train and 100 km truck transportation in China, and transoceanic transport from China to Belgium was estimated to be 19,994 km based on searates.com data. Lorry transport from Antwerp (Belgium) to Cologne (Germany) was estimated to be 500 km. For the solar cells in the VIPV, heterojunction technology (SHJ) was chosen in STREET due

**Table 35.1** Input parameters of the manufacturing process and vehicle integration process.

| Part of system | Parameter |
| --- | --- |
| Wafer | Type: n-type c-Si |
| | Thickness: 180 nm |
| Cell | Technology SHJ c-Si |
| | Area: 239 mm$^2$ |
| | Efficiency: 22.5% |
| Panel | Efficiency: 19.7% |
| | Glass thickness: 2 mm |
| | Back: EVA Back Foil |
| | Cell number per panel: 72 |
| Vehicle integration | Panel number: 3 |
| | Total PV area: 4.8 m$^2$ |
| | PVCU number: 3 |

to the best trade-off between efficiency and costs. The LCI data on material and energy consumption were added for heterojunction cell processing, referring to Louwen et al. (2012, 2016) and Olson et al. (2013).

### 35.2.4 Input Parameters for the Energy Flow Model

The main factor for estimating PV electricity output is the effective solar irradiance, which depends on the orientation, location, season, time, and module configuration. The operation location for the reference case is Cologne, Germany. The hourly global horizontal solar irradiance has been defined by averaging hourly incident global horizontal radiation data extracted from the PVGIS database. The onboard electricity generation was simulated based on degradation, system losses, and shadowing factors (see Table 35.2). A 19.7% module efficiency was assumed (Olson et al. 2013). In line with IEA-PVPS methodology guidelines (Frischknecht et al. 2015), a degradation of 0.7% per year was estimated. Based on data on LUVs in average delivery services, the operation period of the reference case was set to be eight years (Hacker et al. 2015).

The overall efficiency of the VIPV system was calculated according to the literature guidelines. Due to dynamic shading, an average of 70% VIPV performance compared to residential PV can be expected for the use case (NEDO 2018). Furthermore, since generated energy cannot be used directly for traction of the vehicle and must be stored in the battery, DC-charging/discharging loss of 2% has to be taken into account. An additional loss of 5% is considered due to the DC/DC converter. The loss of the MPP tracking additionally limits its efficiency in the model to 95% (NEDO 2018). A performance loss of 9% due to temperature increase and low irradiance is assumed (Araki et al. 2018). The overall average efficiency losses of the VIPV system can be found in Table 35.3.

### 35.2.5 Input Parameters of the Grid Charging

The emissions of the grid can vary massively depending on the different power plants. Fossil power plants still dominate power generation in Germany. Known studies usually state the yearly emission factor of the grid electricity considering the average carbon footprints of each power plant caused by the plant's life cycle (construction, fuel production, operation, EoL) (Icha 2021). Hourly average emissions of the German electricity mix vary depending

**Table 35.2** Input parameters for the VIPV operation.

| Parameter | Value | Unit |
| --- | --- | --- |
| Capacity | 930 | Wp |
| Module efficiency | 19.7 | % |
| Annual degradation | 0.7 | %/a |
| Operation lifetime | 8 | a |
| Location | Cologne [50.938, 6.954] | Lat/Lon |
| Database | PVGIS-CMSAF | / |

**Table 35.3** VIPV system efficiency.

| Loss coefficient | Changes in output | Unit |
|---|---|---|
| MPPT loss | −5 | % |
| Temperature/low irradiance | −9 | % |
| DC/DC conversions | −5 | % |
| DC charging/discharging loss | −2 | % |
| Average shadowing factor | −30 | % |

on the day and night times. This fact is a supporting argument for VIPV since electrically powered vehicles – especially commercially used vehicles – are mainly connected at night, with higher grid emissions per kWh. Thus, the actual GWP associated with the network charging – and thus the potential of VIPV to reduce emissions – is likely to be higher than estimated from the average grid mix. In 2020, partly due to COVID, the German GWP reached a minimum value of 0.087 and a maximum of 0.664 $kgCO_2eq$/kWh, averaging at 0.3 $kgCO_2eq$/kWh. The German electricity mix was modeled using SMARD electricity generation data (Icha 2021) and projected based on future projections. Until 2028, the yearly electricity mix GWP is expected to decrease by 2% annually. With no limitations of a particular driving or charging profile, the average value of 0.4 $kgCO_2eq$/kWh for the German electricity mix is used as the reference in this LCA. The losses due to grid distribution are additionally calculated in the model since the electricity used for LUV charging is based on an energy consumption perspective.

## 35.3 Results

### 35.3.1 Manufacturing Process of the VIPV

The investigation of the manufacturing phase determines the GWP before the start of the operation. The results of the model of the VIPV show a similar GWP to ordinary PV systems. The main contributor, Solar-Grade Process, is responsible for 444.30 $kgCO_2eq$, a third of total VIPV emissions. The process of integration of the cells into the panel emits 235.24 $kgCO_2eq$. The total amount released during the manufacturing process is 1143.12 $kgCO_2eq$ (see Figure 35.4).

### 35.3.2 Operation Phase of the VIPV

The onboard electricity generation was simulated based on the assumptions of degradation, system losses, and shadowing factors, as previously described in Section 35.2.4. While driving the EV, the batteries will discharge and recharge using the onboard PV modules. The degree of VIPV's impact on a battery charge is expected to vary with the usage patterns: various daily driving distances have different depths of discharge corresponding to daily driving durations. In this study, all incoming irradiance during the day is considered to be

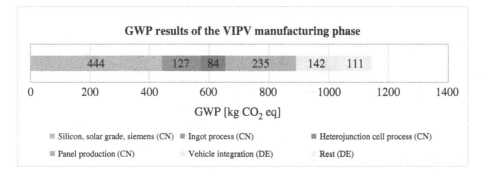

**Figure 35.4** GWP Results of the VIPV manufacturing phase, including all subsystems. The production regions are CN – China and DE – Germany.

used, assuming collecting energy and charging the battery even if not driving. The average yield in the urban area is 936 kWh/kWp annually. The average expected VIPV electricity contributed to the battery is 479 kWh/a.

For the reference scenario of eight years of operation, and shadowing factor of 30%, and calculated degradation, the overall VIPV contribution is 3738 kWh. Prolonged operation of 12 years generates 5527 kWh in total. Based on these numbers and according to calculations described in Kanz et al. (2020) a value of 0.3 kgCO$_2$eq/kWh is calculated for VIPV. For prolonged use of 12 years, the GWP can be reduced to 0.2 kgCO$_2$eq/kWh. For stationary PV, a much lower number of 0.05 kgCO$_2$eq/kWh is reported in the literature (Hsu et al. 2012). These values cannot be compared to VIPV directly. Key performance characteristics such as irradiation of 1700 kWh/m$^2$/yr, system lifetime of 30 years, and a performance ratio of 0.80 are estimated for rooftop systems. Due to shadowing, PV systems on vehicles have a shorter lifetime and receive less irradiation than stationary PV systems.

### 35.3.3 Sensitivity Analysis

The results of the study are wide-ranging. The variations arise mainly from system operating assumptions (e.g. solar irradiation, system lifetime, shadowing factors) and technology improvements (e.g. electricity consumption for manufacturing processes). Some adjustments are considered in this chapter for the reference case of the eight years, 0.7% degradation, and 30% shadowing factor of the operation. PV-generated power is an essential variable for the reduction in emissions. An increasing shadowing factor causes a significant growth of emissions per kWh. The analysis of the increased shadowing factor of 40% results in a higher GWP of 0.435 kgCO$_2$eq/kWh. The ecological benefit over the grid charge disappears completely if this value is reached (see Figure 35.5). Thus, the shadowing and mobility profiles will play a significant role in future LCAs. These factors should be calculated separately for specific use cases to achieve precise GWP analysis.

Lifetime extension of the vehicle operation will automatically result in reduced emissions per produced kWh of VIPV. The sensitivities also show that if VIPV is used for a prolonged timeframe of 12 years, the emission factor of onboard electricity generation

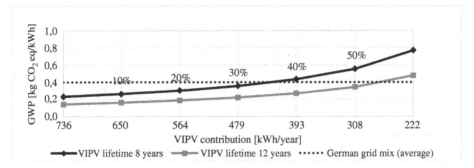

**Figure 35.5** GWP depending on the shadowing factor and "Prolonged Use" Scenario of 12 years of operation. The influence of the shadowing factor (in percentage) is demonstrated. The reference case of 30% shadowing can be compared to the grid.

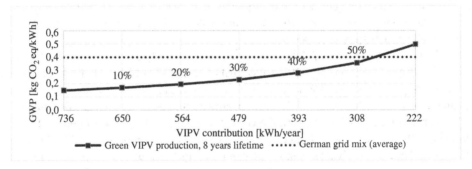

**Figure 35.6** GWP depending on the shadowing factor and "green" VIPV production Scenario of eight years of operation. The influence of the shadowing factor (in percentage) is demonstrated. The reference case of 30% shadowing can be compared to the grid.

decreases to 0.221 kgCO$_2$eq/kWh. A reduction of 38% compared to the reference case of eight years is noted. GWP comparable to the grid emissions can be reached with a shadow factor of 55% or an average annual VIPV generation of about 260 kWh/a. Figure 35.5 demonstrates the potential of the more extended operation phase for different shadowing factors.

Based on the findings of the sensitivity analyses, the highest potential for emission reduction in the manufacturing phase can be confirmed for a "green" electricity scenario, where the manufacturing processes are fully covered by renewable electricity. For China's electricity mix, the emission factor of 0.831 kgCO$_2$-eq/kWh is used for the previous simulation of the reference case. This value is based on the GaBi Education Database 2017. An increased share of renewable energy in the Chinese grid will lower the GWP of the VIPV production. Using green electricity has the potential to achieve a climate-friendly production of PV modules, as is already the case for hydropower in use today. For the "green" electricity scenario, hydro plants with average emissions of 0.003 kgCO$_2$-eq/kWh cover the energy demand of the manufacturing phase in China (Memmler et al. 2018). For this case of 30% shadowing and eight years of operation, the GWP can be reduced from 0.357 to 0.230 kgCO$_2$-eq/kWh (see Figure 35.6).

### 35.3.4  Reliability of the Data

The LCI in this study is based on extracting the data from published reliable literature. A commercial LCA software, GaBi Version 8.7.1.30, is used to model and calculate the impact assessment (Sphera Solutions 2022). Essential materials and electricity mixes are calculated based on data represented by the ecoinvent database unless otherwise noted. Data on production is mainly based on the guidelines for Photovoltaics (Frischknecht et al. 2015), additionally considering the heterojunction process (Olson et al. 2013; Louwen et al. 2012). Some values for the vehicle integration process and PVCU are adjusted after internal communication in STREET. The reason for modification was mainly lacking access to the supply chain model data. The data used for this LCA vary in quality and reliability. To limit the resulting uncertainty, the differences in the data sources were analyzed and scored by referring to the Quality Pedigree Matrix Flow Indicators determined by DIN 14044 (DIN ISO 14040 2006). The overview and detailed description of the LCA data quality can be found in Kanz (Kanz et al. 2020).

## 35.4  Discussion and Conclusions

The study reports a unique observation that integrating a PV system on an LUV StreetScooter can improve the GWP of the vehicle when the reference case of an average shadowing factor of 30% and eight years of operation time are evaluated. Nevertheless, the results of the LCA show that viability is heavily dependent on the vehicle's deployment region and usage scenario. For the functional unit of 1 kWh of onboard generated PV electricity, the emission factor of $0.357 \, kgCO_2$ eq/kWh is calculated. In comparison, the average GWP of the grid charge is expected to be $0.4 \, kgCO_2$ eq/kWh. With increasing the shadowing factor, emissions per kWh grow significantly. The ecological benefit to the grid charge disappears completely when the shadowing factor reaches 40%. However, if the operation time is prolonged to 12 years, the shadowing factor can reach 55% with similar emissions to the grid charge. For the reference case with 30% shadowing, a reduction of GWP of 38% can be noted in the case of prolonged operation. Based on the findings of the sensitivity analyses, the highest potential for emission reduction in the manufacturing phase can be confirmed for a "green" electricity scenario, where hydro plants are used for PV production in China. For the case of 30% shadowing and eight years of operation, the GWP can be reduced from 0.357 to $0.230 \, kgCO_2$-eq/kWh.

The results of this research could be helpful for car manufacturers to calculate emissions per vehicle, for political institutions to estimate environmental impacts for the transport sector, and for business parties in the solar market to identify further application possibilities and critical areas for the improvement of PV technology in-vehicle integrations. One of the critical challenges of this LCA was finding an appropriate vehicle usage model to reproduce the ratio of using VIPV. Tests with radiation sensors investigating shading and reflection conditions would help cover missing data and provide irradiance profiles for the LCA. In addition, no established and reliable routes can be found in the literature for the recycling process. Nevertheless, recycling is crucial, and further research should include material depletion, recyclability options, and removable mounting structures for possible second use. Consistent international methods for evaluating emissions reduction will help communicate the benefits of the different driving and charging behaviors and evaluate the added value of reducing grid power consumption.

| Acronym | Description |
| --- | --- |
| BEV | Battery Electric Vehicle |
| PV | Photovoltaic |
| VIPV | Vehicle Integrated Photovoltaic |
| IEA | International Energy Agency |
| LCA | Life Cycle Assessment |
| GWP | Global Warming Potential |
| LUV | Light Utility Vehicle |
| NEDO | New Energy and Industrial Technology Development Organization |
| $CO_2$ | Carbon Dioxide |
| $CH_4$ | Methane |
| $N_2O$ | Nitrous Oxide |
| CN | China |
| DE | Germany |
| PVCU | Photovoltaic Control Unit |
| EVA | Ethylene Vinyl Acetate |
| FZJ | Forschungszentrum Jülich |
| MPPT | Maximum Power Point Tracking |

## Author Biography

**Olga Kanz** is currently working at the Institute of Energy Research (IEK-5) of Forschungszentrum Jülich. She studied at Technische Hochschule Köln in Cologne and finished with a diploma in Renewable Energy Engineering. Olga is highly interested in sustainable technologies and energy systems. She is currently investigating the environmental impact of different solar-powered hydrogen production methods as a Ph.D. student at the Eindhoven University of Technology.

## References

Araki, K., Ji, L., Kelly, G., and Yamaguchi, M. (2018). To do list for research and development and international standardization to achieve the goal of running a majority of electric vehicles on solar energy. *Coatings* 8 (7): 251.

DIN ISO 14040 (2006). *DIN ISO 14040 Environmental Management-Life Cycle Assessment-Requirements and Guidelines.*

DIN ISO 14044 (2006). *Environmental Management-Life Cycle Assessment-Principles and Framework*.

Frischknecht, R., Itten, R., Sinha, P. et al. (2015). Life Cycle Inventories and Life Cycle Assessments of Photovoltaic Systems. IEA PVPS Task 12, Subtask 2.0, LCA Report IEA-PVPS 12. International Energy Agency.

Hacker, F., Waldenfels, R., and Moschall, M. (2015). *Wirtschaftlichkeit von Elektromobilität in gewerblichen Anwendungen. Betrachtung von Gesamtnutzungskosten, ökonomischen Potenzialen und möglicher $CO_2EQ$ -Minderung im Auftrag der Begleitforschung zum BMWi Förderschwerpunkt IKT für Elektromobilität*.

Helmers, E. (2017). Electric car life cycle assessment based on real-world mileage and the electric conversion scenario. *The International Journal of Life Cycle Assessment* 22: 15–30.

Hsu, D., O'Donoughue, P., Fthenakis, V. et al. (2012). *Life Cycle Greenhouse Gas Emissions of Crystalline Silicon Photovoltaic Electricity Generation*. Wiley Online Library.

Icha, P. (2021). *Entwicklung der spezifischen Kohlendioxid-Emissionen des deutschen Strommix in den Jahren 1990–2020*. Dessau-Roßlau: Umweltbundesamt.

IEA (2021). State-of-the-Art and Expected Benefits of PV-Powered Vehicles. Report IEA-PVPS Task 17. International Energy Agency – Photovoltaic Power Systems Programme.

Kanz, O., Reinders, A., May, J., and Ding, K. (2020). Environmental impacts of integrated photovoltaic modules in light utility electric vehicles. *Energies* 13 (19): 5120. https://doi.org/10.3390/en13195120.

Louwen, A., van Sark, W., Turkenburg, W. et al. (2012). R&D integrated life cycle assessment: a case study on the R&D of silicon heterojunction (SHJ) solar cell based PV systems. *27th European Photovoltaic Solar Energy Conference and Exhibition*. https://doi.org/10.4229/27.

Louwen, A., van Sark, W., Schropp, R., and Faaij, A. (2016). A cost roadmap for silicon heterojunction solar cells. *Solar Energy Materials and Solar Cells* 147: 295–314. https://doi.org/10.1016/j.solmat.2015.12.026.

Memmler, M., Lauf, D.T., and Schneider, S. (2018). Emissionsbilanz erneuerbarer Energieträger – Bestimmung der vermiedenen Emissionen im Jahr 2017. *Climate Change* 156.

NEDO (2018). *PV-Powered Vehicle Strategy Committee Interim Report*. New Energy and Industrial Technology Development Organization.

Olson, C., de Wild-Scholten, M., and Scherff, M. (2013). Life cycle assessment of hetero-junction solar cells. *27th European Photovoltaic Solar Energy Conference and Exhibition*. https://doi.org/10.4229/26thEUPVSEC2011-4AV.3.57.

Peibst, R.F.-H. (2021). Demonstration of feeding vehicle-integrated photovoltaic-converted energy into the high-voltage on-board network of practical light commercial vehicles for range extension. *Solar RRL* 6 (5): 2100516. https://doi.org/10.1002/solr.202100516.

Sphera Solutions (2022). *GaBi Software and Database*.

# Index

*Photovoltaic Solar Energy: From Fundamentals to Applications, Volume 2*, First Edition.
Edited by Wilfried van Sark, Bram Hoex, Angèle Reinders, Pierre Verlinden, and Nicholas J. Ekins-Daukes.
© 2024 John Wiley & Sons Ltd. Published 2024 by John Wiley & Sons Ltd.
Companion website: www.wiley.com/go/PVsolarenergy